METHODS IN MOLECULAR BIOLOGY

For further volumes:
http://www.springer.com/series/7651

The Ubiquitin Proteasome System

Methods and Protocols

Edited by

Thibault Mayor

Department of Biochemistry and Molecular Biology, Michael Smith Laboratories, University of British Columbia, Vancouver, BC, Canada

Gary Kleiger

Department of Chemistry and Biochemistry, University of Nevada Las Vegas, Las Vegas, NV, USA

Humana Press

Editors
Thibault Mayor
Department of Biochemistry
and Molecular Biology
Michael Smith Laboratories
University of British Columbia
Vancouver, BC, Canada

Gary Kleiger
Department of Chemistry
and Biochemistry
University of Nevada Las Vegas
Las Vegas, NV, USA

ISSN 1064-3745 ISSN 1940-6029 (electronic)
Methods in Molecular Biology
ISBN 978-1-4939-8705-4 ISBN 978-1-4939-8706-1 (eBook)
https://doi.org/10.1007/978-1-4939-8706-1

Library of Congress Control Number: 2018952458

This Humana Press imprint is published by the registered company Springer Science+Business Media, LLC part of Springer Nature.
The registered company address is: 233 Spring Street, New York, NY 10013, U.S.A.

Preface

In 1975, a paper was published in the *Proceedings of the National Academy of Sciences* describing a newly discovered polypeptide capable of inducing the differentiation of T cells [1]. The authors named this protein ubiquitous immunopoietic polypeptide (UBIP) "because it is widespread and perhaps even universally represented in living cells. . . ." It soon became clear that this polypeptide, later renamed ubiquitin, is a key signaling molecule in eukaryotic cell biology. Indeed, approximately 5% of human genes are dedicated towards the expression of enzymes and factors that control the ubiquitin proteasome system (UPS), a massive network of integrated pathways responsible for regulating protein degradation, protein localization, and enzyme activity. The pioneering research of Aaron Ciechanover, Avram Hershko, Irwin Rose, and Alex Varshavsky has led to a field that continues to make new and important discoveries regarding how cells function. Greater than 54,000 articles have been indexed in the US National Library of Medicine PubMed database that contain the word "ubiquitin" in either the title or abstract (Fig. 1). The UPS has gifted both biotech and pharmaceutical companies with fresh, new targets for the treatment of various human diseases. Some five drugs have been approved in the United States by the Federal Drug Administration that target enzymes in the UPS, with the promise of many more in the not too distant future. The ubiquitin field continues to grow and change at a dizzying pace with significant advances for both newer methodologies—such as mass spectrometry-based proteomics and cryo-electron microscopy—and the more traditional biochemical, cell biological, and structural analyses and assays. This volume is designed for both veterans in the field wanting to expand their bag of tricks and for novices just getting started.

The methods have been organized in five parts. In the first part, five methods are described that can be applied to the determination of the mechanisms of action for ubiquitin-conjugating enzymes (E2s), ubiquitin ligases (E3s), and de-ubiquitylating

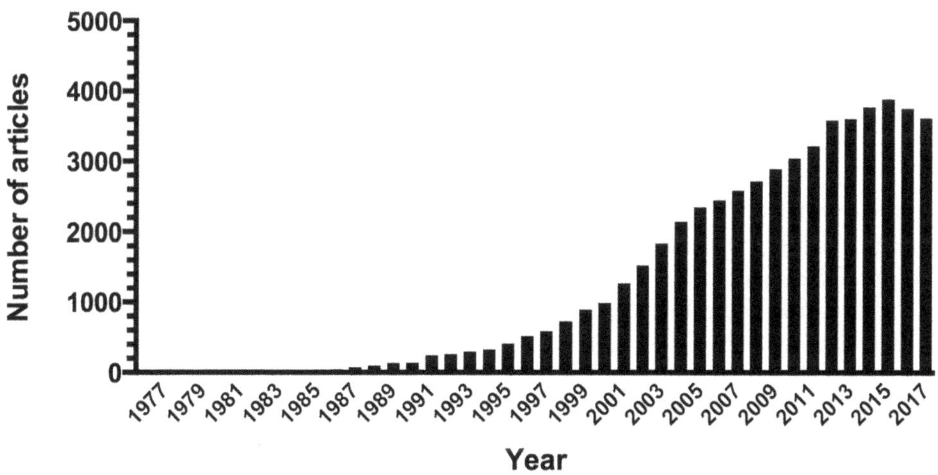

Fig. 1 Publishing in the ubiquitin field: still going strong!

(DUB) enzymes. There are approximately 35 E2s, and perhaps as many as 600 E3s and 90 DUBs in the human proteome, with the vast majority awaiting in-depth characterization. These chapters will be helpful for any researcher aiming to reconstitute in vitro ubiquitylation as a means of uncovering E2 and/or E3 function, or to infer the affinity of a DUB for its substrate.

In the second part of this volume, six methods are described for new technological advances that will enable the researcher to (1) assemble and purify poly-ubiquitin chains that are uniform with respect to linkage type in quantities sufficient for biochemical and structural analyses; (2) discover the binding partners of essentially any protein in the UPS; (3) screen for ubiquitin variants with unique properties that can modulate enzymes in the UPS within the context of living cells; (4) discover the unique amino acid sequence signatures that target proteins to be recognized and degraded by the UPS; (5) use flow cytometry towards identifying the UPS enzymes responsible for protein turnover in a high-throughput manner; and (6) identify and characterize ubiquitylation pathways using the easy to work with model organism *E. coli* that, lacking the UPS, optimizes the researcher's chances for success.

In the third part, cutting-edge methods are described for three areas in structural biology: X-ray crystallography, small-angle X-ray scattering (SAXS), and cryo-electron microscopy. Methods to trap dynamic multi-subunit complexes for structural characterization using protein cross-linking are described (while this method is specific to the ubiquitin-like protein SUMO, the similarities between the ubiquitin and SUMO cascades are so great that these methods will undoubtedly prove useful for members of the ubiquitin field). It is often desirable to characterize the structures of UPS enzymes in solution, and SAXS is a powerful and relatively simple method described here to achieve this goal. While cryo-electron microscopy has enabled the determination of the structures of large, multi-subunit complexes to near or at atomic resolution, one challenge has been the preparation of the grids where macromolecules are placed prior to freezing. A detailed protocol for this procedure has been included.

In the fourth part, seven chapters are dedicated to methodologies to study the 26S proteasome, the multi-subunit complex that recognizes and degrades ubiquitylated proteins. These chapters cover (1) the expression and purification of the 19S proteasome regulatory complex in *E. coli*, as well as the ability to introduce chemical modifications to surface residues on any given subunit; (2) the use of native gel electrophoresis to separate and study the proteasome during various stages of assembly; (3) the use of isotopic pulse-chase to directly measure the rates of protein degradation within living cells; (4) three unique and complementary methods to purify proteasomes from cells either grown in isolation or from intact tissues; (5) several quantitative *in vitro* methods for assaying the sequential activities of the 26S proteasome; (6) methods to assess the functional impact of proteasome phosphorylation; and (7) novel proteasome substrates that enable the measurement of protein degradation in a high-throughput manner.

The fifth part is dedicated to cutting-edge proteomic techniques that are first summarized in a review and then covered in three chapters describing methods to (1) identify ubiquitination sites on endogenous proteins from any eukaryotic organism or tissue; (2) enrich for poly-ubiquitin chains with specific linkage types for mass spectrometry analysis; and (3) employ cross-linking to determine domain topology within a multi-subunit complex such as the proteasome.

We are, first and foremost, most grateful to all of our colleagues that contributed to this volume. While it is undoubtedly true that many researchers consider their research areas as special, we are constantly taken aback by the generosity and good will of the investigators that make the ubiquitin field their home. We are grateful to Springer for the opportunity to feature our work through this book series and to John Walker for his dedicated help in assembling this book. Lastly, we would like to dedicate this volume to our postdoctoral advisor, Ray Deshaies, as well as his long-time research associate, Rati Verma. This volume would not have been possible without their expert mentorship and friendship.

Vancouver, BC, Canada *Thibault Mayor*
Las Vegas, NV, USA *Gary Kleiger*

Reference

1. Goldstein G, Scheid M, Hammerling U, Schlesinger DH, Niall HD, Boyse EA (1975) Isolation of a polypeptide that has lymphocyte-differentiating properties and is probably represented universally in living cells. Proc Natl Acad Sci U S A 72(1):11–15

Contents

Part V Proteomic Methods to Study the Ubiquitin Proteasome System

Contributors

ALON APPLEBOIM • *Department of Biological Chemistry, The Alexander Silberman Institute of Life Sciences, The Hebrew University of Jerusalem, Jerusalem, Israel; School of Computer Science and Engineering, The Hebrew University of Jerusalem, Jerusalem, Israel*

ILAN ATTALI • *Department of Biochemistry and Molecular Biology, Institute of Structural Biology, George S. Wise Faculty of Life Sciences, Tel Aviv University, Tel Aviv, Israel*

JARED A. M. BARD • *Department of Molecular and Cell Biology and California Institute for Quantitative Biosciences, University of California at Berkeley, Berkeley, CA, USA*

DAVID BARFORD • *MRC Laboratory of Molecular Biology, Cambridge, UK*

ERIC J. BENNETT • *Section of Cell and Developmental Biology, Division of Biological Sciences, University of California, San Diego, La Jolla, CA, USA*

LORI BUETOW • *Cancer Research UK Beatson Institute, Glasgow, UK; Institute of Cancer Sciences, University of Glasgow, Glasgow, UK*

LARISSA A. CANADEO • *Department of Molecular Biosciences, University of Texas at Austin, Austin, TX, USA*

ERICK R. CASTELLANOS • *Department of Structural Biology, Genentech Inc., South San Francisco, CA, USA*

LEIFU CHANG • *MRC Laboratory of Molecular Biology, Cambridge, UK*

GALEN ANDREW COLLINS • *Department of Cell Biology, Harvard Medical School, Boston, MA, USA*

SOPHIE A. COMYN • *Department of Biochemistry and Molecular Biology, Michael Smith Laboratories, University of British Columbia, Vancouver, BC, Canada*

PAUL R. ELLIOTT • *Division of Protein and Nucleic Acid Chemistry, MRC Laboratory of Molecular Biology, Cambridge, UK*

AMIT FULZELE • *Section of Cell and Developmental Biology, Division of Biological Sciences, University of California, San Diego, La Jolla, CA, USA*

MADS GABRIELSEN • *Cancer Research UK Beatson Institute, Glasgow, UK; Institute of Cancer Sciences, University of Glasgow, Glasgow, UK*

RICHARD G. GARDNER • *Department of Pharmacology, University of Washington, Seattle, WA, USA*

YIFAT GEFFEN • *Department of Biological Chemistry, The Alexander Silberman Institute of Life Sciences, The Hebrew University of Jerusalem, Jerusalem, Israel; School of Computer Science and Engineering, The Hebrew University of Jerusalem, Jerusalem, Israel*

KALLE GEHRING • *Department of Biochemistry, McGill University, Montréal, QC, Canada*

ALFRED L. GOLDBERG • *Department of Cell Biology, Harvard Medical School, Boston, MA, USA*

CONNOR HILL • *Department of Chemistry and Biochemistry, University of Nevada, Las Vegas, NV, USA*

SPENCER HILL • *Department of Chemistry and Biochemistry, University of Nevada, Las Vegas, NV, USA*

DANNY T. HUANG • *Cancer Research UK Beatson Institute, Glasgow, UK; Institute of Cancer Sciences, University of Glasgow, Glasgow, UK*

LAN HUANG • *Department of Physiology and Biophysics, University of California, Irvine, CA, USA*

Jon M. Huibregtse • *Department of Molecular Biosciences, University of Texas at Austin, Austin, TX, USA*
Tal Keren-Kaplan • *Department of Biochemistry and Molecular Biology, Institute of Structural Biology, George S. Wise Faculty of Life Sciences, Tel Aviv University, Tel Aviv, Israel*
Hyoung Tae Kim • *Department of Cell Biology, Harvard Medical School, Boston, MA, USA*
Donald S. Kirkpatrick • *Department of Microchemistry, Proteomics and Lipidomics, Genentech, Inc., South San Francisco, CA, USA*
Gary Kleiger • *Department of Chemistry and Biochemistry, University of Nevada, Las Vegas, NV, USA*
Rachel E. Klevit • *Department of Biochemistry, University of Washington, Seattle, WA, USA*
David Komander • *Division of Protein and Nucleic Acid Chemistry, MRC Laboratory of Molecular Biology, Cambridge, UK*
Chueh-Ling Kuo • *Department of Cell Biology, Harvard Medical School, Boston, MA, USA*
Olga Levin-Kravets • *Department of Biochemistry and Molecular Biology, Institute of Structural Biology, George S. Wise Faculty of Life Sciences, Tel Aviv University, Tel Aviv, Israel*
Christopher D. Lima • *Structural Biology Program, Sloan Kettering Institute, Memorial Sloan Kettering Cancer Center, New York, NY, USA; Howard Hughes Medical Institute, New York, NY, USA*
Zhao-Qing Luo • *Department of Biological Sciences, Purdue Institute for Inflammation, Immunology and Infectious Disease, Purdue University, West Lafayette, IN, USA*
Andreas Martin • *Department of Molecular and Cell Biology and California Institute for Quantitative Biosciences, University of California at Berkeley, Berkeley, CA, USA; Howard Hughes Medical Institute, University of California at Berkeley, Berkeley, CA, USA*
Kirby Martinez-Fonts • *Department of Molecular Biosciences, The University of Texas at Austin, Austin, TX, USA*
Andreas Matouschek • *Department of Molecular Biosciences, The University of Texas at Austin, Austin, TX, USA*
Marissa L. Matsumoto • *Department of Structural Biology, Genentech Inc., South San Francisco, CA, USA*
Thibault Mayor • *Department of Biochemistry and Molecular Biology, Michael Smith Laboratories, University of British Columbia, Vancouver, BC, Canada*
Martin A. Michel • *Division of Protein and Nucleic Acid Chemistry, MRC Laboratory of Molecular Biology, Cambridge, UK*
Michael T. Morgan • *Department of Biophysics and Biophysical Chemistry, Johns Hopkins University School of Medicine, Baltimore, MD, USA*
Hazel F. O'Connor • *Department of Molecular Biosciences, University of Texas at Austin, Austin, TX, USA*
Soyeon Park • *Department of Molecular, Cellular, and Developmental Biology, University of Colorado Boulder, Boulder, CO, USA*
Avinash Persaud • *Department of Biochemistry and Molecular Biology, Institute of Structural Biology, George S. Wise Faculty of Life Sciences, Tel Aviv University, Tel Aviv, Israel*
Gali Prag • *Department of Biochemistry and Molecular Biology, Institute of Structural Biology, George S. Wise Faculty of Life Sciences, Tel Aviv University, Tel Aviv, Israel*

JIAZHANG QIU • *Key Laboratory of Zoonosis, Ministry of Education, College of Veterinary Medicine, Jilin University, Changchun, China*
RITU RATHI • *Department of Biochemistry and Molecular Biology, Institute of Structural Biology, George S. Wise Faculty of Life Sciences, Tel Aviv University, Tel Aviv, Israel*
TOMMER RAVID • *Department of Biological Chemistry, The Alexander Silberman Institute of Life Sciences, The Hebrew University of Jerusalem, Jerusalem, Israel*
KATHERINE H. REITER • *Department of Biochemistry, University of Washington, Seattle, WA, USA*
JEROEN ROELOFS • *Division of Biology, Kansas State University, Manhattan, KS, USA*
AMALIA ROSE • *Department of Biochemistry and Molecular Biology, Michael Smith Laboratories, University of British Columbia, Vancouver, BC, Canada*
ZHE SHA • *Department of Cell Biology, Harvard Medical School, Boston, MA, USA*
DAR SHAPIRA • *Department of Biochemistry and Molecular Biology, Institute of Structural Biology, George S. Wise Faculty of Life Sciences, Tel Aviv University, Tel Aviv, Israel*
ITAI SHARON • *Department of Biochemistry and Molecular Biology, Institute of Structural Biology, George S. Wise Faculty of Life Sciences, Tel Aviv University, Tel Aviv, Israel*
NOA SHOHAT • *Department of Biochemistry and Molecular Biology, Institute of Structural Biology, George S. Wise Faculty of Life Sciences, Tel Aviv University, Tel Aviv, Israel*
ANNA SHUSTERMAN • *Department of Biochemistry and Molecular Biology, Institute of Structural Biology, George S. Wise Faculty of Life Sciences, Tel Aviv University, Tel Aviv, Israel*
SACHDEV S. SIDHU • *Donnelly Centre for Cellular and Biomolecular Research, University of Toronto, Toronto, ON, Canada; Banting and Best Department of Medical Research, University of Toronto, Toronto, ON, Canada; Department of Molecular Genetics, University of Toronto, Toronto, ON, Canada*
AMIT KUMAR SINGH GAUTAM • *Department of Molecular Biosciences, The University of Texas at Austin, Austin, TX, USA*
FREDERICK C. STREICH JR. • *Structural Biology Program, Sloan Kettering Institute, Memorial Sloan Kettering Cancer Center, New York, NY, USA*
ANJANA SUPPAHIA • *Division of Biology, Kansas State University, Manhattan, KS, USA*
CALEB D. SWAIM • *Department of Molecular Biosciences, University of Texas at Austin, Austin, TX, USA*
NETA TANNER • *Department of Biochemistry and Molecular Biology, Institute of Structural Biology, George S. Wise Faculty of Life Sciences, Tel Aviv University, Tel Aviv, Israel*
JEAN-FRANÇOIS TREMPE • *Department of Pharmacology and Therapeutics, McGill University, Montréal, QC, Canada*
JORDAN J. S. VERPLANK • *Harvard Medical School, Boston, MA, USA*
KENRICK A. WAITE • *Division of Biology, Kansas State University, Manhattan, KS, USA*
XIAORONG WANG • *Department of Physiology and Biophysics, University of California, Irvine, CA, USA*
CYNTHIA WOLBERGER • *Department of Biophysics and Biophysical Chemistry, Johns Hopkins University School of Medicine, Baltimore, MD, USA*
YI JIMMY ZENG • *Department of Microchemistry, Proteomics and Lipidomics, Genentech, Inc., South San Francisco, CA, USA*
WEI ZHANG • *Donnelly Centre for Cellular and Biomolecular Research, University of Toronto, Toronto, ON, Canada*
JINGHUI ZHAO • *Department of Cell Biology, Harvard Medical School, Boston, MA, USA; AbbVie, Cambridge, MA, USA*

Part I

Methods to Uncover the Mechanisms of Action for Enzymes that Assemble or Disassemble Poly-Ubiquitin Chains

Chapter 1

Characterization of RING-Between-RING E3 Ubiquitin Transfer Mechanisms

Katherine H. Reiter and Rachel E. Klevit

Abstract

Protein ubiquitination is an essential posttranslational modification that regulates nearly all cellular processes. E3 ligases catalyze the final transfer of ubiquitin (Ub) onto substrates and thus are important temporal regulators of ubiquitin modifications in the cell. E3s are classified by their distinct transfer mechanisms. RING E3s act as scaffolds to facilitate the transfer of Ub from E2-conjugating enzymes directly onto substrates, while HECT E3s form an E3~Ub thioester intermediate prior to Ub transfer. A third class, RING-Between-RING (RBR) E3s, are classified as RING/HECT hybrids based on their ability to engage the E2~Ub conjugate via a RING1 domain while subsequently forming an obligate E3~Ub intermediate prior to substrate modification. RBRs comprise the smallest class of E3s, consisting of only 14 family members in humans, yet their dysfunction has been associated with neurodegenerative diseases, susceptibility to infection, inflammation, and cancer. Additionally, their activity is suppressed by autoinhibitory domains that block their catalytic activity, suggesting their regulation has important cellular consequences. Here, we identify technical hurdles faced in studying RBR E3s and provide protocols and guidelines to overcome these challenges.

Key words Ubiquitin, Transthiolation, Ubiquitin E3 ligase, RBR, HHARI, Parkin, HOIP, RING-Between-RING

1 Introduction

Ubiquitin (Ub) modifications are generated by the sequential transfer of Ub between a trio of enzymes (E1-activating, E2-conjugating, and E3-ligase) onto substrates. The resulting products have dramatic effects on protein function and half-life, and therefore their generation must be tightly regulated. E3 ligases provide both spatial and temporal control over the final ubiquitin transfer step by acting as scaffolds that recruit both E2~Ub (~denotes thioester) and substrates. E3s are classified into three families based on their distinct transfer mechanisms: RING (*R*eally *I*nteresting *N*ew *G*ene), HECT (*h*omologous to *E*6AP *c*arboxyl *t*erminus), and RBR (RING-Between-RING). RING E3s use an eponymous RING domain to engage E2~Ub and promote Ub

Thibault Mayor and Gary Kleiger (eds.), *The Ubiquitin Proteasome System: Methods and Protocols*, Methods in Molecular Biology, vol. 1844, https://doi.org/10.1007/978-1-4939-8706-1_1,

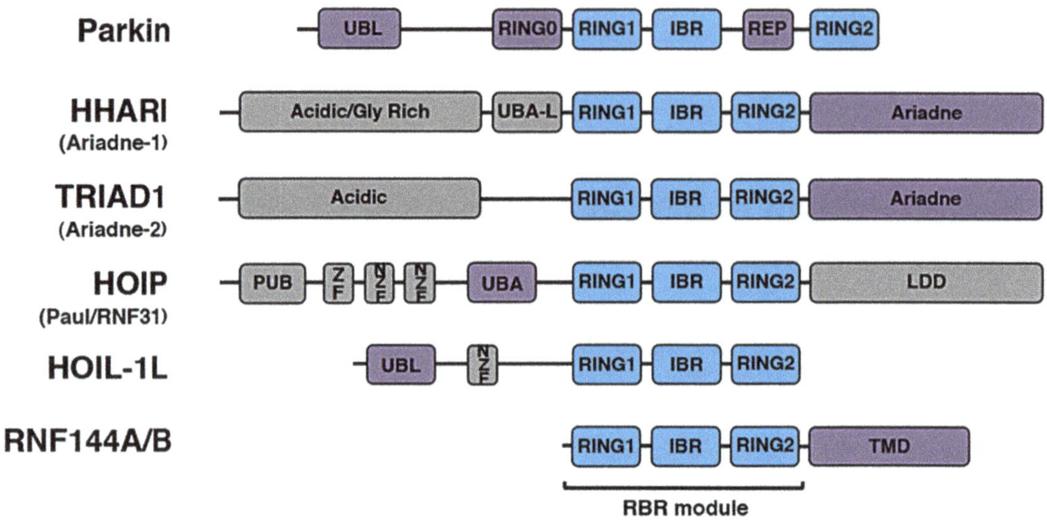

Fig. 1 Domain architecture of RBR enzymes. Domain colors represent: RING1-IBR-RING2 module (blue), autoinhibitory domains (purple), accessory domains (grey)

transfer directly onto substrate residues, most often a lysine. HECT E3s form an obligate E3~Ub thioester intermediate prior to Ub transfer. RBRs have been dubbed as RING/HECT hybrids based on their use of a RING1 domain to bind E2~Ub and an active-site cysteine to mediate transfer of ubiquitin through an E3~Ub intermediate [1–5]. All RBRs share a structurally similar catalytic module, composed of an E2-binding domain (RING1), an in-between-RING domain (IBR) of unknown function, and a catalytic domain (RING2) (Fig. 1). Despite their shared catalytic core, it is becoming increasingly clear that RBR E3s possess distinct features and properties associated with their unique functions that demand further investigation.

RBR E3 ligases comprise an important class of enzymes involved in neurodegeneration (Parkin), organogenesis (HHARI), and inflammation (HOIP), yet little is known about how they perform their cellular functions [6–10]. Study of RBR E3s has been greatly facilitated by the discovery that these enzymes contain an active-site cysteine residue, allowing for the straightforward generation of structurally intact ligase-dead forms of the RBRs [1]. Subsequent structural and biochemical studies of several RBR E3s (Parkin, HHARI, and HOIP) provide a mechanistic foundation on which to further understand RBR function, but much remains to be learned from these well-studied cases as well as for other members of the family [11–19]. As outlined below, basic information is lacking for virtually every member of the RBR family. In this article, we describe available approaches and tools to address the enzymatic E2:RBR pairs and their activity.

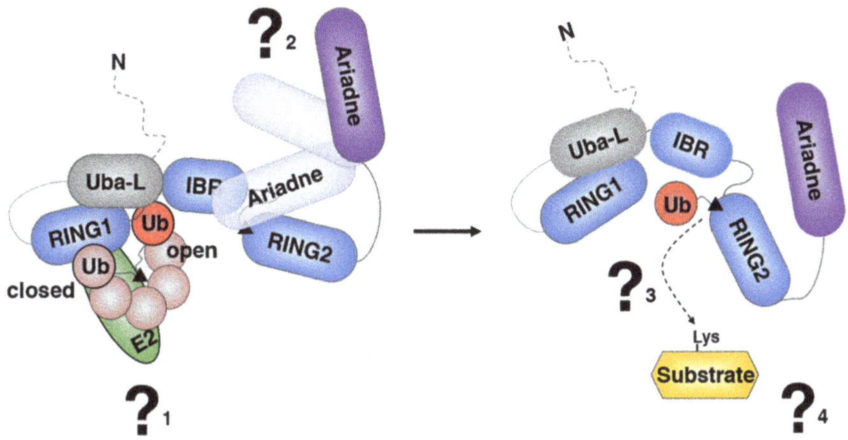

Fig. 2 Fundamental questions surrounding RBR ubiquitin transfer. HHARI domain structure shown as an example. (1) What E2s work with each RBR? (2) How are RBR enzymes activated? (3) What ubiquitin products does each RBR form? (4) What substrates do RBRs target?

To fully investigate the cellular function of a given RBR E3, answers to four questions are needed (Fig. 2):

Which E2s work with each RBR?

How are RBR enzymes activated?

What ubiquitin products does an RBR E3 generate?

What are the substrates for a given RBR E3?

Each question poses its own set of challenges and technical hurdles, as discussed below. Furthermore, there is a chicken and egg situation when none of the information is available: for example, how can one determine the products of an RBR-catalyzed reaction if one does not know with which of the ~36 human E2s it works. An additional layer of complexity and challenge arises from the fact that most RBR E3s are auto-inhibited. Thus, one must first discover ways in which auto-inhibition can be released before other in vitro investigations can be carried out.

1.1 RBR Enzymes Are Auto-inhibited

Ub transfer by RBR E3s is generally suppressed by an auto-inhibitory mechanism (for a recent review, *see* [20]). Of the human RBR E3s whose auto-inhibition has been studied, a majority involve domains outside the RING1-IBR-RING2 catalytic module that physically occlude the RING2 catalytic site. These include Parkin, HHARI, TRIAD1, and, possibly HOIP [2, 4, 5, 16, 21]. Thus, in vitro studies of RBR activity require the release of auto-inhibition. In some cases, such as HHARI (*see* below), inhibition can be released by removal of the inhibitory domain [1, 16]. Such an approach can generate constructs of an RBR E3 that can (1) bind E2~Ub and (2) generate the E3~Ub intermediate, but it is not currently known whether removal of a non-RBR domain will have an impact on either the types of products generated or on the ability of the RBR E3 to transfer Ub to a substrate. We briefly review what is currently known about how to generate activated forms of RBR E3s, as these may serve as guides for how to proceed in systems for which the information is lacking.

HHARI auto-inhibition is mediated by its C-terminal Ariadne domain, which blocks the RING2 active site (Fig. 1) [16]. HHARI retains its ability to bind an E2 from the auto-inhibited state, but that binding does not release the inhibition [15, 22]. This observation is consistent with recent structural studies in which UbcH7~Ub bound to auto-inhibited HHARI still has ~50 Å separation between E2 and E3 active sites, suggesting additional conformational changes are required for transthiolation to occur [15–17]. One mechanism by which inhibition is released involves a second E3 ligase, a cullin-RING ligase (CRL) [2, 22]. Binding of HHARI to a neddylated form of either CUL-1 or CUL-3 activates HHARI to allow transfer of Ub via its RING2 active site [2]. This, in turn, promotes HHARI-mediated mono-ubiquitination of cullin substrates [22]. This mode of activation has been demonstrated both in cellular and in vitro contexts for known cullin substrates, making it a viable, if technically difficult, method to generate active HHARI [2, 7, 22]. However, it is not yet known whether HHARI activated in this manner is capable of modifying other (non-cullin-specific) substrates. Activation of HHARI can also be achieved by disrupting the intramolecular interaction between the Ariadne domain and RING2, either by deletion of the Ariadne domain, or by point mutations at the RING2:Ariadne interface (Table 1) [16, 22].

Table 1
Active RBR constructs that retain activity in the absence of activating partners

RBR	Construct	Residues	Mechanism of activation	Ref
HHARI	RING1-IBR-RING2	177-395	Catalytic module. Deletion of auto-inhibitory Ariadne domain	[1, 15, 37]
	Open A	F430A, E431A, E503A	Point mutations disrupt the RING2: Ariadne binding interface	[16, 22]
	Open B	R420A, N423A, E503A	Point mutations disrupt the RING2: Ariadne binding interface	[15, 16]
	Δariadne	1-400	Deletion of auto-inhibitory Ariadne domain	[16, 17]
Parkin	RING1-IBR-REP-RING2	217-465	Catalytic module. Deletion of auto-inhibitory N-terminus	[1, 13, 37]
HOIP	RING1-IBR-RING2-LDD	697-1072	Catalytic module. Deletion of auto-inhibitory N-terminus	[4, 18, 37]
TRIAD1	Δariadne	1-348	Deletion of auto-inhibitory Ariadne domain	[37]

Although all RBR E3s have a RING1-IBR-RING2 module in common, flexible linkers and additional idiosyncratic elements give rise to unique inter-domain orientations and relationships and, therefore, distinct activation requirements. Some of these inter-domain interactions are the source of auto-inhibition: for example, the Parkin RING0 domain occludes its RING2 active site, while a small repressor domain (REP) that is located between the IBR and RING2 blocks the E2 binding site in RING1 (Fig. 1) [11, 14, 21]. Parkin activation is coupled to phosphorylation events that lead to the onset of mitophagy [23, 24]. The mitochondrial-associated kinase PTEN-induced kinase (PINK1) activates Parkin by two distinct mechanisms. In response to mitochondrial damage, PINK1 phosphorylates the Parkin UBL domain, releasing it from the RBR module and increasing Parkin ligase activity [25–28]. Additionally, ubiquitin is phosphorylated by PINK1, and binding of phospho-Ub to Parkin allosterically activates the E3 [29, 30]. These phosphorylation events lead to the translocation of Parkin to damaged mitochondria, where it can tag them for removal. The study of Parkin is an active area of investigation, given its immediate medical relevance (Parkinson's disease), and this is now aided by the commercial availability of recombinant phospho-Ub and PINK1.

RBR activation by E3:E3 complex formation is not limited to HHARI, as the linear ubiquitin chain assembly complex (LUBAC) forms linear chains through the action of two RBR-containing subunits: HOIP, which harbors catalytic activity, and HOIL-1L. Interactions between HOIP's ubiquitin-associated domain and the ubiquitin-like domain (UBL) of either HOIL-1L or an additional subunit, Shank-associated RH domain-interacting protein (SHARPIN), promote complex formation and activation [31–34]. Both HOIP and HOIL-1L contain catalytic cysteine residues; however, linear chain activity resides in the HOIP RING2 domain and a C-terminal linear ubiquitin chain determining domain (LDD). HOIP activity is inhibited by domains adjacent to the HOIP RBR module. While the exact mechanism of auto-inhibition is unknown, HOIP catalytic activity is restored through binding to the HOIL-1 UBL domain and is further enhanced through interactions with full-length HOIL-1 [5]. This enhanced activity is dependent on the HOIL-1 catalytic cysteine, even though HOIL-1 is not intrinsically active on its own [5]. To date, in vitro studies of HOIP activity have relied on "active" constructs containing the RBR and C-terminal LDD domain (*see* Table 1). While a structure of full-length HOIP has remained elusive, several crystal structures have captured snapshots of the HOIP catalytic domain poised for linear chain assembly [18, 19]. These structures helped to define the essential role of the LDD domain in linear ubiquitin chain formation by orienting an acceptor ubiquitin N-terminus proximal to a donor ubiquitin at the HOIP active site. Additionally, a structure of $\mathrm{HOIP}^{\mathrm{RBR\text{-}LDD}}$ in complex with UbcH5~Ub captured the E2 in an elongated open conformation [19]. Notably, there is an E2 active site in close proximity to an E3 active site in this crystal, suggesting a model for E2-to-RBR ubiquitin transfer [19]. In the future, revealing whether these mechanisms extend to all RBRs and identifying idiosyncratic details of individual RBRs will be of great interest to the field.

1.2 RBR E3 and Their E2s

Like other E3 families, RBRs will likely interact with multiple E2 enzymes in response to distinct cellular signals and enzyme expression levels. To date, most RBRs studied are functional with UbcH7 and UbcH5, yet their affinities for the E2s vary substantially, a factor which may have important ramifications in the cell. Additionally, while UbcH5 is both lysine and cysteine reactive, the reactivity profile of UbcH7 is limited to only cysteines, making it a dedicated E2 for RBR and HECT ligases. To date, high-throughput E2 screens have been limited to RING and HECT ligases; thus the general interaction and reactivity of E2s with RBRs require further analysis. There are reports of other E2s working in concert with RBR E3s (e.g., UBE2K works with LUBAC [35]), but the complete cadre of E2s for a given RBR E3 has yet to be defined. Additionally, the sequence of events by which RBRs are activated and E2~Ub binds is unknown, and therefore the ability to detect E2 interactions may require active RBR constructs, as described above. Understanding how RBRs interact and function with a select set of E2s is a fundamental question that must be addressed as we begin to probe RBR ubiquitin transfer mechanisms.

1.3 RBR E2~Ub:E3 Transfer Mechanisms

Mechanistically, RING1 domains are distinct from traditional RINGs in their ability to restrict E2~Ub transfer to a catalytic cysteine. Canonical RING domains generally employ a "linchpin" residue that restricts the Ub of a bound E2~Ub conjugate to be in a "closed state," which is intrinsically more active toward lysines [36]. RBR RING1s, however, lack a linchpin residue and instead promote an open conformation of E2~Ub to prevent aminolysis [15, 17, 19]. In the case of HHARI, an extended loop in RING1 serves as a wedge that disfavors the closed E2~Ub conjugate, thereby promoting an open state and transthiolation with the E3 active site [15, 17]. However, HOIP RING1 does not contain an extended loop, yet the HOIP:UbcH5~Ub structure depicts an open conjugate, suggesting additional interactions may restrict closed E2~Ub conformations [19]. Recent studies have shown that HHARI, Parkin, and HOIP RING2 domains contain a weak ubiquitin binding surface that is important for transthiolation [37]. A donor ubiquitin binding site on the HOIP IBR-RING2 linker identified in the HOIP:UbcH5-Ub crystal structure suggests this interaction may be a common feature required for Ub transfer to the RBR active site [19].

1.4 RBR Product Formation

It is now a generally accepted tenet that the last moiety holding an activated Ub prior to its transfer to substrate determines the type of ubiquitin product formed on the substrate. In the case of canonical RING E3s, it is the E2 that determines product, while the formation of an RBR~Ub intermediate shifts this important function to the RBR E3. This "rule" predicts that an E2's activity with traditional RING E3s may not reflect the final product formed when that E2 works with RBR E3s. Such a relationship is illustrated by the Ube2K:LUBAC interaction. In the absence of any E3, or in the presence of RING E3s, Ube2K builds K48-linked poly-Ub chains, but when paired with LUBAC, the RBR HOIP uses the Ub provided by Ube2K to assemble linear Ub chains despite the intrinsic Ube2K preference [35]. The type of ubiquitin product formed is not a conserved characteristic among RBRs: HHARI modifies substrates with mono-ubiquitin, HOIP forms linear ubiquitin chains, and Parkin forms poly-ubiquitin chains [15, 22, 38, 39]. In many cases, our understanding of the RBR E3~Ub transfer mechanism has been limited to auto-ubiquitination of the E3, but whether product preferences hold true for bona fide substrates remains to be determined. Product type can be determined in vitro in several ways. Ubiquitin variants, such as lysine-less Ub (K0) that is unable to form lysine-linked chains, distinguish poly-Ub chains from multiple-mono-Ub events, an important distinction as both products may appear as high-molecular-weight ubiquitin smears on reducing SDS-PAGE gels. Additionally, chain type may be parsed out using Ub mutants that carry a single lysine (with all others mutated) to determine linkage specificity. As noted above, ubiquitin interactions with RBRs play an important role in the transfer mechanism, so any variant of Ub should be used with caution, as mutations that disrupt these interactions may confound interpretation of results obtained with them. Therefore, mass spectrometric analysis of products should be performed for a rigorous identification of product type.

1.5 RBR Substrates

Interactions between E3s and substrates are often transient. Consequently, the identification of E3:substrate pairs by common binding techniques such as yeast two hybrids and co-immunoprecipitation can be challenging and are often not fruitful. Quantitative proteomics studies have revealed numerous binding partners for HHARI and Parkin, yet verification of actual substrates has been slow [22, 40]. Akin to studying E2~Ub transthiolation, the study of RBR substrate interactions and modifications is complicated by auto-inhibition and a lack of information regarding the substrate binding domains within RBR E3s. For the most commonly studied RBRs (Parkin, HHARI, HOIP), it has become routine to use the core catalytic module (RING1-IBR-RING2) to generate E3 activity in assays; however, whether these domains are sufficient for specific substrate modification has yet to be determined.

1.6 Future Approaches

Although structural information revealing an active state of a full-length RBR has been elusive, numerous structures of inhibited states reveal that in all cases significant conformational changes are necessary to bridge the gap between E2~Ub and E3 active sites. Thus, a clear goal is to determine how the active sites are brought together to generate the E3~Ub intermediate that is required for RBR function. In addition to purely structural approaches, more sensitive tools are being developed for the detection of E3~Ub conjugates. Electrophilic activity-based probes, such as Ub-vinyl methyl ester (VME) have been widely used in the field to trap thioesters [41]. However, these probes are stoichiometric suicide inhibitors and therefore limit the ability to assess turnover. Recently, a fluorescent Ub probe (UbFluor) was used to report on the transthiolation activity of Parkin [42]. Termed a "bypass system" (ByS), given that E1, ATP, and E2 are not required for the probe's modification of the RING2 active site, this probe shows great promise for use with other RBRs to simplify the study of the catalytically important intermediate state.

The biochemical assays described in this chapter are a starting point to further define RBR Ub transfer steps. Here, we describe methods to screen for RBR-reactive E2s as well as to characterize RBR product formation, the output of which requires an active RBR construct. We provide a summary of active constructs described in the literature (Table 1) as a basis for designing active constructs for closely related RBRs. These studies will naturally be iterative at first and limited to the use of known RBR E2s (UbcH5 and UbcH7). Once an E2/RBR molecular network is established, one can begin to uncover the details of how these E3s orchestrate Ub transfer.

2 Materials

Prepare all solutions using ultrapure water (resistivity of 18.2 MΩ cm at 25 °C). All stock solutions should be filtered using 0.22 μm filter units.

2.1 Auto-ubiquitination Components

1. Reaction buffer: 0.02 M HEPES pH 7.5, 0.1 M NaCl, 0.5 mM DTT.
2. Recombinant human HA-Ub (*see* **Notes 1** and **2**).
3. Recombinant human E1 (*see* **Notes 1** and **2**).
4. Recombinant human UbcH5 and UbcH7 (*see* **Notes 1** and **2**).
5. Recombinant human RBR enzymes (*see* **Notes 1** and **2**).
6. ATP stock solution: 0.1 M adenosine triphosphate (ATP), 0.1 M $MgCl_2$. Dissolve ATP in H_2O and adjust pH to 7.0 with sodium bicarbonate. Prepare immediately before use.

7. 1 M DTT stock solution. Dissolve DTT in H_2O and store small aliquots at −20 °C.
8. 2× reducing SDS sample buffer.
9. PVDF filter for immunoblotting.

2.2 Yeast Two-Hybrid (Y2H) Assays

1. Y2H vectors: Bait and prey expression vectors designed to express fusion proteins of the GAL4 DNA-binding domain or -activating domain with your bait and prey protein, respectively. Vector backbones are available commercially, and E2 prey clones have been deposited in Addgene [43].
2. Yeast host strain: AH109 (Clonetech).
3. YPAD medium: Dissolve 20 g bacto peptone, 10 g bacto yeast extract, 20 g glucose, and 40 mg adenine sulfate dihydrate in 1000 mL H_2O. Autoclave and store at room temperature or 4 °C.
4. Synthetic dextrose minimal media (SD) -Leu/-Trp plates. SD minimal media packets are available commercially (*see* **Note 3**). Dissolve packets in H_2O and autoclave. Pour plates and let cool. Store at 4 °C.
5. Selective media SD -His/-Leu/-Trp plates. SD minimal media packets are available commercially (*see* **Note 3**). Prepare as SD -Leu/-Trp plates above. Once media is cool to the touch, supplement with filter sterilized 3-amino-1,2,4-triazole (3AT) to 0–10 mM (optimized to reduce false positives). Pour plates and let cool. Store at 4 °C.

3 Methods

3.1 Designing an Active RBR Construct

The study of RBR ubiquitin transfer requires active E3 constructs, yet assessment of activity requires a functional E2:RBR pair. Two human E2s, UbcH7 and UbcH5, appear to be active with every RBR E3 tested to date, making them good choices for initial studies. Given the limited knowledge of RBR activation, active constructs are typically designed to lack the auto-inhibition domain and, at a minimum, contain the RBR module. Due to interactions between the RBR module and auxiliary domains, the presence of additional domains may be required to achieve solubility. Table 1 includes a list of commonly used active RBR constructs that have been successfully purified in the literature. For less studied RBRs, activated constructs will need to be designed de novo. Each RBR seems to have its own mechanisms for activation; thus, inhibitory domains surrounding each RBR module will need to be identified by domain deletion. Additionally, the identification of point mutations aimed at disrupting inhibitory domain interactions can provide a means for working with more complete constructs (*see*

Note 4). RBR enzymes are highly active; thus the types of products assembled during auto-ubiquitination assays will need to be further validated (*see* below).

3.2 In Vitro Auto-ubiquitination Assay

The following method is modified from our studies on HHARI auto-ubiquitination:

1. Set up a 50 μL reaction on ice: Mix 1 μM E1 enzyme, 5 μM UbcH7, 20 μM HA-Ub, and 2 μM active RBR enzyme in 47.5 μL 20 mM HEPES pH 7.5, 100 mM NaCl, and 0.5 mM DTT (*see* **Note 5**). Incubate reaction mixture and 100 mM ATP/$MgCl_2$ stock at 37 °C for 1 min.
2. Initiate reaction by adding 2.5 μL ATP stock and incubate reaction at 37 °C for 10 min.
3. Take 10 μL aliquots at 0, 1, 5, and 10 min time-points, and add 10 μL of 2× reducing SDS sample buffer to stop the reaction. Boil for 5 min.
4. Separate the reaction mixture by 15% SDS-PAGE.
5. Transfer to a PVDF filter and perform standard immunoblotting analysis with anti-HA antibody (*see* **Note 6**).

3.3 Defining E2:RBR Pairs: Yeast Two-Hybrid Screening

Our understanding of RBR transthiolation and substrate modification has been limited to the study of individual E2:E3 pairs. To gain insight into RING1 E2 recruitment and functional transthiolation reactions, high-throughput studies and/or use of a complete E2 panel can identify which E2s are capable of pairing with an RBR E3. These types of screens have been performed for RING and HECT E3s but have yet to be pursued for RBRs. Many E2:E3 interactions are of modest affinity and are therefore difficult to capture by pull-down assays. Given there are only 14 human RBRs, yeast two-hybrid (Y2H) assays provide a tractable way to explore the range of E2 interactions, using previously created E2 Y2H libraries available on Addgene [43]. Targeted Y2H assays have been used to identify E2:E3 pairs but may require the use of low-stringency conditions to identify potential positive hits [43]. These must be validated biochemically, as false positives are likely to appear under such liberal conditions.

Generate Y2H clones as described [43]. Briefly:

1. Design primers for the full-length and active deletion coding sequences for each RBR of interest (*see* **Note 7**). Include the endogenous stop codon and restriction enzyme sites for ligation into the Gal4-activation domain (GAD) or Gal4 DNA-binding domain (GBD) plasmids.
2. PCR amplify the RBR coding sequence from human cDNA (*see* **Note 8**).

3. Digest RBR amplification and pGAD/pGBD plasmid with the appropriate restriction enzymes.
4. Ligate RBR insert into pGAD/pGBD plasmid.
5. Sequence verify final constructs.
6. Co-transform E2 bait clones (available from Addgene) and RBR prey clones into yeast strain AH109 and grow in YPAD media for 12 h.
7. Select for positive transformants on minimal SD -Leu/-Trp plates for 24 h.
8. Inoculate 100 μL of sterile water with a single colony and spot via serial dilutions onto selective media (SD -His/-Leu/-Trp) supplemented with 0–10 mM 3-amino-1,2,4-triazole (3AT).
9. Incubate yeast at 30 °C and score growth for 7 days.
10. Verify positive interactions via additional binding studies (e.g., co-immunoprecipitation, pulldowns, NMR) and assess activity with in vitro auto-ubiquitination assays.

3.4 Determining the Types of RBR Ub Products

Each RBR dictates the types of Ub products it forms on substrates. Thus, knowledge of the associated E2(s) is not sufficient to predict the RBR product formation. Additionally, RBRs are highly active, and as seen with HHARI, can form multi-mono-Ub products during auto-ubiquitination assays. Multiple mono- vs. poly-Ub can be distinguished using lysine-less ubiquitin mutants (Ub^{K0}), while distinct chain linkages can be assessed using lysine-only ubiquitin mutants (ex: Ub^{K48}), in which all other lysine residues are mutated to arginine. Rigorous characterization of linkage type should include mass spectrometric analysis of products. Additionally, whether substrates themselves play a role in RBR product formation is not yet known and will need to be assessed on a substrate basis.

1. Perform auto-ubiquitination assays, as described above, in the presence of one of the following ubiquitin variants: HA-Ub^{K0}, -Ub^{K6}, -Ub^{K11}, -Ub^{27}, -Ub^{29}, -Ub^{33}, -Ub^{48}, or -Ub^{63}.
2. Multiple mono-ubiquitination will be revealed as collapsed chain formation (1–2× ubiquitin modifications) in the presence of Ub^{K0} (*see* Fig. 4e [7] for an example Ub pattern).

4 Notes

1. Recombinant HA-Ub, human and wheat E1 enzymes, UbcH5, and UbcH7 are commercially available. In our lab, we purify these enzymes using an *E. Coli* BL21 (λDE3) expression system [1, 15].

2. Auto-ubiquitination assays require activation of RBRs. *See* Table 1 for a list of commonly used active RBR constructs. When making RBRs in-house, we supplement our RBR expression media with 0.2 mM $ZnCl_2$.
3. SD minimal dropout media can be made in-house if your lab has access to individual amino acid stock reagents.
4. RBR structural studies have revealed several interaction surfaces between the RBR module and auxiliary domains, which likely affect the stability of the protein. For example, residues C-terminal to the HOIP RBR module are required for solubility. Additionally, when selecting active RBR constructs for substrate modification assays, one must take into consideration that RBR substrate binding domains are unknown.
5. We use human E1 and UbcH7 in our HHARI ubiquitination assays; however, wheat E1 and UbcH5 can also be used with minimal impact to auto-ubiquitination results.
6. When possible, auto-ubiquitination assays should include a comparison to auto-inhibited and catalytically dead E3s. Additionally, the use of T7-tagged RBR constructs allows for the direct assessment of modified E3 species.
7. Several auto-inhibited and active constructs should be pursued for each RBR, as the sequence of E2 binding and RBR release from auto-inhibition has yet to be parsed out for all RBRs.
8. Human cDNA libraries are commercially available or can be purified from cell lines. Y2H kits with premade cDNA libraries are available through commercial vendors.

References

1. Wenzel DM, Lissounov A, Brzovic PS, Klevit RE (2011) UBCH7 reactivity profile reveals parkin and HHARI to be RING/HECT hybrids. Nature 474(7349):105–108. https://doi.org/10.1038/nature09966
2. Kelsall IR, Duda DM, Olszewski JL, Hofmann K, Knebel A, Langevin F, Wood N, Wightman M, Schulman BA, Alpi AF (2013) TRIAD1 and HHARI bind to and are activated by distinct neddylated Cullin-RING ligase complexes. EMBO J 32(21):2848–2860. https://doi.org/10.1038/emboj.2013.209
3. Ho SR, Mahanic CS, Lee YJ, Lin WC (2014) RNF144A, an E3 ubiquitin ligase for DNA-PKcs, promotes apoptosis during DNA damage. Proc Natl Acad Sci U S A 111(26): E2646–E2655. https://doi.org/10.1073/pnas.1323107111
4. Smit JJ, Monteferrario D, Noordermeer SM, van Dijk WJ, van der Reijden BA, Sixma TK (2012) The E3 ligase HOIP specifies linear ubiquitin chain assembly through its RING-IBR-RING domain and the unique LDD extension. EMBO J 31(19):3833–3844. https://doi.org/10.1038/emboj.2012.217
5. Stieglitz B, Morris-Davies AC, Koliopoulos MG, Christodoulou E, Rittinger K (2012) LUBAC synthesizes linear ubiquitin chains via a thioester intermediate. EMBO Rep 13 (9):840–846. https://doi.org/10.1038/embor.2012.105
6. Kitada T, Asakawa S, Hattori N, Matsumine H, Yamamura Y, Minoshima S, Yokochi M, Mizuno Y, Shimizu N (1998) Mutations in the parkin gene cause autosomal recessive juvenile parkinsonism. Nature 392 (6676):605–608. https://doi.org/10.1038/33416
7. Dove KK, Kemp HA, Di Bona KR, Reiter KH, Milburn LJ, Camacho D, Fay DS, Miller DL,

Klevit RE (2017) Two functionally distinct E2/E3 pairs coordinate sequential ubiquitination of a common substrate in Caenorhabditis elegans development. Proc Natl Acad Sci U S A 114(32):E6576–E6584. https://doi.org/10.1073/pnas.1705060114

8. Qiu X, Fay DS (2006) ARI-1, an RBR family ubiquitin-ligase, functions with UBC-18 to regulate pharyngeal development in C. elegans. Dev Biol 291(2):239–252. https://doi.org/10.1016/j.ydbio.2005.11.045
9. Iwai K, Fujita H, Sasaki Y (2014) Linear ubiquitin chains: NF-kappaB signalling, cell death and beyond. Nat Rev Mol Cell Biol 15 (8):503–508. https://doi.org/10.1038/nrm3836
10. Ikeda F (2015) Linear ubiquitination signals in adaptive immune responses. Immunol Rev 266 (1):222–236. https://doi.org/10.1111/imr.12300
11. Kumar A, Aguirre JD, Condos TE, Martinez-Torres RJ, Chaugule VK, Toth R, Sundaramoorthy R, Mercier P, Knebel A, Spratt DE, Barber KR, Shaw GS, Walden H (2015) Disruption of the autoinhibited state primes the E3 ligase parkin for activation and catalysis. EMBO J 34(20):2506–2521. https://doi.org/10.15252/embj.201592337
12. Riley BE, Lougheed JC, Callaway K, Velasquez M, Brecht E, Nguyen L, Shaler T, Walker D, Yang Y, Regnstrom K, Diep L, Zhang Z, Chiou S, Bova M, Artis DR, Yao N, Baker J, Yednock T, Johnston JA (2013) Structure and function of Parkin E3 ubiquitin ligase reveals aspects of RING and HECT ligases. Nat Commun 4:1982. https://doi.org/10.1038/ncomms2982
13. Trempe JF, Sauve V, Grenier K, Seirafi M, Tang MY, Menade M, Al-Abdul-Wahid S, Krett J, Wong K, Kozlov G, Nagar B, Fon EA, Gehring K (2013) Structure of parkin reveals mechanisms for ubiquitin ligase activation. Science 340 (6139):1451–1455. https://doi.org/10.1126/science.1237908
14. Wauer T, Komander D (2013) Structure of the human Parkin ligase domain in an autoinhibited state. EMBO J 32(15):2099–2112. https://doi.org/10.1038/emboj.2013.125
15. Dove KK, Olszewski JL, Martino L, Duda DM, Wu XS, Miller DJ, Reiter KH, Rittinger K, Schulman BA, Klevit RE (2017) Structural studies of HHARI/UbcH7 approximately Ub reveal unique E2 approximately Ub conformational restriction by RBR RING1. Structure 25(6):890–900.e5. https://doi.org/10.1016/j.str.2017.04.013
16. Duda DM, Olszewski JL, Schuermann JP, Kurinov I, Miller DJ, Nourse A, Alpi AF, Schulman BA (2013) Structure of HHARI, a RING-IBR-RING ubiquitin ligase: autoinhibition of an Ariadne-family E3 and insights into ligation mechanism. Structure 21 (6):1030–1041. https://doi.org/10.1016/j.str.2013.04.019
17. Yuan L, Lv Z, Atkison JH, Olsen SK (2017) Structural insights into the mechanism and E2 specificity of the RBR E3 ubiquitin ligase HHARI. Nat Commun 8(1):211. https://doi.org/10.1038/s41467-017-00272-6
18. Stieglitz B, Rana RR, Koliopoulos MG, Morris-Davies AC, Schaeffer V, Christodoulou E, Howell S, Brown NR, Dikic I, Rittinger K (2013) Structural basis for ligase-specific conjugation of linear ubiquitin chains by HOIP. Nature 503 (7476):422–426. https://doi.org/10.1038/nature12638
19. Lechtenberg BC, Rajput A, Sanishvili R, Dobaczewska MK, Ware CF, Mace PD, Riedl SJ (2016) Structure of a HOIP/E2~ubiquitin complex reveals RBR E3 ligase mechanism and regulation. Nature 529(7587):546–550. https://doi.org/10.1038/nature16511
20. Dove KK, Klevit RE (2017) RING-between-RING E3s ligases: emerging themes amid the variations. J Mol Biol 429:3363. https://doi.org/10.1016/j.jmb.2017.08.008
21. Chaugule VK, Burchell L, Barber KR, Sidhu A, Leslie SJ, Shaw GS, Walden H (2011) Autoregulation of Parkin activity through its ubiquitin-like domain. EMBO J 30 (14):2853–2867. https://doi.org/10.1038/emboj.2011.204
22. Scott DC, Rhee DY, Duda DM, Kelsall IR, Olszewski JL, Paulo JA, de Jong A, Ovaa H, Alpi AF, Harper JW, Schulman BA (2016) Two distinct types of E3 ligases work in unison to regulate substrate Ubiquitylation. Cell 166 (5):1198–1214.e24. https://doi.org/10.1016/j.cell.2016.07.027
23. Pickrell AM, Youle RJ (2015) The roles of PINK1, parkin, and mitochondrial fidelity in Parkinson's disease. Neuron 85(2):257–273. https://doi.org/10.1016/j.neuron.2014.12.007
24. Eiyama A, Okamoto K (2015) PINK1/Parkin-mediated mitophagy in mammalian cells. Curr Opin Cell Biol 33:95–101. https://doi.org/10.1016/j.ceb.2015.01.002
25. Kondapalli C, Kazlauskaite A, Zhang N, Woodroof HI, Campbell DG, Gourlay R, Burchell L, Walden H, Macartney TJ, Deak M, Knebel A, Alessi DR, Muqit MM (2012) PINK1 is activated by mitochondrial

membrane potential depolarization and stimulates Parkin E3 ligase activity by phosphorylating serine 65. Open Biol 2(5):120080. https://doi.org/10.1098/rsob.120080

26. Shiba-Fukushima K, Imai Y, Yoshida S, Ishihama Y, Kanao T, Sato S, Hattori N (2012) PINK1-mediated phosphorylation of the Parkin ubiquitin-like domain primes mitochondrial translocation of Parkin and regulates mitophagy. Sci Rep 2:1002. https://doi.org/10.1038/srep01002
27. Kane LA, Lazarou M, Fogel AI, Li Y, Yamano K, Sarraf SA, Banerjee S, Youle RJ (2014) PINK1 phosphorylates ubiquitin to activate Parkin E3 ubiquitin ligase activity. J Cell Biol 205(2):143–153. https://doi.org/10.1083/jcb.201402104
28. Kazlauskaite A, Kondapalli C, Gourlay R, Campbell DG, Ritorto MS, Hofmann K, Alessi DR, Knebel A, Trost M, Muqit MM (2014) Parkin is activated by PINK1-dependent phosphorylation of ubiquitin at Ser65. Biochem J 460(1):127–139. https://doi.org/10.1042/BJ20140334
29. Koyano F, Okatsu K, Kosako H, Tamura Y, Go E, Kimura M, Kimura Y, Tsuchiya H, Yoshihara H, Hirokawa T, Endo T, Fon EA, Trempe JF, Saeki Y, Tanaka K, Matsuda N (2014) Ubiquitin is phosphorylated by PINK1 to activate parkin. Nature 510 (7503):162–166. https://doi.org/10.1038/nature13392
30. Wauer T, Simicek M, Schubert A, Komander D (2015) Mechanism of phospho-ubiquitin-induced PARKIN activation. Nature 524 (7565):370–374. https://doi.org/10.1038/nature14879
31. Ikeda F, Deribe YL, Skanland SS, Stieglitz B, Grabbe C, Franz-Wachtel M, van Wijk SJ, Goswami P, Nagy V, Terzic J, Tokunaga F, Androulidaki A, Nakagawa T, Pasparakis M, Iwai K, Sundberg JP, Schaefer L, Rittinger K, Macek B, Dikic I (2011) SHARPIN forms a linear ubiquitin ligase complex regulating NF-kappaB activity and apoptosis. Nature 471 (7340):637–641. https://doi.org/10.1038/nature09814
32. Gerlach B, Cordier SM, Schmukle AC, Emmerich CH, Rieser E, Haas TL, Webb AI, Rickard JA, Anderton H, Wong WW, Nachbur U, Gangoda L, Warnken U, Purcell AW, Silke J, Walczak H (2011) Linear ubiquitination prevents inflammation and regulates immune signalling. Nature 471 (7340):591–596. https://doi.org/10.1038/nature09816
33. Tokunaga F, Nakagawa T, Nakahara M, Saeki Y, Taniguchi M, Sakata S, Tanaka K, Nakano H, Iwai K (2011) SHARPIN is a component of the NF-kappaB-activating linear ubiquitin chain assembly complex. Nature 471 (7340):633–636. https://doi.org/10.1038/nature09815
34. Yagi H, Ishimoto K, Hiromoto T, Fujita H, Mizushima T, Uekusa Y, Yagi-Utsumi M, Kurimoto E, Noda M, Uchiyama S, Tokunaga F, Iwai K, Kato K (2012) A non-canonical UBA-UBL interaction forms the linear-ubiquitin-chain assembly complex. EMBO Rep 13(5):462–468. https://doi.org/10.1038/embor.2012.24
35. Kirisako T, Kamei K, Murata S, Kato M, Fukumoto H, Kanie M, Sano S, Tokunaga F, Tanaka K, Iwai K (2006) A ubiquitin ligase complex assembles linear polyubiquitin chains. EMBO J 25(20):4877–4887. https://doi.org/10.1038/sj.emboj.7601360
36. Pruneda JN, Littlefield PJ, Soss SE, Nordquist KA, Chazin WJ, Brzovic PS, Klevit RE (2012) Structure of an E3:E2~Ub complex reveals an allosteric mechanism shared among RING/U-box ligases. Mol Cell 47(6):933–942. https://doi.org/10.1016/j.molcel.2012.07.001
37. Dove KK, Stieglitz B, Duncan ED, Rittinger K, Klevit RE (2016) Molecular insights into RBR E3 ligase ubiquitin transfer mechanisms. EMBO Rep 17(8):1221–1235. https://doi.org/10.15252/embr.201642641
38. Ordureau A, Sarraf SA, Duda DM, Heo JM, Jedrychowski MP, Sviderskiy VO, Olszewski JL, Koerber JT, Xie T, Beausoleil SA, Wells JA, Gygi SP, Schulman BA, Harper JW (2014) Quantitative proteomics reveal a feedforward mechanism for mitochondrial PARKIN translocation and ubiquitin chain synthesis. Mol Cell 56(3):360–375. https://doi.org/10.1016/j.molcel.2014.09.007
39. Tokunaga F, Sakata S, Saeki Y, Satomi Y, Kirisako T, Kamei K, Nakagawa T, Kato M, Murata S, Yamaoka S, Yamamoto M, Akira S, Takao T, Tanaka K, Iwai K (2009) Involvement of linear polyubiquitylation of NEMO in NF-kappaB activation. Nat Cell Biol 11 (2):123–132. https://doi.org/10.1038/ncb1821
40. Sarraf SA, Raman M, Guarani-Pereira V, Sowa ME, Huttlin EL, Gygi SP, Harper JW (2013) Landscape of the PARKIN-dependent ubiquitylome in response to mitochondrial depolarization. Nature 496(7445):372–376. https://doi.org/10.1038/nature12043
41. Borodovsky A, Ovaa H, Kolli N, Gan-Erdene T, Wilkinson KD, Ploegh HL, Kessler BM (2002) Chemistry-based functional proteomics reveals novel members of

the deubiquitinating enzyme family. Chem Biol 9(10):1149–1159
42. Park S, Foote PK, Krist DT, Rice SE, Statsyuk AV (2017) UbMES and UbFluor: novel probes for RBR E3 ubiquitin ligase PARKIN. J Biol Chem 292:16539. https://doi.org/10.1074/jbc.M116.773200
43. Christensen DE, Brzovic PS, Klevit RE (2007) E2-BRCA1 RING interactions dictate synthesis of mono- or specific polyubiquitin chain linkages. Nat Struct Mol Biol 14(10):941–948. https://doi.org/10.1038/nsmb1295

Chapter 2

Single-Turnover RING/U-Box E3-Mediated Lysine Discharge Assays

Lori Buetow, Mads Gabrielsen, and Danny T. Huang

Abstract

RING and U-box ubiquitin ligases promote ubiquitin (Ub) transfer by priming Ub-conjugated E2 in a closed conformation to optimize the thioester bond for nucleophilic attack by substrate lysine. Here, we describe a single-turnover lysine discharge assay for direct assessment of the activity of any RING/U-box E3-E2~Ub complex.

Key words Ubiquitination, Lysine discharge, RING or U-box E3 ligases, Radiolabeling

1 Introduction

Ubiquitination is a posttranslational modification whereby the small 76 amino acid protein modifier ubiquitin (Ub) is covalently attached to a protein substrate [1]. Ubiquitination is achieved by the sequential actions of three enzymes: Ub-activating enzyme (E1), a Ub-conjugating enzyme (E2), and a Ub ligase (E3) [2, 3]. Initially, E1 activates Ub in a Mg^{2+}-ATP-dependent manner, and a thioester bond is formed between the catalytic cysteine of E1 and the diglycine motif at the C-terminus of Ub to produce E1~Ub complex, where ~ denotes a thioester (Fig. 1a). E1~Ub subsequently recruits an E2, and Ub is transferred to the active site cysteine of E2 via its tail, thereby forming an E2~Ub intermediate (Fig. 1b). Next, an E3 recruits E2~Ub and substrate to mediate the transfer of Ub to substrate (Fig. 1c). In this final step, an isopeptide/amide link is formed between Gly 76 of Ub and a free amino group from a lysine side chain or the N-terminus of a substrate protein (Fig. 1d).

The human genome encodes two E1s, approximately 40 E2s and over 600 putative E3s [4]. E3s are categorized into three families based on their Ub transfer mechanism: HECT, RING-between-RING (RBR), and RING/U-box [5]. HECT and RBR

Thibault Mayor and Gary Kleiger (eds.), *The Ubiquitin Proteasome System: Methods and Protocols*, Methods in Molecular Biology, vol. 1844, https://doi.org/10.1007/978-1-4939-8706-1_2, © Springer Science+Business Media, LLC, part of Springer Nature 2018

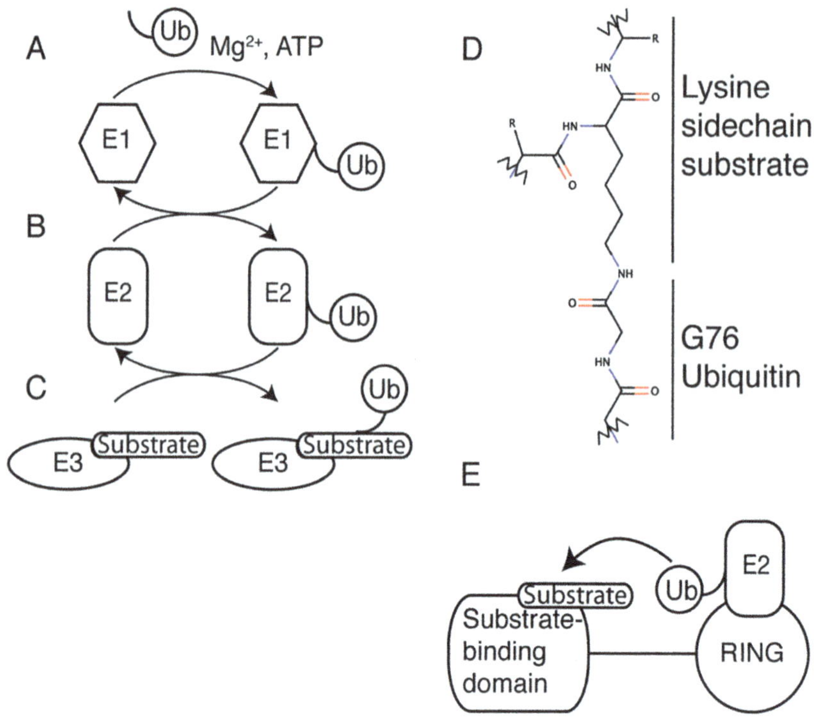

Fig. 1 Schematic overview of the ubiquitination cascade. (**a**) Ub is activated by E1 and (**b**) transferred to E2-conjugating enzyme. (**c**) E3 ligase mediates the transfer of Ub from E2~Ub, onto substrate. (**d**) Close-up of final isopeptide bond between Gly 76 of Ub and a Lys side chain on substrate. (**e**) RING E3 recruits E2~Ub and substrate to facilitate direct transfer of Ub from E2 onto the substrate

E3s form an E3~Ub complex prior to transferring Ub to substrate, whereas RING/U-box E3s catalyze the transfer of Ub directly from E2~Ub to substrate (Fig. 1e). A shared feature among all RING/U-box E3s is the presence of a RING or U-box domain that recruits E2~Ub and promotes Ub transfer. Simple RING/U-box E3s encode a RING/U-box domain and substrate-binding domain within a single polypeptide chain, whereas multisubunit E3s like cullin-RING ligases encode these domains on multiple polypeptide chains. Some simple RING E3s like RNF38 function as monomers, but others, like XIAP or BRCA1/BARD, are only functional as dimers [5].

Although binding is observed for many E2-E3 pairs in vitro, this does not always translate to function. For example, BRCA1/BARD binds the E2 UbcH7 [6] but does not enhance its reactivity [7]. E2s have an intrinsic reactivity that generally involves Ub transfer to a thiol as in a cysteine residue or to an amine as in a lysine residue or the N-terminus of a protein [8]. Based on these properties, it is essential to select an E2 for lysine discharge assays that has intrinsic reactivity toward lysine. Some E2s, like UbcH7, are only reactive toward cysteine [7], whereas others, like Ube2W,

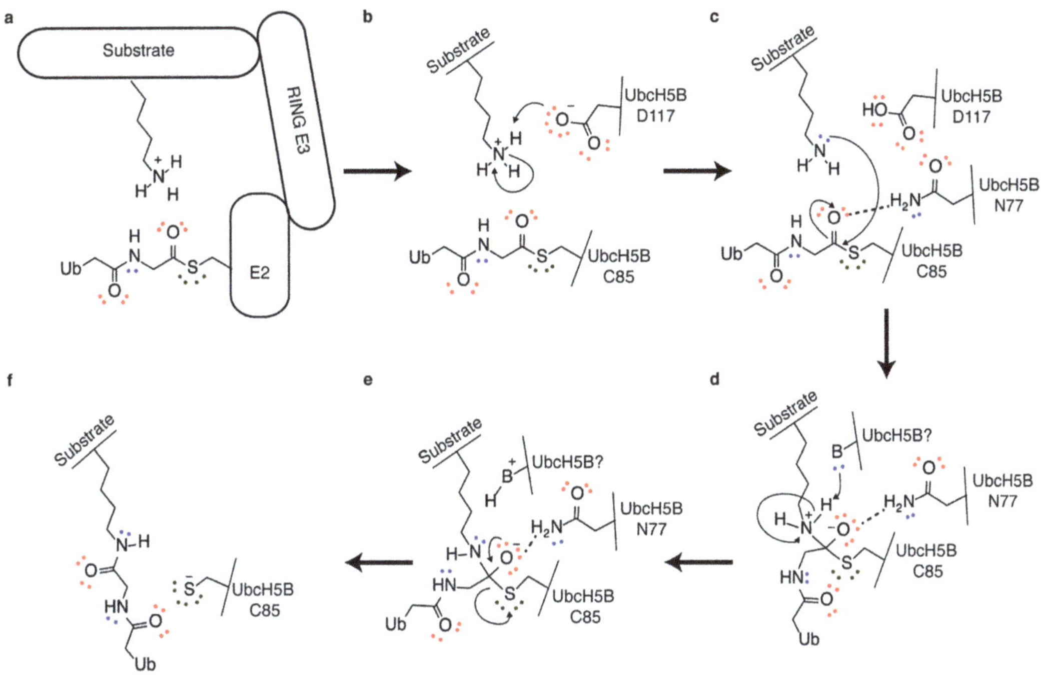

Fig. 2 Enzymatic mechanism of substrate ubiquitination by RING/U-box E3 ligases. (**a**) RING/U-box E3 ligases recruit substrate and E2~Ub, thereby forming a RING/U-box-substrate-E2~Ub complex; spatial restraints imposed by the E3 frequently play a key role in substrate lysine selectivity. (**b**) A residue from the E2 (D117 in UbcH5B) deprotonates the lysine side chain amine, thereby converting it to a nucleophile. (**c**) This nucleophile attacks the carbonyl of the E2~Ub thioester, forming an oxyanion intermediate in which covalent interactions are present between substrate, the catalytic Cys of E2 (C85 in UbcH5B) and Ub. The oxyanion intermediate is stabilized by N77 in UbcH5B. (**d**) Deprotonation of the intermediate, which may be performed by a residue from the E2, followed by (**e**) elimination of a thiol, (**f**) thereby freeing the catalytic Cys of E2 to accept another Ub and producing an isopeptide bond between the side chain of a substrate lysine and Gly 76 of Ub. MarvinSketch was used for drawing chemical structures and reactions [19]

only react with the N-terminal amino group of a protein or peptide [9–11].

The chemical reaction profile of RING/U-box mediated Ub transfer involves a number of steps with several species of intermediates. Initially, both E2~Ub and a protein substrate are recruited to form a RING/U-box-substrate-E2~Ub complex (Fig. 2a). The substrate lysine side chain or N-terminal amino group is then deprotonated to produce a nucleophile (Fig. 2b). This activated nucleophile subsequently attacks the carbonyl group of the E2~Ub thioester to form an oxyanion intermediate in which E2 is covalently linked to both Ub and substrate (Fig. 2c). In the final steps, a thiol group is eliminated (Fig. 2d, e), thereby forming an amide/isopeptide bond between substrate and ubiquitin (Fig. 2f).

To promote Ub transfer, RING/U-box E3s bring E2~Ub together with substrate and optimize the orientation of the E2~Ub thioester for nucleophilic attack [5]. In addition, some

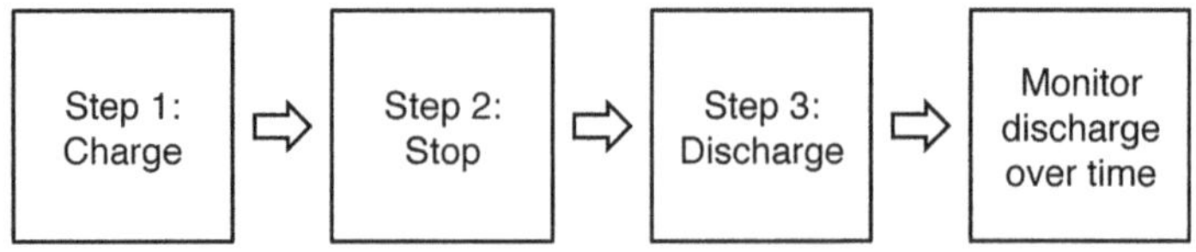

Fig. 3 Schematic overview a lysine discharge reaction. (Step 1) Charge: E1, E2, and Ub are mixed to produce E2~Ub. (Step 2) Stop: chemical and/or enzymatic reagents are added to inactivate E1, thereby preventing production of more E2~Ub or recharging of the E2 after discharge. (Step 3) Discharge: L-lysine, E3, and/or additional components are added and the discharge of E2~Ub over time is monitored by SDS-PAGE

RING/U-box E3s have been shown to impose conformational restraints that regulate which substrate lysines can access the active site of E2~Ub [12–14]. In this chapter, we describe how to perform a single-turnover lysine discharge assay in which we use SDS-PAGE to follow Ub transfer from E2~Ub to free lysine in solution (Fig. 3). This removes any dependency on substrate recognition and binding and facilitates analysis of the mechanism used by RING/U-box E3s to prime E2~Ub and promote Ub transfer. By using this reductionist approach combined with mutagenesis, we can explore the role of the E2 in coordinating donor Ub (the thioesterified Ub) in the presence and absence of E3, the role of the RING or U-box domain in coordinating donor Ub and priming E2~Ub for transfer, and the role of additional components in stimulating or inhibiting the actions of the RING or U-box domain in mediating Ub transfer.

A single-turnover lysine discharge reaction consists of a charge step in which Ub is loaded onto E2 by E1, a stop step in which reagents are added to inactivate E1 and prevent additional loading of Ub onto E2, and a final step in which lysine alone or a mixture of lysine with E3 and/or additional components is added to initiate discharge of E2~Ub (Fig. 3). Each step requires empirical optimization. E2-charging efficiency depends on the E1-E2 pair and can be influenced by a number of factors including changing the duration and/or temperature of the charging step, the source organism for E1, and relative concentrations of E1, E2, and Ub. In most cases, adding EDTA quenches the charge reaction. However, for some E1-E2 pairs, inclusion of apyrase together with EDTA seems to work more effectively. For a clear, visually convincing result, the disappearance of the E2~Ub band should be readily apparent over the time points selected and can be affected by adjusting factors like the concentration of lysine and/or E3 or running the assay at a different temperature for the discharge step.

We routinely use *Arabidopsis thaliana* Uba1 as our E1, human UbcH5B as our E2, and radiolabeled Ub (^{32}P-Ub). We have devised a method to purify relatively large quantities of active *Arabidopsis* Uba1 from *E. coli* compared to other species [15] and found it to be fairly robust with respect to storage, handling, and E2-charging efficiency. We use UbcH5B when possible/relevant because almost 100% of UbcH5B can be loaded with Ub in our charge reaction. Additionally, most RING/U-box E3s react with UbcH5B~Ub, and substrate lysine modification is fairly promiscuous with this E2 in vitro [16]. In cases where we want to avoid stimulation of Ub transfer by non-covalent binding of Ub to the surface of UbcH5B opposite the active site Cys [17, 18], we use an S22R substitution on UbcH5B. Although lysine discharge assays can be performed using Coomassie staining alone, radiolabeled assays are more sensitive and prevent detection/interpretation problems that arise due to overlapping bands on a gel, whether these bands arise from similarly sized assay components or contaminants. Other methods like fluorescent labeling of Ub are also suitable, but the labeling and detection processes require additional optimization and/or standardization such as what fraction of Ub requires labeling to obtain a detectable signal, how stable is the label, and how much gel-to-gel variation is present. Hence, when possible, we use ^{32}P-Ub in our assays.

In the following method, we describe the procedure used to scout for suitable lysine discharge conditions to investigate Ub transfer mediated by the RING domain of RNF38 ($RNF38_{RING}$, residues 389–C). "Suitable conditions" for these assays depend on the type of comparison being made. Previously, to demonstrate that $RNF38_{RING}$-mediated Ub transfer proceeded via the same general mechanism as other RING E3s, we introduced mutations into $RNF38_{RING}$, UbcH5B, or Ub and investigated their effect on lysine discharge reactions compared to wild-type [18]. For this type of comparison in which the mutations are expected to slow discharge, "suitable conditions" involve lysine/E3 concentrations and a time frame where UbcH5B~Ub is discharged within the first few time points. In contrast, when we used $RNF38_{RING}$ to demonstrate that non-covalent binding of Ub to UbcH5B stimulates E3-mediated Ub transfer, the lysine/E3 concentrations and time frame were adjusted so that UbcH5B~Ub was incomplete at the later time points in the absence of excess non-covalent Ub.

2 Materials

Prepare all solutions with ultrapure water. Purification protocols for *Arabidopsis* E1, 2TK-Ub (*see* **Note 1**), UbcH5B, and $RNF38_{RING}$ are available in the indicated references [15, 18].

2.1 Preparation of Buffers and Solutions

1. 5 M NaCl.
2. 1 M Tris–HCl, adjusted to pH 7.6.
3. 2 M $MgCl_2$.

4. 1 M NaOH.
5. 0.5 M ATP, adjusted to pH 7–8 with NaOH and stored at −20 °C in 100 μL aliquots.
6. 10× charge buffer: 500 mM Tris–HCl, pH 7.6, 50 mM ATP, 50 mM $MgCl_2$. When mixing the buffer, $MgCl_2$ is added after all the other components including water to avoid producing an insoluble Mg-ATP complex. This buffer is prepared fresh weekly and stored at 4 °C.
7. 1.5 M L-lysine, pH 7.6: Add 2.46 g L-lysine monohydrate (Formedium) to ~4–5 mL water, adjust pH to 7.6 with NaOH, and then add water to a final volume of 10 mL. This solution is stored in 0.5–1 mL aliquots at −20 °C and can be repeatedly frozen and thawed.
8. TBS buffer: 150 mM NaCl, 25 mM Tris–HCl, pH 7.6.
9. 0.5 M EDTA, pH 8.0.
10. 1 U/μL apyrase (Sigma A6535) solubilized in a buffer containing 50 mM NaCl, 25 mM HEPES, pH 7.0, 1 mM $MgCl_2$, and 1 mM DTT. Snap freeze in liquid nitrogen, and store at −80 °C in 5–10 μL aliquots. Each aliquot is thawed only once and used fresh on the day of the experiment.
11. 20 mg/mL BSA (Sigma A3294) in water. Store at −20 °C in 200 μL aliquots. Can be repeatedly frozen and thawed.
12. 10× γ-^{32}P-ATP labeling buffer: 1 M NaCl, 150 mM Tris–HCl, pH 7.6, 120 mM $MgCl_2$, 10 mM DTT. Store at −20 °C in 100 μL aliquots, and re-freeze/thaw as needed.

2.2 Sample Preparation and Analysis

1. 1.5 mL Eppendorf tubes.
2. NuPAGE 4–12% Bis–Tris gels (Invitrogen Thermo Fisher Scientific) or similar.
3. 4× NuPAGE LDS sample buffer (Invitrogen Thermo Fisher Scientific).
4. 2× SDS-PAGE sample buffer (nonreducing): Equal volumes of nonreducing 4× NuPAGE LDS sample buffer and water.
5. 1× MES running buffer: 50 mM MES, 50 mM Tris–HCl, pH 7.3, 0.1% SDS, 1 mM EDTA (in our case, prepared from a 20× solution purchased from Invitrogen Thermo Fisher Scientific).
6. Plastic containers with dimensions suitable to accommodate the gel.
7. Fixing/gel drying buffer: 5% methanol, 8% acetic acid, and 5% glycerol.
8. Whatman paper, cut into pieces slightly bigger than gel.
9. Whatman paper, cut to the size of the gel dryer.

10. Gel dryer.
11. Film or phosphor storage cassette.
12. Film developer or phosphor imager.

3 Methods

3.1 Preparation of 100 μM Radioactively Labeled ^{32}P-Ub (See Note 2)

1. Calculate the volume of 2TK-Ub (*see* **Note 1**) to achieve a concentration of 100 μM Ub in 200 μL labeling buffer. This volume is 2.9 μL based on our current 2TK-Ub stock concentration of 6.96 mM.
2. Mix 20 μL 10× γ-^{32}P-ATP labeling buffer with 169.9 μL water, 2.9 μL 2TK-Ub, 6.2 μL γ-^{32}P-ATP (6000 Ci/mmol 10 mCi/mL), and 1 μL of cAMP-dependent protein kinase, catalytic subunit (New England Biolabs, P6000L). Incubate at room temperature for 1–4 h. Store at 4 °C.

3.2 Preparation of Lysine Discharge Solutions

Prepare discharge solution(s) just prior to running assays. With the exception of BSA (*see* **Note 3**), the concentration of lysine, E3 and/or additional components should be twice the final desired concentration (*see* **Note 4**). The stability of the E3 and any additional components will determine the concentrations of NaCl, buffer, and DTT required in the discharge solution. For $RNF38_{RING}$, NaCl and buffer (Tris–HCl, pH 7.6) concentrations can range between 50 and 150 mM and 10–50 mM, respectively, and no DTT is required:

1. Lysine dilution buffer/lysine-only discharge buffer: 20 mM lysine, 1 mg/mL BSA, 50 mM NaCl, 25 mM Tris–HCl, pH 7.6 (200 μL).
2. Lysine/E3 discharge solution: 20 mM lysine, 1 mg/mL BSA, and 800 nM $RNF38_{RING}$ (100 μL, *see* **Note 5**).

3.3 Preparation of Stop Solution

Stop solution should be prepared just prior to running assays (*see* **Note 6**):

1. Stop solution: 250 mM NaCl, 500 mM Tris–HCl, pH 7.6, 150 mM EDTA, and 0.1 U/μL apyrase. After charging and stopping, a reaction contains 0.5 units of apyrase for every 0.1 μmol of ATP.

3.4 Preparation of Charge Stock Solution

Charge stock solution should be prepared just prior to running assays:

1. Per 25 μL charge reaction: 1× charge buffer, 12.5 μM UbcH5B, 12.5 μM ^{32}P-Ub (*see* **Note 7**), ~75–100 mM NaCl, minimal amounts of reducing agent, i.e., only what is carried over from stock solutions (*see* **Note 8**).

2. Calculate the volume of each reagent required per reaction. The relative amounts of water and TBS buffer will vary depending on the stock concentrations of E1, E2, and ^{32}P-Ub and the NaCl content of each of these solutions. For these assays, both the stocks of UbcH5B and ^{32}P-Ub are at 100 μM and contain 150 mM NaCl. The stock of *Arabidopsis* E1 is at 30 μM and contains ~200 mM NaCl. In this case, the volume required to bring the solution to 25 μL can be split equally between water and TBS buffer. If one of the components is very dilute or very concentrated, then the ratio will need adjusting. Each 25 μL reaction requires 2.5 μL 10× charge buffer, 3.125 μL 100 μM UbcH5B, 3.125 μL 100 μM ^{32}P-Ub, 0.833 μL 30 μM *Arabidopsis* Uba1, 1.25 μL BSA (20 mg/mL), 7.1 μL water, and 7.1 μL TBS buffer.
3. There are six reactions in total (*see* below), but the stock volume is calculated based on eight to account for handling errors. Mix the components in the following order:
   (a) 20 μL 10× charge buffer.
   (b) 56.8 μL water.
   (c) 56.8 μL TBS buffer.
   (d) 10 μL BSA (20 mg/mL).
   (e) 25 μL 100 μM UbcH5B.
   (f) 6.7 μL 30 μM *Arabidopsis* Uba1 (*see* **Note 9**).
4. Flick tube and microfuge briefly (~10 s). Add 21.9 μL to each of six Eppendorf tubes.

3.5 Assay Assembly and Execution

1. Prepare sample tubes. Based on an assay with five time points including a zero time point (T_0), each assay requires 25 μL of a charged E2~Ub solution, 6.125 μL of a stop solution, and 30 μL of a discharge solution.
2. Add 6 μL 2× SDS-PAGE sample buffer to Eppendorf tubes for T_0 for each reaction. Make sure there is no reducing agent in the sample buffer.
3. Add 4 μL of 4× sample buffer to the tubes for assay time points following addition of discharge solution. For the assays described here, four time points are taken per reaction.
4. Dilute 30 μL of lysine/E3 discharge solution from Subheading 3.2, **step 1** with 30 μL of lysine dilution buffer prepared in Subheading 3.2, **step 2**.
5. Dilute 30 μL of lysine/E3 discharge solution from **step 4** with 30 μL of lysine dilution buffer prepared in Subheading 3.2, **step 1**. Repeat twofold E3/lysine discharge buffer serial dilution a total of four times. The discharge solutions for the reactions are as follows:

(a) Reaction 1: 20 mM lysine, 1 mg/mL BSA.

(b) Reaction 2: 20 mM lysine, 800 nM $RNF38_{RING}$, 1 mg/mL BSA.

(c) Reaction 3: 20 mM lysine, 400 nM $RNF38_{RING}$, 1 mg/mL BSA.

(d) Reaction 4: 20 mM lysine, 200 nM $RNF38_{RING}$, 1 mg/mL BSA.

(e) Reaction 5: 20 mM lysine, 100 nM $RNF38_{RING}$, 1 mg/mL BSA.

(f) Reaction 6: 20 mM lysine, 50 nM $RNF38_{RING}$, 1 mg/mL BSA.

6. Add 3.125 μL 100 μM ^{32}P-Ub to 21.9 μL charge stock aliquoted in Subheading 3.4, **step 4**. Start the timer.
7. Flick tube to mix and spin down briefly (~5 s) in a benchtop mini centrifuge.
8. Incubate for 15 min (*see* **Note 10**).
9. Add 6.125 μL stop solution. Flick the tube to mix, and spin down briefly (~5 s) in a benchtop mini centrifuge (*see* **Note 11**).
10. About 15–30 s after adding stop solution, remove 6 μL and add to T_0 sample tube and vortex (*see* **Note 12**).
11. About 1–2 min after adding stop solution, add 25.125 μL discharge solution from Subheading 3.5, **step 5**. Pipette up and down 5–10 times.
12. Flick the tube to mix and spin down briefly (~5 s) in a benchtop mini centrifuge.
13. Take 12 μL time points at 30, 60, 90, and 120 s after adding discharge solution for each reaction. Vortex immediately upon adding to gel sample buffer (*see* **Note 13**).
14. Microfuge all samples for ~15 s prior to loading onto SDS-PAGE gels (*see* **Notes 14** and **15**).
15. Load samples onto SDS-PAGE gels by reaction with T_0 followed by the time course. Use nonreducing buffers when running gels.
16. Run gels at 200 V for approximately 35 min so that the ^{32}P-Ub band almost reaches the bottom of the gel.
17. Transfer the gels to plastic containers filled with fixing/gel drying buffer. Trim off the foot and most of the stacker, leaving ~3 mm of each well. Leave in buffer for 2–5 min.
18. Prepare a stack of at least three pieces of gel-sized Whatman filter paper. Label top piece with a unique gel identifier.
19. Pre-wet the labeled filter paper in fixing/gel drying buffer, and return to top of stack. Place gel on the labeled filter paper.

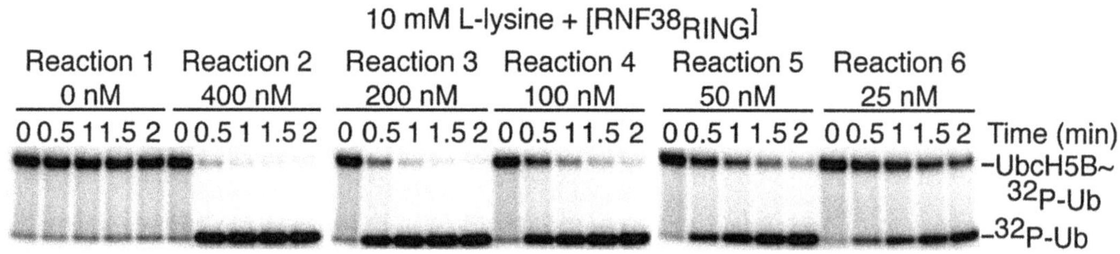

Fig. 4 Nonreduced autoradiograms of reactions 1–6 as described in the methods section showing the discharge of UbcH5B~^{32}P-Ub over 2 min with final concentrations of 10 mM L-lysine and 0–400 nM $RNF38_{RING}$ as indicated. For assays in which comparisons are made with a stimulating component like non-covalent Ub [18], 25–50 nM $RNF38_{RING}$ and 10 mM L-lysine are a suitable starting point provided that the stimulatory component/effect is strong enough to drive the reaction to completion in 2 min. For comparisons across reactions involving UbcH5B mutations that slow discharge, a suitable starting concentration of $RNF38_{RING}$ would be between 200 and 400 nM, depending on the extent that mutants disrupt discharge. A possible next step in the optimization process would be to test a narrower concentration range of $RNF38_{RING}$ with a mutation that abolishes E3-mediated discharge

20. Place gel stack on three pieces of Whatman filter paper that are cut to a size slightly smaller than your gel dryer.
21. Dry gels. We use cling film to avoid contaminating the gel-drying equipment with radioactive material.
22. Expose gels to film in a darkroom or a phosphor storage cassette to capture the signal from ^{32}P-Ub. Try different exposure times to ensure an optimal and accurate visual representation of the results (*see* **Notes 16** and **17**). The exposure time is dependent upon the concentration of the radioactive material and the age of the ^{32}P used (half-life is about 14 days).
23. Analyze results based upon the disappearance of the E2~Ub band over time (*see* Fig. 4) to determine if a suitable condition has been found or further screening is required.

4 Notes

1. Labeling in our experiment relies on a 2TK tag (RRASV) at the N-terminus of Ub. The gene encoding Ub has been cloned into the BamHI and Not sites of a modified pGEX-2TK GST expression vector (GE Healthcare) in which the thrombin cleavage site has been replaced with TEV and a non-cleavable His_6 tag precedes the GST tag.
2. The final labeled concentration of 2TK-Ub and amount of γ-^{32}P-ATP can be varied.
3. Some E3s adhere to the surface of the Eppendorf tubes, which reduces apparent activity and affects assay reproducibility. When possible, BSA is added to discharge solutions prior to

adding E3 or additional protein/peptide components to prevent adherence.

4. Although we do not fit kinetic parameters with lysine discharge assays, the concentrations of various reactants are determined as if kinetic analyses were intended. Under ideal conditions, the maximum concentration of E3 (the enzyme) used in any reaction is at least tenfold lower than E2~Ub (the substrate).
5. Protein stocks are stored at −80 °C at a minimum concentration of 1 mg/mL. Diluted working E3 stocks are prepared in TBS containing 1 mg/mL BSA to minimize losses from surface adherence. For the discharge assays presented here, the −80 °C stock of $RNF38_{RING}$ was at 8.7 mg/mL (937 μM). A 5 μM working stock in TBS containing 1 mg/mL BSA was prepared to make the discharge solution.
6. Although apyrase is inhibited by EDTA, we have found that using a combination of the two is slightly more effective at stopping charging for certain E1–E2 pairs. The concentrations of the other reagents in the stop solution are unchanged regardless of whether apyrase is used.
7. When using a new E2, initial experiments need to be done to characterize the level of ubiquitin charging onto E2 that can be achieved with a given E1. An ideal charge reaction will convert all the E2 to E2~Ub without producing by-products like diUb, E2–Ub, or E1–Ub, where—denotes an isopeptide bond between the C-terminal diglycine motif of Ub and the side chain of a lysine residue on the protein. The species of origin of the E1 plays a significant role in the E2-conjugating efficiency as well as the duration, temperature of the charge reaction, and molar ratio of ^{32}P-Ub to E2. For a given E2, we optimize charging conditions by using SDS-PAGE to analyze time courses of several charge reactions in which ^{32}P-Ub:E2:E1 ratios are varied. Almost all of UbcH5B can reliably be converted to UbcH5B~Ub up to concentrations of 20 μM UbcH5B using molar ratios of ^{32}P-Ub:UbcH5B between 1:1 and 2:1 with 1 μM *Arabidopsis* E1.
8. Because the E2~Ub thioester is labile, it is important that the concentration of reducing agent is kept to a minimum.
9. At concentrations less than ~100 nM, there is a pronounced reduction in Uba1 activity due to surface adherence. Hence, when possible, BSA is included in the charge reaction and Uba1 is added after BSA. If BSA is not used, Uba1 is always added after E2.
10. Ub mutants can influence the rate of the E1-E2 transthiolation reaction. In reference [15], some charge reactions with Ub mutants were run for 30 min instead of 15 min to produce

equivalent amounts of E2~Ub to the corresponding wild-type reaction.

11. Stopping the reaction is critical. New E1-E2 combinations should be tested by setting up a charge reaction, adding the stop solution, and collecting samples over 10 min to check that the reaction has stopped charging and to investigate the stability of E2~Ub. For UbcH5B, once the charge reaction is stopped, a reduction in UbcH5B~Ub is evident in as little as 5 min.
12. For each reaction in an assay, the time between adding stop, removing T_0, and adding discharge solution should be the same and kept to a minimum.
13. Thioester bonds are labile at higher temperature, so do *not* boil the samples.
14. The viscosity of the gel sample buffer from Invitrogen frequently causes problems with reproducibly pipetting accurate volumes. Hence, all the sample volume for each time point is loaded onto the gel.
15. Because of the labile nature of E2~Ub, samples are sensitive to discharge even after denaturation with gel loading buffer. Hence, gels should be run promptly after completing an assay, and there are no benefits to holding back "extra" sample in the event that there is a problem with a gel. In addition, when running multiple reactions in an assay, it is best to stagger the reactions to minimize the time between completing the first reaction and loading the gels.
16. When working with new systems, overexposure to check for by-products is advised to ensure that the concentration of lysine is saturating under the conditions tested. When working with a new system, films should be exposed for a long time to check that Ub transfer occurs to free lysine and not lysine residues on the surface of any of the proteins in the system. We have previously observed formation of diUb and E3–Ub (where—indicates an isopeptide linkage) when lysine concentrations were too low.
17. When multiple gels are run to compare reactions in an assay, the gels should be exposed to film at the same time and onto the same film.

Acknowledgments

This work was supported by Cancer Research UK (A23278) and European Research Council (ERC) under the European Union's Horizon 2020 research and innovation program (grant agreement n° 647849).

References

1. Hershko A, Ciechanover A (1998) The ubiquitin system. Annu Rev Biochem 67:425–479
2. Pickart CM, Eddins MJ (2004) Ubiquitin: structures, functions, mechanisms. Biochim Biophys Acta 1695:55–72
3. Dye BT, Schulman BA (2007) Structural mechanisms underlying posttranslational modification by ubiquitin-like proteins. Annu Rev Biophys Biomol Struct 36:131–150
4. Komander D, Rape M (2012) The ubiquitin code. Annu Rev Biochem 81:203–229
5. Buetow L, Huang DT (2016) Structural insights into the catalysis and regulation of E3 ubiquitin ligases. Nat Rev Mol Cell Biol 17:626–642
6. Brzovic PS, Keeffe JR, Nishikawa H et al (2003) Binding and recognition in the assembly of an active BRCA1/BARD1 ubiquitin-ligase complex. Proc Natl Acad Sci U S A 100:5646–5651
7. Wenzel DM, Lissounov A, Brzovic PS et al (2011) UBCH7 reactivity profile reveals parkin and HHARI to be RING/HECT hybrids. Nature 474:105–108
8. Stewart MD, Ritterhoff T, Klevit RE et al (2016) E2 enzymes: more than just middle men. Cell Res 26:423–440
9. Tatham MH, Plechanovova A, Jaffray EG et al (2013) Ube2W conjugates ubiquitin to alpha-amino groups of protein N-termini. Biochem J 453:137–145
10. Scaglione KM, Basrur V, Ashraf NS et al (2013) The ubiquitin-conjugating enzyme (E2) Ube2w ubiquitinates the N terminus of substrates. J Biol Chem 288:18784–18788
11. Vittal V, Shi L, Wenzel DM et al (2015) Intrinsic disorder drives N-terminal ubiquitination by Ube2w. Nat Chem Biol 11:83–89
12. Scott DC, Sviderskiy VO, Monda JK et al (2014) Structure of a RING E3 trapped in action reveals ligation mechanism for the ubiquitin-like protein NEDD8. Cell 157:1671–1684
13. Mcginty RK, Henrici RC, Tan S (2014) Crystal structure of the PRC1 ubiquitylation module bound to the nucleosome. Nature 514:591–596
14. Chang L, Zhang Z, Yang J et al (2015) Atomic structure of the APC/C and its mechanism of protein ubiquitination. Nature 522:450–454
15. Dou H, Buetow L, Sibbet GJ et al (2012) BIRC7-E2 ubiquitin conjugate structure reveals the mechanism of ubiquitin transfer by a RING dimer. Nat Struct Mol Biol 19:876–883
16. Brzovic PS, Klevit RE (2006) Ubiquitin transfer from the E2 perspective: why is UbcH5 so promiscuous? Cell Cycle 5:2867–2873
17. Brzovic PS, Lissounov A, Christensen DE et al (2006) A UbcH5/ubiquitin noncovalent complex is required for processive BRCA1-directed ubiquitination. Mol Cell 21:873–880
18. Buetow L, Gabrielsen M, Anthony NG et al (2015) Activation of a primed RING E3-E2-ubiquitin complex by non-covalent ubiquitin. Mol Cell 58:297–310
19. Marvin (version 17.21.0) (2017) ChemAxon. http://www.chemaxon.com

Chapter 3

Methods for NAD-Dependent Ubiquitination Catalyzed by *Legionella pneumophila* Effector Proteins

Jiazhang Qiu and Zhao-Qing Luo

Abstract

Ubiquitination is one of the most important posttranslational modifications in eukaryotic cells where it regulates the activity, cellular localization, and half-life of proteins. Ubiquitination thus affects many essential cellular processes, including vesicle trafficking, cell cycle, DNA repair, immune response, and protein homeostasis. The ubiquitin system is exclusive to eukaryotes; however, pathogenic bacteria have developed effective strategies to influence the host ubiquitin system for their own benefit. *Legionella pneumophila* is the causative agent of Legionnaires' disease, a severe form of pneumonia. This bacterium utilizes a type IV secretion system to translocate more than 300 effector proteins into host cells. These virulence factors modulate a wide spectrum of host processes to support its intracellular survival and replication. Hijacking of host ubiquitin system is an important theme in Legionella virulence, and a number of *L. pneumophila* effector proteins have been shown to possess ubiquitin ligase or deubiquitinase activity. Among these, members of the SidE family effector proteins (SidEs) catalyze ubiquitination of several ER-associated Rab small GTPases by a mechanism that bypasses the requirement of ATP and the E1, E2 enzymes. Here, we summarize the experimental details of Rab small GTPases ubiquitination catalyzed by SdeA, a member of the SidE family.

Key words Dot/Icm, mART, ADP-ribosylation, Rab small GTPases

1 Introduction

Ubiquitination, the biochemical process of covalent attachment of the 76-residue ubiquitin modifier to a substrate protein, is a prevalent protein modification conserved among all eukaryotic organisms [1]. Ubiquitination leads to degradation of proteins, alterations in protein function, or cellular localization and thereby regulates virtually all key cellular processes, including vesicle trafficking, cell cycle, DNA repair, and innate and adaptive immunity [2]. Conventional ubiquitination occurs via a universally conserved three-enzyme cascade that involves the ubiquitin-activating enzyme (E1), the ubiquitin-conjugating enzyme (E2), and the ubiquitin ligase (E3) [1]. Ubiquitin is first activated by E1 with the consumption of ATP, leading to the formation of a thioester

Thibault Mayor and Gary Kleiger (eds.), *The Ubiquitin Proteasome System: Methods and Protocols*, Methods in Molecular Biology, vol. 1844, https://doi.org/10.1007/978-1-4939-8706-1_3,

linkage between the carboxyl terminus of ubiquitin and a cysteine residue in E1; E2 catalyzes the transfer of ubiquitin from E1 to its catalytic cysteine residue via a transesterification reaction; the ubiquitin moiety is eventually transferred to substrate by the activity of the E3 enzyme, leading to the formation of an isopeptide bond between mostly the ε-amino group of a lysine residue on the substrate and the carboxyl end of ubiquitin.

Bacteria do not possess genes coding for the classic ubiquitin system [3]. However, many bacterial pathogens have developed effective strategies to co-opt the host ubiquitin system to benefit their own life cycle [3]. This is mainly achieved by specialized protein secretion systems, such as type III and IV secretion systems, which allow bacterial pathogens to deliver virulence factors called effectors into host cells. Many bacterial effectors are functional mimicry of eukaryotic E3 ligases or deubiquitinases, which interfere with host cellular processes by altering ubiquitin signaling [4]. The facultative intracellular bacterial pathogen *Legionella pneumophila* infects a wide spectrum of host cells, ranging from freshwater amoebae to mammalian macrophages [5]. *L. pneumophila* delivers more than 300 effector proteins into host cells via the Dot/Icm type IV secretion system to subvert diverse host processes for the creation of an intracellular niche permissive for its replication [6]. The ubiquitin network is important for *L. pneumophila* virulence [7], and active modulation of this machinery has emerged as an important theme in the field of *L. pneumophila* pathogenesis [8]. In particular, our group recently discovered an unprecedented mechanism of ubiquitination catalyzed by the SidE family effectors such as SdeA of *L. pneumophila* [9]. These proteins catalyze ubiquitination of several ER-associated Rab small GTPases by a mechanism in which ubiquitin is activated by ADP-ribosylation, thus bypassing the requirement of ATP, the E1 and E2 enzymes [9]. Here we summarize the procedure for SdeA-catalyzed ubiquitination of Rab33b under different experimental conditions.

2 Materials

2.1 L. Pneumophila Infection

1. Charcoal-yeast extract (CYE) plates.
2. (*N*-(2-Acetamido)-2-aminoethanesulfonic acid) ACES-buffered yeast extract (AYE) broth.
3. HEK293 cells.
4. Dulbecco's modified minimum Eagle's medium (DMEM) supplemented with 10% FBS.
5. Lipofectamine 3000.
6. Plasmids encoding 4xFlag-Rab33b and FCγRII.
7. Anti-*L. pneumophila* serum produced in rabbit (*see* **Note 1**).

8. Phosphate-buffered saline (PBS).
9. RIPA buffer: 25 mM Tris–HCl (pH 7.6), 150 mM NaCl, 1% NP-40, 1% sodium deoxycholate, 0.1% SDS, and complete protease inhibitors.
10. Anti-Flag M2 beads.
11. Anti-Flag antibody.
12. 5 mg/mL 3xFlag peptide in 50 mM Tris–HCl (pH 7.4), 150 mM NaCl.
13. 5× SDS sample buffer: 0.3 M Tris–HCl (pH 6.8), 10% SDS, 50% glycerol, 20% 2-mercaptoethanol, 0.01% bromophenol blue.
14. Coomassie brilliant blue staining solution: coomassie brilliant blue 0.1%, 50% methanol (v/v), 10% glacial acetic acid (v/v) in H_2O.

2.2 In Vitro Ubiquitination Assay

1. *E. coli* strain BL21(DE3).
2. pQE30-4xFlag-Rab33b and pQE30-SdeA.
3. Luria-Bertani (LB) medium.
4. 1 M Isopropyl-D-thiogalactoside (IPTG).
5. 100 mg/mL ampicillin.
6. Ni^{2+}-NTA beads.
7. 1 M imidazole.
8. Phosphate-buffered saline (PBS).
9. Dialysis buffer: 25 mM Tris–HCl (pH 7.5), 150 mM NaCl, 5% glycerol, 1 mM DTT.
10. 100 mM β-nicotinamide adenine dinucleotide (β-NAD) in deionized water.
11. 1 M Tris–HCl (pH 7.5).
12. Human ubiquitin (Boston Biochem), stored at −20 °C.
13. Anti-ubiquitin antibody.

3 Methods

3.1 Ubiquitination of Rab33b During L. Pneumophila Infection

3.1.1 Preparation of L. Pneumophila for Infection

1. Streak out the required *L. pneumophila* strains from −80 °C stocks onto CYE plates; culture the bacteria in AYE broth to the postexponential phase (OD_{600} = 3.3–3.8).
2. Dilute *L. pneumophila* to 1 × 10^8 cfu/mL with fresh AYE broth, and opsonize the bacteria with rabbit anti-*L. pneumophila* serum at 1:500 ratio (v/v) for 30 min at 37 °C before infection.

3.1.2 Preparation of Mammalian Cells Expressing 4xFlag-Rab33b and FCγRII

1. Culture HEK293 cells in the DMEM medium supplemented with 10% FBS (*see* **Note 2**).
2. Grow the cells to about 80% confluence, and transfect them with 4xFlag-Rab33b and FCγRII plasmids (at 1:1 ratio) using Lipofectamine 3000 following manufacturer's instructions.
3. 24 h post-transfection, change the culture medium with fresh DMEM supplemented with 10% FBS.

3.1.3 Bacterial Infection and Immunoprecipitation (IP)

1. Infect HEK293 cells with opsonized *L. pneumophila* at an MOI of 10 for 2 h.
2. Use a cell scraper to detach the cells, and collect infected cells by centrifugation at 1200 × *g* for 5 min.
3. Wash cells twice with PBS.
4. Lyse the cells with the RIPA buffer containing protease inhibitors for 10 min at 4 °C.
5. Remove insoluble cell debris by centrifugation at 12,000 × *g* for 10 min at 4 °C.
6. Add 20 μL Flag beads to the cell lysates, and allow the immunoprecipitation reaction to proceed for 4 h at 4 °C on a rotatory rack.
7. Collect Flag beads by centrifugation at 12,000 × *g* for 5 min, and wash three times with RIPA buffer.
8. Add 50 μL 1× SDS sample buffer to washed Flag beads, and boil at 100 °C for 5 min.
9. Resolve the released proteins by SDS-PAGE, and transfer to nitrocellulose membrane for detection of 4xFlag-Rab33b using the Flag-specific antibody.

3.1.4 Analyze the Modification of Rab33b During L. Pneumophila Infection

1. Infect HEK293 cells expressing 4xFlag-Rab33b and FCγRII with *L. pneumophila*, and perform immunoprecipitation with Flag beads as describe above (*see* **Note 3**).
2. Elute 4xFlag-Rab33b from Flag beads 3xFlag peptide at 150 μg/mL following manufacturer's instructions.
3. Add 5× SDS loading buffer to the eluates, boil the samples at 100 °C for 5 min, and resolve the proteins by SDS-PAGE; stain the gels with Coomassie brilliant blue.
4. Protein bands corresponding to ubiquitinated 4xFlag-Rab33b (with higher molecular weight) are excised from the gels to be analyzed by mass spectrometry (*see* **Note 4**).

3.2 In Vitro Ubiquitination of Rab33b by SdeA

3.2.1 Protein Purification

1. The *E. coli* BL21(DE3) strains harboring pQE30-4xFlag-Rab33b or pQE30-SdeA are grown in LB broth containing 100 μg/mL of ampicillin at 37 °C; transfer 40 mL of overnight cultures to 1 L of LB broth, and grow the cultures to OD_{600} of 0.6–0.8.
2. Add IPTG to a final concentration of 0.2 mM; bacterial cultures are further incubated in a shaker at 18 °C for 16–18 h.
3. Harvest bacterial cells by centrifugation at 12,000 × *g* for 10 min; lyse cells by sonication in PBS buffer containing protease inhibitors. Remove insoluble debris by centrifugation twice at 12,000 × *g* for 10 min at 4 °C.
4. Incubate the cleared cell lysates with 1.5 mL of Ni^{2+}-NTA beads for 2 h at 4 °C with rotation. After washing the beads with 20 bed volumes of PBS buffer containing 20 mM imidazole, bound proteins are eluted with 5 mL of PBS containing 300 mM imidazole.
5. Dialyze proteins with dialysis buffer. Protein concentrations can be determined by a Bradford assay, and the quality of proteins can be assessed by Coomassie brilliant blue staining.

3.2.2 Ubiquitination Reaction

1. In a 50 μL reaction that contains 50 mM Tris–HCl (pH 7.5), add 5 μg of SdeA, 5 μg of Rab33b (*see* **Note 5**) and 10 μg of ubiquitin (*see* **Note 6**), and 0.4 mM β-NAD. Reaction is allowed to proceed at 37 °C for 2 h.
2. Terminate the reaction by adding 12.5 μL of 5× SDS loading buffer following boiling at 100 °C for 5 min.
3. Resolve the samples by SDS-PAGE; ubiquitinated Rab33b and SdeA can be probed by Coomassie staining or by immunoblotting with antibodies specific for ubiquitin or Flag.

4 Notes

1. *L. pneumophila* cells from 10 mL culture were washed with PBS twice before being fixed with 4% formaldehyde for 30 min. After washing with 15 mL PBS for five times, fixed cells resuspended in 2 mL PBS were used to immunized a rabbit following a standard protocol (Pocono Rabbit Farm and Laboratory, Canadensis, PA). Sera that react well with *L. pneumophila* (up to 1:100,000 dilution for immunofluorescence staining of *L. pneumophila* bacteria internalized by macrophages) were used.
2. HEK293 cells are seeded on a 6-well plate, and 2×10^6 cells are required for Western blot.
3. 2×10^8 cells are required for a good Coomassie staining.

4. Protein bands corresponding to a shifted Rab33b were digested with trypsin, and the resulting protein fragments were identified by mass spectrometry. The ubiquitin tryptic fragments were identified in bands corresponding to shifted Rab33b.
5. Ubiquitination of Rab33b catalyzed by SdeA requires the mART motif located in the middle of the protein; a negative control can be established using the SdeA$_{E/A}$ mutant that is defective in the mART motif.
6. All lysine variants of ubiquitin, as well as the ubiquitin derivative containing an alanine substitution in the last two glycine residues (Ub-AA), still can be used in the SdeA-catalyzed ubiquitination.

References

1. Hershko A, Ciechanover A (1998) The ubiquitin system. Annu Rev Biochem 67:425–479. https://doi.org/10.1146/annurev.biochem.67.1.425
2. Haglund K, Dikic I (2005) Ubiquitylation and cell signaling. EMBO J 24(19):3353–3359. https://doi.org/10.1038/sj.emboj.7600808
3. Maculins T, Fiskin E, Bhogaraju S, Dikic I (2016) Bacteria-host relationship: ubiquitin ligases as weapons of invasion. Cell Res 26 (4):499–510. https://doi.org/10.1038/cr.2016.30
4. Zhou Y, Zhu YQ (2015) Diversity of bacterial manipulation of the host ubiquitin pathways. Cell Microbiol 17(1):26–34. https://doi.org/10.1111/cmi.12384
5. Newton HJ, Ang DKY, van Driel IR, Hartland EL (2010) Molecular pathogenesis of infections caused by Legionella pneumophila. Clin Microbiol Rev 23(2):274–298. https://doi.org/10.1128/CMR.00052-09
6. Qiu JZ, Luo ZQ (2017) Legionella and Coxiella effectors: strength in diversity and activity. Nat Rev Microbiol 15(10):591–605. https://doi.org/10.1038/nrmicro.2017.67
7. Dorer MS, Kirton D, Bader JS, Isberg RR (2006) RNA interference analysis of Legionella in Drosophila cells: exploitation of early secretory apparatus dynamics. PLoS Pathog 2 (4):315–327. https://doi.org/10.1371/journal.ppat.0020034
8. Hubber A, Kubori T, Nagai H (2014) Modulation of the ubiquitination machinery by Legionella. Curr Top Microbiol Immunol 376:227–247. https://doi.org/10.1007/82_2013_343
9. Qiu J, Sheedlo MJ, Yu K, Tan Y, Nakayasu ES, Das C, Liu X, Luo ZQ (2016) Ubiquitination independent of E1 and E2 enzymes by bacterial effectors. Nature 533(7601):120–124. https://doi.org/10.1038/nature17657

Chapter 4

Using In Vitro Ubiquitylation Assays to Estimate the Affinities of Ubiquitin-Conjugating Enzymes for Their Ubiquitin Ligase Partners

Spencer Hill, Connor Hill, and Gary Kleiger

Abstract

Ubiquitin ligases (E3s) function by binding to both a protein substrate and to ubiquitin-conjugating enzymes (E2s) bound to ubiquitin. E3s facilitate the transfer of ubiquitin from the E2 active site to an E3-bound substrate. Thus, the affinity of the interaction of an E2 with its E3 partner is of considerable interest. The purpose of this work is to (1) provide protocols for the purification of the human E2 Cdc34, as well as for some additional protein components needed for the assays described here whose purification protocols haven't been described elsewhere in detail; (2) provide the researcher with critical information regarding the proper long-term storage of these enzymes to retain maximal activity; (3) provide a protocol to benchmark Cdc34 activity with previously described activity levels in the literature; and (4) provide a simple and rapid means of measuring E2 affinity for an E3.

Key words Ubiquitin ligase, Ubiquitin-conjugating enzyme, Protein-protein interaction, Affinity, Cdc34, Ube2R1/2, UbcH5, Skp1-Cullin Fbox (SCF) ubiquitin ligase , In vitro ubiquitylation

1 Introduction

Protein ubiquitylation involves the coordinated action of three classes of enzymes: ubiquitin-activating enzyme (E1), ubiquitin-conjugating enzyme (E2), and ubiquitin ligase (E3). Ubiquitin is first activated by the E1, forming a covalent complex between E1 and the C-terminus of ubiquitin. The E2 then binds to the E1-ubiquitin complex, and ubiquitin is transferred from E1 to the E2 active site. E3s act as scaffolds, binding both to the protein substrate and to the E2-ubiquitin complex [1]. In most cases, E3s also stimulate the transfer of ubiquitin from the E2 onto a lysine residue on either the E3-bound substrate or ubiquitin(s) covalently attached to the substrate [2]. Multiple lysine residues on the substrate may become modified with ubiquitins prior to substrate dissociation from the E3 (termed multi-mono-ubiquitylation). In other cases, ubiquitins can be transferred from E2 to ubiquitins on

Thibault Mayor and Gary Kleiger (eds.), *The Ubiquitin Proteasome System: Methods and Protocols*, Methods in Molecular Biology, vol. 1844, https://doi.org/10.1007/978-1-4939-8706-1_4,

the substrate forming a chain (termed poly-ubiquitylation). In either case, the ubiquitylated protein is typically degraded by the 26S proteasome [3].

Since E2 interaction with E3 is required for substrate ubiquitylation, it is often an advantage for the researcher to know the affinity of an E2 for its cognate E3. For instance, having an estimate of the affinity (e.g., equilibrium dissociation constant for E2 binding to E3) is useful when designing in vitro ubiquitylation assays. The affinity informs the researcher of the minimal concentration of E2 that is necessary to fully saturate the E3. Knowing the affinity of E2 for E3 is also relevant since proteomic methods are now available to estimate the concentrations of proteins inside the cell [4]. Comparison of cellular E2 and E3 concentrations with their affinity for each other provides insight into whether E3 activity is maximal within the cell. Lastly, both E2s and E3s have been found to contain mutations in humans that may lead to disease, and the ability to measure E2 affinity for E3 will enable the determination of whether these mutations disrupt the E2-E3 complex.

A variety of biophysical methods have been adapted for measuring the equilibrium dissociation constant of an E2 for an E3, including surface plasmon resonance [5, 6], isothermal calorimetry [7, 8], and fluorescence resonance energy transfer (FRET) [9] as well as fluorescence polarization [10]. However, the development of these assays is nearly always expensive and time-consuming. It is also common that these methodologies will not adapt to specific pairs of E2s and E3s, meaning that the researcher will have to scout through several before successfully developing the binding assay. Finally, most of these methods require substantial technical and theoretical know-how in order to employ them. Thus, it would be desirable to have a method that is conceptually simple, easy to use, and inexpensive to implement.

E2 affinity for E3 has also been estimated by measuring the Michaelis-Menten kinetics of ubiquitylation reactions containing increasing amounts of E2 and then using the reaction velocities to estimate the Michaelis constant, K_m, of E2 for E3. Previous studies have shown that the K_m is often consistent with estimates of the affinity of an E2 for E3 derived from traditional biophysical methods [9]. Measuring the Michaelis-Menten kinetics is particularly powerful because all one needs to perform the experiments are highly purified proteins and a substrate that can be detected quantitatively. Here we describe the purification of the human E2 Cdc34 (also referred to as Ube2R1/2) as well its physiological E3 partner Skp1-Cullin Fbox (SCF) ubiquitin ligase. The purification of a synthetic peptide derived from the bona fide SCF substrate β-catenin modified with a single ubiquitin is also described. All of these proteins retain their activities during long-term storage. However, it is critical to follow specific procedures during storage, and attention is drawn to these points within the protocols. An assay is described to facilitate comparison of the activities of the freshly prepared enzymes with benchmark levels from the literature. Finally, the procedure for measuring the K_m of Cdc34 or another human E2, UbcH5, for SCF is described, and several important considerations regarding both the implementation and the analysis are noted.

2 Materials

Prepare all solutions using ultrapure water (prepared by purifying deionized water, to attain a sensitivity of 18 MΩ-cm at 25 °C) and analytical grade reagents.

2.1 Buffers and Solutions

1. Luria broth (LB): 10 g/L tryptone, 10 g/L NaCl, 5 g/L yeast extract.
2. 100 mg/mL ampicillin stock.
3. 34 mg/mL chloramphenicol stock (in ethanol).
4. 0.8 M isopropyl β-D-1-thiogalactopyranoside (IPTG) stock.
5. 1× phosphate-buffered saline with Tween 20 (PBS-T): 12 mM phosphate (pH 7.4), 137 mM NaCl, 2.7 mM KCl, 0.1% (v/v) Tween 20.
6. Lysis buffer: 30 mM Tris–HCl (pH 7.5), 250 mM NaCl, 20 mM imidazole-HCl (pH 8.0), 0.1% IgePal, 5% glycerol, 1 mM β-mercaptoetanol, commercial protease inhibitor cocktail lacking EDTA (PIC) (*see* **Note 1**).
7. Wash buffer: 30 mM Tris–HCl (pH 7.5), 250 mM NaCl, 20 mM imidazole-HCl (pH 8.0), 5% glycerol, 1 mM β-mercaptoethanol.
8. Tobacco etch virus protease (TEV) buffer: 50 mM Tris–HCl (pH 8.0), 50 mM NaCl, 5% glycerol, 1 mM DTT, 0.5 mM EDTA.
9. Nickel elution buffer: 50 mM HEPES-Na (pH 7.5), 250 mM NaCl, 300 mM imidazole-HCl.
10. 0.5 M imidazole-HCl (pH 8.0).
11. Coomassie blue solution: 40% (v/v) methanol, 10% glacial acetic acid (v/v), 0.1% Coomassie Brilliant Blue R-250 (w/v).
12. Storage buffer: 30 mM Tris–HCl (pH 7.5), 100 mM NaCl, 10% glycerol, 1 mM DTT.
13. Ion exchange buffer A: 25 mM HEPES-Na (pH 6.8), 5% glycerol, 1 mM β-mercaptoethanol.
14. Ion exchange buffer B: 25 mM HEPES-Na (pH 6.8), 500 mM NaCl, 5% glycerol, 1 mM β-mercaptoethanol.

15. 10x neddylation buffer: 300 mM Tris–HCl (pH 7.5), 50 mM $MgCl_2$, 20 mM ATP.
16. 1 M dithiothreitol (DTT).
17. 10× reaction buffer: 300 mM Tris–HCl (pH 7.5), 1 M NaCl, 50 mM $MgCl_2$, 20 mM DTT, 20 mM ATP.
18. 2× reducing quench buffer: 100 mM Tris–HCl (pH 6.8), 30 mM EDTA, 20% (v/v) glycerol, 2% (w/v) SDS, 4% (v/v) β-mercaptoethanol, 0.1% (w/v) bromophenol blue.

2.2 Reagents and Materials

All proteins and enzymes described are of human origin unless otherwise stated.

1. 1.5 mL Eppendorf tubes.
2. Rosetta BL21(DE3) chemically competent cells.
3. 10 cm LB-agar plates: 10 g/L tryptone, 10 g/L NaCl, 5 g/L yeast extract, 15 g/L agar.
4. 15 and 50 mL conical tubes.
5. Ni-NTA agarose resin.
6. Disposable 25 mL columns.
7. Commercial TEV protease (must possess His_6 tag).
8. Protein molecular weight marker.
9. Disposable desalting columns (*see* **Note 2**).
10. 3 and 10 kDa cutoff centrifugal concentrators.
11. 0.22 μm spin filters.
12. Liquid nitrogen.
13. Commercial His_6-Nedd8.
14. APPBP1/UBA3, bacterially expressed as a GST fusion [11].
15. Ubc12, bacterially expressed as a GST fusion [12].
16. Commercial His_6-ubiquitin.
17. Ube1 (E1), expressed in insect cells as a GST fusion or in bacterial cells with a His_6 affinity tag as described [13, 14].
18. UbcH5, bacterially expressed as a GST fusion [15].
19. $Cul1^{Lys\ 720\ Arg}$-Rbx1 complex (C/R^{K720R}), using the Split-n-Coexpress method [16] (*see* **Note 3**).
20. Skp1/βTRCP complex, expressed in insect cells as described [16].
21. Commercial β-catenin substrate peptide (β-cat) [17] (*see* **Note 4**).
22. γ-^{32}P ATP.
23. Commercial protein kinase A (PKA) enzyme.
24. Commercial ubiquitin protein.

25. Lys 48 Arg (K48R) ubiquitin with N-terminal PKA phosphorylation site (*see* **Note 5**).
26. Ubiquitin containing C-terminal aspartic acid residue (D77 ubiquitin) (*see* **Note 5**); preparation described in Subheading 3.1.
27. Cdc34, bacterially expressed; preparation described in Subheading 3.2.
28. Cul1-Rbx1 complex (C/R), using the Split-n-Coexpress method (*see* **Note 6**).
29. Cul1-Rbx1-Nedd8 (C/R^{Nedd8}); preparation described in Subheading 3.3.
30. Mono-ubiquitin β-catenin substrate (β-cat-$(Ub)_1$) [17]; preparation described in Subheading 3.4.
31. SDS-PAGE gels, 18% acrylamide.
32. 3 mm chromatography paper.
33. Ultra clear cellophane (*see* **Note 7**).

2.3 Equipment

1. Refrigerated protein purification workstation including:
   (a) Gradient pump with UV detector.
   (b) Size exclusion chromatography column optimized for resolving proteins between 3 kDa and 75 kDa.
   (c) Cation exchange column.
   (d) Fraction collector.
2. UV/vis spectrophotometer.
3. Tabletop centrifuge:
   (a) Rotor compatible with 1.5 mL Eppendorf tubes.
4. Large centrifuge:
   (a) Rotor capable of processing 50 mL volumes at 32,000 × *g*.
   (b) Rotor capable of processing 1 L volumes at 5000 × *g*.
5. Gel dryer (*see* **Note 8**).
6. Storage phosphor screen.
7. Molecular imager capable of autoradiography.

3 Methods

3.1 Expression and Purification of Human Ubiquitin with C-Terminal Aspartic Acid (D77 Ubiquitin)

1. Clone the gene for human ubiquitin and containing an N-terminal His_6 affinity tag and C-terminal aspartic acid residue (D77 ubiquitin) into pET-11b or an equivalent bacterial protein expression vector.
2. Transform 100 ng of plasmid into 50 μL of chemically competent Rosetta(DE3) cells, and recover after heat shock in 350 μL

of LB for 1 h with mild shaking at 37 °C. Deposit 150 μL of the transformants onto a 10 cm agar plate containing 100 μg/mL ampicillin and 34 μg/mL chloramphenicol, and let grow overnight.

3. Wash the bacterial colonies from the plate with 6 mL of LB. Add this equally to 2 Fernbach flasks each containing 1 L of LB supplemented with 100 μg/mL ampicillin, 34 μg/mL chloramphenicol, and 10 g/L dextrose (*see* **Note 9**).
4. Shake the cultures at 120 rpm at 37 °C until they reach an OD_{600} of 0.8, which should take approximately 4.5 h.
5. Pellet cultures at 5000 × *g* for 10 min in 1 L centrifuge bottles, discard the old media, and transfer the cells to fresh LB supplemented with only 0.4 mM IPTG.
6. Shake the cultures at 30 °C for 4 h, pellet them at 5000 × *g* for 10 min, and discard the media.
7. Resuspend each pellet with 25 mL of 1× PBS-T, and then pellet each in a 50 mL conical at 5000 × *g* for 10 min. Discard supernatant, and then drop freeze pellets in liquid nitrogen before long-term storage at −80 °C.
8. Equilibrate a size exclusion column with a molecular weight range of 3 kDa to 75 kDa with at least three column volumes of ddH_2O. Repeat with storage buffer.
9. Thaw cells on ice and add 25 mL of cold lysis buffer to each bacterial pellet.
10. Sonicate each pellet three times for 90 s using 0.5 s pulses, with a 50% duty cycle. Keep lysates in an ice-water bath at all times. Provide at least 5 min for lysates to cool immediately after each round of sonication. After sonication each lysate should be approximately 30 mL.
11. Equilibrate 0.5 mL of Ni-NTA agarose resin by first resuspending in 10 mL wash buffer followed by centrifugation at 1000 × *g* for 2 min. Remove the supernatant, and then repeat with another 10 mL of wash buffer.
12. Transfer lysates to centrifuge tubes and spin at 32,000 × *g* for 60 min. Transfer each 30 mL of lysate to 0.5 mL of the Ni-NTA resin. Rotate the lysate-resin mixture for 1 h at 4 °C.
13. Centrifuge the lysate-resin mixture at 1000 × *g* for 2 min. Discard the supernatant. Now wash the resin with 10 mL of wash buffer by resuspending the resin thoroughly. Repeat three times before transferring the final resin wash buffer mixture to a disposable 25 mL column where the filter has been pre-equilibrated in wash buffer.

14. Wash the resin in the disposable column twice each time with 20 mL of wash buffer, waiting for the flow of buffer to stop completely before continuing. Apply a stopcock to the column.
15. Add 2.7 mL of nickel elution buffer to the resin and incubate for 10 min. Elute into a 3 kDa cutoff centrifugal filter.
16. Concentrate to a final volume to achieve the best resolution with your size exclusion column. We always pass the protein solution through a 0.22 μm centrifugal filter unit prior to loading onto the column. Check the fractions with a UV signal by SDS-PAGE and Coomassie blue solution to ascertain the final level of protein purity.
17. Collect all fractions containing pure D77 ubiquitin. Measure the absorbance of the solution at 280 nm, using an extinction coefficient of 1280 $M^{-1}cm^{-1}$ to determine the initial protein concentration. Concentrate the sample using a 3 kDa cutoff concentrator to at least 500 μM, ideally 1 mM or higher (*see* **Note 10**). The typical final yield of purified D77 ubiquitin is 8 mg per 1 L of starting culture.
18. Aliquot protein and drop freeze in liquid nitrogen prior to long-term storage at −80 °C.

3.2 Expression and Purification of Human Cdc34

1. Clone the gene for human Cdc34B/Ube2R2 (hereafter referred to as Cdc34), transform into chemically competent cells, and prepare bacterial cultures, as described in Subheading 3.1, **steps 1** through **3**.
2. Shake the cultures at 120 rpm at 37 °C until they reach an OD_{600} of 1.0–1.2, which should take approximately 4.5 h.
3. Follow **steps 5** through 7 in Subheading 3.1.
4. Follow **steps 9** through **14** in Subheading 3.1.
5. Begin equilibrating a commercial desalting column into TEV buffer according to the manufacturer.
6. Add 2.7 mL of nickel elution buffer to the Ni-NTA resin, and incubate for 10 min.
7. Collect the eluate, and add an appropriate volume to the desalting column according to the manufacturer. Elute the protein from the desalting column using TEV buffer, and collect in a 15 mL conical tube.
8. The protein should be pure enough to approximate the yield by measuring the absorbance at 280 nm. Use an extinction coefficient of 37,200 $M^{-1}cm^{-1}$. Add additional TEV buffer to bring the final volume of the eluate to 6 mL.
9. Collect a 20 μL sample prior to addition of TEV protease, and dilute in 2× reducing quench buffer. Add 1 mg of TEV per

25 mg of protein in the eluate. Incubate at 4 °C overnight without rotating.

10. Equilibrate a size exclusion column as described in Subheading 3.1, **step 8**.
11. The next morning, load a SDS-PAGE gel with protein molecular weight marker, 6 μL pre-TEV sample from **step 9**, and 6 μL of post-TEV sample. Run the gel and stain in Coomassie blue solution. If TEV digestion of Cdc34 was efficient, a slight downward shift of a 30 kDa band in the pre-TEV lane, corresponding to loss of the His_6 tag, should be visible in the post-TEV lane. Digestion need not be 100% efficient before proceeding to the next step (*see* **Note 11**).
12. Equilibrate another 0.5 mL of Ni-NTA resin in wash buffer as described in Subheading 3.1, **step 11**. Use a fresh desalting column to equilibrate Cdc34 into wash buffer and to remove the EDTA. Add the Cdc34 eluate to the Ni-NTA resin and rotate for 1 h at 4 °C. Add the slurry to a 25 mL disposable column where the filter has been pre-equilibrated in wash buffer, this time collecting the flow-through. His_6-TEV and Cdc34 still containing the His_6 tag will remain bound to the resin (*see* **Note 11**).
13. Concentrate the protein using a centrifugal concentrator with a 10 kDa cutoff as necessary according to the maximum load volume of the size exclusion column. Filter the protein in a microcentrifuge using 0.22 μm spin filters immediately prior to loading onto the size exclusion column.
14. After separation, load 6 μL of all fractions corresponding to the strongest peak by UV absorbance on a SDS-PAGE gel. Stain the gel in Coomassie blue solution to verify the product and purity.
15. Combine all fractions containing Cdc34. Check the protein concentration by measuring the absorbance at 280 nm using an extinction coefficient of 37,200 $M^{-1}cm^{-1}$. Concentrate the protein to 100 μM (*see* **Note 12**). The typical final yield of purified Cdc34 is approximately 2 mg per 1 L of culture.
16. Aliquot the sample between as many tubes as possible (*see* **Note 12**), and then drop freeze in liquid nitrogen prior to long-term storage at −80 °C. Cdc34 can be safely stored for at least 3 years this way.

3.3 Neddylation of Cul1-Rbx1 (C/R^{Nedd8}) and Purification

1. Our protocol is a modified version of that described previously [17].
2. Equilibrate a cation exchange column with five column volumes of ion exchange buffer A, then five column volumes of ion exchange buffer B, and then five column volumes of

buffer A again. Reset the UV absorbance setting on your chromatography instrument. Equilibrate 0.5 mL of Ni-NTA resin in wash buffer as described in Subheading 3.1, **step 11**.

3. Add the following components in sequential order (values for the final concentrations of each are shown). While the volume of the reaction can vary, we recommend at minimum a reaction volume of 1 mL to recover a sufficient amount of product, and 2.5 mL is preferable if sufficient reagents are available.
   (a) ddH_2O.
   (b) 10× neddylation buffer.
   (c) 6 μM His_6-Nedd8.
   (d) 1 μM APPBP1/UBA3. Incubate for 1 min before proceeding to next step.
   (e) 5 μM Ubc12. Incubate for 1 min before proceeding to next step.
   (f) 4 μM C/R.
4. After addition of C/R, incubate reaction for 12 min at room temperature, and then quench by adding 1 M DTT to achieve a final concentration of 10 mM. Aliquot 5 μL of the reaction into 5 μL of 2× reducing quench buffer, and keep for analysis of the reaction efficiency (*see* **step 6** below).
5. Immediately load the full reaction onto the cation exchange column, and wash with buffer A at 0.3 mL/min until the UV signal is back to zero. Recover the flow-through in case the protein does not bind to the column.
6. Initiate a salt gradient with ion exchange buffers A and B with a flow rate of 0.3 mL/min. We typically collect fraction volumes of 0.5 mL. C/R^{Nedd8} should begin eluting from the column at a concentration of 150 mM NaCl (~30% buffer B). After separation is complete, load 6 μL of the fraction corresponding to the strongest peak on an SDS-PAGE gel, along with the sample from **step 4** as well as the flow-through sample from **step 5**. Stain in Coomassie blue solution to verify product formation and purity.
7. Collect all fractions containing C/R^{Nedd8}, and supplement with 0.5 M imidazole-HCl (pH 8.0) to a final concentration of 20 mM. Incubate the eluate on the Ni-NTA resin with rotating for 1 h at 4 °C.
8. Pour eluate over a disposable column where the filter has been pre-equilibrated in wash buffer, discarding the flow-through. Wash the resin thoroughly with 10 mL wash buffer, waiting for the flow of buffer to stop completely before continuing. Repeat with another 10 mL of wash buffer.

9. Place a stopcock on the exit port of the column, and add 2.5 mL of nickel elution buffer. Incubate for 10 min.
10. Collect the eluate, and then exchange the protein into storage buffer by size exclusion chromatography, dialysis, or repeated cycles of reducing sample volume followed by dilution into storage buffer using a centrifugal concentrator (*see* **Note 13**). The extinction coefficient for C/R^{Nedd8} is 115,000 $M^{-1}cm^{-1}$. Concentrate to at least 4 μM for use in biochemical assays.
11. Aliquot the protein into several microcentrifuge tubes, and then drop freeze into liquid nitrogen, and store at −80 °C (*see* **Note 14**). Typical yield of purified C/R^{Nedd8} is approximately 50% of starting C/R.

3.4 Preparation and Purification of Mono-ubiquitin β-Catenin Substrate (β-Cat-(Ub)1)

1. Our protocol is a modified version of that described previously [17].
2. Add the following components in sequential order (values for the final concentrations of each are shown), with a total reaction volume of at least 1 mL to recover a sufficient amount of product.
   (a) ddH_2O.
   (b) 10× reaction buffer.
   (c) 130 μM His_6-ubiquitin.
   (d) 125 nM human E1, incubate for 1 min.
   (e) 2 μM human UbcH5, incubate for 1 min.
   (f) 450 nM C/R^{K720R}.
   (g) 450 nM Skp1/βTRCP.
   (h) 100 μM β-cat peptide.
3. Allow reaction to proceed overnight at approximately 22 °C (about 12 h), and quench with the addition of 1 M DTT to a final concentration of 10 mM.
4. Filter the reaction contents using a 0.22 μm spin filter, and then load onto a size exclusion column with a molecular weight range of 3 kDa to 75 kDa. β-cat-$(Ub)_1$ should be the third-to-last peak to elute, followed by a peak with weak UV absorbance corresponding to ubiquitin and a peak with large UV absorbance corresponding to unmodified β-cat. Load samples from these peaks onto a SDS-PAGE gel, and stain with Coomassie blue solution to confirm the fractions containing the desired product and to assess purity.
5. Equilibrate 0.5 mL of Ni-NTA resin in wash buffer as described in Subheading 3.1, **step 11**.
6. Follow **steps 7** through **9** in Subheading 3.3 (*see* **Note 15**).
7. Collect the eluate, and concentrate to a final volume that is appropriate for your size exclusion chromatography column,

using a 3 kDa cutoff concentrator. Filter the sample using a 0.22 μm spin filter, and load onto the column collecting 0.5 mL fractions during the entire run.

8. Concentrate fractions containing the final product to 50 μM. The extinction coefficient used for β-cat-$(Ub)_1$ is 9500 $M^{-1}cm^{-1}$. The typical yield of β-cat-$(Ub)_1$ relative to starting substrate is 15%. Drop freeze aliquots in liquid nitrogen and store at −80 °C.

3.5 Di-ubiquitin Synthesis Assay

1. Thaw K48R ubiquitin, PKA kit reagents, and γ-^{32}P ATP on ice. Follow the kit manufacturer's instructions to prepare a ^{32}P K48R ubiquitin stock of 100 μM. ^{32}P K48R can go through at least 10 freeze-thaw cycles.
2. Prepare a reaction mixture for a final volume of 20 μL with the sequential addition of ddH_2O, 10× reaction buffer, and ^{32}P-labeled K48R ubiquitin to a final concentration of 10 μM.
3. Prepare six 1.5 mL Eppendorf tubes, each containing 3 μL of 2× reducing quench buffer.
4. Add human E1 to a final concentration of 0.25 μM to the reaction mixture, lightly vortex, and incubate for 1 min to achieve thioester formation between E1 and ubiquitin.
5. Add human Cdc34 to a final concentration of 2 μM to the reaction mixture, lightly vortex, and incubate for 2 min.
6. If measuring E3 activation, add C/R^{Nedd8} to a final concentration of 0.1 μM. Lightly vortex, and then briefly spin down contents using a tabletop microcentrifuge to ensure that the reaction contents are entirely contained in the bottom of the tube.
7. Immediately add D77 ubiquitin to a final concentration of 100 μM to initiate the reaction. Lightly vortex, and begin collecting time points by transferring 3 μL of the reaction to the respective tubes containing quench buffer. If measuring E3 activation, time points every 10 s are required. If measuring E3-independent activity, 2 min per time point should be sufficient to achieve adequate product formation for the quantitation step (*see* **Notes 16** and **17**).
8. Load the substrates and products from each time point onto a reducing SDS-PAGE gel. A reaction lacking human E1 typically serves as the negative control.
9. Place your SDS-PAGE gel onto a filter paper that has already been soaked in ddH_2O, and then overlay a wet sheet of cellophane onto the gel. Thoroughly dry the gel, and then image using a phosphor screen, being careful not to overexpose (*see* **Note 8**). Visualize the gel using any molecular imager capable of detecting phosphorescence (Fig. 1a, b).

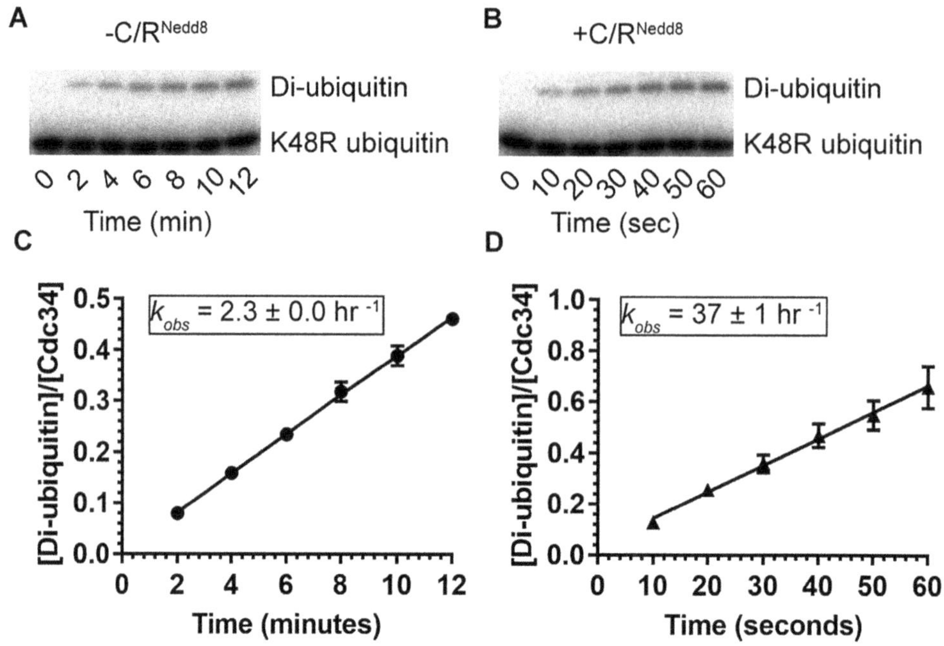

Fig. 1 A simple Cdc34-dependent di-ubiquitin synthesis assay to measure Cdc34 activity after purification. (**a**) Autoradiogram of a time course for a reaction containing human Cdc34, K48R donor ubiquitin, and D77 acceptor ubiquitin. (**b**) Same as (**a**) but with the addition of C/R^{Nedd8}. Notice that the presence of C/R^{Nedd8} enhances the rate of di-ubiquitin formation by 16-fold. (**c**) Linear regression of product formation versus time for the E3-independent reaction in (**a**). Notice that product formation is highly linear with respect to time. (**d**) Linear regression of product formation versus time for the E3-dependent reaction in (**b**). All data points represent the mean of values collected in duplicate. Error bars represent the standard error of measurement

10. Using the software with your molecular imager, quantitate the amount of both unmodified substrate and di-ubiquitin per reaction time point. Use the area where product normally appears in the minus E3 negative control as a measure of background levels, using object average subtraction. Subtract this from all substrate and product bands, and record the intensities of each.
11. Divide the amount of di-ubiquitin product by the total signal produced by both product and substrate per lane to calculate the percentage of substrate converted to product. Next, multiply the percentage by the starting concentration of K48R ubiquitin substrate (10 μM), and divide by the concentration of Cdc34 (2 μM) to find the normalized values.
12. Using a graphing software package, input the normalized values for product into the Y-axis and time (hr) into the X-axis. Use linear regression to determine the best fit to the data. The slope of the line provides Cdc34 activity (k_{obs}) in units of hr^{-1}. Typical rates for both the E3-independent and E3-dependent assays are 2.3 hr^{-1} and 37 hr^{-1}, respectively.

13. Use the r^2 value as a basis to determine quality of the fit. In our experience, a value of less than 0.98 indicates the possibility of excessive product formation for the later time points or insufficient product at earlier ones (Fig. 1c, d).

3.6 Using Michaelis-Menten Kinetics to Estimate the Affinity of Cdc34 and UbcH5 for Their E3 Partner SCF(βTRCP)

1. Thaw β-cat or β-cat-$(Ub)_1$ substrate, PKA kit reagents, and γ-^{32}P ATP on ice. Following the kit manufacturer's instructions, prepare a ^{32}P-labeled substrate stock of at least 50 μM (*see* **Note 18**).
2. Thaw 10× reaction buffer, ubiquitin, E1, Cdc34, UbcH5, C/R^{Nedd8}, and Skp1/βTRCP complex on ice, and then spin down all aliquots briefly using a microcentrifuge to pellet any precipitate that may have formed.
3. Prepare a twofold dilution series of Cdc34 and UbcH5 using storage buffer and extending to at least nine samples. For wild-type Cdc34 and UbcH5, 100 μM stocks are a good starting point for the twofold dilution series such that the E2s will achieve saturation of SCF(βTRCP). If assaying mutant E2s that may compromise binding to E3, concentrate the E2 stocks above 100 μM immediately prior to performing the experiment, and do not freeze these samples for long-term storage (*see* **Note 12**). Since E2s are diluted in storage buffer, it is critical that each reaction contain the same volume of E2 over the entire dilution series (*see* **Note 19**).
4. Prepare an Eppendorf tube containing equimolar amounts of C/R^{Nedd8} and Skp1/βTRCP complex (together referred to as SCF(βTRCP).
5. Prepare a stock of ddH_2O, 10× reaction buffer, ubiquitin, and E1 in an Eppendorf tube. Prepare enough stock to be equally distributed to nine Eppendorf tubes for the E2 titration series. Add enough ubiquitin and E1 such that their final concentrations in each reaction will be 60 μM and 1.0 μM, respectively. We typically use a final reaction volume of 10 μL.
6. Add the E2s from the dilution series sample to its respective reaction tube, vortexing briefly after each addition. We've done this assay by diluting the E2 samples either 1:10 or 1:5 into the reaction mixture. Allow the E2s to charge with ubiquitin for 2 min (slightly longer is acceptable if necessary to fully pipette all reactions).
7. Add SCF(βTRCP) to a final concentration of 0.1 μM, vortexing briefly. Then briefly spin down reaction contents in microcentrifuge. Do not allow SCF(βTRCP) to incubate with the other components prior to initiating the reaction for more than 1 min (*see* **Note 20**).
8. Start a timer when adding radiolabeled substrate at a final concentration of 5 μM to the reactions, lightly vortexing each

reaction after the addition of substrate. If you are working with Cdc34 and β-cat, an 8 min reaction duration is sufficient to convert approximately 15% of substrate to product. If you are working with Cdc34 and β-cat-(Ub)$_1$ or UbcH5 and β-cat, a 2.5 min time point should be sufficient to also observe 15% substrate conversion to product. Quench each reaction in 2× reducing quench buffer at the desired time.

9. Resolve substrates and products on an 18% reducing SDS-PAGE gel (*see* **Note 21**). Include a minus E1 reaction as a negative control. Place the finished gel onto filter paper that has just been soaked in ddH_2O, and then cover the gel with ddH_2O soaked cellophane. Thoroughly dry the gel prior to autoradiography (*see* **Note 8**).
10. Expose the dried gel to a phosphor screen, and then visualize with a molecular imager (Fig. 2a, b).

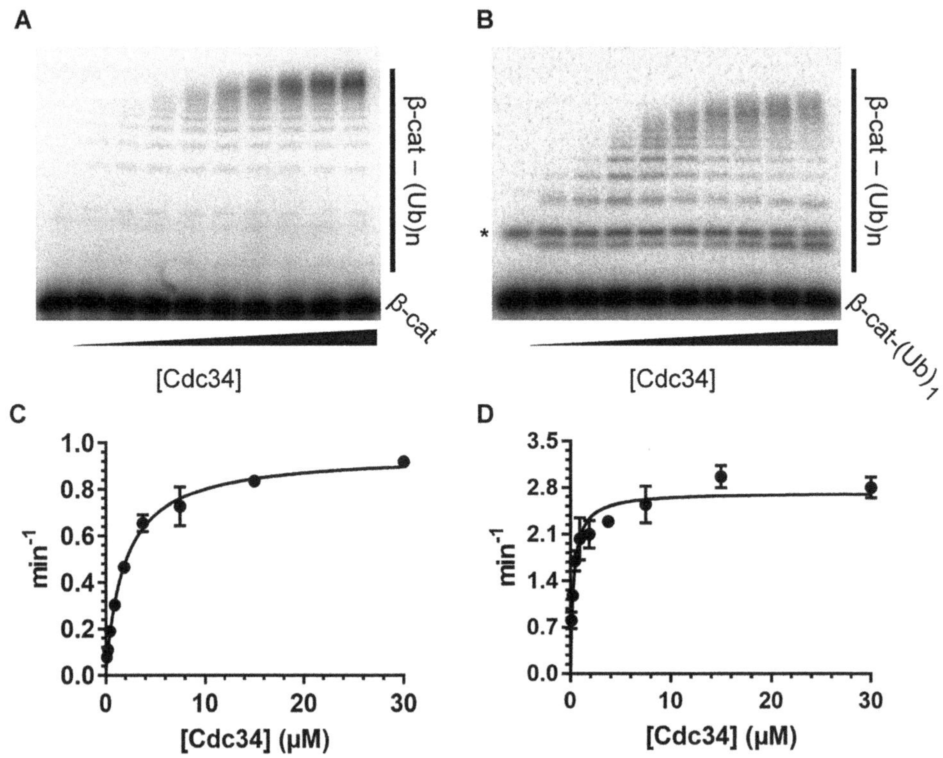

Fig. 2 Determining the affinity of an E2 for an E3 using Michaelis-Menten kinetics. (**a**) Autoradiogram of ubiquitylation reactions containing a twofold dilution series of Cdc34. Each lane corresponds to a reaction incubated for 8 min prior to quenching in 2× reducing quench buffer. (**b**) Same as (**a**) but using β-cat-(Ub)$_1$ substrate. Each reaction was incubated for 2.5 min prior to quenching in 2× reducing quench buffer. The asterisk indicates a contaminant band whose signal was removed during background correction. (**c**) Nonlinear regression of the reaction velocities from (**a**) fit to the Michaelis-Menten equation. (**d**) Same as (**c**) except the reaction velocities from (**b**) were used. All K_m values were derived from reaction velocities from duplicate data points. Error bars represent the standard error of measurement

11. Using your molecular imager's software for quantitation, determine the amount of substrate and product (defined as any ^{32}P-labeled band that migrates through the gel slower than the substrate) for each reaction. Use the area where product normally appears in the minus E3 negative control as a measure of background levels using object average subtraction. Subtract this from all substrate and product bands, and record the intensities of each.
12. Divide the amount of product by the total signal for all products as well as substrate to calculate the percentage of substrate modified for each reaction. Next multiply the percentage by the starting concentration of substrate (5 μM), and then divide by both the concentration of SCF(βTRCP) (0.1 μM) and the time of the reaction. This provides the normalized rate of each reaction in units of min^{-1}.
13. Using graphing software capable of fitting data to nonlinear equations, fit the data to the Michaelis-Menten equation where the E2 concentrations have been plotted on the X-axis and the normalized reaction rates have been plotted on the Y-axis (Fig. 2c, d). This procedure results in an estimate of the Michaelis constant, K_m, as well as V_{max}, the maximal reaction velocity under saturating conditions of E2 for SCF(βTRCP) (Table 1). Note that the K_m may vary both with storage conditions depending on the E2 used as well as with the nature of

Table 1
The K_m values of Cdc34 or UbcH5 for SCF were measured as a function of both the ionic strength of the reaction buffer and E2 exposure to multiple freeze-thaw cycles

E2	K_m
Cdc34, 100 μM	1.9 μM ± 0.1 μM
Cdc34, 100 μM, 40% ddH$_2$O	6.7 μM ± 1.0 μM
Cdc34, 300 μM	4.1 μM ± 0.4 μM
Cdc34, 1 mM	3.7 μM ± 0.2 μM
Cdc34, 1 mM, single freeze-thaw cycle	4.2 μM ± 0.3 μM
Cdc34, 1 mM, five freeze-thaw cycles	8.0 μM ± 0.7 μM
Cdc34, 100 μM, β-cat-(Ub)$_1$	320 nM ± 50 nM
UbcH5, 100 μM	420 nM ± 30 nM
UbcH5, 1 mM	330 nM ± 50 nM
UbcH5, 1 mM, five freeze-thaw cycles	660 nM ± 70 nM

the substrate (Table 1; *see* **Notes 12** and **22**). If all steps are followed carefully, an r^2 of 0.9 for duplicate reactions should be attainable (*see* **Note 23**).

4 Notes

1. Purifying ubiquitin does not require the use of PIC.
2. These protocols were written for the use of gravity flow disposable desalting columns with a bed volume of 5 mL. Volumes may be adjusted as needed to support the use of other products such as buffer exchange spin columns or dialysis kits.
3. Cullin-RING ligases (CRLs) are modified by the ubiquitin-like protein Nedd8, which enhances both E2 binding and the rate of ubiquitin transfer from E2- to E3-bound substrate. In CRLs containing the Cul1 subunit (e.g., SCF), modification with Nedd8 occurs on Lys 720 on Cul1. Since UbcH5 can ubiquitylate Lys 720 in vitro that also results in activation of the E3, the Lys 720 to Arg mutation has been incorporated to suppress this.
4. The β-cat peptide sequence is N-term KAWQQQSYLD-phosphoS-GIH-phosphoS-GATTTAPRRASY and contains a C-terminal PKA phosphorylation site for radiolabeling [17]. While there are multiple commercial vendors that can, in principle, produce this peptide, our experience has been that the peptide is not trivial to synthesize. We have had good results with New England Peptide.
5. K48R and D77 ubiquitin are described as donor and acceptor ubiquitins in the context of Cdc34 reactions because the E2 has a strong preference for forming poly-ubiquitin chains that are covalently linked through Lys 48 on ubiquitin. Therefore, K48R ubiquitin may still be thioesterified to Cdc34 but inhibits poly-ubiquitin chain formation. While D77 ubiquitin has an intact Lys 48 residue, the additional aspartate residue at the C-terminus eliminates the possibility of charging by E1 onto the Cdc34 active site. Thus, only di-ubiquitin product will be formed when assays are performed with both K48R donor and D77 acceptor ubiquitins. This makes quantitation of substrate and product far simpler for this assay.
6. It is critical that the overnight thrombin digestion be performed on the eluate (e.g., not while still attached to resin) on ice. Any agitation, such as rotating, will likely lead to precipitation. Also, we recommend using Sigma bovine plasma thrombin for the digest (T7513), as seemingly similar products have resulted in less efficient cleavage.

7. While normal plastic cellophane may be adequate, we use ultra clear cellophane from Research Products International Corp. (order number 1080), which we believe reduces the likelihood of gel cracking during the drying procedure.
8. It is critical to evenly dry the gel such that no cracks form. In our experience, it is critical to use a gel dryer that is connected to a dedicated vacuum system.
9. Overnight cultures can be used instead of inoculating from a selective media plate. However, in our experience, the overnight culture must be optimized to still be in log phase growth immediately prior to inoculation of the 1 L cultures, and starting from fresh bacteria on an agar plate has been much easier.
10. Ubiquitin is well-known for its remarkable stability. We have solubilized lyophilized commercial ubiquitin at concentrations of 3.5 mM for long-term storage without any notable reduction in activity after multiple freeze-thaw cycles.
11. It is okay if the TEV digestion of Cdc34 does not go to completion. The additional Ni-NTA step will remove His_6-TEV as well as any undigested His_6-Cdc34 from the elution. If Cdc34 digestion appears 100% complete by Coomassie blue staining, we then skip the Ni-NTA incubation step. However, only do this if your size exclusion column has sufficient resolution to separate Cdc34 from TEV protease.
12. Human Cdc34 activity is sensitive to the concentration of stocks frozen for long-term storage as well as the number of freeze-thaw cycles. Do not freeze Cdc34 protein at a concentration higher than 150 μM. Additionally, we recommend using fresh Cdc34 protein that has only been frozen once for results featured in a publication. UbcH5 is significantly more resistant to these effects than Cdc34 (*see* Table 1). However, when stored near 1 mM concentrations for years, the corresponding author has anecdotally observed a significant decline in activity as well (data not shown).
13. In the past, a size exclusion chromatography step was used for polishing the purification of C/R^{Nedd8} and for exchanging the protein into storage buffer. In our experience, this step doesn't measurably improve the activity or the purity. However, it will significantly reduce the final yield of product, and for this reason we've eliminated this step in the purification.
14. In our experience, C/R^{Nedd8} is fairly resistant to both multiple freeze-thaw cycles and long-term storage at −80 °C. However, an excess of 3 years of long-term storage may result in qualitative changes in activity including the extent of poly-ubiquitin chain lengths achieved during the reaction. While this issue may not have an effect on the multi-turnover reaction described here (which considers any substrate modified by

one or more ubiquitins as product), it could affect the results from experiments that quantify individual rates of ubiquitin transfer such as single-encounter reactions under pre-steady-state conditions [18].

15. Repeated rounds of size exclusion chromatography may result in pure product in the absence of a His_6-tagged peptide, although shorter peptide substrates may result in insufficient separation of unmodified substrate from the mono-ubiquitylated product.
16. The amount of K48R ubiquitin (e.g., donor) should be in at least twofold excess over the E2 concentration to ensure that the E2 remains charged with ubiquitin throughout the time course, and in the presence of E3, a fivefold excess should be considered minimum. However, changing from a fivefold to a tenfold excess does not substantially affect the rate both in the absence and presence of E3.
17. The rates (k_{obs}) determined from this method do not represent the maximal rate of di-ubiquitin formation since acceptor ubiquitin has a K_m for Cdc34 greater than 1 mM. Thus, changing acceptor ubiquitin concentration by two- or threefold should result in linear changes to k_{obs}. This note is useful if the researcher wants to either reduce or increase the length of the time points for any reason while still maintaining adequate product formation for quantitation.
18. β-cat peptides contain two phosphoserines that are critical for binding to βTRCP. In our experience, these peptides are relatively impervious to freeze-thaw cycles, but their phosphate modifications may slowly hydrolyze in solution, so samples should be kept on ice and frozen immediately after use.
19. Cdc34's affinity for SCF(βTRCP) is mediated through electrostatic interactions, causing it to be sensitive to changes in the ionic strength of the final reaction conditions after all components have been added [9]. All experiments were performed with a consistent ratio of ddH_2O to other reaction components (55% of the final reaction volume), unless otherwise noted. It is important to be consistent and consider the amount of storage buffer being added to the reaction from other protein components when designing the conditions for these assays. For instance, note that replacing 15% of the total water in the reaction with storage buffer results in a significant increase in the K_m (Table 1). It is important to consider that some E2s other than Cdc34 may also be sensitive to reaction conditions such as the ionic strength, concentration of glycerol, etc.
20. The combination of Cdc34 with SCF(βTRCP) results in the rapid production of di-ubiquitin product. Thus, allowing a reaction to sit for too long prior to adding substrate can result

in a significant amount of free ubiquitin being consumed in the formation of di-ubiquitin, potentially interfering with the reaction.

21. An 18% polyacrylamide gel is critical for resolving the substrate peptide from a mono-ubiquitylated product.
22. The identity of the substrate and the rate of ubiquitin transfer to that substrate can have substantial effects on the value of K_m. For instance, note that the values for the K_m of Cdc34 for SCF do not agree when comparing reactions containing β-cat or β-cat-$(Ub)_1$ as substrates (Fig. 2 and Table 1). Thus, great care should be taken in selecting a substrate that is appropriate for Michaelis-Menten kinetics.
23. For assays that estimate the K_m of Cdc34 for SCF(βTRCP) and containing Cdc34 and β-cat substrate, the E1 concentration must be at least 1.0 μM. When the E1 concentration is lower than 1.0 μM, reactions containing the highest concentrations of Cdc34 will have higher ratios of uncharged Cdc34 to Cdc34-ubiquitin complex. This will lower the reaction velocities and results in the false appearance that E2 saturation of E3 has been achieved. This is often indicated during the fitting procedure by an r^2 value much lower than 0.9. For reasons unknown, reactions containing UbcH5 at concentrations up to 30 μM can tolerate E1 concentrations substantially below 1.0 μM. Similarly, Cdc34 can tolerate lower E1 concentrations with β-cat-$(Ub)_1$ substrate. In our experience, it is best to scout for conditions where the amount of ubiquitin or E1 doesn't limit E2 or E3 activity.

Acknowledgments

This work, including the efforts of Spencer Hill, Connor Hill, and Gary Kleiger, was funded by HHS | National Institutes of Health (NIH) (R15 GM117555- 01).

References

1. Yau R, Rape M (2016) The increasing complexity of the ubiquitin code. Nat Cell Biol 18 (6):579–586. https://doi.org/10.1038/ncb3358
2. Zheng N, Shabek N (2017) Ubiquitin ligases: structure, function, and regulation. Annu Rev Biochem 86:129–157. https://doi.org/10.1146/annurev-biochem-060815-014922
3. Swatek KN, Komander D (2016) Ubiquitin modifications. Cell Res 26(4):399–422. https://doi.org/10.1038/cr.2016.39
4. Mirzaei H, Rogers RS, Grimes B, Eng J, Aderem A, Aebersold R (2010) Characterizing the connectivity of poly-ubiquitin chains by selected reaction monitoring mass spectrometry. Mol BioSyst 6(10):2004–2014. https://doi.org/10.1039/c005242f
5. Das R, Liang YH, Mariano J, Li J, Huang T, King A, Tarasov SG, Weissman AM, Ji X, Byrd RA (2013) Allosteric regulation of E2:E3 interactions promote a processive ubiquitination machine. EMBO J 32(18):2504–2516. https://doi.org/10.1038/emboj.2013.174

6. Buetow L, Gabrielsen M, Anthony NG, Dou H, Patel A, Aitkenhead H, Sibbet GJ, Smith BO, Huang DT (2015) Activation of a primed RING E3-E2-ubiquitin complex by non-covalent ubiquitin. Mol Cell 58 (2):297–310. https://doi.org/10.1016/j.molcel.2015.02.017
7. Eletr ZM, Kuhlman B (2007) Sequence determinants of E2-E6AP binding affinity and specificity. J Mol Biol 369(2):419–428. https://doi.org/10.1016/j.jmb.2007.03.026
8. Wright JD, Mace PD, Day CL (2016) Secondary ubiquitin-RING docking enhances Arkadia and Ark2C E3 ligase activity. Nat Struct Mol Biol 23(1):45–52. https://doi.org/10.1038/nsmb.3142
9. Kleiger G, Saha A, Lewis S, Kuhlman B, Deshaies RJ (2009) Rapid E2-E3 assembly and disassembly enable processive ubiquitylation of cullin-RING ubiquitin ligase substrates. Cell 139(5):957–968. https://doi.org/10.1016/j.cell.2009.10.030
10. Eletr ZM, Huang DT, Duda DM, Schulman BA, Kuhlman B (2005) E2 conjugating enzymes must disengage from their E1 enzymes before E3-dependent ubiquitin and ubiquitin-like transfer. Nat Struct Mol Biol 12 (10):933–934. https://doi.org/10.1038/nsmb984
11. Huang DT, Schulman BA (2005) Expression, purification, and characterization of the E1 for human NEDD8, the heterodimeric APPBP1-UBA3 complex. Methods Enzymol 398:9–20. https://doi.org/10.1016/S0076-6879(05)98002-6
12. Chiba T (2005) In vitro systems for NEDD8 conjugation by Ubc12. Methods Enzymol 398:68–73. https://doi.org/10.1016/S0076-6879(05)98007-5
13. Beaudenon S, Huibregtse JM (2005) High-level expression and purification of recombinant E1 enzyme. Methods Enzymol 398:3–8. https://doi.org/10.1016/S0076-6879(05)98001-4
14. Zheng M, Liu J, Yang Z, Gu X, Li F, Lou T, Ji C, Mao Y (2010) Expression, purification and characterization of human ubiquitin-activating enzyme, UBE1. Mol Biol Rep 37 (3):1413–1419. https://doi.org/10.1007/s11033-009-9525-3
15. Lorick KL, Jensen JP, Weissman AM (2005) Expression, purification, and properties of the Ubc4/5 family of E2 enzymes. Methods Enzymol 398:54–68. https://doi.org/10.1016/S0076-6879(05)98006-3
16. Li T, Pavletich NP, Schulman BA, Zheng N (2005) High-level expression and purification of recombinant SCF ubiquitin ligases. Methods Enzymol 398:125–142. https://doi.org/10.1016/S0076-6879(05)98012-9
17. Saha A, Deshaies RJ (2008) Multimodal activation of the ubiquitin ligase SCF by Nedd8 conjugation. Mol Cell 32(1):21–31. https://doi.org/10.1016/j.molcel.2008.08.021
18. Pierce NW, Kleiger G, Shan SO, Deshaies RJ (2009) Detection of sequential polyubiquitylation on a millisecond timescale. Nature 462 (7273):615–619. https://doi.org/10.1038/nature08595

Chapter 5

Competition Assay for Measuring Deubiquitinating Enzyme Substrate Affinity

Michael T. Morgan and Cynthia Wolberger

Abstract

Assays of the affinity of a deubiquitinating enzyme for substrate, either through binding studies or determination of the Michaelis constant, K_M, can shed light on substrate selectivity and the effects of mutations on substrate interactions. The difficulty in generating sufficient quantities of ubiquitinated substrate frequently presents a barrier to these studies. We describe here an alternative approach that can be used in cases where non-hydrolyzable, chemically ubiquitinated substrate analogs can be more readily generated. The substrate analog can be utilized as a competitive inhibitor in kinetics experiments monitoring cleavage of ubiquitin-AMC (Ub-AMC) by the deubiquitinating enzyme. The resulting inhibitory constant, K_i, provides a reliable approximation of the K_d for ubiquitinated substrate. We show how this approach can be used to assay the affinity of the yeast SAGA DUB module for nucleosomes containing monoubiquitinated H2B.

Key words Enzyme kinetics, Deubiquitinating enzymes, Ubiquitin, Enzyme inhibition, Equilibrium binding

1 Introduction

Deubiquitinating enzymes (DUBs) remove ubiquitin from substrate proteins and are involved in virtually every signaling process in eukaryotes. The approximately 90 DUBs encoded by the human genome target a vast variety of substrates and serve distinct roles in biology. There are six distinct classes of DUBs: the USP, UCH, OTU, MJD, and recently discovered MINDY cysteine protease DUBS and the JAMM metalloproteases [1]. All of these enzymes catalyze hydrolysis of the isopeptide linkage that joins the ubiquitin C-terminus to a lysine or, in some cases, to a protein amino terminus [2]. A detailed understanding of the substrate specificity of a particular DUB or the impact of mutations entails in vitro assays of purified components. Methods for measuring specificity include determining the kinetic constants, k_{cat} and K_M, or assaying binding of a catalytic mutant to substrate using methods such as Förster

Thibault Mayor and Gary Kleiger (eds.), *The Ubiquitin Proteasome System: Methods and Protocols*, Methods in Molecular Biology, vol. 1844, https://doi.org/10.1007/978-1-4939-8706-1_5, © Springer Science+Business Media, LLC, part of Springer Nature 2018

resonance energy transfer (FRET) or isothermal titration calorimetry (ITC). In many cases, however, obtaining sufficient quantities of ubiquitinated substrate is difficult, if not impossible, thus presenting an obstacle to quantitative measures of substrate affinity and specificity.

We describe an alternative approach to estimating the affinity of a DUB for its substrate that can be implemented in cases where a non-hydrolyzable analog of the ubiquitinated substrate is available. The non-hydrolyzable substrate can be used as a competitive inhibitor in assays monitoring cleavage of the fluorogenic substrate, ubiquitin-AMC (Ub-AMC). The inhibition constant, K_i, which is determined by measuring Ub-AMC cleavage activity as a function of increasing competitor concentration, provides an estimate of the affinity of the DUB for the ubiquitinated substrate.

The protocol described here is for the yeast Spt-Ada-Gcn5 acetyltransferase (SAGA) DUB module, a four-protein subcomplex of the 1.8 MDa SAGA transcriptional coactivator [3]. The DUB module cleaves ubiquitin from Lys123 of yeast histone H2B in the context of the nucleosome core particle (NCP-Ub) [4–6]. Binding assays and kinetic studies of the DUB module are challenging due to the difficulty in generating sufficient quantities of nucleosomes that are specifically monoubiquitinated at H2B-K123. H2B-Ub substrate can be made by a semisynthetic approach that utilizes H2B and ubiquitin fragments generated by native chemical ligation [7, 8], which can be incorporated it into nucleosomes. The yield, however, is limited, and generating significant quantities is highly labor-intensive, making this approach impractical for kinetic or binding studies. While FRET methods require less material, it is first necessary to identify appropriate sites on both the enzyme and nucleosome for the donor and acceptor fluorophores.

We devised an approach for approximating the affinity of the DUB module for NCP-Ub that takes advantage of a method for generating nucleosomes with a non-hydrolyzable linkage between the ubiquitin C-terminus and Lys123 of histone H2B [9, 10]. Cysteine residues substituted for ubiquitin residue G76 and H2B residue K123 (K120 in *Xenopus* H2B) can be cross-linked with dichloroacetone (DCA) [9, 10], thus approximating the native isopeptide linkage. Nucleosomes containing H2B with the cross-linked ubiquitin (H2B-Ub_{DCA}) are then used to inhibit the cleavage of Ub-AMC by the DUB module [10]. The point at which rates of Ub-AMC cleavage are reduced by 50% is described as the inhibition constant, K_i. If NCP-Ub_{DCA} behaves as a competitive inhibitor, then $K_i \approx K_d$.

In this protocol, we describe how to identify the appropriate enzyme concentration for Ub-AMC cleavage assays, determine the Michaelis-Menten constants for Ub-AMC cleavage, and approximate the K_i of the non-hydrolyzable NCP-Ub_{DCA} analog. This method can be generalized to any DUB for which a non-hydrolyzable ubiquitinated substrate can be generated.

2 Materials

All solutions should be prepared with ultra-pure water and analytical-grade materials. Buffers can be prepared in advance, aliquoted in convenient volumes, and frozen at −20 °C for several months. *See* **Note 1** for details of materials used for data acquisition and analysis. The use of a multichannel pipettor is optional.

2.1 Kinetics of Ubiquitin-AMC Cleavage

1. 384-well, low-volume, flat bottom, black polystyrene plates.
2. Purified SAGA deubiquitinating module (DUBm) (as described in [11]) in DUBm storage buffer: 10 mM HEPES pH 8, 150 mM NaCl, 20 μM $ZnCl_2$, and 5 mM dithiothreitol (DTT).
3. Ubiquitin-AMC, a fluorogenic substrate in which the ubiquitin C-terminus is covalently linked to an AMC molecule via a peptide bond (Boston Biochem). When this bond is cleaved by the DUB, there is an increase in fluorescence.
4. Dimethyl sulfoxide (DMSO).
5. Assay buffer: 20 mM HEPES buffer (pH 7.5), 150 mM NaCl, 20 μM $ZnCl_2$, 1 mM DTT, 0.1 mg/mL bovine serum albumin (BSA).

2.2 Determination of Inhibition Constant by Non-hydrolyzable Substrate

1. Nucleosome storage buffer: 10 mM Tris–HCl (pH 7.5), 50 mM KCl, and 1 mM DTT.
2. Non-hydrolyzable substrate, nucleosome core particle (NCP) with ubiquitin C-terminus conjugated to H2B residue 123 via a DCA linkage (NCP-Ub_{DCA}), prepared as previously described [10], in nucleosome storage buffer.
3. Serial dilutions of Ub-AMC in DMSO (*see* Subheading 3.3.1).

3 Methods

3.1 Determine Range of Enzyme Concentrations Over Which Reaction Velocity Is Linear

It is first necessary to determine the concentration range of enzyme in which the rate of Ub-AMC cleavage increases linearly with increasing enzyme concentration. In practice, enzymes can adhere to surfaces, lose activity during handling, or exhibit prep-to-prep variability in a way that may significantly alter the amount of active enzyme. Measurements of initial rates are most easily obtained when the reaction is slowest, which corresponds to the lowest reasonable concentration of enzyme. It is therefore important to determine the lowest enzyme concentrations at which the activity of the DUB in question is linear. Such an approach ensures that the experimenter is working at DUB concentrations at which the assumptions of kinetic analysis hold and that results are reproducible. In this section, we describe a workflow for determining the linear range of the DUB module.

3.2 Determining the Linear Range of Ub-AMC Cleavage

1. Prepare 60 μL of 10 μM Ubiquitin-AMC (Ub-AMC) substrate in assay buffer; this will be a 10× stock of Ub-AMC (final concentration of Ub-AMC was chosen to be at least 5× the highest concentration of DUBm used in **step 2** below).
2. Prepare 90 μL each of 10 nM, 25 nM, 50 nM, 100 nM, 150 nM, and 200 nM DUBm in assay buffer. This volume was chosen to slightly exceed the total amount needed for the measurements below.
3. To perform the first replicate, pipette 27 μL of each of the six DUBm dilutions into six wells of a microplate. Cover with tape.
4. Incubate the plate for 20 min in the plate reader, which has been pre-equilibrated at the assay temperature (30 °C for the yeast DUB module).
5. Initiate the cleavage reaction by adding 3 μL of the 10× Ub-AMC stock to each well containing DUB enzyme, giving a final reaction concentration of 1 μM Ub-AMC (*see* **Note 2**). The pipetting can be done either in rapid succession with a repeater pipette or at once with a multichannel pipette. For cases in which the reaction rate is quite fast, such that a significant amount of Ub-AMC is consumed during the time it takes the plate reader to monitor multiple wells, it will be necessary to assay one concentration at a time (*see* **Notes 3** and **4** on gain setting).
6. Record the fluorescence increase in each well as a function of time (*see* **Note 5**).
7. Repeat **steps 2–6** two additional times such that three measurements of each enzyme concentration are made.
8. Identify the linear range of the reaction, which is typically over the period in which approximately 10% of substrate is consumed.
9. Determine the initial velocity (v_i) for each enzyme concentration by measuring the slope of the line in the linear range of the reaction. This is done by plotting the increasing fluorescence values as a function of time and using linear regression in Excel (or similar software) to fit a line to the data points.
10. Using Prism, Excel, or other data analysis software, plot initial velocity (v_i) divided by the corresponding enzyme concentration ($v_i/[E]$) as a function of the enzyme concentration ($[E]$). This is in essence a plot of the first derivative of the rate as a function of enzyme concentration. An example is shown in Fig. 1
11. The linear range of enzyme activity corresponds to the concentration range over which the slope of the plot of $v_i/[E]$ versus $[E]$ is approximately 0. For the data shown in Fig. 1, the

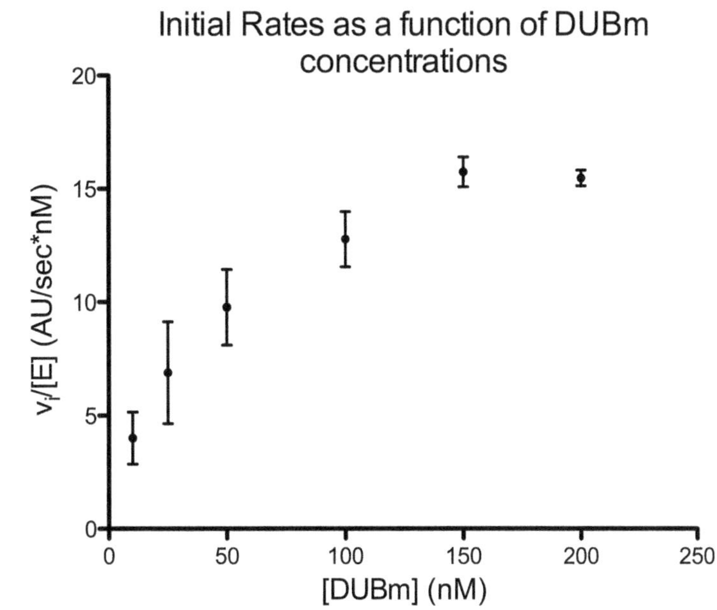

Fig. 1 Identifying the enzyme concentration to use in assays. To find the minimal concentration of enzyme for use in kinetic assays, the initial rate of Ub-AMC cleavage is plotted as a function of enzyme concentration. The same concentration of Ub-AMC is the same in each assay

activity of the DUB module is linear after an enzyme concentration of 150 nM (*see* **Note 6**).

12. If the plot does not reveal a concentration regime that reflects the desired stability of $v_i/[E]$, consider altering the buffer conditions, exploring a different concentration range of enzyme, or repurifying the enzyme.

3.3 Determining Michaelis-Menten Kinetic Constants for DUB Module Cleavage of Ub-AMC

Classical enzyme kinetic assays can be used to measure the Michaelis-Menten constants that define the enzyme's activity on Ub-AMC. Since commercially available Ub-AMC is typically dissolved in DMSO, an important consideration is maintaining constant solvent content while increasing the concentration of the Ub-AMC substrate. It is important to keep the volume of substrate added to the reaction as small as possible because high DMSO concentrations can inhibit enzyme activity. Since the concentration of substrate required to reach maximum enzyme velocity (V_{max}) differs between DUBs, the need to minimize the proportion of DMSO in the reaction can, in practice, limit the maximum amount of substrate that can be used.

In order to accurately determine the relative amount of ubiquitin that has been cleaved as a function of time, it is first necessary to generate a standard curve, which is used to convert arbitrary fluorescence signal into units of concentration.

3.3.1 Generating a Ub-AMC Standard Curve

1. Make 10 μL stocks of 5, 10, and 20 μM Ub-AMC.
2. Make 300 μL of a 1 μM stock of DUB module in assay buffer (*see* **Note 7**).
3. Pipette 27 μL of the DUB stock into each of 9 microplate wells, forming three sets of three wells.
4. Add 3 μL of 5 μM Ub-AMC to each of the first three wells. Repeat with the 10 and 20 μM UB-AMC stocks with the next two sets of wells, such that there are three cleavage reactions for each concentration of Ub-AMC (now at 0.5, 1, and 2 μM for each respective Ub-AMC stock).
5. Monitor the fluorescence in each well with the plate reader until there is no further increase in signal in any of the wells (*see* **Note 8**). At this point, the Ub-AMC in each well is assumed to be fully digested.
6. Record the signal from each well. Each set of three readings should then be averaged.
7. Plot the average fluorescence values and their corresponding errors as a function of Ub-AMC concentration, and use Excel to fit the values to a straight line.
8. The algebraic expression of the line, $y = mx + b$, can then be used to convert fluorescence units into concentration values; given any fluorescence value (y), solving for x gives the concentration of Ub-AMC that has been cleaved. This equation will be used to measure the amount of Ub-AMC consumed in all experiments (Fig. 2).

3.3.2 Measuring Reaction Velocities, K_M, and V_{max}

1. Make a series of 10 μL stocks of Ub-AMC diluted in pure DMSO, at the following concentrations: 10, 20, 30, 40, 50, 70, 100, and 192 μM Ub-AMC (in this case, the Ub-AMC obtained from the vendor was at a concentration of 192 μM).
2. Make a working stock of DUBm by diluting with assay buffer to 1.1× the optimal enzyme concentration as determined in Subheading 3.1 (165 nM in this case).
3. Pipette 27 μL of the working enzyme stock into three wells of the microplate. Each of the three wells will constitute a triplicate measurement of each substrate concentration.
4. Incubate the microplate containing the enzyme in the plate reader at the desired temperature (30 °C in this case) for 20 min.
5. Add 3 μL of the 10 μM Ub-AMC stock to each well to initiate each reaction replicate.
6. Immediately begin monitoring the cleavage of Ub-AMC with the plate reader once the substrate is added.

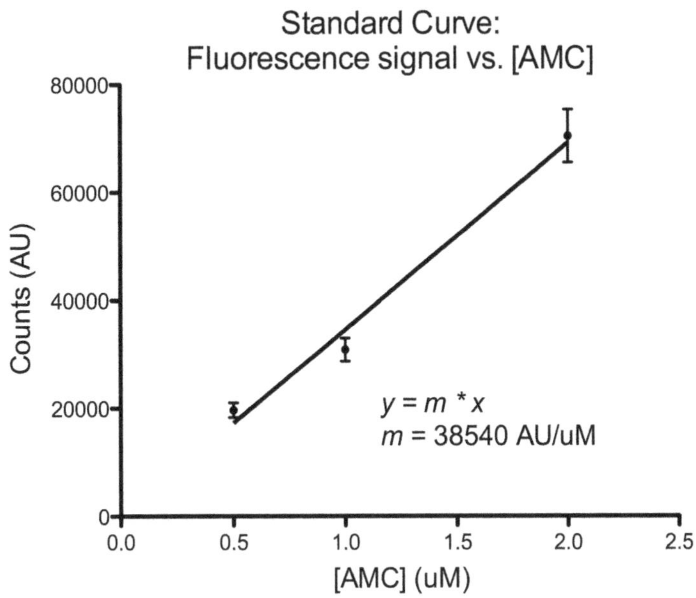

Fig. 2 The standard curve of fluorescence as a function of cleaved Ub-AMC concentration. A plot of fully digested Ub-AMC at different concentrations is used to convert fluorescence units (AU) into concentration units (μM)

7. For each well, determine the interval during which the fluorescence increases linearly as a function of time (*see* **Note 9**).
8. Repeat **steps 3**–**7** for 20, 30, 40, 50, 70, 100, and 192 μM Ub-AMC concentrations, until triplicate measurements of each desired substrate concentration are obtained.
9. Determine the slope of the linear region of the cleavage reaction for each well as described in **step 9** of Subheading 3.2, and convert the fluorescence units into concentration of product as described in **step 8** of Subheading 3.3.1. The result will be the reaction velocity in units of μM/s. Average the values for each triplicate point to obtain the average value of v_i for each concentration.
10. Plot the average initial velocities as a function of Ub-AMC concentration as shown in Fig. 3. These values should be plotted during the course of the experiment in order to monitor whether the reaction is approaching saturation, meaning that the reaction velocity does not change appreciably as the Ub-AMC concentration is increased. Since the K_M of the DUB for Ub-AMC can be high, making it impractical to approach saturation, data analysis software such as Prism can be used to fit the data to the Michaelis-Menten equation and monitor the range of 95% confidence intervals for V_{max} and K_M values to determine when the values are sufficiently constrained.

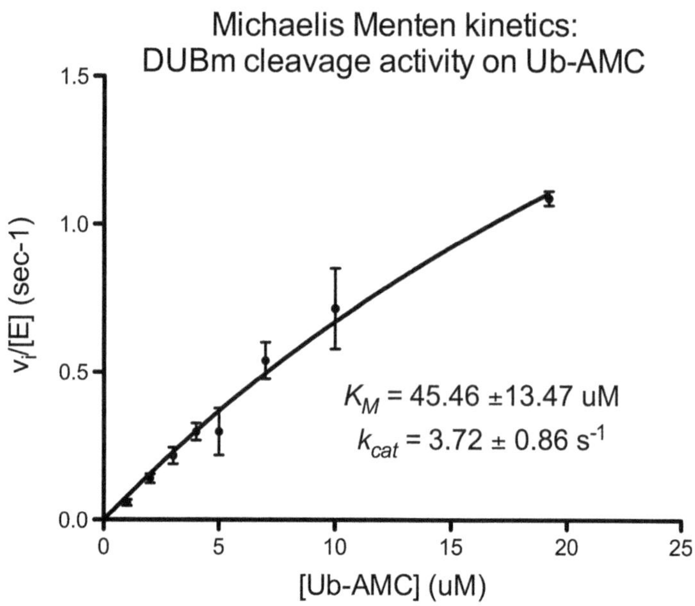

Fig. 3 Michaelis-Menten kinetics of DUB module cleavage of Ub-AMC. Plot of initial reaction velocity as a function of Ub-AMC concentration can be fit to the Michaelis-Menten equation to determine values for K_M and k_{cat}.

3.4 Measure the Ki of Nucleosome Containing Non-hydrolyzable H2B-Ub Substrate Analog

The kinetic constants of the deubiquitinating enzyme are used to determine the K_i of a non-cleavable substrate analog, which in this application is the ubiquitinated nucleosome containing a non-hydrolyzable linkage between the ubiquitin C-terminus and H2B (NCP-Ub_{DCA}). The cleavage of Ub-AMC is monitored in the presence of increasing concentrations of NCP-Ub_{DCA}, which binds to the DUBm and competes for substrate cleavage. The effect of increasing the concentration of competitor should be assayed for at least three different Ub-AMC concentrations. Substrate analog concentration should span a range of both above and below the estimated K_d of the substrate analog in order to derive well-determined values. In this case, we used electrophoretic mobility shift assays (EMSA) to estimate the K_d of the DUB module affinity for NCP-Ub_{DCA} and inferred that the K_d likely approximates the K_i (*see* **Note 10**). If the K_d is not known, it may be necessary to assay a range in order to identify a set of concentrations over which the apparent K_M of the reaction is reduced by greater than 50% (*see* **Note 11**).

1. Prepare a 187.5 nM stock solution of the DUB module in assay buffer (1.25× the final enzyme concentration in each reaction). For the experiment described here, 500 μL of enzyme stock is a sufficient volume to measure Ub-AMC cleavage at five

different Ub-AMC concentrations in the presence of three different NCP-Ub_{DCA} concentrations, as well as in the absence of NCP-Ub_{DCA} (storage buffer only added).

2. Make 15 μL Ub-AMC stock solutions at concentrations of 10 μM, 20 μM, 30 μM, 50 μM, and 100 μM in DMSO, each representing a 10× stock solution of the final substrate concentration in each reaction.
3. Prepare 15 μL each of 10× NCP-Ub_{DCA} stock solutions at concentrations of 10, 25, and 50 μM. Add 15 μL of NCP storage buffer to a fourth tube (to serve as the no-inhibitor control).
4. Add 120 μL DUB stock to the 10 μM NCP-Ub_{DCA} stock, giving a total volume of 135 μL. This is the exact volume to be used, whereas previous steps provided a slight excess, so pipette with care or make a slightly larger volume. Incubate for at least 20 min.
5. Aliquot 27 μL of the DUBm-NCP-Ub mixture into each of five wells in the microplate, and incubate for 20 min at 30 °C.
6. To ensure that early time points are captured, initiate one reaction at a time, and monitor fluorescence with the plate reader. To initiate the first reaction, add 3 μL of the 10 μM Ub-AMC stock to the first well (final concentration 1 μM). Monitor the increase in fluorescence as a function of time.
7. Repeat **step 6** with the 20 μM, 30 μM, 50 μM, and 100 μM Ub-AMC stocks (giving final substrate concentrations of 2, 3, 5, and 10 μM).
8. Verify that the initial rates are linear by monitoring the fluorescence signal.
9. Repeat **steps 4–8** for each concentration of the inhibitor, NCP-UB_{DCA}, as well as for the NCP storage buffer (no-inhibitor control).
10. Use the standard curve determined in Subheading 3.2.1 to convert initial rates of fluorescence evolution (AU/s) into initial rates of product formation (μM/s).
11. For each inhibitor concentration, plot the initial rates as a function of substrate concentration, and determine the apparent K_M:

$$v_i = V_{max} \times \frac{[S]}{K_M^{obs}(I) + [S]}$$

The values of the apparent K_M for each inhibitor concentration are then used to determine the K_i from the following equation:

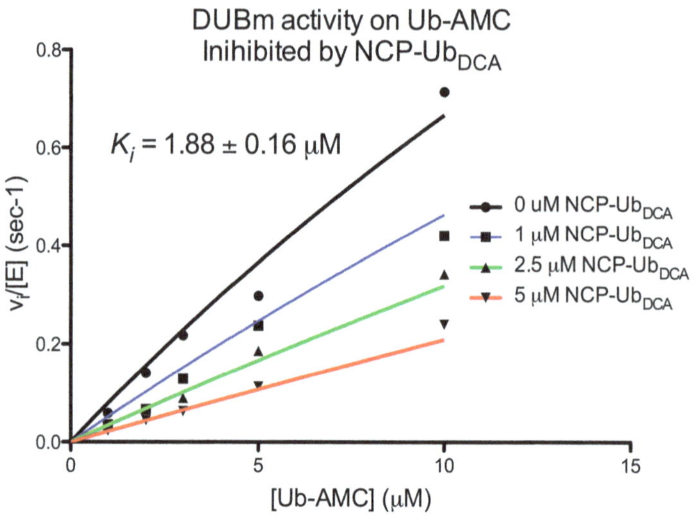

Fig. 4 Initial velocity of Ub-AMC cleavage in the presence of increasing concentrations of non-hydrolyzable inhibitor. DUB module cleavage of Ub-AMC is plotted for different concentrations of NCP-Ub$_{DCA}$, which acts as a substrate analog. The K_i is derived as described in the text

$$K_M^{obs}(I) = K_M \times \left(1 + \frac{[I]}{K_i}\right)$$

Data analysis software such as GraphPad Prism have the option to determine the K_i directly from a spreadsheet containing the values for the reaction velocity as a function of substrate and inhibitor concentration, as demonstrated in Fig. 4.

4 Notes

1. The experiments described here were performed on a BMG Labtech POLARstar Omega plate reader. Data are fitted with software for analyzing enzyme kinetics, such as Prism (GraphPad). The plate reader or fluorometer used to measure fluorescence should be equipped with a temperature control mechanism, and the instrument should be set to equilibrate at the desired temperature 20 min before use; for yeast enzymes, 30 °C is appropriate. If the fluorescence detection instrument does not have monochromators, filters should be used (345 nm for excitation, 445 nm for emission, both with a 10-nm bandpass to ensure sufficient signal).
2. Throughout this protocol, thorough mixing is necessary to produce high-quality data. We recommend using a second pipette with a fresh tip to mix each reaction immediately after adding the substrate.

3. It is critical to ensure that the rate of the reaction is observable on a time scale matching the instrument's read time. If the reaction is too fast to measure multiple wells at once, one should measure one reaction at a time. We suggest pilot experiments to estimate the speed of the reaction.
4. For all steps in this protocol, we assume that the maximum fluorescence one would want to measure is 10% of the highest overall Ub-AMC concentration used throughout (in this case, 2 μM, or roughly 10% of the highest concentration used in Subheading 3.2.2). Thus, a 2 μM standard of Ub-AMC precleaved with 1 μM DUBm for 30 min should be used to determine the instrument's gain setting, and the same gain value should be used throughout the experiments described.
5. Typically, 10–20 points collected within 30 s of initiation are sufficient for a robust measurement. Longer monitoring windows may be needed for slow reactions. As substrate is consumed, the rate of the first 10% of product formation will be approximately linear with respect to substrate concentration, before curving toward the plateau corresponding to full cleavage. It is the initial, linear phase of the reaction that is used to fit initial velocity values (v_i).
6. Slower reactions are easier to monitor, especially if monitoring multiple wells in a plate reader format. For this reason, it is advantageous to use the lowest enzyme concentration within the linear range. Alternatively, one can lower the temperature of the reaction to slow the kinetics.
7. This enzyme concentration is one that will rapidly digest each sample of Ub-AMC to completion.
8. Progress of the reaction can be periodically checked; however typically the Ub-AMC will be fully digested within 20 min of initiating the reaction.
9. The Michaelis-Menten kinetic approximations hold only when a small proportion (<10%) of substrate has been consumed.
10. Reliability of K_i estimates depends on how well-inhibited the enzyme is. We recommend that substrate analog concentrations cover a range at which the enzyme is (1) <50% inhibited in its initial velocity, (2) approximately 50% inhibited, and (3) >50% inhibited.
11. Note that the protocol below utilizes a single replicate of each condition due to material limitations. Given the high ratio of observations (20) to the single parameter to be determined, K_i, this approach can still yield an accurate measurement. If possible, triplicate measurements of each concentration would be optimal.

Acknowledgments

Supported by the National Institute of General Medical Sciences (GM095822).

References

1. Mevissen TET, Komander D (2017) Mechanisms of Deubiquitinase specificity and regulation. Annu Rev Biochem 86:159–192. https://doi.org/10.1146/annurev-biochem-061516-044916
2. Keusekotten K, Elliott PR, Glockner L, Fiil BK, Damgaard RB, Kulathu Y, Wauer T, Hospenthal MK, Gyrd-Hansen M, Krappmann D, Hofmann K, Komander D (2013) OTULIN antagonizes LUBAC signaling by specifically hydrolyzing Met1-linked polyubiquitin. Cell 153(6):1312–1326. https://doi.org/10.1016/j.cell.2013.05.014
3. Spedale G, Timmers HT, Pijnappel WW (2012) ATAC-king the complexity of SAGA during evolution. Genes Dev 26(6):527–541. https://doi.org/10.1101/gad.184705.111
4. Lee KK, Florens L, Swanson SK, Washburn MP, Workman JL (2005) The deubiquitylation activity of Ubp8 is dependent upon Sgf11 and its association with the SAGA complex. Mol Cell Biol 25(3):1173–1182. https://doi.org/10.1128/MCB.25.3.1173-1182.2005
5. Lee KK, Swanson SK, Florens L, Washburn MP, Workman JL (2009) Yeast Sgf73/Ataxin-7 serves to anchor the deubiquitination module into both SAGA and Slik(SALSA) HAT complexes. Epigenetics Chromatin 2(1):2. https://doi.org/10.1186/1756-8935-2-2
6. Kohler A, Schneider M, Cabal GG, Nehrbass U, Hurt E (2008) Yeast Ataxin-7 links histone deubiquitination with gene gating and mRNA export. Nat Cell Biol 10 (6):707–715. doi:ncb1733 [pii]. https://doi.org/10.1038/ncb1733
7. Kumar KS, Spasser L, Ohayon S, Erlich LA, Brik A (2011) Expeditious chemical synthesis of ubiquitinated peptides employing orthogonal protection and native chemical ligation. Bioconjug Chem 22(2):137–143. https://doi.org/10.1021/bc1004735
8. McGinty RK, Kim J, Chatterjee C, Roeder RG, Muir TW (2008) Chemically ubiquitylated histone H2B stimulates hDot1L-mediated intranucleosomal methylation. Nature 453 (7196):812–816. https://doi.org/10.1038/nature06906
9. Yin L, Krantz B, Russell NS, Deshpande S, Wilkinson KD (2000) Nonhydrolyzable diubiquitin analogues are inhibitors of ubiquitin conjugation and deconjugation. Biochemistry 39(32):10001–10010
10. Morgan MT, Haj-Yahya M, Ringel AE, Bandi P, Brik A, Wolberger C (2016) Structural basis for histone H2B deubiquitination by the SAGA DUB module. Science 351 (6274):725–728. https://doi.org/10.1126/science.aac5681
11. Samara NL, Datta AB, Berndsen CE, Zhang X, Yao T, Cohen RE, Wolberger C (2010) Structural insights into the assembly and function of the SAGA deubiquitinating module. Science 328(5981):1025–1029. https://doi.org/10.1126/science.1190049

Part II

Tools to Study Ubiquitylation

Chapter 6

Enzymatic Assembly of Ubiquitin Chains

Martin A. Michel, David Komander, and Paul R. Elliott

Abstract

The availability of different polyubiquitin chains of specific linkage types has changed the appreciation of the specificity in the ubiquitin (Ub) system. Numerous E2 Ub-conjugating enzymes and E3 Ub ligases, Ub-binding domains (UBDs), and deubiquitinases (DUBs) are now known to assemble, bind, or hydrolyze individual linkage types, respectively. Biochemical and structural studies of these processes require milligram quantities of pure polyUb. Here we describe protocols that allow the enzymatic synthesis and purification of six of the eight homotypic polyUb chains through the use of chain-specific Ub ligases and DUBs.

Key words Ubiquitin, Deubiquitinase, Ligase, Linkage-specific, Enzymatic assembly, Protein purification

1 Introduction

Protein ubiquitination orchestrates a spectrum of different signaling events that include substrate degradation, protein localization, and enzyme activation or inactivation [1, 2]. This remarkable diversity in the outcome of Ub signaling is achieved through functionally distinct Ub signals. Monoubiquitination on one or multiple substrate Lys residues contrasts with polyubiquitination, in which Ub chains are attached to proteins. The latter can be assembled via eight internal sites in Ub including seven Lys residues and the N-terminus. In addition, chains extending beyond diUb can be composed of multiple linkage types and achieve complex architectures. These are collectively referred to as heterotypic chains and can include mixed linkage chains without branches, as well as branched architectures, in which individual Ub molecules are ubiquitinated on multiple positions. Finally, other types of posttranslational modifications can occur on Ub, such as phosphorylation or acetylation, and each may alter its signaling properties [1, 2]. Most importantly, it is now clear that cells use the plethora of Ub signals extensively in defined functional pathways. The Ub system equips

Thibault Mayor and Gary Kleiger (eds.), *The Ubiquitin Proteasome System: Methods and Protocols*, Methods in Molecular Biology, vol. 1844, https://doi.org/10.1007/978-1-4939-8706-1_6, © Springer Science+Business Media, LLC, part of Springer Nature 2018

cells with a highly tunable signaling capacity, and disruptions to the system are emerging to be deleterious for cells [3].

Such a complex system requires specificity of its components, from the generation of defined polyUb signals by linkage- and architecture-specific Ub assembly systems to the recognition of defined polyUb chains by specific UBDs and to the linkage-selective hydrolysis of polyUb chains by DUBs [4]. The last decade has seen an explosion of information on the specificity of the Ub system, and revealed mechanistic insights into how this specificity is achieved, as well as linking particular Ub signals to defined cellular pathways. For example, recent results highlight how mitophagy, the targeted destruction of damaged parts of mitochondria by autophagy, is regulated by at least four different chain types including the previously uncharacterized Lys6-linked Ub chains, as well as by phosphorylation on Ub Ser65 [5].

Insights into the specificity of the Ub system have largely emerged from biochemical and biophysical experiments that comprehensively compare all eight linkage types. Key methodologies for this are quantitative mass spectrometry techniques [6], but lower-throughput gel-based studies are also widely used [7]. A large number of linkage-specific proteins have been analyzed structurally, generating a wealth of mechanistic molecular information on linkage specificity. A key requirement for the latter studies is access to milligram amounts of pure polyUb chains of defined linkage and length.

Significant advances have been made to generate polyUb chains and modified Ub moieties through various synthetic or semisynthetic strategies [8, 9]. While these approaches allow for the production of milligram quantities of Ub chains, only a subset of the methods achieve chemically identical isopeptide bonds between Ub monomers, which may negatively affect chain recognition by UBDs or DUBs. More importantly, while these methods have made much needed reagents commercially available, they are not easily reproduced in biochemical laboratories.

Enzymatic assembly protocols for polyUb chains present a more accessible and cheaper alternative. Here, Ub variants, assembly enzymes, and DUBs are all separately expressed in *E. coli* and purified. Since most enzymatic assembly reactions generate a mixture of linkage types and chain lengths, unwanted linkage types are removed with the help of linkage-specific DUBs. Individual chain lengths can then be purified by cation exchange or size exclusion chromatography. Unlike in chemical methods, there is no unfolding of Ub during the process, which is beneficial as partially misfolded polyUb may result in either artifactual binding, or lack of binding, to UBDs, as well as misrepresentative DUB activity.

Protocols detailing the enzymatic synthesis of Lys48 and Lys63 polyUb chains have existed for more than a decade [10, 11]. While Met1-linked "linear" chains can be expressed in *E. coli* and purified just like monoUb, from constructs expressing head-to-tail Ub fusions [12]. More recently, strategies have emerged for the assembly of Lys6-linked [13], Lys11-linked [14], as well as Lys29- and Lys33-linked polyUb chains [15], which are provided in detail below. Each method relies on specific E2 enzymes or HECT/HECT-like E3 ligases for assembly, and some require incorporation of DUB-based purification steps to eliminate unwanted chain types.

The protocols presented below enable production of milligram amounts of homotypic polyUb for seven linkage types. We describe how linkage-specific DUBs can be used to ensure that only the desired polyUb chain is produced. A general procedure is also described for the purification of polyUb chains. Finally, we describe strategies for how more complicated polyUb topologies can be generated through the inclusion of Ub point mutations that direct the assembly of defined polyUb architectures.

While an enzymatic in vitro assembly system has not yet been described for Lys27-linked polyUb chains, the outlined strategies should be easily adaptable once an enzyme with this activity is identified.

2 Materials

2.1 Purified Recombinant Ub and Ub-Activating Enzyme (E1)

1. Purified Ub from recombinant overexpression (*see* **Note 1**) or purchased from Sigma. **Note 1** also applies to expression of Met1-linked "linear" Ub chains.
2. Purified recombinant Ub-activating enzyme (E1). Mouse E1 works well with human E2 Ub-conjugating enzymes. E1 can be purchased from commercial sources or purified from *E. coli* (*see* **Note 2**).

2.2 Purified Recombinant E2 Ub-Conjugating Enzymes

1. UBE2L3 (*see* **Note 3**).
2. UBE2R1 (*see* **Note 3**).
3. UBE2V1 (*see* **Note 3**).
4. UBE2N (*see* **Note 3**).
5. UBE2S-UBD (*see* **Note 3**) [14].

2.3 Purified Recombinant HECT/HECT-Like E3 Ub Ligases

1. NleL (*see* **Note 4**) [13].
2. UBE3C (*see* **Note 4**) [15].
3. AREL1 (*see* **Note 4**) [15].

2.4 Purified Recombinant Deubiquitinating Enzymes

1. Cezanne (*see* **Note 5**) [16].
2. OTUB1* (*see* **Note 5**) [15].
3. AMSH* (*see* **Note 5**) [15].

2.5 Buffers, Reagents, and Materials

1. 20× ligation buffer: 800 mM Tris–HCl, pH 8.5, 200 mM $MgCl_2$, 10 mM dithiothreitol (DTT).
2. 100 mM ATP (adjust pH to ~8.0 with sodium bicarbonate). Freshly prepare and store at −20 °C in small aliquots.
3. 1 M DTT. Freshly prepare and store at −20 °C in small aliquots.
4. Buffer A: 50 mM sodium acetate, pH 4.5.
5. Buffer B: 50 mM sodium acetate, pH 4.5, 1 M NaCl.
6. Storage buffer: 20 mM Tris–HCl, pH 7.4.
7. 1 M Tris–HCl, pH 8.0.
8. Spin concentrator with a molecular weight cutoff of less than 5000 Da.
9. 0.22 μm filter.
10. 4–12% Bis-Tris SDS-PAGE gels.

2.6 Chromatography and Protein Purification

1. 6 mL Resource S cation exchange column (or, alternatively, a 1 mL Resource S column, a 1 mL Mono S 5/50 GL column, or equivalent).
2. A protein purification workstation (e.g., AKTA pure).

3 Methods

Here, we outline the enzymatic assembly of homotypic polyUb chains of defined linkage and length using wild-type Ub to yield milligram quantities suitable for biochemistry and structural biology applications. Six of the seven isopeptide-linked polyUb chain types can be synthesized enzymatically, with the exception being Lys27 chains. Met1-linked "linear" polyUb can be assembled enzymatically by the E3 ubiquitin ligase HOIP [17], yet we prefer to directly express untagged Met1-linked polyUb chains of defined lengths in *E. coli* and purify them according to a described protocol for monoUb (*see* **Note 1**) [10].

3.1 Assembly Strategy

Table 1 summarizes the linkage-specific assembly enzymes used to generate different polyUb chains, the linkage composition immediately after the reaction, and the final composition after treatment with linkage-specific DUBs.

3.2 Assembly of PolyUb Chains

1. Mix the components for the particular chain type according to the table below (Table 2), and incubate at 37 °C for 3 h or overnight, as indicated (*see* **Notes 6** and **7**). For large-scale assemblies, we typically use ~20 mg of Ub in a final reaction volume of 1 mL. The result of a typical assembly reaction is shown in Fig. 1a.

Table 1
Estimated purity of different polyUb chains using wild-type Ub in the assembly reaction before and after incubation with linkage-specific DUBs

	Lys6	Lys11	Lys29	Lys33	Lys48	Lys63
E2 /E3	UBE2L3	UBE2S-UBD	UBE2L3	UBE2L3	UBE2R1	UBE2V1 UBE2N
E3	NleL	–	UBE3C	AREL1	–	–
Composition (untreated)	50% K6 50% K48	90% K11 10% K63	60% K48 25% K29 10% K11 5% other	35% K33 35% K11 20% K48 5% K63 5% other	>95% K48	>95% K63
Composition (DUB-treated)	>95% K6	>95% K11	>95% K29	>95% K33	N.A	N.A.

2. After the assembly reaction is completed, add 1 M DTT to a final concentration of 10 mM to stop the reaction and to discharge any thioester-linked Ub conjugates off of the E1, E2, or E3 enzymes.
3. Add the appropriate DUBs according to Table 2 (*see* **Note 9**). Linkage-specific DUBs are added to cleave unwanted chain types that are generated as by-products during the assembly reaction.

3.3 Purification of PolyUb Chains

1. Prepare the sample for cation exchange chromatography by dialyzing overnight into buffer A or by diluting the sample tenfold in ice-cold buffer A. Assembly reaction enzymes will precipitate during the pH change. However, adding strong acid directly to the reaction could result in polyUb precipitation and should be avoided.
2. Filter the sample using a 0.22 μm filter to remove precipitated enzymes.
3. Load the sample onto a 6 mL Resource S column (*see* **Note 10**).
4. Elute the polyUb chains in buffer A with a linear gradient of 0–50% buffer B over 350 mL. A typical elution profile is shown in Fig. 1b.
5. Analyze the assembly reaction and quality of the cation exchange chromatography separation by SDS-PAGE gel electrophoresis using 4–12% Bis-Tris protein gels. Do not boil the polyUb samples prior to SDS-PAGE analysis (*see* **Note 11**).
6. Pool peak fractions and adjust the pH to ~7.0 using 1 M Tris–HCl (pH 8.0) (*see* **Note 12**).

Table 2
Overview of assembly conditions for differently linked Ub chains. For assembly of Lys11-linked polyUb chains beyond triUb, please *see* Note 8

	Lys6	Lys11	Lys29	Lys33	Lys48	Lys63
Mix						
E1	100 nM	250 nM	1 μM	1 μM	1 μM	1 μM
E2	600 nM UBE2L3	5 μM UBE2S-UBD	10 μM UBE2L3	10 μM UBE2L3	25 μM UBE2R1	8 μM UBE2V1 8 μM UBE2N
E3	10 μM NleL	–	30 μM UBE3C	30 μM AREL1		
Total Ub	2.5 mM	2.5 mM	2.5 mM	2.5 mM	2.5 mM	1.5 mM
ATP	10 mM	10 mM	10 mM	10 mM	10 mM	10 mM
Ligation buffer	1×	1×	1×	1×	1×	1×
H_2O	To 1 mL	To 1 mL	To 1 mL	To 1 mL	To 1 mL	To 1 mL
Reaction conditions	3 h at 37 °C	3 h at 37 °C	ON at 37 °C	ON at 25 °C	3 h at 37 °C	3 h at 37 °C
DUB treatment (at 37 °C)	≥3 h 2 μM OTUB1*	≥3 h 2 μM AMSH1*	≥3 h 2 μM OTUB1* AMSH* 1 h 400 nM Cezanne	≥3 h 2 μM OTUB1* 1 h 400 nM Cezanne	–	–

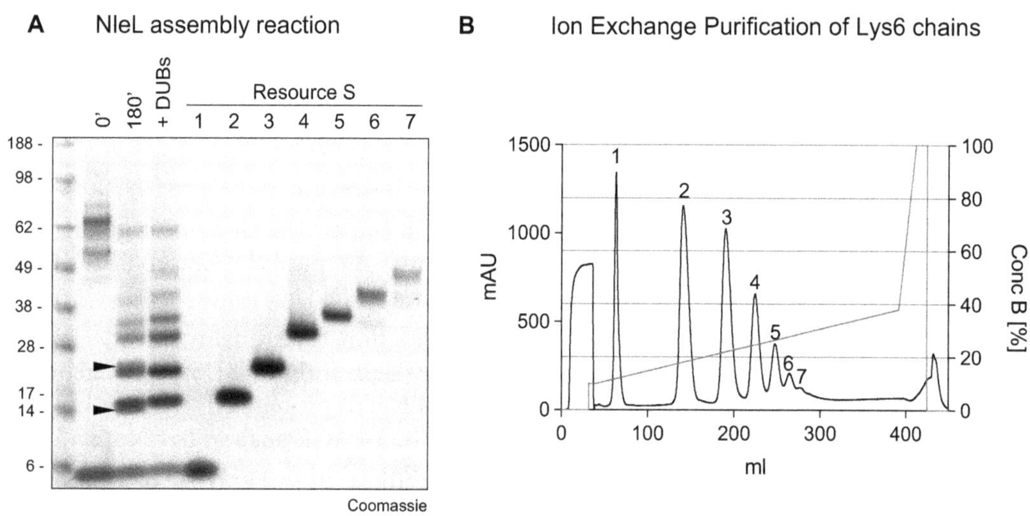

Fig. 1 Enzymatic assembly and purification of Lys6 chains. (**a**) SDS-PAGE of an assembly reaction of Lys6 chains with the resulting peak fractions from a Resource S ion exchange chromatography run. (**b**) An elution profile from a Resource S run for Lys6 chains. Peak fractions used in A are indicated

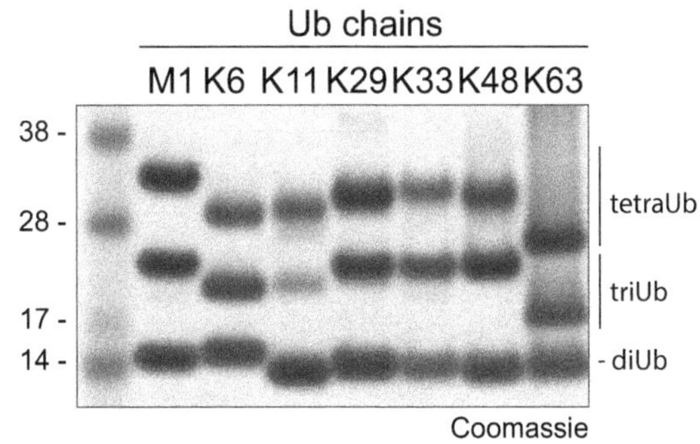

Fig. 2 SDS-PAGE showing the purities of purified di- to tetraUb chains of different Ub linkage types

7. Concentrate the pH-adjusted polyUb chains using 5000 Da cutoff spin concentrators.
8. *Optional*: if further purity is required (e.g., for structural studies) an additional purification step using size exclusion chromatography (e.g., Superdex 75 10/300 or similar, equilibrated in storage buffer) can be used (*see* **Note 13**).
9. Concentrated Ub chains can be flash-frozen in liquid nitrogen and kept indefinitely at −20 °C (*see* **Note 14**).

Differently linked Ub chains have different electrophoretic mobilities during separation by SDS-PAGE. A gel summarizing the seven Ub linkages for diUb to tetraUb is shown in Fig. 2. This can be used as a visual diagnostic to assess chain purity but also enables competition experiments, whereby tetraUb of different linkage types can be mixed and used in simple gel-based competition pull-down experiments [18].

3.4 Quality Control

The gold standard for assessing Ub linkage composition of enzymatically assembled polyUb chains is to perform AQUA mass spectrometry that produces absolute values of the abundance of different chain types in the sample [19]. However, this technique is not readily available in most laboratories. Alternatively, the purified DUBs used in these protocols can be used diagnostically in UbiCRest analysis [20]. For instance, incubating purified Lys48-linked chains with AMSH* (specific for Lys63-linked chains) should not produce any monoUb, while OTUB1* (specific for Lys48-linked chains) should result in a complete collapse of the chains into monoUb.

3.5 Using Ub Mutants to Simplify the Protocol and Increase Yields

Assembly reactions can also be performed with Ub Lys-to-Arg mutants [10] that obviate the need for treatment with linkage-specific DUBs. This simplifies the protocol and typically increases yields as no Ub is used for the assembly of undesired linkages (Fig. 3). For instance, Ub K48R can be used in an assembly

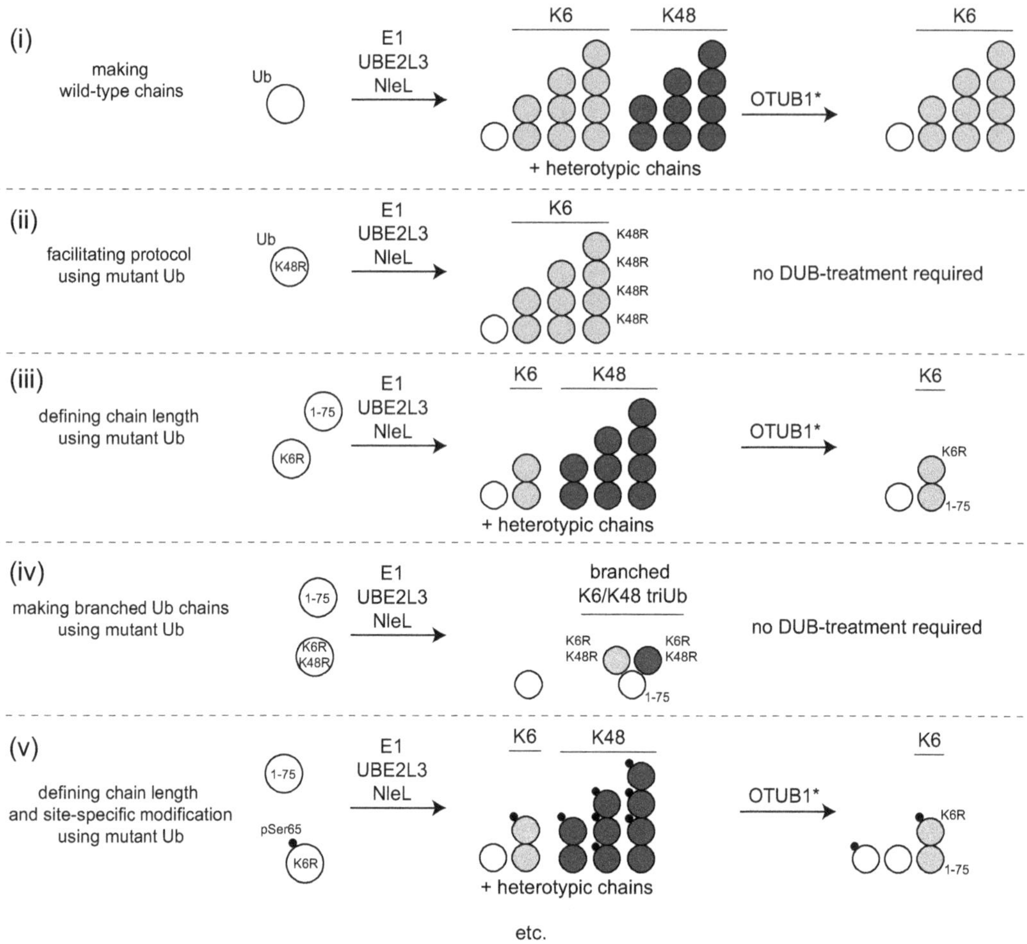

Fig. 3 Using Ub mutants to expand the complexity of enzymatically assembled Ub chains exemplified for Lys63 chains. (i) If Ub chain assembly reactions are performed as described in Subheading 3.2 with the E2 UBE2L3 and the E3 NleL, both Lys6- and Lys48-linked chains are formed. To obtain pure Lys6 chains, treatment with OTUB1* is necessary to hydrolyze Lys48-linked chains. (ii) Instead of using OTUB1* treatment, Ub K48R can be used to prevent the formation of Lys48-linked chains. (iii) Chain length can be controlled using Ub_{1-75}. (iv) Both chain length and linkage type can be controlled using Ub_{1-75} and Ub K6R/K48R. (v) Additional Ub variants can also be incorporated into chains. Light gray circles, polyUb chains linked through Lys6. Dark gray circles, polyUb chains linked through Lys48. Small black circles, phosphate on Ser65

reaction using NleL to produce Lys6 chains, without the need for OTUB1* treatment.

Other Ub mutants such as Ub_{1-75} can also be useful for the controlled assembly of Ub chains since this Ub cannot be thioesterified to E2 or E3 enzymes and acts as a proximal chain terminator [10]. If assembly reactions are performed with more than one Ub mutant, mix them at the ratio in which they appear in the desired product and with the total Ub concentration equaling those listed in Table 2. As an example, consider the desired product of Lys6 diUb that is to be assembled from Ub_{1-75} K48R and Ub

K6R/K48R. Mix 1.25 mM of Ub_{1-75} K48R and 1.25 mM Ub K6R/K48R in the assembly reaction. The only product should be the desired Lys6 diUb and unassembled monoUb.

3.6 Extending the Complexity

As discussed in the introduction, further complexity exists within the Ub signal: polyUb chains can be mixed or branched, and individual Ub moieties can be modified by posttranslational modifications. By using Ub mutants, more complex chain topologies are also amenable to enzymatic synthesis, and even heterotypic mixed and branched polyUb chains can be assembled in a controlled manner (Fig. 3). For example, mixing Ub_{1-75} and Ub K6R/K48R at a 1:2 ratio in an assembly reaction using NleL (without OTUB1* treatment) will yield defined, branched Lys6-/Lys48-linked triUb [13].

This methodology can be further combined with using differentially modified monoUb in the assembly reaction. For example, monoUb can be isotopically labeled (for NMR studies [13, 14]) or phosphorylated on Ser65 by PINK1 [21] (Fig. 3). For example, Ub K6R phosphorylated on Ser65 by the protein kinase PINK1 can be used in an assembly reaction with Ub_{1-75} using NleL (and OTUB1*) to generate specific Lys6 diUb where Ser65 is phosphorylated only on the Ub in the distal position of the chain [21].

4 Notes

1. We routinely express Ub in *E. coli* (Rosetta II DE3) grown in auto-induction medium [22] at 37 °C for 24 h. Ub is purified using perchloric acid precipitation (0.5–1%) followed by cation exchange chromatography using a protocol derived from [10]. This protocol can also be used for expression and purification of Met1-linked polyUb.
2. Plasmids for the overexpression of His_6-tagged mouse E1 can be obtained from Addgene (plasmid #32534).
3. Plasmids encoding GST-tagged E2s can be obtained through Addgene (UBE2S-UBD: #66713; the remaining constructs were being provided to Addgene during the preparation of the manuscript and should be available at the time of publication). GST-E2s are purified by glutathione Sepharose affinity chromatography and, following cleavage of GST, seldom require further purification.
4. Plasmids for bacterial expression of E3 enzymes can be found on Addgene (AREL1, #66710; NleL, #66716; UBE3C, #66711). UBE3C and AREL1 are purified by immobilized metal affinity chromatography and subsequent size exclusion chromatography [15], whereas NleL is purified analogously to GST-E2s [13].

5. All plasmids for recombinant overexpression of the linkage-specific DUBs can be purchased from Addgene (AMSH*, #66712; Cezanne, #61581; OTUB1*, #65441). OTUB1* and AMSH* are enhanced forms of OTUB1 and AMSH, respectively, that are more active while still retaining linkage specificity [15].
6. The stocks of the proteins used in the assembly reactions should be of sufficient concentration (2–10 mg/mL), so that in the final assembly reaction mix, the DTT concentration is below ~1 mM. This is to ensure that thioester-linked Ub is not reduced off of the E1, E2, and E3 enzymes.
7. The assembly conditions in Table 2 are chosen to produce maximum yields of diUb to tetraUb chains. If other chain lengths are desired, the assembly time can be increased or decreased to shift the products to shorter and longer chains, respectively. NleL, UBE2S, UBE2R1, and UBE2N/UBE2V1 assemble chains efficiently, whereas AREL1 and UBE3C require longer incubations and tend to assemble only shorter chains.
8. UBE2S-UBD efficiently assembles diUb chains but does not generate longer chains in large quantities. If longer chains are desired, we recommend using Lys11-linked diUb as the starting material in an assembly reaction using UBE2S to generate Lys11-linked tetraUb.
9. The treatment with OTUB1* and AMSH* can be extended to overnight with improved product purity. Cezanne will begin to cleave other chain types upon prolonged incubation, and therefore, it is advisable to follow the indicated times and concentrations in Table 2 (400 nM for 1 h at 37 °C).
10. A 1 mL Resource S or Mono S 5/50 GL column can be used for cation exchange, although we found the separation of different Ub chain lengths to be superior on a 6 mL Resource S. Column capacity is crucial for achieving maximum resolution. The use of smaller columns tends to result in monoUb co-eluting with some polyUb fractions. MonoUb can be removed from polyUb through an additional size exclusion chromatography step.
11. Overloading of Ub samples on SDS-PAGE gels or boiling of polyUb samples tends to result with bands with a smeared appearance.
12. PolyUb chains can precipitate at low pH. Avoid keeping fractions in the cation exchange buffer for extended periods of time. We observe significant loss of polyUb chains if the cation exchange fractions are concentrated at pH 4.5. Adjusting the pH to neutrality reduces sample loss.

13. Longer chains are harder to resolve on both cation exchange and size exclusion chromatography columns. If longer chains are desired, we recommend using shorter chains as inputs in the assembly reactions. For example, the use of triUb instead of monoUb will produce hexa, nona, dodeca, etc. polyUb chains.
14. Avoid multiple freeze-thaw cycles for longer polyUb chains.

Acknowledgments

We would like to thank present and past members of the DK laboratory who have helped to develop and refine described protocols by experimentation and discussions. Work in the D.K. lab is funded by the Medical Research Council [U105192732], the European Research Council [309756, 724804], and the Lister Institute for Preventive Medicine. M.A.M. was supported by a PhD fellowship of the Boehringer Ingelheim Fonds and a Doc. Mobility fellowship of the Swiss National Science Foundation.

References

1. Swatek KN, Komander D (2016) Ubiquitin modifications. Cell Res 26:399–422. https://doi.org/10.1038/cr.2016.39
2. Yau R, Rape M (2016) The increasing complexity of the ubiquitin code. Nat Cell Biol 18:579–586. https://doi.org/10.1038/ncb3358
3. Rape M (2018) Ubiquitylation at the crossroads of development and disease. Nat Rev Mol Cell Biol 19:59–70. https://doi.org/10.1038/nrm.2017.83
4. Komander D, Rape M (2012) The ubiquitin code. Annu Rev Biochem 81:203–229. https://doi.org/10.1146/annurev-biochem-060310-170328
5. Harper JW, Ordureau A, Heo J-M (2018) Building and decoding ubiquitin chains for mitophagy. Nat Rev Mol Cell Biol 19:93–108. https://doi.org/10.1038/nrm.2017.129
6. Ordureau A, Münch C, Harper JW (2015) Quantifying ubiquitin signaling. Mol Cell 58:660–676. https://doi.org/10.1016/j.molcel.2015.02.020
7. Mevissen TET, Hospenthal MK, Geurink PP et al (2013) OTU deubiquitinases reveal mechanisms of linkage specificity and enable ubiquitin chain restriction analysis. Cell 154:169–184. https://doi.org/10.1016/j.cell.2013.05.046
8. Hameed DS, Sapmaz A, Ovaa H (2017) How chemical synthesis of ubiquitin conjugates helps to understand ubiquitin signal transduction. Bioconjug Chem 28:805–815. https://doi.org/10.1021/acs.bioconjchem.6b00140
9. Mali SM, Singh SK, Eid E, Brik A (2017) Ubiquitin signaling: chemistry comes to the rescue. J Am Chem Soc 139:4971–4986. https://doi.org/10.1021/jacs.7b00089
10. Pickart CM, Raasi S (2005) Controlled synthesis of Polyubiquitin chains. In: Ubiquitin and protein degradation, Part B. Elsevier, Amsterdam, pp 21–36
11. Dong KC, Helgason E, Yu C et al (2011) Preparation of distinct ubiquitin chain reagents of high purity and yield. Structure 19:1053–1063. https://doi.org/10.1016/j.str.2011.06.010
12. Reyes-Turcu FE, Shanks JR, Komander D, Wilkinson KD (2008) Recognition of polyubiquitin isoforms by the multiple ubiquitin binding modules of isopeptidase T. J Biol Chem 283:19581–19592. https://doi.org/10.1074/jbc.M800947200
13. Hospenthal MK, Freund SMV, Komander D (2013) Assembly, analysis and architecture of atypical ubiquitin chains. Nat Struct Mol Biol 20:555–565. https://doi.org/10.1038/nsmb.2547
14. Bremm A, Freund SMV, Komander D (2010) Lys11-linked ubiquitin chains adopt compact

conformations and are preferentially hydrolyzed by the deubiquitinase Cezanne. Nat Struct Mol Biol 17:939–947. https://doi.org/10.1038/nsmb.1873
15. Michel MA, Elliott PR, Swatek KN et al (2015) Assembly and specific recognition of K29- and K33-linked polyubiquitin. Mol Cell 58:95–109. https://doi.org/10.1016/j.molcel.2015.01.042
16. Mevissen TET, Kulathu Y, Mulder MPC et al (2016) Molecular basis of Lys11-polyubiquitin specificity in the deubiquitinase Cezanne. Nature 538:402–405. https://doi.org/10.1038/nature19836
17. Stieglitz B, Morris-Davies AC, Koliopoulos MG et al (2012) LUBAC synthesizes linear ubiquitin chains via a thioester intermediate. EMBO Rep 13:840–846. https://doi.org/10.1038/embor.2012.105
18. Komander D, Reyes-Turcu F, Licchesi JDF et al (2009) Molecular discrimination of structurally equivalent Lys 63-linked and linear polyubiquitin chains. EMBO Rep 10:466–473. https://doi.org/10.1038/embor.2009.55
19. Kirkpatrick DS, Gerber SA, Gygi SP (2005) The absolute quantification strategy: a general procedure for the quantification of proteins and post-translational modifications. Methods 35:265–273. https://doi.org/10.1016/j.ymeth.2004.08.018
20. Hospenthal MK, Mevissen TET, Komander D (2015) Deubiquitinase-based analysis of ubiquitin chain architecture using ubiquitin chain restriction (UbiCRest). Nat Protoc 10:349–361. https://doi.org/10.1038/nprot.2015.018
21. Gersch M, Gladkova C, Schubert AF et al (2017) Mechanism and regulation of the Lys6-selective deubiquitinase USP30. Nat Struct Mol Biol 24:920–930. https://doi.org/10.1038/nsmb.3475
22. Studier FW (2005) Protein production by auto-induction in high density shaking cultures. Protein Expr Purif 41:207–234

Chapter 7

Ubiquitin-Activated Interaction Traps (UBAITs): Tools for Capturing Protein-Protein Interactions

Hazel F. O'Connor, Caleb D. Swaim, Larissa A. Canadeo, and Jon M. Huibregtse

Abstract

UBAITs (*Ub*iquitin-*A*ctivated *I*nteraction *T*raps) are reagents that capitalize on the biochemistry of the ubiquitin system to covalently trap transient protein-protein interactions. UBAITs consist of an affinity-tagged protein of interest fused to a short linker followed by a C-terminal ubiquitin moiety. When charged in an E1- and E2-dependent manner, the C-terminal ubiquitin moiety of the UBAIT has the potential to form an amide linkage with lysine side chains of a protein that interacts transiently with the protein of interest, thereby covalently trapping the protein-protein interaction. The partner protein can then be identified by affinity-based purification of the UBAIT coupled with mass spectroscopy methods. While originally designed to identify substrates of ubiquitin ligases, UBAITs can, in principle, be used for identifying interaction partners of virtually any protein of interest. Here we describe methods for utilizing UBAITs in both cell-based and in vitro experiments.

Key words UBAIT, Protein-protein interactions, Ubiquitin ligases, HECT E3, RING E3s

1 Introduction

Many genetic and biochemical approaches have been employed to identify substrates and regulatory proteins of ubiquitin ligases, including yeast two-hybrid assays, co-immunoprecipitation approaches, and protein-protein interaction arrays [1–3]. While each of these has its own advantages and disadvantages, we sought a method that would be applicable to a wide range of E3s and would overcome the challenges posed by potentially weak or transient enzyme-substrate interactions. The UBAIT method was developed with ubiquitin ligase substrate identification in mind [4], but as described below it is a generally applicable method for identifying proteins that interact with almost any protein of interest.

UBAITs take advantage of the first two steps of the ubiquitin conjugation pathway: the E1 activation reaction and the transthiolation reaction to an E2 enzyme. Artificial ubiquitin proteins with

Thibault Mayor and Gary Kleiger (eds.), *The Ubiquitin Proteasome System: Methods and Protocols*, Methods in Molecular Biology, vol. 1844, https://doi.org/10.1007/978-1-4939-8706-1_7, © Springer Science+Business Media, LLC, part of Springer Nature 2018

small N-terminal affinity tags—such as FLAG-tagged or His6-tagged ubiquitin—have been used routinely for many years, both in cells and in vitro, for monitoring ubiquitylation reactions and purifying conjugates [5, 6]. In this sense, UBAITs are simply N-terminally tagged ubiquitin proteins, with the N-terminal extension consisting of an affinity-tagged protein of interest followed by a short linker to ubiquitin. We have incorporated proteins of 100 kDa or larger (e.g., yeast Rsp5, human Itch) into UBAITs, and remarkably the E1 enzyme can use these fusion proteins as substrates, forming UBAIT~E1 thioester-linked conjugates. E2 enzymes can react with the UBAIT- E1 thioesters, releasing the E1 enzyme and forming a UBAIT~E2 thioester. Finally, in cells or in vitro, if a partner protein interacts transiently with the protein of interest within the UBAIT, a lysine side chain of the partner protein can react with the Ub~E2 thioester of the UBAIT; this will displace the E2 and form a stable amide bond between the UBAIT and the protein of interest (Fig. 1a). Affinity purification of the UBAIT, even under denaturing conditions, will co-purify the interacting protein(s), which can then be identified by standard mass spectrometry methods.

Important controls to incorporate into a UBAIT experiment include parallel expression of a "ΔGG" mutant UBAIT, where the two terminal glycine residues of ubiquitin are deleted. The ΔGG UBAIT cannot be charged by the E1 enzyme and is a negative control for conjugate formation. Additional negative controls are an empty vector control and a control in which the affinity tagged protein of interest is expressed without any ubiquitin sequences. Further controls, based on the known biochemistry of the particular protein of interest, can also be incorporated into the experiment. For example, if a specific mutation within the protein of interest is known to either positively or negatively influence a specific function, it may be useful to express and purify a UBAIT containing that mutation.

We have primarily used affinity purification tags (e.g., TAP tags) [4] at the N-terminus of the UBAIT; however it is possible to also incorporate the affinity tag into the linker sequence between the protein of interest and ubiquitin (e.g., FLAG tag) [7]. The latter may be particularly advantageous if the goal is to create a UBAIT by altering an endogenous chromosomal locus by a single-step gene modification. A consideration in choosing the affinity tag is whether the user wishes to incorporate a denaturation step into purification of the UBAIT conjugates (as described in the large-scale protocols, below), which can decrease background protein contamination and takes full advantage of the stable covalent bond formed between UBAITs and their interacting proteins. With respect to the linker, we have so far observed few restraints on the identity of the sequence or linker length (varying between 4 and 20 residues) in the UBAITs we have analyzed, although we

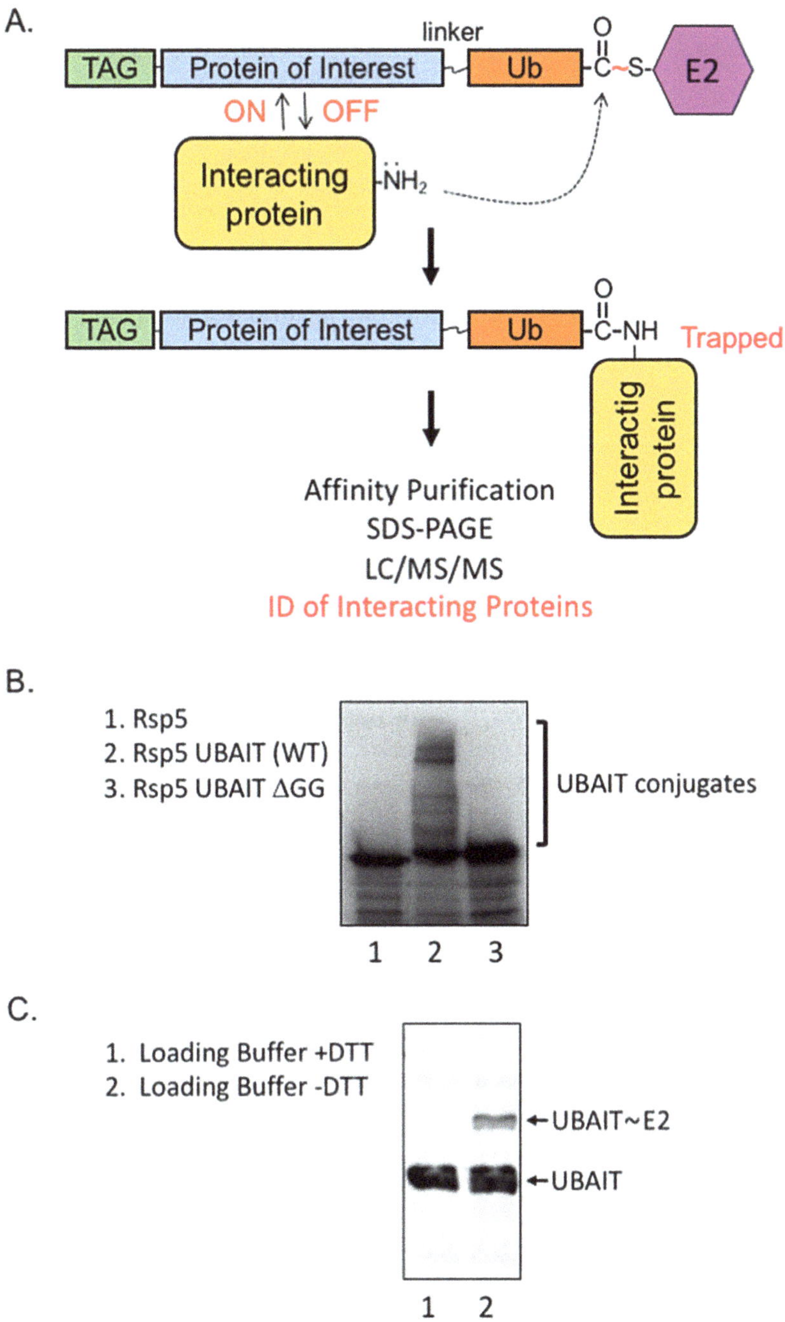

Fig. 1 The UBAIT strategy for trapping interacting proteins. (**a**) An N-terminally tagged protein of interest is expressed as a fusion protein to a short linker and ubiquitin. The UBAIT is activated by charging it to an E2 enzyme in an ATP- and E1-dependent reaction. In the presence of an interacting partner protein, amino groups on lysine side chains can react, by proximity, with the C-terminus of the UBAIT, displacing the E2 and trapping the interacting protein via an amide bond. UBAIT conjugates are affinity purified, size-fractionated by SDS-PAGE, and subject to LC-MS/MS protein identification to identifying the interacting proteins. (**b**) A representative example of UBAIT expression and conjugate formation. NTAP-tagged Rsp5 (lane 1), NTAP-

can imagine cases where this might be an important variable. The linker sequences we have chosen are predicted to be unstructured peptides (e.g., GSGG repeats). In some cases we have found UBAIT expression and stability can be enhanced by incorporating either K48R or lysine-less ubiquitin in the fusion proteins, which may prevent polyubiquitylation and proteasomal degradation of the UBAIT.

As described previously [4], the UBAIT method was originally designed to identify substrates of ubiquitin ligases, but the fact that it is essentially a proximity ligation reaction based on the chemically activated C-terminus of the UBAIT (and independent of whether the protein of interest has E3 ligase activity) makes it applicable to almost any type of protein. The protocols below describe the isolation of interacting proteins using UBAITs expressed in yeast or mammalian cells, as well as a protocol for using purified UBAITs for in vitro trapping reactions. For expression in cells, we describe protocols for initial small-scale characterization of UBAITs and larger-scale protocols for expression and purification of UBAIT conjugates that are carried forward for mass spectrometry analyses. Figure 1b shows an example of a small-scale UBAIT analysis (where the protein of interest was yeast Rsp5) that met the criteria for moving forward with the large-scale prep: both the WT and ΔGG UBAITs were expressed similarly to TAP-tagged Rsp5 (with no ubiquitin sequences), and conjugates were only apparent with the WT UBAIT. The protocol described here for using UBAITs for in vitro experiments focuses on the generation of a purified E2-charged UBAIT that can then be used to covalently trap-binding partners. Figure 1c demonstrates thioester formation between a purified UBAIT and E2, indicating the UBAIT is charged and ready to be used for interacting proteins that might, for example, be present in a total cell lysate. Recently, the in vitro UBAIT system was used to trap and identify the cell surface receptor (LFA-1) for ISG15 [7], a protein that can function as an extracellular signaling molecule. In that case, the E2-charged ISG15 UBAIT was added directly to intact cells, and covalent conjugates were affinity purified and subject to mass spectrometry-based protein identification. Other examples of using UBAITs for in vitro trapping reactions were described previously [4].

Fig. 1 (continued) Rsp5 UBAIT (WT ubiquitin), and NTAP-Rsp5 UBAIT ΔGG (ΔGG ubiquitin) were expressed and purified from yeast and analyzed by immunoblotting with anti-TAP antibody. The three proteins were expressed at similar levels, and higher molecular weight conjugates were apparent with only the WT UBAIT. (**c**) In vitro charging of UBAITs, monitored by Western blot. A FLAG-tagged UBAIT was charged in the presence of ATP, E1, and E2 (UbcH5a) proteins. The reaction products were separated by SDS-PAGE using loading buffer with DTT (lane 1) or without DTT (lane 2). Successful charging is indicated by the formation of a DTT-sensitive UBAIT~E2 thioester conjugate (lane 2), which is hydrolyzed in the presence of DTT (lane 1)

There are a wide variety of options for characterization of affinity-purified UBAIT conjugates and identification of the trapped proteins. The protocols below describe a single approach, which is to isolate the conjugates by excision from SDS-PAGE gels, followed by in-gel tryptic digest and MS/MS protein identification. This simple approach is advantageous since the SDS-PAGE gel size fractionates the UBAIT conjugates, and only proteins of a higher molecular weight than the UBAIT, itself, need to be analyzed.

2 Materials

Stock solutions should be prepared in 18.2 MΩ-cm ultrapure water for all solutions unless otherwise noted.

2.1 UBAIT Expression Vectors: Yeast, Mammalian, and Bacterial

Yeast Plasmids

1. pYES2-TAP (no ubiquitin sequences) (Fig. 2a).
2. pYES2-TAP WT UBAIT (WT ubiquitin) (Fig. 2a).
3. pYES2-TAP ΔGG UBAIT (ΔGG ubiquitin) (Fig. 2a).

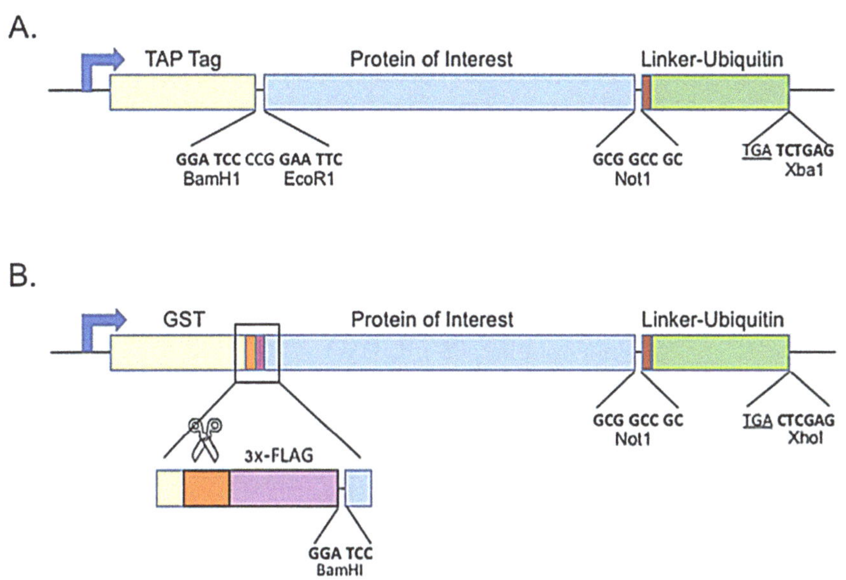

Fig. 2 Representative UBAIT expression vectors. (**a**) Vector design for yeast and mammalian expression vectors. ORFs are cloned into TAP-tagged UBAITs with the indicated restriction sites. The reading frames are indicated by the spaces between triplet nucleotides. Yeast expression vectors are based on the pYES2 vector, and mammalian vectors are based on pcDNA3. (**b**) Vector design for expression of UBAITs in *E. coli* for in vitro trapping reactions. GST fusion proteins are expression in a pGEX-6p-1 vector with a 3X-FLAG sequence downstream of a PreScission Protease cleavage site (scissors icon). Cloning sites and reading frames are indicated as in (**a**)

Mammalian Plasmids

4. pcDNA3.1-NTAP (no ubiquitin sequences) (Fig. 2a).
5. pcDNA3.1-NTAP WT UBAIT (WT ubiquitin) (Fig. 2a).
6. pcDNA3.1-NTAP ΔGG UBAIT (ΔGG ubiquitin) (Fig. 2a).

Bacterial Plasmids

7. pGEX-6p-1-3xFLAG WT UBAIT (WT ubiquitin) (Fig. 2b).
8. pGEX-6p-1-3XFLAG ΔGG UBAIT (ΔGG ubiquitin) (Fig. 2b).

2.2 Growth and Lysis of Yeast Cells

1. *Saccharomyces cerevisiae* FY56 (MATα, *his4-912 ΔR5*, *lys2-128Δ*, *ura3-52*) or other strain allowing for selection of *URA3*-based pYES2 plasmids.
2. 20% sucrose (filter sterilized).
3. 20% galactose (filter sterilized).
4. 1 L synthetic complete medium without uracil (SC-Ura): 6.67 g yeast nitrogen base without amino acids (Difco cat no. 291940), 0.77 g Ura dropout supplement (Clontech cat no. 630416), in H_2O, 2% sucrose (added after autoclaving and cooling).
5. 30 °C plate and shaking incubators.
6. Disruptor Beads 0.5 mm/yeast (Electron Microscopy Sciences cat no. 72408-05).
7. Bead beater 2.0 mL screw cap tubes (Genesee Scientific).
8. Bead beater O-Ring screw caps (Genesee Scientific).
9. Collection tubes: Conical bottom culture tubes, 5 mL.
10. 27G × ½ needle (0.4 mm × 13 mm).
11. Biospec Products Mini-BeadBeater-96 (cat no. 1001).
12. SDS-PAGE loading buffer (4×): 25 mL 1 M Tris–HCl, pH 6.8, 40 mL glycerol, 8 g SDS, 6.17 g DTT, 4 mg bromophenol blue, 35 mL H_2O.
13. Coomassie stain: 50% H_2O, 40% MeOH, 10% acetic acid, 0.25% Coomassie Brilliant Blue R-250.
14. Destaining solution: 50% H_2O, 40% MeOH, 10% acetic acid.
15. TAP antibody: Anti-protein A antibody (Sigma cat no. P3775).
16. Goat anti-rabbit IgG secondary antibody (LI-COR cat no. 926-32211).
17. Phosphate-buffered saline (PBS): 10 mM Na_2HPO_4, 1.8 mM KH_2PO_4, 137 mM NaCl, 2.7 mM KCl (pH 7.4).
18. Odyssey Blocking Buffer in PBS (LI-COR cat no. 927-40000).
19. TNET: 10 mM Tris–HCl, pH 7.5, 2.5 mM EDTA, 50 mM NaCl, 0.1% Tween 20.

20. NP40 lysis buffer: 1% Nonidet P-40, 100 mM Tris–HCl, pH 7.9, 100 mM NaCl, 1 mM DTT, 100 μM phenylmethylsulfonyl fluoride, 4 μM leupeptin, 0.3 μM aprotinin, and 10 mM N-ethylmaleimide.
21. RIPA buffer: 50 mM Tris–HCl, pH 7.4, 150 mM NaCl, 1% Nonidet P-40, 0.1% SDS, 0.5% sodium deoxycholate, 1 mM DTT, 100 μM phenylmethylsulfonyl fluoride, 4 μM leupeptin, 0.3 μM aprotinin, and 10 mM N-ethylmaleimide.
22. IgG Sepharose beads (GE Healthcare).
23. Denaturation buffer: 20 mM Tris–HCl, pH 8, 50 mM NaCl, 5 mM DTT, 1% SDS.
24. NP40 wash buffer: 0.1% Nonidet P-40, 100 mM Tris–HCl, pH 7.9, 100 mM NaCl.
25. 50 mL Nalgene Oak Ridge High-Speed PPCO Centrifuge Tubes.
26. LI-COR® Odyssey® Detection System or similar system for Western blot analyses.

2.3 Culturing and Lysis of Mammalian Cells

1. HEK 293T cells (ATCC).
2. Standard tissue culture plates (6-well tissue culture plates and 100 mm × 20 mm tissue culture dishes).
3. Tissue culture media for HEK 293T cells: Dulbecco's Modification of Eagle's Medium (DMEM) with L-glutamine and 4.5 g/L glucose and without sodium pyruvate (Corning cat no. 10017CV), 10% fetal bovine serum (Sigma cat no. F4135), penicillin-streptomycin solution.
4. Transfection media: Dulbecco's Modification of Eagle's Medium (DMEM) with L-glutamine and 4.5 g/L glucose and without sodium pyruvate (Corning cat no. 10017CV), 10% fetal bovine serum (Sigma cat no. F4135).
5. X-tremeGENE HP DNA Transfection Reagent (Sigma cat no. 06366546001).
6. High salt RIPA buffer: RIPA buffer: 50 mM Tris–HCl, pH 7.4, 400 mM NaCl, 1% Nonidet P-40, 0.1% SDS, 0.5% sodium deoxycholate, 1 mM DTT, 100 μM phenylmethylsulfonyl fluoride, 4 μM leupeptin, 0.3 μM aprotinin, and 10 μM N-ethylmaleimide.

2.4 In Vitro UBAIT Charging

1. *Escherichia coli* BL21 DE3 cells.
2. 1 L LB broth with 100 μg/mL ampicillin: 10 g tryptone, 5 g yeast extract, 10 g NaCl in 1 L H_2O. Autoclave, cool, and add ampicillin to 100 μg/mL.
3. LB agar plates with 100 μg/mL ampicillin.
4. 100 mM isopropyl β-D-1-thiogalactopyranoside (IPTG).

5. Sonicator with microtip.
6. PreScission Protease.
7. PreScission Protease cleavage (PC) buffer: 50 mM Tris–HCl, pH 8.0, 150 mM NaCl, 0.01% triton, 2.5 mM EDTA.
8. PBS-Triton: PBS, 1% Triton x-100.
9. Ube1 (Boston Biochem cat no. E-305).
10. UbcH5a (commercial or in-house preparation; *see* Subheading 3.5, **step 12**).
11. M2 FLAG affinity gel (Sigma cat no. A2220).
12. Glutathione Sepharose (GE Healthcare cat no. 17075601).
13. 1 M DTT.
14. 100 mM ATP.
15. 40 mM $MgCl_2$.
16. 8 M urea.
17. $T_{25}N_{50}$: 25 mM Tris–HCl, pH 7.5, 50 mM NaCl.
18. Bradford reagent.
19. Bovine serum albumin (BSA).

3 Methods

3.1 Small-Scale UBAIT Analysis in Yeast

1. Clone open reading frame (ORF) of interest into pYES2-NTAP and -NTAP UBAIT vectors (*see* **Notes 1** and **2**), and transform into yeast (*URA3* selection).
2. Inoculate 5 mL of SC-Ura with *S. cerevisiae* transformants (pYES2-NTAP, NTAP WT UBAIT, NTAP ΔGG UBAIT) in a 30 mL glass culture tube, and incubate at 30 °C for 14–16 h, shaking at 250 rpm.
3. Dilute starter culture to 10 mL with fresh media and add 1 mL 20% galactose. Incubate cultures for a further 3 h.
4. Harvest cells by centrifugation at 2053 × *g* for 5 min (*see* **Note 3**).
5. Resuspend pellet in 150 μL NP40 lysis buffer. Transfer the resuspended cells to a bead beater tube, and add glass beads until the beads are visible above the sample. Cap tubes tightly.
6. Lyse cells by alternating bead beating for 2 min, followed by 2 min on ice, for 6 rounds (*see* **Note 4**).
7. To collect the lysate, puncture the bottom of the bead beater tubes with an 18-gauge needle (*see* **Note 5**). Loosen the cap and place in collection tube.
8. Centrifuge at 228 × *g* for 1 min. Discard bead beater tube (the beads should be dry), and transfer lysate to a 1.5 mL tube, along with 500 μL fresh NP40 lysis buffer.

9. Clear the lysate by centrifugation at 18,407 × *g* for 10 min at 4 °C.
10. Transfer lysate to a new 1.5 mL tube, and remove a 5% aliquot (input) for a Western blot (save on ice).
11. Add 20 μL IgG Sepharose bead slurry for immunoprecipitation of UBAIT proteins (*see* **Note 6**), and place on a rotator for 2 h at 4 °C.
12. Collect beads by gentle centrifugation. Remove supernatant and wash three times with 100 μL RIPA buffer. After last wash, remove as much supernatant as possible.
13. Add SDS loading dye to beads (30 μL 1×) and input aliquot (11 μL 4×). Boil samples at 95 °C for 5 min.
14. Analyze by SDS-PAGE and immunoblotting (*see* **Note 7**). For TAP-tagged UBAITs, probe with anti-protein A (1:1,000,000 in 1× TNET), followed by the corresponding LI-COR secondary antibody (1:15,000 in Odyssey Blocking Buffer). Visualize results using the LI-COR Odyssey Detection System (*see* **Notes 8** and **9**).

3.2 Large-Scale Expression and Analysis of UBAIT Targets in Yeast

1. Inoculate 200 mL SC-Ura starter cultures with *S. cerevisiae* transformants containing the relevant plasmids (pYES2-NTAP, pYES2-NTAP WT UBAIT, pYES2-NTAP ΔGG UBAIT), and incubate at 30 °C for 14–16 h, shaking at 250 rpm.
2. Dilute starter cultures to 1.5 L with fresh media and incubate at 30 °C. Once cells reach OD_{600nm} 0.5, add galactose to 2%. Incubate cultures for a further 3 h.
3. Harvest cells by centrifugation at 3210 × *g* for 5 min (*see* **Note 3**).
4. Resuspend cells in 4.8 mL NP40 lysis buffer. Transfer approximately 300 μL resuspended cells to bead beater tubes (16 tubes per sample). Add glass beads to each tube until the beads are visible above the sample, and cap tubes tightly.
5. Lyse cells by alternating bead beating for 2 min, followed by 2 min on ice, for a total of six rounds (*see* **Note 4**).
6. To collect the lysate, puncture the bottom of the bead beater tube with an 18-gauge needle (*see* **Note 5**). Loosen the cap and place in a collection tube.
7. Centrifuge at 228 × *g* for 1 min. Discard the bead beater tubes (the beads should be dry), and pool lysate from each collection tube into a 50 mL Oak Ridge tube. Add 10 mL RIPA buffer to the lysate.
8. Clear the lysate by centrifugation at 13,750 × *g* for 20 min at 4 °C.

9. Transfer lysate to a 50 mL conical tube, and add 300 μL IgG Sepharose bead slurry for immunoprecipitation of UBAIT proteins (*see* **Note 6**). Place on a rotator at for 2 h at 4 °C.
10. Collect beads by gentle centrifugation. Transfer to 1.5 mL tubes and wash three times with 600 μL RIPA buffer. After last wash, remove as much supernatant as possible.
11. Add 200 μL denaturation buffer to the beads and boil samples at 95 °C for 5 min.
12. Add 1 mL of NP40 wash buffer to the tube, and centrifuge at 18,407 × *g* for 1 min.
13. Transfer eluate to a new 15 mL tube.
14. Add 10 mL NP40 wash buffer and 300 μL IgG Sepharose bead slurry (*see* **Notes 6** and **10**). Place on a rotator at for 2 h at 4 °C.
15. Collect beads by gentle centrifugation. Transfer to 1.5 mL tubes and wash three times with 600 μL RIPA buffer. After last wash, remove as much supernatant as possible.
16. Add 75 μL SDS-PAGE loading buffer (1×) to the beads, and boil samples at 95 °C for 5 min.
17. Analyze samples by SDS-PAGE (*see* **Note 11**) followed by Coomassie staining for 3 h and destaining overnight (*see* **Note 12**).
18. Excise a gel slice above the unconjugated UBAIT for WT and ΔGG control and the corresponding section of gel for the empty vector control. Send samples for in-gel tryptic digestion and LC-MS/MS to identify peptides.
19. Analyze data by searching against the UniProt yeast database along with decoy databases using SEQUEST HT (latest version of Proteome Discoverer, Thermo Scientific). Allow for up to two missed cleavage sites in the fully tryptic peptides. Filter the peptide-spectrum matches (PSMs) using Percolator (a part of Proteome Discoverer). Apply a false discovery rate (FDR) of <1% to ensure that only reliable PSMs are retained. Potential protein of interest (POI)-interacting proteins are identified as proteins with greater than or equal to fivefold higher spectral counts in the WT UBAIT sample compared to the empty vector control and those with two or more spectral counts in the WT UBAIT sample (*see* **Note 13**).

3.3 Small-Scale UBAIT Analysis in Mammalian Cells

1. Clone open reading frame (ORF) of interest into pcDNA3.1-NTAP and -NTAP UBAIT vectors, and transform into *E. coli* DH10 cells to generate and purify microgram quantities of each plasmid.
2. Plate HEK293T cells in a 6-well tissue culture plate so that cells are ~70% confluent at time of transfection (*see* **Note 14**).

Remove media and replace with transfection media. Transfect 1 μg UBAIT-expressing plasmid DNA per well, using X-tremeGENE HP DNA Transfection Reagent (*see* **Note 15**).

3. After approximately 36 h, harvest the cells (*see* **Note 16**). Remove media and wash cells gently with 1 mL ice-cold PBS. Add 1 mL ice-cold PBS to the plate, resuspend the cells, and transfer to a 1.5 mL tube on ice.
4. Harvest cells by centrifugation at 4500 × *g* for 4 min at 4 °C.
5. To lyse cells, resuspend pellet in 200 μL NP40 lysis buffer and incubate on ice for 5 min.
6. Clear the lysate by centrifugation at 18,407 × *g* for 10 min at 4 °C.
7. Transfer lysate to a new 1.5 mL tube, and remove a 10% aliquot (input) to a new tube for Western blot analysis (save on ice).
8. Add 800 μL RIPA buffer and 40 μL IgG Sepharose bead slurry for immunoprecipitation of UBAIT proteins (*see* **Note 6**). Place on a rotator at 4 °C for 2 h.
9. Collect beads by gentle centrifugation. Remove supernatant and wash three times with 100 μL RIPA buffer. After last wash, remove as much supernatant as possible.
10. Add SDS loading dye to beads (30 μL 1×) and input aliquot (6.7 μL 4×). Boil samples at 95 °C for 5 min.
11. Analyze by SDS-PAGE gel and Western blot (*see* **Note 8**). For TAP-tagged UBAITs, probe with anti-protein A (1:1,000,000 in 1× TNET), followed by the corresponding LI-COR secondary antibody (1:15,000 in Odyssey Blocking Buffer). Visualize results using the LI-COR Odyssey Detection System (*see* **Notes 8** and **9**).

3.4 Large-Scale Expression and Analysis of UBAIT Targets in Mammalian Cells

1. Plate HEK293T cells in 100 mm tissue culture dishes so that cells are ~70% confluent at time of transfection (start with ten 100 mm dishes per plasmid). Transfect 3 μg UBAIT-expressing plasmid DNA per dish using X-tremeGENE HP DNA Transfection Reagent (*see* **Note 15**).
2. After approx. 36 h, harvest the cells (*see* **Note 16**). Remove media and wash cells gently with 2 mL ice-cold PBS. Add 1 mL ice-cold PBS to the plate, and using a cell scraper, scrape the cells into the liquid, and transfer to a 15 mL conical tube on ice. Collect cells from all ten plates in the 15 mL conical tube.
3. Pellet cells by centrifugation at 197 × *g* at 4 °C for 5 min (*see* **Note 3**).
4. To lyse cells, resuspend pellet in 8 mL NP40 lysis buffer, and transfer to a 50 mL Oak Ridge tube. Incubate on ice for 15 min.

5. Clear the lysate by centrifugation at 13,750 × *g* at 4 °C for 10 min.
6. Transfer the cleared lysate to a 15 mL conical tube, and add 4 mL RIPA buffer and 200 μL IgG Sepharose bead slurry (*see* **Note 6**).
7. Collect beads by gentle centrifugation, transfer to 1.5 mL tubes, and wash three times with 600 μL high salt RIPA buffer. After last wash remove as much supernatant as possible.
8. Add 75 μL SDS-PAGE loading buffer (1×) to the beads and boil samples at 95 °C for 5 min.
9. Analyze samples by SDS-PAGE (*see* **Note** 7), followed by Coomassie staining for 3 h and destaining overnight (*see* **Note 12**).
10. Excise a gel slice above the unconjugated UBAIT for WT and ΔGG control and the corresponding section of gel for the empty vector control. Send samples for in-gel tryptic digestion and LC-MS/MS to identify peptides.
11. Analyze data by searching against the UniProt human database along with decoy databases using SEQUEST HT (latest version of Proteome Discoverer, Thermo Scientific). Allow for up to two missed cleavage sites in the fully tryptic peptides. Filter the peptide-spectrum matches (PSMs) using Percolator (part of Proteome Discoverer). Apply a false discovery rate (FDR) of <1% to ensure that only reliable PSMs are retained. Potential POI-interacting proteins are identified as proteins with greater than or equal to fivefold higher spectral counts in the WT UBAIT sample compared to the empty vector control and those with two or more spectral counts in the WT UBAIT sample (*see* **Note 13**).

3.5 In Vitro UBAIT Charging and Trapping

1. Clone open reading frame (ORF) of interest into pGEX-6p-1-3xFLAG to construct GST-3xFLAG UBAIT and GST-3xFLAG UBAIT ΔGG. Transform into *E. coli* BL21 cells.
2. Inoculate a 150 mL LB starter culture with BL21 pGEX-6p-1-3xFLAG UBAIT from LB ampicillin 100 μg/mL agar plates, and incubate at 37 °C for 14–16 h.
3. Dilute to 3 L with fresh media and incubate at 37 °C for 2 h. Add 3 mL of 100 mM IPTG, and incubate cultures for a further 3 h at 30 °C.
4. Harvest by centrifugation at 3210 × *g* for 5 min (*see* **Note 3**).
5. Resuspend the cells in 30 mL PBS-Triton buffer on ice. Sonicate on ice at 50% duty cycle, output control 5, for 30 s.
6. Clear the lysate by centrifugation at 15,000 × *g* for 10 min at 4 °C. Transfer to a 15 mL tube, and add 200 μL Glutathione Sepharose slurry (*see* **Note 6**). Place on a rotator for 2 h at 4 °C.

7. Collect beads by gentle centrifugation and transfer to a 1.5 mL tube.
8. Wash three times with 500 μL PBS-Triton buffer followed by three washes with 500 μL PC buffer. Resuspend the beads in a 50:50 beads/PC buffer slurry.
9. Perform site-specific cleavage of the protein from the beads by adding 2 μL PC enzyme, and rotate for 14–16 h at 4 °C.
10. Collect beads by gentle centrifugation. Transfer the protein-containing supernatant to a new 1.5 mL tube.
11. Quantify the protein purity by analyzing 2 μL by SDS-PAGE along with the protein concentration using Bradford Reagent and comparison with BSA protein standards.
12. Production of E2 (UbcH5a) enzyme is similar to expression and purification of the UBAIT proteins described above, with the following differences: Inoculate a 25 mL LB starter culture with BL21-expressing GST-UbcH5a. Dilute culture to 250 mL with fresh media. After incubating 2 h at 37 °C, induce with 250 μL of 100 mM IPTG, and incubate cultures further for 2 h at 37 °C. Harvest by centrifugation and resuspend the cell pellet in 10 mL of PBS-Triton buffer. Following sonication and centrifugation, transfer the cleared lysate to a 15 mL conical, and add 100 μL Glutathione Sepharose slurry. Finish purification as above (*see* **Note 17**).
13. Mix UBAIT (~10 μM) and excess molar E2 with 239 nM UBE1 in 5 mM DTT, 5 mM $MgCl_2$, and 10 mM ATP (*see* **Note 18**). Bring up the volume to 200 μL with $T_{25}N_{50}$. Incubate for 1 h at 25 °C to complete the charging reaction.
14. To confirm charging of UbcH5a with the UBAIT, take 20% of your charging reaction, and divide into two 1.5 mL tubes (20 μL per tube).
15. Add SDS loading buffer (6.7 μL 4×) to one sample and SDS loading buffer without DTT (6.7 μL 4X-DTT) to the other.
16. Run samples on the appropriate percentage SDS-PAGE gel with an empty lane between each sample and immunoblot for the FLAG tag-UBAIT.
17. Successful charging is indicated by a higher molecular weight band equivalent to UBAIT +E2 present in the sample that was run in SDS loading buffer without DTT, but not present in the sample run in SDS loading buffer with DTT.
18. Add the charged UBAIT to the sample of interest (250–1000 μL) for 1 h at 37 °C (*see* **Note 19**).
19. Denature sample by bringing solution to a final concentration of 6 M urea; incubate for 20 min at 25 °C.

20. Transfer to a 15 mL conical and dilute to 1 M urea with 7 mL RIPA buffer, and add 100 μL M2 FLAG bead slurry (*see* **Note 6**) for 14–16 h at 4 °C on a rotator.
21. Collect beads by gentle centrifugation, and wash three times with 500 μL RIPA buffer. After the last wash, remove as much supernatant as possible.
22. Add SDS-PAGE loading buffer (40 μL 1×), and boil samples at 95 °C for 5 min. Analyze samples by SDS-PAGE, followed by Coomassie staining for 3 h and destaining overnight.
23. Excise a gel slice above the unconjugated UBAIT for WT and ΔGG control and the corresponding section of gel for the empty vector control. Send samples for in-gel tryptic digestion and LC-MS/MS to identify peptides.
24. Analyze data by searching against the appropriate database along with decoy databases using SEQUEST HT (latest version of Proteome Discoverer, Thermo Scientific). Allow for up to two missed cleavage sites in the fully tryptic peptides. Filter the peptide-spectrum matches (PSMs) using Percolator (part of Proteome Discoverer). Apply a false discovery rate (FDR) of <1% to ensure that only reliable PSMs are retained. Potential POI-interacting proteins are identified as proteins with greater than or equal to twofold higher spectral counts in the WT UBAIT sample compared to the ΔGG UBAIT control across three biological replicates (*see* **Note 13**).

4 Notes

1. Overlapping extension PCR has been used for the construction of previous UBAITs [4]; other cloning techniques may also be used.
2. Other affinity tags besides TAP may be built into the UBAIT construct (e.g., FLAG or 3X-FLAG tag within the linker instead of an N-terminal tag or an N-terminal 3X-FLAG instead of TAP).
3. Cells can be frozen at −80 °C until lysis if necessary.
4. Although six rounds of bead beating are usually sufficient for lysis, lysis efficiency should be monitored under a light microscope. Dip a p200 pipette tip into a sample tube, and smear the sample from the pipette tip onto a glass slide, along with 5 μL of water. Cover with a coverslip. The un-lysed cells will appear bright and reflective in comparison to lysed cells. Perform additional cycles if necessary.
5. Be cautious with the needle. The needle quickly becomes blunt; changing the needle often is recommended.

6. IgG Sepharose beads are washed in NP40 wash buffer, Glutathione Sepharose beads are washed in PBS-Triton, and M2 FLAG resin is washed in RIPA buffer. All beads were washed three times before use and used as a 50:50 beads-to-buffer slurry.
7. The percentage SDS-PAGE gel will depend on the expected size of the UBAIT protein.
8. Other detection systems will also work, such as HRP-linked secondary antibodies for detection on X-ray film.
9. In rare instances, the WT UBAIT for some proteins of interest does not express as well or sometimes, not at all, whereas the ΔGG counterpart may be expressed at normal levels compared to the non-UBAIT protein. This suggests that the ubiquitin sequence is destabilizing the UBAIT, and in such instances the UBAIT approach may not be useful.
10. UBAITs were bound to IgG beads in a first non-denaturing purification and then boiled to denature. After collecting the supernatant, fresh IgG beads were added to the sample for a second round of binding. The two-step process reduces background binding and ensures that only covalently trapped proteins are co-purified with the UBAITs (as opposed to non-covalently bound proteins).
11. For in-gel tryptic digestion, less than or equal to 10% is recommended. Leave an empty space between lanes to minimize the chance of cross-contamination of samples during the gel-loading process.
12. A strong band corresponding to the expected molecular weight for the WT UBAIT, and a smear upward from the unconjugated UBAIT, is indicative of a well-functioning UBAIT (Fig. 1b). The ΔGG UBAIT may have a stronger band at the expected molecular weight, and there should be no higher molecular weight conjugates. Sometimes, despite a positive outcome in the Western blot for the small-scale experiment, there is no smear visible for the WT UBAIT in the Coomassie-stained gel when the experiment is scaled up. This may indicate that further scale-up is necessary. To help with the problem-solving process, take samples at different stages of the experiment for Western blot analysis.
13. Common background proteins identified by unrelated UBAITs include the yeast proteins Cpr1, Ddi1, Ecm21, histone H2B, Met4, Pal1, Pgk1, Uba1, Ubc1, Ubi4, and Yhr097c, as well as the human proteins Amot, Fam115a, and Spg20. In addition, UBAITs commonly identify components of the ubiquitin system. Further filtering can also be carried out to address a particular biological question.

14. HEK293T cells are easy to grow and transfect and therefore work well for the expression of UBAITs; however, this protocol can be adapted to other transfectable cell lines.
15. Follow manufacturer's guidelines for use of transfection reagent and the recommended confluency for efficient transfection.
16. Cells should have reached full confluency when they are harvested; more or less time may be required. One well of a 6-well plate is ~1.2×10^6 HEK293T cells. Ten 100 mm dishes are ~ 1×10^8 HEK293T cells.
17. We have found that human UbcH5a and yeast Ubc4 work well for most in vitro UBAIT experiments; however preliminary experiments may be useful for determining which E2 works best for UBAIT charging in individual experiments. In addition, the protein yield is dependent on the culture size. Increase or decrease the culture size as appropriate.
18. Use excess molar amount of E2 protein to UBAIT to encourage maximum charging. It is critical to have enough UBAIT to detect your conjugates on a Coomassie-stained gel. Scale up as appropriate.
19. The protocol was developed with NK-92 cells [7] but can be adapted for other cell lines, cell lysates, or purified protein. The amount of material needed to get sufficient hits by mass spectrometry will be specific to the experiment and may need to be optimized.

References

1. O'Connor HF, Huibregtse JM (2017) Enzyme--substrate relationships in the ubiquitin system: approaches for identifying substrates of ubiquitin ligases. Cell Mol Life Sci 74:3363–3375
2. O'Leary CE, Lewis EL, Oliver PM (2015) Ubiquitylation as a rheostat for TCR signaling: from targeted approaches toward global profiling. Front Immunol 6:618
3. Iconomou M, Saunders DN (2016) Systematic approaches to identify E3 ligase substrates. Biochem J 473:4083–4101. https://doi.org/10.1042/BCJ20160719
4. O'Connor HF, Lyon N, Leung JW et al (2015) Ubiquitin-activated interaction traps (UBAITs) identify E3 ligase binding partners. EMBO Rep 16:1699–1712. https://doi.org/10.15252/embr.201540620
5. Ling R, Colón E, Dahmus ME, Callis J (2000) Histidine-tagged ubiquitin substitutes for wild-type ubiquitin in Saccharomyces cerevisiae and facilitates isolation and identification of in vivo substrates of the ubiquitin pathway. Anal Biochem 282:54–64. https://doi.org/10.1006/abio.2000.4586
6. Peng J, Schwartz D, Elias JE et al (2003) A proteomics approach to understanding protein ubiquitination. Nat Biotechnol 21:921–926. https://doi.org/10.1038/nbt849
7. Swaim CD, Scott AF, Canadeo LA, Huibregtse JM (2017) Extracellular ISG15 signals cytokine secretion through the LFA-1 integrin receptor. Mol Cell 68:581–590.e5. https://doi.org/10.1016/j.molcel.2017.10.003

Chapter 8

Generating Intracellular Modulators of E3 Ligases and Deubiquitinases from Phage-Displayed Ubiquitin Variant Libraries

Wei Zhang and Sachdev S. Sidhu

Abstract

Ubiquitination is a posttranslational protein modification pathway regulating diverse cellular processes that are implicated in numerous human diseases. However, targeting the enzymes in the ubiquitination cascade potently and selectively remains a major challenge. Recently we devised a methodology to generate ubiquitin-based modulators for E3 ligases and deubiquitinases, enzymes that control the specificity of protein ubiquitination and deubiquitination, respectively. Here, we describe methods to generate libraries of ubiquitin variants and perform phage display selections to isolate high-affinity binders for target proteins. Importantly, the strategy introduced here can be applied to other small protein domains mediating protein-protein interactions to engineer tools for target validation and potential therapeutic development.

Key words Phage display, Combinatorial library, Ubiquitin, Inhibitor, Activator, E3 ligase, Deubiquitinase

1 Introduction

Ubiquitination is a highly orchestrated posttranslational modification pathway essential for many critical cellular processes including DNA replication and repair, endocytic trafficking, and immune response [1]. In a multistep enzymatic cascade, ubiquitination covalently attaches one or more copies of the small, evolutionarily conserved protein ubiquitin (Ub) to substrate proteins through the coordinated actions of E1 Ub-activating enzymes, E2 Ub-conjugating enzymes, and E3 Ub ligases [2–5] (Fig. 1). Notably, the process is reversed by deubiquitinases (DUBs), which remove Ub moieties from substrates to prevent protein degradation or to alter signaling [6]. The thousands of ligases, DUBs, and adapter proteins that mediate ubiquitination together comprise the Ub-proteasome system (UPS) [7].

Many E3 ligases and DUBs are potential targets for treatment of numerous human diseases including cancer, neuronal disorders,

Thibault Mayor and Gary Kleiger (eds.), *The Ubiquitin Proteasome System: Methods and Protocols*, Methods in Molecular Biology, vol. 1844, https://doi.org/10.1007/978-1-4939-8706-1_8,

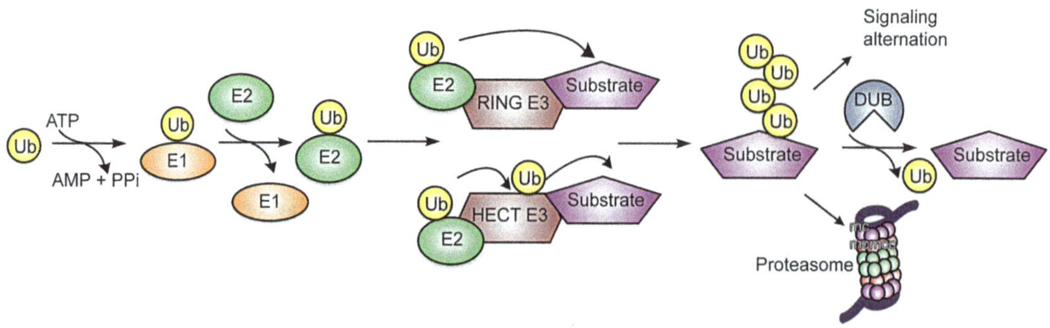

Fig. 1 The ubiquitination pathway. Ubiquitination involves a three-step enzymatic cascade. *Step 1*: E1 employs ATP to adenylate Ub to form a high-energy thioester bond between the Ub C-terminal carboxyl group and the thiol group of the active site cysteine residue. *Step 2*: The activated Ub is then transferred to the cysteine residue of an E2 enzyme through a thioester linkage. *Step 3*: An E3 ligase recruits a charged E2 and helps transfer Ub (RING-type E3s mediate direct transfer of Ub from E2 to substrate, whereas HECT-type E3s covalently bind Ub and then transfer to substrate proteins) to specific protein substrates to form mono- (not shown) or poly-Ub chains, which will lead to different cellular responses, including altered signaling events or protein degradation through the proteasome. Finally, deubiquitinases (DUBs) catalyze the cleavage of Ub from polypeptides to reverse the ubiquitination process

and infectious diseases [2, 7–13]. Therefore, the development of specific and potent inhibitors and activators of E3 ligases and DUBs would have significant therapeutic potential. This has been demonstrated by the clinical success of the proteasome inhibitor bortezomib, which was approved for treatment of relapsed or refractory multiple myeloma [14, 15]. However, bortezomib is a global inhibitor of the proteasome; therefore, the efficacy of this approach is limited by toxicity and drug resistance. Consequently, inhibitors of specific components of the ubiquitination/deubiquitination cascade may provide routes to better therapeutics with fewer side effects.

We have devised a technology based on protein engineering and phage display to rapidly and effectively select protein modulators for virtually any of the thousands of Ub-interacting proteins [16, 17]. Our approach is based on the fact that most proteins that interact with Ub recognize a common surface with low affinity but high specificity, mediating weak interactions through large and diverse surfaces [18]. The weak interactions may be required because Ub has to be released quickly after the enzymatic reaction has occurred. We hypothesized that introduction of mutations in Ub could enhance affinity for a particular protein while maintaining selectivity. Importantly, these variants of Ub (UbVs) may act as potent and competitive inhibitors of enzymatic activity by blocking the binding of Ub substrates.

We thus created a phage-displayed library of ten billion unique UbVs with mutations at positions across the Ub surface that binds to target proteins [16] (Fig. 2). In our first study, we generated

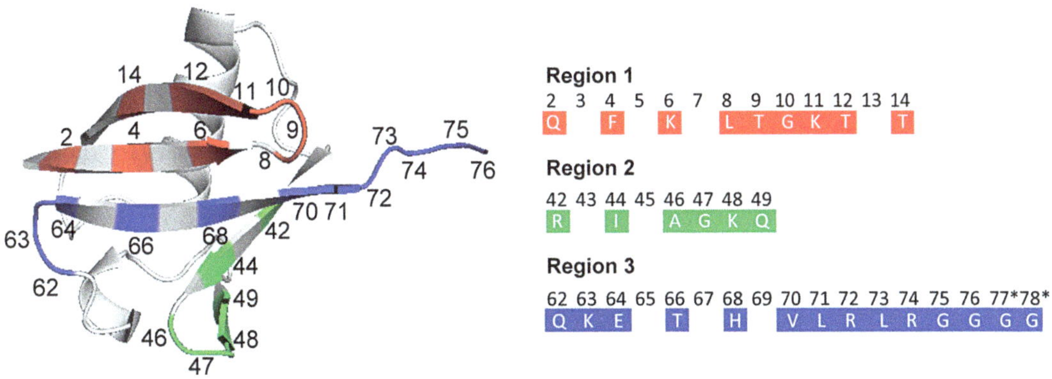

Fig. 2 Design of the ubiquitin variant library. The ubiquitin (Ub) structure (PDB entry 1UBQ) is shown as a ribbon (gray), and positions diversified in the library are colored as follows: Region 1 (red), Region 2 (green), and Region 3 (blue). The sequences of the regions targeted in the library design are shown to the right and are colored as in the structure

UbV binders for several human DUBs and showed inhibition of catalytic activity both in vitro and in cells [16]. Structural analysis of UbVs in complex with DUBs validated our strategy for the generation of specific inhibitors by optimization of intermolecular contacts through a combinatorial approach, which small molecules are unable to achieve.

To date, we have generated UbV modulators for >100 UPS components ([16, 19–25] and unpublished work). We have confirmed selectivity and potency of UbVs in vitro, and, importantly, for many UbVs we have confirmed activity and specificity in cells [20, 22, 24–26] and three-dimensional organoids [22]. We have shown that UbVs can bind their targets with high affinity, and small amounts are often sufficient to cause biological effects. Indeed, in our first study [16], we showed that UbV.21.4 blocked the deubiquitination of RIP1 by USP21 and restored NF-κB activation, whereas UbV.8.2 interacted with USP8 in cells and caused ubiquitination, lysosomal localization, and downregulation of activated EGFR, which is the expected outcome of USP8 inhibition. For E3 ligases, we obtained specific and potent UbV inhibitors and/or activators for the whole family of human HECT E3 ligases [22] and important RING E3 ligases [19]. Similarly, we generated UbV binders for F-box proteins that impair SCF E3 ligase complex formation and inhibit activity in cells [20].

There are five features that make UbVs unique among the small molecule and chemical compound inhibitors that have thus far been reported to target the UPS. (1) The facile and versatile nature of the protein engineering process enables rapid and cost-effective production of inhibitors and activators. (2) UbVs enable inhibition and activation modes that cannot be accessed by genetic technologies such as RNA interference or CRISPR (clustered regularly interspaced short palindromic repeats). (3) UbVs can target families of proteins at sites that small molecules may fail to reach (e.g., E2-binding sites on HECT and RING E3 ligases). (4) The structural details of the interactions between UbVs and their targets can be used to guide the design of small molecule inhibitors. (5) UbVs are amenable to systematic optimization to increase efficacy and reduce off-target effects. In summary, our technology has tremendous transformative capability to empower UPS drug discovery. In this chapter, we describe step-by-step procedures to enable the research community to use our technology to develop high-affinity UbV binders for E3 ligases, DUBs, and other components of the UPS.

2 Materials and Reagents

2.1 Stock Solutions

1. 10% BSA (w/v) in water.
2. 10 mM ATP.
3. 5 mg/mL carbenicillin. Dissolve in water and filter-sterilize.
4. 100 mg/mL chloramphenicol in ethanol.
5. 100 mM dithiothreitol (DTT).
6. 100 mM dNTP mix (solution containing 25 mM each of dATP, dCTP, dGTP, and dTTP).
7. 40 mM dNTP mix (10 mM of each dATP, dCTP, dGTP, and dTTP).
8. GST-tagged recombinant protein and GST alone (*see* **Note 1**).
9. 100 mM HCl.
10. HRP/anti-M13 antibody conjugate (GE Healthcare, 27-9421-01).
11. 5 mg/mL kanamycin in water, filter-sterilize.
12. 1 M phosphoric acid (H_3PO_4).
13. QIAprep Spin M13 Kit.
14. QIAquick Gel Extraction Kit.
15. QIAquick PCR Purification Kit.
16. T4 DNA ligase.
17. T4 DNA polymerase.
18. T4 polynucleotide kinase.
19. T7 DNA polymerase.
20. Taq DNA polymerase and reaction buffer.
21. 5 mg/mL tetracycline. Dissolve in water and filter-sterilize.
22. 1 M Tris-HCl, pH 11.0.
23. Ultrapure irrigation USP water (*see* **Note 2**).
24. 25 mg/mL uridine in ethanol.

2.2 Equipment

1. 96-well MaxiSorp immunoplates (Nunc, 436954).
2. 96-well clear flat bottom polystyrene non-binding surface microplate (Corning, 3641).
3. 96-well PCR plate.
4. 96-well mini tubes (Axygen, MTS-06-C-R-S).
5. 384-well clear MaxiSorp plate (Nunc, 460518).
6. Baffled flasks.
7. Benchtop centrifuge.
8. ECM-600 electroporator (BTX).
9. Shaker-incubator.
10. Centrifuge bottles.
11. Sorvall centrifuge (SS34 rotor for 50 mL tube).
12. 0.2 cm gap electroporation cuvette.
13. Water bath (12–25 °C).
14. Two-inch magnetic stirbars, soaked in ethanol.
15. Titer Plate Shaker.
16. PowerWave XS microplate spectrophotometer (BioTek).

2.3 Buffers and Media

1. 2YT medium: 10 g yeast extract, 16 g tryptone, 5 g NaCl. Add water to make up the volume to 1.0 L, adjust pH to 7.0 with NaOH, and autoclave.
2. 2YT/carb/cmp medium: 2YT, 50 μg/mL carbenicillin, 10 μg/mL chloramphenicol.
3. 2YT/carb/kan medium: 2YT, 50 μg/mL carbenicillin, 25 μg/mL kanamycin.
4. 2YT/carb/kan/uridine medium: 2YT, 50 μg/mL carbenicillin, 25 μg/mL kanamycin, 0.25 μg/mL uridine.
5. 2YT/carb/KO7 medium: 2YT, 50 μg/mL carbenicillin, 10^{10} M13KO7 phage per mL.
6. 2YT/tet medium: 2YT, 10 μg/mL tetracycline.
7. 80% (v/v) glycerol: 800 mL ultrapure glycerol in 200 mL ultrapure irrigation USP water, filter-sterilize.
8. LB medium: 5 g yeast extract, 10 g tryptone, 10 g NaCl. Add water to make up the volume to 1.0 L, adjust pH to 7.0 with NaOH, and autoclave. For LB agar, add 20 g of granulated agar.
9. LB/carb plates: LB agar, 50 μg/mL carbenicillin.
10. LB/kan plates: LB agar, 25 μg/mL kanamycin.
11. LB/tet plates: LB agar, 10 μg/mL tetracycline.

12. PBS buffer: 137 mM NaCl, 3 mM KCl, 8 mM Na_2HPO_4, and 1.5 mM KH_2PO_4. Adjust pH to 7.2 with HCl and autoclave.
13. PBT buffer: PBS buffer, 0.05% Tween 20, 1% BSA. Filter-sterilize.
14. PCR cleanup mix: add 20 μL exonuclease I (200 units) and 20 μL shrimp alkaline phosphatase (20 units) to 960 μL H_2O.
15. PEG/NaCl: 20% PEG 8000 (w/v), 2.5 M NaCl. Mix and autoclave.
16. PT buffer: PBS buffer, 0.05% Tween 20. Filter-sterilize.
17. SOC medium: 5 g bacto-yeast extract, 20 g bacto-tryptone, 0.5 g NaCl, 0.2 g KCl. Add water to make up the volume to 1.0 L, and adjust pH to 7.0 with NaOH, and autoclave; add 5.0 mL of autoclaved 2.0 M $MgCl_2$ and 20 mL of filter-sterilized 1.0 M glucose.
18. TAE buffer: 40 mM Tris-acetate, 1.0 mM EDTA, adjust pH to 8.0; autoclave.
19. 1% TAE/agarose gel: TAE buffer, 1.0% (w/v) agarose, 1:5000 (v/v) 10% (w/w) ethidium bromide.
20. 10× TM buffer: 0.1 M $MgCl_2$, 0.5 M Tris-HCl (pH 7.5).
21. TMB substrate: Mix equal volumes of 3,3′,5,5′-tetramethylbenzidine (TMB) and H_2O_2 peroxidase substrate solution B.

2.4 Bacterial and Phage Strains Used in the Protocol

1. *Escherichia coli* CJ236 (New England Biolabs).
2. *E. coli* OmniMAX (Thermo Fisher Scientific).
3. *E. coli* SS320 (Sidhu Lab).
4. M13KO7 helper phage (Sidhu Lab, *see* **Note 3**).
5. p3-Ub (Sidhu Lab).

2.5 Oligonucleotides Used in the Protocol

To construct the Ub variant library as described in Subheading 3, the following oligonucleotides were used [16]. The nucleotide ratio in degenerate oligonucleotides was adjusted to 70% of wild type represented by 5 = A, 6 = C, 7 = G, and 8 = T and 10% of each of the other three nucleotides. Also "N" denotes an equimolar mixture of all four nucleotides; "K" denotes an equimolar mixture of G and T (*see* **Note 4**).

1. pdl1: GAC GAT GAC AAA ATG (6)(5)(7) ATT (8)(8)(6) GTG (5)(5)(5) ACC (6)(8)(8) (5)(6)(7) (7)(7)(7) (5)(5)(7) (5)(6)(6) ATC (5)(6)(6) CTC GAG GTT GAA CCC.
2. pdl2a: CCT CCT GAT CAG CAG (5)(7)(5) CTG (5)(8)(6) TTT (7)(6)(8) (7)(7)(6) (5)(5)(7)(6)(5)(7) CTG GAA GAT GGA CGT.

3. pdl3a: TCT GAC TAC AAT ATT (6)(5)(5) (5)(5)(7) (7)(5) (7) TCT (5)(6)(8) CTT (6)(5)(8) CTT (7)(8)(7) (8)(8)(7) (5) (7)(5) (6)(8)(8) (6)(7)(8) (7)(7)(8) NNK NNK NNK GGC GGA TCC GGT GGC CTC.

   In addition, the following primers were required for cloning the UbV from phagemid vectors and for sequencing.

4. WZ107: Forward primer, 5'-GTAAAACGACGGCCAGTT-CATCGGCTCGTATAATGTG-3′ (used to PCR-amplify DNA fragment for sequencing).
5. WZ108: Reverse primer, 5'-CAGGAAACAGCTATGACGC-GACAGAATCAAGTTTGCC-3′ (used to PCR-amplify DNA fragment for sequencing).
6. M13 forward (−20): Primer used for sequencing, 5′-GTA AAA CGA CGG CCA GTG-3′.

3 Methods

3.1 Isolation of dU-ssDNA Template

1. A phagemid was designed for the display of Ub with the FLAG (DYKDDDDK) peptide fused to its N terminus (p3-Ub wt, [16]) (*see* **Note 5**). Transform phagemid p3-Ub into competent *E. coli* CJ236 cells following standard plasmid transformation protocols. Plate the transformed cells on LB/carb plate; incubate overnight at 37 °C.
2. Pick a single colony harboring the phagemid into 1 mL 2YT media supplemented with M13KO7 helper phage (10^{10} PFU/mL) and appropriate antibiotics to maintain the host F′ episome and the phagemid. For example, 2YT/carb/cmp media contains carbenicillin to select for phagemids that carry the β-lactamase gene and chloramphenicol to select for the F′ episome of *E. coli* CJ236.
3. Incubate with shaking at 200 rpm (revolutions per minute) for 2 h at 37 °C.
4. Add kanamycin (25 μg/mL) to select clones that have been coinfected with M13KO7, which carries a kanamycin resistance gene.
5. Incubate with shaking at 200 rpm for 6 h at 37 °C, and transfer the culture to 30 mL 2YT/carb/kan/uridine media.
6. Incubate with shaking at 200 rpm for 20 h at 37 °C.
7. Centrifuge at 27,000 × *g* for 10 min at 4 °C.
8. Precipitate phage particles by transferring the supernatant to a new tube containing 1/5 volume of PEG/NaCl and incubate on ice for 10 min.

9. Centrifuge at 12,000 × *g* for 10 min at 4 °C. Decant the supernatant, centrifuge at 12,000 × *g* for 2 min, and aspirate the remaining supernatant.
10. Resuspend the phage pellet in 0.5 mL PBS and transfer to a 1.5 mL tube.
11. Centrifuge at 16,000 × *g* for 5 min at 4 °C to pellet insoluble matter, and transfer the supernatant to a new 1.5 mL tube.
12. The following steps (**steps 12–22**) are modified from the QIAprep Spin M13 Kit protocol and are performed at room temperature. Add 7.0 μL of buffer MP (from Spin M13 Kit) and mix. Incubate at room temperature for at least 2 min.
13. Apply the sample to a QIAprep Spin M13 column in a 2 mL tube. Centrifuge at 6000 × *g* for 30 s. Discard the flow-through. The phage particles remain bound to the column matrix.
14. Add 0.7 mL buffer MLB (from the Spin M13 Kit) to the column. Centrifuge at 6000 × *g* for 30 s and discard the flow-through.
15. Add 0.7 mL buffer MLB. Incubate at room temperature for at least 1 min.
16. Centrifuge at 6000 × *g* for 30 s. Discard the flow-through. The DNA is separated from the protein coat and remains adsorbed to the matrix.
17. Add 0.7 mL of buffer PE (from the Spin M13 Kit). Centrifuge at 6000 × *g* for 30 s and discard the flow-through.
18. Repeat **steps 15–17** to remove residual proteins and salt.
19. Centrifuge at 6000 × *g* for 30 s in a fresh 1.5 mL tube to remove residual PE buffer from column.
20. Transfer the column to a 1.5 mL tube.
21. Add 100 μL of buffer EB (from Spin M13 Kit; 10 mM Tris-HCl, pH 8.5) to the center of the column membrane. Incubate at room temperature for 10 min.
22. Centrifuge at 6000 × *g* for 30 s. Save the eluant, which contains the purified dU-ssDNA.
23. Analyze the dU-ssDNA by electrophoresing 1.0 μL on a 1% TAE/agarose gel at 100 V for 1 h. The DNA should appear as a predominant single band, but faint bands with lower electrophoretic mobility may be visible (Fig. 3). These are likely caused by secondary structure in the dU-ssDNA.
24. Determine the DNA concentration by measuring absorbance at 260 nm ($A_{260} = 1.0$ for 33 ng/μL of ssDNA). Typical DNA concentrations range from 200 to 500 ng/μL.

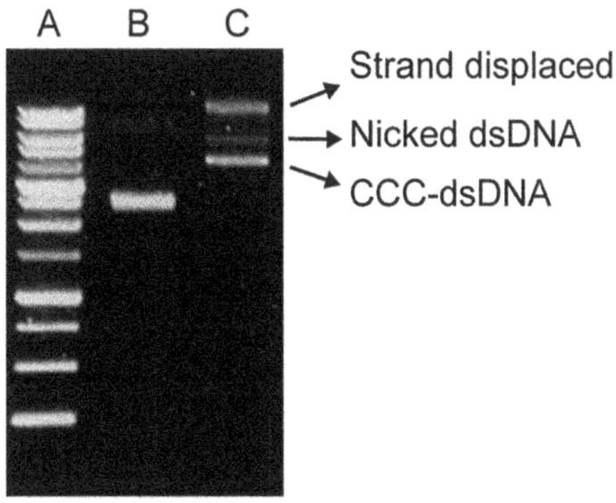

Fig. 3 Synthesis of heteroduplex CCC-dsDNA. The reaction products were electrophoresed on a 1.0% agarose gel for DNA visualization. The lanes were loaded as follows: (**a**) 1 kb DNA marker, (**b**) dU-ssDNA template, and (**c**) reaction product from the heteroduplex CCC-dsDNA synthesis reaction. The lower band is correctly extended and ligated CCC-dsDNA. The middle band is nicked dsDNA. The upper band is strand-displaced DNA

3.2 In Vitro Synthesis of Heteroduplex CCC-dsDNA

1. The first step in the synthesis of heteroduplex CCC-dsDNA is the phosphorylation of the mutagenic oligonucleotide. The mutagenic oligonucleotide must be 5′-phosphorylated to enable ligation by T4 DNA ligase. Set up the phosphorylation reaction in a 1.5 mL tube as shown in Table 1.
2. Incubate for 1.0 h at 37 °C and use immediately for annealing.
3. Anneal the phosphorylated oligonucleotide generated as above to the dU-ssDNA template. Set up the annealing reaction in a 1.5 mL tube as shown in Table 2.
4. Anneal the phosphorylated mutagenic oligonucleotide to the dU-ssDNA template, according to the following protocol: 90 °C for 3 min, 50 °C for 5 min, and 20 °C for 5 min. The reaction is incubated at the DNA melting temperature (90 °C) for only a brief period to minimize dephosphorylation of the mutagenic oligonucleotide.
5. Following annealing, add T7 DNA polymerase and T4 DNA ligase as detailed in Table 3 to complete the enzymatic synthesis of CCC-dsDNA. T7 DNA polymerase uses the annealed mutagenic oligonucleotide as a primer for the template-directed synthesis of a complementary DNA strand. Subsequently, T4 DNA ligase seals the nick to form heteroduplex CCC-dsDNA.
6. Incubate overnight at room temperature.
7. Affinity-purify and desalt the heteroduplex CCC-dsDNA using a modification of the protocol described in the QIAquick PCR Purification Kit (Qiagen). To the reaction mixture from **step 6**, add 1.0 mL buffer QG (from the PCR Kit) and mix.

Table 1
Oligonucleotide phosphorylation reaction

Component	Amount	Final
Mutagenic oligonucleotide	Variable	0.6 μg
10× TM buffer	2 μL	1×
10 mM ATP	2 μL	1 mM
100 mM DTT	1 μL	5 mM
T4 polynucleotide kinase (10 U/μL)	2 μL	20 U
ddH_2O	Up to 20 μL	

Table 2
Annealing reaction

Component	Amount	Final
dU-ssDNA template	Variable	20 μg
10× TM buffer	25 μL	1×
Phosphorylated mutagenic oligonucleotide	20 μL	0.6 μg
ddH_2O	Up to 250 μL	

Table 3
Enzymatic synthesis of CCC-dsDNA

Component	Amount	Final
Annealed oligonucleotide/template mixture	250 μL	
10 mM ATP	10 μL	346 μM
100 mM dNTP mix (25 mM of each nucleotide)	10 μL	865 μM of each nucleotide
100 mM DTT	15 μL	5.19 mM
T4 DNA ligase	1 μL	2000 ligation units/ 30 Weiss units
T7 DNA polymerase (10 U/μL)	3 μL	30 U

8. Apply the sample to two QIAquick PCR Purification spin columns placed in 2 mL microcentrifuge tubes. Centrifuge at 16,000 × *g* for 1 min in a microcentrifuge. Discard the flow-through. **Steps 9–13** are carried out at room temperature.

9. Add 750 μL buffer PE (from the PCR Kit) to each column and centrifuge at 16,000 × *g* for 1 min.
10. Transfer the column to a fresh 1.5 mL tube and centrifuge at 16,000 × *g* for 1 min.
11. Transfer the column to a fresh 1.5 mL tube, and add 35 μL of ultrapure irrigation USP water to the center of the membrane. Incubate at room temperature for 10 min.
12. Centrifuge at 16,000 × *g* for 1 min to elute the DNA. Combine the eluants from the two columns (combined volume will be approximately 60 μL owing to the dead volume in the column). The DNA can be used immediately for *E. coli* electroporation or be frozen at −20 °C for up to 1 month for later use.
13. Electrophorese 1.0 μL of the eluted reaction product alongside the dU-ssDNA (Fig. 3).

3.3 *E. coli* Electroporation and Phage Propagation

1. Chill the purified CCC-dsDNA (20 μg in a minimum volume) and a 0.2 cm gap electroporation cuvette on ice.
2. Thaw a 350 μL aliquot of electrocompetent *E. coli* SS320 on ice. Add the cells to the DNA and mix by pipetting several times (avoid introducing bubbles).
3. Transfer the mixture to the cuvette and electroporate. Follow the manufacturer's instructions (*see* **Note 6**).
4. Immediately transfer the electroporated cells to 15 mL SOC media in a 250 mL baffled flask, rinse the cuvette several times with a total of 10 mL pre-warmed SOC media, and transfer to the flask (*see* **Note** 7).
5. Incubate with shaking at 200 rpm for 30 min at 37 °C.
6. Make serial dilutions to determine the library diversity. Add 90 μL 2YT medium to each well of a single column of a 96-well plate. Remove 10 μL from the culture flask (**step 4**), and add it to the first well of the plate. Make eight tenfold serial dilutions by taking 10 μL from a single well and adding it to the well in the next row of the same column. Plate 5 μL of the serial dilutions on LB/carb plates (selects for the phagemid) to determine the library diversity. Incubate the plates overnight at 37 °C (*see* **Note 8**).
7. Transfer the culture to a 2 L baffled flask containing 500 mL 2YT media, supplemented with antibiotics for phagemid and M13KO7 helper phage selection (e.g., 2YT/carb/kan media).
8. Incubate overnight with shaking at 200 rpm at 37 °C.
9. Centrifuge the culture at 16,000 × *g* for 10 min at 4 °C.

10. Transfer the supernatant to a fresh bottle, and add 1/5 volume of PEG/NaCl solution to precipitate the phage. Incubate for 15 min on ice.
11. Centrifuge at 16,000 × *g* for 20 min at 4 °C. Decant the supernatant. Respin briefly and remove the remaining supernatant with a pipette.
12. Resuspend the phage pellet in 5 mL PBT buffer and transfer to a 50 mL Falcon tube.
13. Pellet insoluble matters by centrifuging at 27,000 × *g* for 5 min at 4 °C. Transfer the supernatant to a clean tube.
14. Make serial dilutions as described in **step 6** to titer the library to determine concentration.
15. The library can be used immediately for selection experiments or stored at 4 °C for up to 6 months. For long-term storage, add 5 mL 80% ultrapure glycerol to 5 mL of the library phage and store at −80 °C.

3.4 Preparation of the UbV Library for Selection

The phage-displayed library from Subheading 3.3, **step 15**, should be diluted to a concentration representing 100× the library diversity (*see* **Note 9**). We typically prepare 800 μL of library phage per recombinant protein to screen against.

1. Determine how much of the original library is needed for selection according to the library diversity and phage concentration (*see* **Note 9**).
2. Dilute original library at least tenfold with PBT.
3. Add 1/5 volume of PEG/NaCl and incubate on ice for 20 min to precipitate phage.
4. Centrifuge at 16,000 × *g* for 20 min at 4 °C. Remove the supernatant.
5. Centrifuge at 16,000 × *g* for 2 min at 4 °C. Remove the remaining supernatant.
6. Resuspend the phage pellet in appropriate volume of PBT at a concentration of 100× of library diversity per mL. Prepare 800 μL of diluted library for each UbV to screen.

3.5 UbV Library Selection for Binding to GST-Fusion Target Proteins

1. An overview of the procedures for phage display is illustrated in Fig. 4. First, coat 8 wells of a 96-well MaxiSorp immunoplate with 100 μL each of 1 μM protein in 1× PBS (*see* **Note 1**). Shake on the Titer Plate Shaker overnight at 4 °C (25 rpm). This is the target plate.
2. Coat 8 wells of another MaxiSorp immunoplate with 100 μL each of 1 μM GST in 1× PBS. Shake on the Titer Plate Shaker overnight at 4 °C (25 rpm). This is the negative control plate.

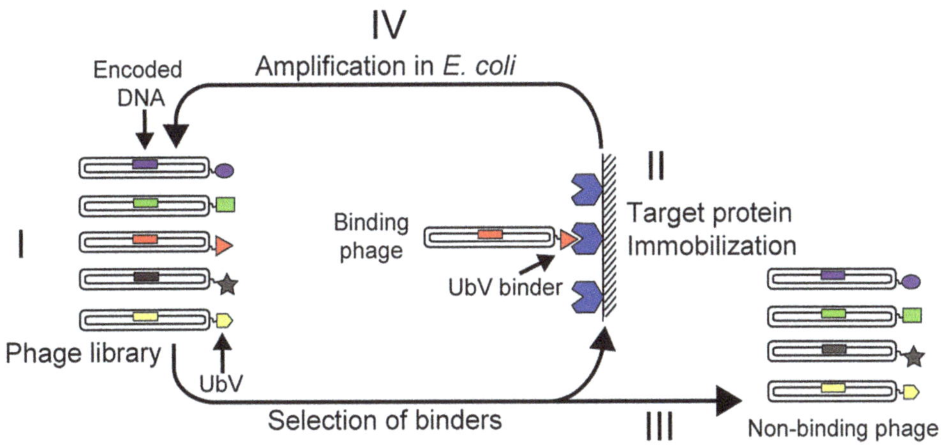

Fig. 4 The phage display cycle. (I) Within the phage pool, each phage particle displays a unique UbV and encapsulates the encoding DNA. (II) Protein-binding phage is captured with immobilized proteins. (III) Non-binding phage is washed away [34]. Bound phage is amplified by infection of bacterial host *E. coli*. The amplified pool can be cycled through additional rounds of selection to further enrich for protein-binding UbVs, and individual clones can be subjected to DNA sequencing to decode sequences of the displayed UbVs

3. From a fresh LB/tet plate, inoculate a single colony of *E. coli* OmniMAX into 5 mL 2YT/tet. Incubate overnight at 37 °C with shaking at 200 rpm.
4. Next day, transfer the culture into 50 mL 2YT/tet and incubate at 37 °C with shaking at 200 rpm until OD_{600} reaches 0.6–0.8.
5. Remove the coating solution from the target and control plates.
6. Add 200 μL/well of PBT blocking buffer (that contains 1% BSA) into each well, shake out, and add another 200 μL. Incubate at room temperature for 1 h with shaking at 25 rpm.
7. Remove the blocking buffer from control plate; add 100 μL of library phage into each well. Shake on the Titer Plate Shaker at 4 °C for 1 h (25 rpm).
8. Remove the blocking buffer from target plate, transfer library phage from wells of control plate into corresponding wells of target plate, and shake on the Titer Plate Shaker at 4 °C for 1.5 h (25 rpm).
9. Remove library phage from target plate; wash the plate with PT buffer as follows: round 1, 4 times; round 2, 6 times; round 3, 8 times; round 4, 10 times; and round 5, 12 times (*see* **Note 10**).
10. Bang the plate upside down onto a paper towel several times to remove any remaining liquid.

11. To elute the bound phage, add 100 μL 100 mM HCl to each well and incubate at room temperature with shaking at 25 rpm for 5 min.
12. Neutralize the pH by adding 12.5 μL 1.0 M Tris-HCl, pH 11, to every 100 μL of eluate. Pool eluted phage from 8 wells in a 1.5 mL microcentrifuge tube, and add 10% BSA to a final concentration of 1%. This is the output from round 1.
13. Use half (500 μL, *see* **Note 11**) of the output to inoculate 4 mL *E. coli* OmniMAX culture ($OD_{600} = 0.6–0.8$) from **step 4** and incubate at 37 °C with shaking at 200 rpm for 30 min.
14. Add M13K07 helper phage to a final concentration of 10^{10} PFU/mL and incubate for 45 min at 37 °C. Transfer it to 40 mL 2YT/carb/kan media and incubate overnight at 37 °C with shaking at 200 rpm.
15. Determine the titer by performing six tenfold serial dilutions of output phage from **step 14**. Spot 5 μL of each dilution on LB/carb and LB/kan plates (*see* **Note 12**).
16. This completes round 1 of the selection. Repeat **steps 1–15** to complete rounds 2–5, using 4 wells (still 100 μL per well) for each round.

3.6 Analysis of Binding Clones by Phage ELISA

1. Calculate the titer of R4-O (*round 4 output*) and R5-O (*round 5 output*). Mix appropriated dilutions ($<10^3$) of R4-O and R5-O (10 μL) with *E. coli* OmniMAX cells (90 μL, $OD_{600} = 0.8$) and plate on LB/carb plates. Incubate overnight at 37 °C.
2. Pick 48 single colonies each from R4-O and R5-O plates, and transfer each colony to 450 μL 2YT/carb/KO7 media in 96-well Axygen mini tubes (Blue Box) and incubate overnight with shaking at 200 rpm at 37 °C.
3. Coat 3 sets of 96 wells in a 384-well MaxiSorp immunoplates overnight at 4 °C with 30 μL/well of 1 μM target protein (in PBS buffer), 30 μL/well of 1 μM GST (in PBS buffer), or 30 μL/well of PBT.
4. Discard coating solution by turning the plate upside down over the sink. Bang the plate upside down onto a paper towel several times to remove any remaining liquid.
5. Add 60 μL/well of blocking buffer (PBT, 1% BSA) and incubate at room temperature with shaking at 25 rpm for 1 h.
6. While blocking, spin down cells in the Blue Box at 2000 × *g* for 10 min at 4 °C. Phage remains in the supernatant. Dilute phage threefold with PBT in a 96-well non-sticky plate (70 μL phage and 140 μL PBT).

7. Remove the blocking buffer from the 384-well plate, add 30 μL of each diluted phage to three different types of wells (containing target proteins, GST, or BSA only), and incubate at room temperature with shaking at 25 rpm for 1 h.
8. Remove the phage and wash the plate four times with PT buffer.
9. Add 30 μL M13-HRP antibody to each well (1:5000 in PBT buffer) and incubate at room temperature with shaking at 25 rpm for 30 min.
10. Wash the plate four times with PT buffer. Add 25 μL TMB substrate to each well.
11. Develop with shaking for a few minutes, and stop the reaction by adding 25 μL of 1.0 M phosphoric acid.
12. Read the OD of each well at 450 nm in a BioTek plate reader (*see* **Note 13**).

3.7 DNA Sequencing of UbV Phagemids

1. In a 96-well PCR plate, set up PCR reactions to amplify UbV-encoding DNA fragments as follows: 10× Taq reaction buffer (2 μL), 40 mM dNTP (0.4 μL), 100 μM FWD primer (WZ107, 0.1 μL), 100 μM REV primer (WZ108, 0.1 μL), Taq polymerase (0.1 μL), ddH_2O (16.3 μL), and phage supernatant (1 μL).
2. Perform PCR program as follows: 95 °C, 5 min; 95 °C, 30 s*; 57 °C, 30 s*; 72 °C, 30 s*; 72 °C, 10 min; and 4 °C for storage (* 26 cycles).
3. Analyze amplified UbV-encoding DNA fragments by electrophoresis using 1% TAE/agarose gel (*see* **Note 14**).
4. Prepare 1 mL of fresh cleanup mix for PCR products. Dispense 10 μL of cleanup mix into each well of a new 96-well PCR plate.
5. Transfer 3 μL of PCR product to each well and mix carefully.
6. Incubate the cleanup reactions at 37 °C for 15 min and at 80 °C for 15 min.
7. Sequence the PCR products with M13 forward primer.
8. Analyze the sequencing results (*see* **Note 15**).

3.8 Follow-Up Studies of UbVs

1. Subcloning of UbV-encoding DNA. We routinely amplify DNA fragments encoding UbVs of interest with an N-terminal FLAG epitope tag and clone these into Gateway Entry vector pDONR221 according to the manufacturer's instructions. Subsequently, the UbV clones can be transferred into Gateway Destination vectors. We usually use pET53 as the bacterial expression vector for biochemical assays or pcDNA3.1 for mammalian cellular assays.

2. Functional characterization. A variety of biochemical, biophysical, and cellular assays can be conducted according to protocols here in this book and published *Methods in Molecular Biology* chapters [27–31] to assess the activities of UbVs as intracellular probes of E3 ligases and DUBs.

4 Notes

1. To conduct selections and following binding and specificity ELISA experiments, we require at least 1 mL proteins with concentration of 20 μM (e.g., for a 50 kDa protein, 1 mg). There is no specific requirement for storage buffer (e.g., Tris or phosphate), but glycerol should be less than 5%, and reducing agent (e.g., DTT) should be below 1 mM. We recommend that the GST tag is not cleaved from proteins. Alternatively, proteins can be biotinylated and immobilized to plates through pre-coating with neutravidin or streptavidin.
2. Unless noted otherwise in Subheading 3, laboratory research grade type 1 water (e.g., after Milli-Q™ purification) was used.
3. Helper phage M13KO7 (carries a kanamycin resistance marker) was prepared as previously described [32]. The final concentration is 10^{13} phage per mL. The helper phage can be used immediately for selection experiments and stored at 4 °C for up to 6 months. For long-term storage, add sterile glycerol to a final concentration of 10% and store at −80 °C.
4. The residues chosen for randomization were identified from structure-function analysis as Ub interaction sites [16]. In most cases, the soft randomization (adjusting the nucleotide ratio in degenerate oligonucleotides to 70% of wild type and 10% of each of the other three nucleotides) method allowed the recovery of wild-type sequence in each place when a mutagenic oligonucleotide was not incorporated into that region (approximately 30% of the population) [33]. This strategy is usually applied to a defined protein domain (e.g., Ub) to constrain the level of variation and mutation frequency. On the other hand, at the three C-terminal end residues of Ub, a hard randomization strategy was applied [32] to maximize the flexibility of Ub tail to contact target proteins.
5. We prefer to use a phagemid to construct phage-displayed libraries [32]. Phagemid is a plasmid that can be introduced to bacterial host by transformation or electroporation and contains the components required for packaging viral DNA into phage particles upon superinfection of the bacterial host with helper phage [32].

6. We use a BTX ECM-600 electroporation system with the following settings: 2.5 kV field strength, 125 Ω resistance, and 50 μF capacitance. Alternatively, a Bio-Rad Gene Pulser can be used with the following settings: 2.5 kV field strength, 200 Ω resistance, and 25 μF capacitance.
7. Pre-warming the SOC media is critical to avoid cold shocking the electroporated cells.
8. It is crucial to use pipette tips with filter to manipulate phage and also remember to change pipette tips between dilutions.
9. For instance, if the library diversity is 2×10^{10}, and the concentration is 2×10^{13} CFU/mL, then dilute the library to 2×10^{12} CFU/mL.
10. Since the selection procedures are based on enrichment of bound phage, durations of washes are gradually increased to maintain selection stringency.
11. The other half is usually kept at 4 °C as backup stocks and future studies, including deep sequencing to collect information of all binding sequences.
12. LB/carb plates select for phagemid to monitor bound phage enrichment across five selection rounds. On the other hand, LB/kan plates select for helper phage M13KO7 as a control.
13. A successful UbV selection process results in majority of phage clones from round 5 (>75%; to a lesser extent for round 4, ~50%) showing high ELISA signals relative to GST alone or other control proteins. If no signal was observed, increase the concentration of target proteins by tenfold and increase the number of wells coated with target proteins by twofold and repeat selections.
14. Typically, if binding is confirmed by ELISA, the PCR will yield a ~300 bp fragment (containing UbV and flanked sequences) for each phage clone.
15. Following a successful UbV selection, a variable number (1–20) of unique UbV sequences are expected to be isolated from the 96 picked clones from round 4 and round 5.

Acknowledgments

We acknowledge the technical assistance and work summary from Mr. Jun Gu. We greatly appreciate the help from past and present collaborators, including Drs. Andreas Ernst, Jason Moffat, Brenda A. Schulman, J. Wade Harper, Daniela Rotin, Danny T. Huang, Brian L Mark, and Marjolein Kikkert. This work is supported by the Canadian Institutes of Health Research (CIHR) project grant (#0000303157) awarded to S.S.S.

References

1. Al-Hakim A, Escribano-Diaz C, Landry MC, O'Donnell L, Panier S, Szilard RK, Durocher D (2010) The ubiquitous role of ubiquitin in the DNA damage response. DNA Repair 9 (12):1229–1240. https://doi.org/10.1016/j.dnarep.2010.09.011
2. Bedford L, Lowe J, Dick LR, Mayer RJ, Brownell JE (2011) Ubiquitin-like protein conjugation and the ubiquitin-proteasome system as drug targets. Nat Rev Drug Discov 10 (1):29–46. https://doi.org/10.1038/nrd3321
3. Komander D, Rape M (2012) The ubiquitin code. Annu Rev Biochem 81:203–229. https://doi.org/10.1146/annurev-biochem-060310-170328
4. Clague MJ, Liu H, Urbe S (2012) Governance of endocytic trafficking and signaling by reversible ubiquitylation. Dev Cell 23(3):457–467. https://doi.org/10.1016/j.devcel.2012.08.011
5. Hershko A, Ciechanover A (1998) The ubiquitin system. Annu Rev Biochem 67:425–479. https://doi.org/10.1146/annurev.biochem.67.1.425
6. Nijman SM, Luna-Vargas MP, Velds A, Brummelkamp TR, Dirac AM, Sixma TK, Bernards R (2005) A genomic and functional inventory of deubiquitinating enzymes. Cell 123 (5):773–786. https://doi.org/10.1016/j.cell.2005.11.007
7. Nalepa G, Rolfe M, Harper JW (2006) Drug discovery in the ubiquitin-proteasome system. Nat Rev Drug Discov 5(7):596–613. https://doi.org/10.1038/nrd2056
8. Petroski MD (2008) The ubiquitin system, disease, and drug discovery. BMC Biochem 9 (Suppl 1):S7. https://doi.org/10.1186/1471-2091-9-S1-S7
9. Cohen P, Tcherpakov M (2010) Will the ubiquitin system furnish as many drug targets as protein kinases? Cell 143(5):686–693. https://doi.org/10.1016/j.cell.2010.11.016
10. Harper JW, King RW (2011) Stuck in the middle: drugging the ubiquitin system at the e2 step. Cell 145(7):1007–1009. https://doi.org/10.1016/j.cell.2011.06.002
11. Maculins T, Fiskin E, Bhogaraju S, Dikic I (2016) Bacteria-host relationship: ubiquitin ligases as weapons of invasion. Cell Res 26 (4):499–510. https://doi.org/10.1038/cr.2016.30
12. Lin YH, Machner MP (2017) Exploitation of the host cell ubiquitin machinery by microbial effector proteins. J Cell Sci 130 (12):1985–1996. https://doi.org/10.1242/jcs.188482
13. Bailey-Elkin BA, Knaap RCM, Kikkert M, Mark BL (2017) Structure and function of viral deubiquitinating enzymes. J Mol Biol 429:3441. https://doi.org/10.1016/j.jmb.2017.06.010
14. Richardson PG, Barlogie B, Berenson J, Singhal S, Jagannath S, Irwin D, Rajkumar SV, Srkalovic G, Alsina M, Alexanian R, Siegel D, Orlowski RZ, Kuter D, Limentani SA, Lee S, Hideshima T, Esseltine DL, Kauffman M, Adams J, Schenkein DP, Anderson KC (2003) A phase 2 study of bortezomib in relapsed, refractory myeloma. N Engl J Med 348(26):2609–2617. https://doi.org/10.1056/NEJMoa030288
15. Richardson PG, Sonneveld P, Schuster MW, Irwin D, Stadtmauer EA, Facon T, Harousseau JL, Ben-Yehuda D, Lonial S, Goldschmidt H, Reece D, San-Miguel JF, Blade J, Boccadoro M, Cavenagh J, Dalton WS, Boral AL, Esseltine DL, Porter JB, Schenkein D, Anderson KC (2005) Bortezomib or high-dose dexamethasone for relapsed multiple myeloma. N Engl J Med 352 (24):2487–2498. https://doi.org/10.1056/NEJMoa043445
16. Ernst A, Avvakumov G, Tong J, Fan Y, Zhao Y, Alberts P, Persaud A, Walker JR, Neculai AM, Neculai D, Vorobyov A, Garg P, Beatty L, Chan PK, Juang YC, Landry MC, Yeh C, Zeqiraj E, Karamboulas K, Allali-Hassani A, Vedadi M, Tyers M, Moffat J, Sicheri F, Pelletier L, Durocher D, Raught B, Rotin D, Yang J, Moran MF, Dhe-Paganon S, Sidhu SS (2013) A strategy for modulation of enzymes in the ubiquitin system. Science 339 (6119):590–595. https://doi.org/10.1126/science.1230161
17. Zhang W, Sidhu SS (2014) Development of inhibitors in the ubiquitination cascade. FEBS Lett 588(2):356–367. https://doi.org/10.1016/j.febslet.2013.11.003
18. Husnjak K, Dikic I (2012) Ubiquitin-binding proteins: decoders of ubiquitin-mediated cellular functions. Annu Rev Biochem 81:291–322. https://doi.org/10.1146/annurev-biochem-051810-094654
19. Brown NG, VanderLinden R, Watson ER, Weissmann F, Ordureau A, Wu KP, Zhang W, Yu S, Mercredi PY, Harrison JS, Davidson IF, Qiao R, Lu Y, Dube P, Brunner MR, Grace CR, Miller DJ, Haselbach D, Jarvis MA, Yamaguchi M, Yanishevski D, Petzold G, Sidhu SS, Kuhlman B, Kirschner MW, Harper

JW, Peters JM, Stark H, Schulman BA (2016) Dual RING E3 architectures regulate multiubiquitination and ubiquitin chain elongation by APC/C. Cell 165(6):1440–1453. https://doi.org/10.1016/j.cell.2016.05.037

20. Gorelik M, Orlicky S, Sartori MA, Tang X, Marcon E, Kurinov I, Greenblatt JF, Tyers M, Moffat J, Sicheri F, Sidhu SS (2016) Inhibition of SCF ubiquitin ligases by engineered ubiquitin variants that target the Cul1 binding site on the Skp1-F-box interface. Proc Natl Acad Sci U S A 113(13):3527–3532. https://doi.org/10.1073/pnas.1519389113
21. Leung I, Dekel A, Shifman JM, Sidhu SS (2016) Saturation scanning of ubiquitin variants reveals a common hot spot for binding to USP2 and USP21. Proc Natl Acad Sci U S A 113:8705. https://doi.org/10.1073/pnas.1524648113
22. Zhang W, Wu KP, Sartori MA, Kamadurai HB, Ordureau A, Jiang C, Mercredi PY, Murchie R, Hu J, Persaud A, Mukherjee M, Li N, Doye A, Walker JR, Sheng Y, Hao Z, Li Y, Brown KR, Lemichez E, Chen J, Tong Y, Harper JW, Moffat J, Rotin D, Schulman BA, Sidhu SS (2016) System-wide modulation of HECT E3 ligases with selective ubiquitin variant probes. Mol Cell 62(1):121–136. https://doi.org/10.1016/j.molcel.2016.02.005
23. Manczyk N, Yates BP, Veggiani G, Ernst A, Sicheri F, Sidhu SS (2017) Structural and functional characterization of a ubiquitin variant engineered for tight and specific binding to an alpha-helical ubiquitin interacting motif. Protein Sci 26:1060. https://doi.org/10.1002/pro.3155
24. Zhang W, Bailey-Elkin BA, Knaap RCM, Khare B, Dalebout TJ, Johnson GG, van Kasteren PB, McLeish NJ, Gu J, He W, Kikkert M, Mark BL, Sidhu SS (2017) Potent and selective inhibition of pathogenic viruses by engineered ubiquitin variants. PLoS Pathog 13(5):e1006372. https://doi.org/10.1371/journal.ppat.1006372
25. Zhang W, Sartori MA, Makhnevych T, Federowicz KE, Dong X, Liu L, Nim S, Dong A, Yang J, Li Y, Haddad D, Ernst A, Heerding D, Tong Y, Moffat J, Sidhu SS (2017) Generation and validation of intracellular ubiquitin variant inhibitors for USP7 and USP10. J Mol Biol 429:3546. https://doi.org/10.1016/j.jmb.2017.05.025
26. Ernst A, Gfeller D, Kan Z, Seshagiri S, Kim PM, Bader GD, Sidhu SS (2010) Coevolution of PDZ domain-ligand interactions analyzed by high-throughput phage display and deep sequencing. Mol BioSyst 6(10):1782–1790. https://doi.org/10.1039/c0mb00061b
27. Maspero E, Polo S (2016) In vitro ubiquitination: self-ubiquitination, chain formation, and substrate ubiquitination assays. Methods Mol Biol 1449:153–160. https://doi.org/10.1007/978-1-4939-3756-1_7
28. Azkargorta M, Escobes I, Elortza F, Matthiesen R, Rodriguez MS (2016) TUBEs-mass spectrometry for identification and analysis of the ubiquitin-proteome. Methods Mol Biol 1449:177–192. https://doi.org/10.1007/978-1-4939-3756-1_9
29. Sigismund S, Polo S (2016) Strategies to detect endogenous ubiquitination of a target mammalian protein. Methods Mol Biol 1449:143–151. https://doi.org/10.1007/978-1-4939-3756-1_6
30. Furukawa M, Andrews PS, Xiong Y (2005) Assays for RING family ubiquitin ligases. Methods Mol Biol 301:37–46. https://doi.org/10.1385/1-59259-895-1:037
31. Russell NS, Wilkinson KD (2005) Deubiquitinating enzyme purification, assay inhibitors, and characterization. Methods Mol Biol 301:207–219. https://doi.org/10.1385/1-59259-895-1:207
32. Tonikian R, Zhang Y, Boone C, Sidhu SS (2007) Identifying specificity profiles for peptide recognition modules from phage-displayed peptide libraries. Nat Protoc 2(6):1368–1386. https://doi.org/10.1038/nprot.2007.151
33. Sidhu SS, Lowman HB, Cunningham BC, Wells JA (2000) Phage display for selection of novel binding peptides. Methods Enzymol 328:333–363
34. Thalassinos K, Grabenauer M, Slade SE, Hilton GR, Bowers MT, Scrivens JH (2009) Characterization of phosphorylated peptides using traveling wave-based and drift cell ion mobility mass spectrometry. Anal Chem 81(1):248–254. https://doi.org/10.1021/ac801916h

Chapter 9

Integrated Proteogenomic Approach for Identifying Degradation Motifs in Eukaryotic Cells

Yifat Geffen, Alon Appleboim, Richard G. Gardner, and Tommer Ravid

Abstract

Since its discovery nearly 40 years ago, many components of the ubiquitin-proteasome system (UPS) have been identified and characterized in detail. However, a key aspect of the UPS that remains largely obscure is the signals that initiate the interaction of a substrate with enzymes of the UPS machinery. Understanding these signals is of particular interest for studies that examine the mechanism of substrate recognition for proteins that have adopted a non-native structure, as part of the cellular protein quality control (PQC) defense mechanism. Such studies are quite salient as the entire proteome makes up the potential battery of PQC substrates, and yet only a limited number of ubiquitination pathways are known to handle misfolded proteins. Our current research aims at understanding how a small number of PQC ubiquitin-protein ligases specifically recognize and ubiquitinate the overwhelming assortment of misfolded proteins. Here, we present a new proteogenomic approach for identifying and characterizing recognition motifs within degradation elements (degrons) in a high-throughput manner. The method utilizes yeast growth under restrictive conditions for selecting protein fragments that confer instability. The corresponding cDNA fragments are analyzed by next-generation sequencing (NGS) that provides information about each fragment's identity, reading frame, and abundance over time. This method was used by us to identify PQC-specific and compartment-specific degrons. It can readily be modified to study protein degradation signals and pathways in other organisms and in various settings, such as different strain backgrounds and under various cell conditions, all of which can be sequenced and analyzed simultaneously.

Key words Ubiquitin-proteasome system, Protein quality control, Degradation signals, Proteogenomics, Next-generation sequencing, Yeast competition assay

1 Introduction

The vitality of every living cell depends on the integrity of its proteome [1]. Therefore, biological mechanisms have evolved to survey the proteome and either repair, eliminate, or sequester defective proteins. Overall, these processes are collectively termed protein quality control (PQC) pathways [2]. The PQC network entails the combined activity of molecular chaperones, ubiquitin conjugation enzymatic complexes, and protein-destructive machines, mainly the proteasome and the lysosome/vacuole

Thibault Mayor and Gary Kleiger (eds.), *The Ubiquitin Proteasome System: Methods and Protocols*, Methods in Molecular Biology, vol. 1844, https://doi.org/10.1007/978-1-4939-8706-1_9,

[3–6]. While many of the enzymatic machineries that lead to misfolded protein ubiquitination and degradation by the proteasome have been identified [7, 8], the mechanisms underlying selection of misfolded proteins for their ultimate demise are still not fully understood [9]. This shortcoming is largely due to our limited knowledge about the state of the eukaryotic proteome and the different ways proteins are affected by stress in living cells. To date, contemporary methods for determining protein degradation kinetics heavily rely on protein extraction and immunoblotting. These techniques do not have the capacity to deal with rapid changes in the entire proteome and so far have failed to provide general rules governing misfolded protein degradation by the UPS.

To overcome these limitations, we developed a method for systematic and quantitative measurement of relative protein degradation kinetics from the entire yeast proteome [10]. The method employs a fusion between *URA3*, which encodes for the essential enzyme orotidine-5′-phosphate (OMP) decarboxylase (Ura3), and a library of protein fragments, generated from a yeast cDNA library. The fusion proteins are expressed in yeast cells auxotrophic for uracil, by deletion of the endogenous *URA3* gene. Constitutive expression of stable Ura3 fusions rescues growth of *ura3Δ* cells in uracil-deficient medium. Consequently, destabilizing Ura3 through the fusion of a protein fragment that functions as a degron (Fig. 1a) can diminish cell growth on media lacking uracil (SD-Ura), due to its reduced steady-state protein levels (Fig. 1b). We previously termed this selection method that translates yeast growth kinetics to protein degradation rates as GiLS (*G*rowth kinetics *I*n *L*iquid culture under *S*elective conditions) [11]. For the purpose of positive screening, cells are counter-selected for Ura3 using 5-fluoroorotic acid (5FOA) (Fig. 1c) [12, 13]. Ura3 converts 5FOA into the toxic compound 5-fluorouracil, causing cell death (Fig. 1b) [14]. Thus, destabilization of Ura3 by a degron leads to faster cell growth on media containing 5FOA, which is predicted to correlate with the degron's potency for recognition by the degradation machinery (Fig. 1d). We further adapted the GiLS assay to fit a proteome-wide degron search, in which translated DNA fragments that confer protein instability are identified by NGS [10]. This new method enabled us, for the first time, to simultaneously identify thousands of degrons from the entire yeast proteome, many of which are in their native protein context.

One of the advantages of this method is its flexibility, since the same selection principles can be used for comparing degrons in different cell compartments, under various stress conditions, or in yeast strains lacking key components of the degradation machinery, such as ubiquitin-protein ligases and molecular chaperones. Yet, this system is by no means restricted to the study of recognition motifs, as it can also be used for studying other protein elements that may confer protein instability, such as degradation initiation

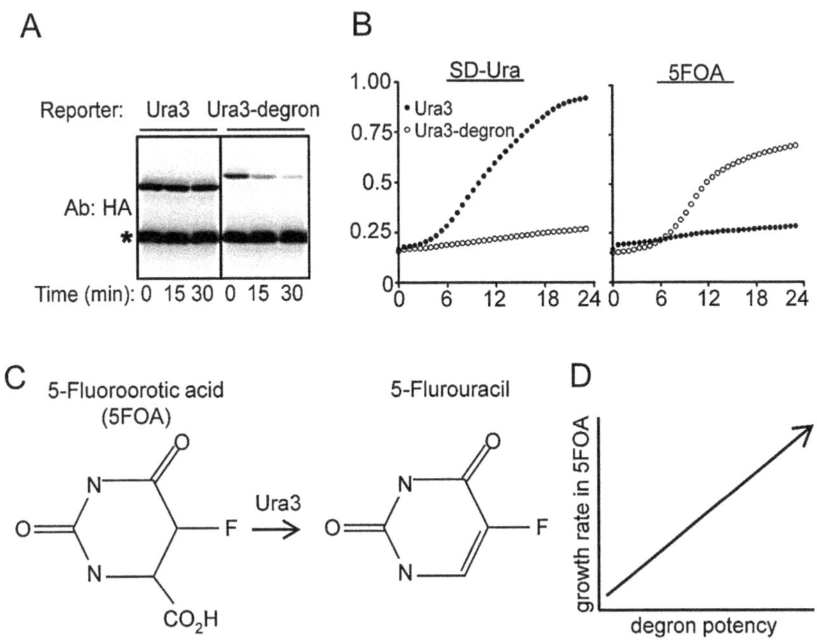

Fig. 1 Principles of Ura3-dependent selection. (**a**) Degradation kinetics of Ura3 with and without a degron. Cells were treated with the translation inhibitor cycloheximide, and samples were collected at the indicated time points. Immunoblotting was done using anti-HA antibodies for the detection of the fusion proteins. *Asterisks* represent a nonspecific band that serves as a loading control. (**b**) Growth curves of cells expressing plasmids with either Ura3 or Ura3 fused to a degron. Cells were incubated in SD-Ura or in SD-Leu media, supplemented with 100 μM 5FOA, for 24 h, and OD_{600} was measured every 30 min. (**c**) Ura3 removes carboxylic group from 5FOA to yield the toxic substance 5-fluorouracil. (**d**) Destabilization of Ura3 by a degron is expected to correlate with enhanced growth rates of cells treated with 5FOA: The higher a degron's potency, the faster the cell's growth kinetics

sequences and ubiquitin conjugation sites, as well as noncoding regions, such as translation termination sequences. We anticipate that the use of this method will help bridging the gap between proteome- and genome-wide studies in the post-genomic era.

2 Materials

2.1 General Strains and Growth Media

All bacteria and yeast media are prepared according to standard protocols.

1. Backbone and library-containing plasmids are listed in Table 1.
2. The yeast strain used in this study was created by switching the mating type of strain BY4741 (open biosystems) from "a" to "alpha" by mating and sporulation. The new strain is listed as TRy481. Its genotype is *MATα*, *his3Δ1*, *met15Δ0*, *ura3Δ0*, and *leu2Δ0* (*see* **Note 1**).

Table 1
List of plasmids

Plasmid	Features	Source
pTR1412	pRS315-CUP1p-URA3orf-HA-GFP-Bam HI Lib cloning site-3x stop (org frame)	[10]
pTR1413	pRS315-CUP1p-URA3orf-HA-GFP-Bam HI Lib cloning site-3x stop (+1 frame)	[10]
pTR1414	pRS315-CUP1p-URA3orf-HA-GFP-Bam HI Lib cloning site-3x stop (+2 frame)	[10]
pTR1804	pRS315-CUP1p-URA3orf-HA-GFP-cDNA library Lib-3x stop (org frame)	[10]
pTR1805	pRS315-CUP1p-URA3orf-HA-GFP-cDNA library-3x stop (+1 frame)	[10]
pTR1806	pRS315-CUP1p-URA3orf-HA-GFP-cDNA library-3x stop (+2 frame)	[10]

3. Electrocompetent bacterial cells for DNA modification and propagation.
4. Yeast-rich media (YPD): 1% yeast extract, 2% peptone, 2% glucose.
5. Minimal synthetic-defined (SD) media: 0.67% yeast nitrogen base (without amino acid), 2% glucose, appropriate dropout mix (typically 0.2%).
6. Lysogeny broth (LB) bacterial media with 100 μg/mL ampicillin (LB-AMP).

2.2 Fusion Library Construction

1. Standard RNA extraction kit.
2. Standard cDNA library construction kit.
3. Sau3AI and BamHI restriction enzymes with corresponding buffers.
4. DNA ligase with the corresponding buffer.
5. Sterile 75% glycerol solution.
6. Standard DNA midi prep kit.

2.3 Library Selection in Yeast and Library Preparation for NGS

1. 5-Fluoroorotic acid (5FOA) is dissolved in the appropriate media at the concentration of 100 μM, filter sterilized, and stored in the dark at 4 °C.
2. 100 mM $CuSO_4$ in H_2O is stored in the dark at room temperature and used at the concentration of 100 μM.
3. Standard DNA miniprep kit.
4. Glass beads, 425–600 μm
5. Library primers (final concentration of 0.5 μM for each primer) (*see* Table 2).
6. High fidelity Taq (final concentration of 0.02 U/μL) with buffer and dNTPs (final concentration of 200 μM).

Table 2
List of primer sequences

Primer	Sequences
Forward primer mix	1. TCGTCGGCAGCGTCAGATGTGTATAAGAGACA**NNNA** *GCCTACGATCGACGAA* 2. TCGTCGGCAGCGTCAGATGTGTATAAGAGACA**NNNCA** *GCCTACGATCGACGAA* 3. TCGTCGGCAGCGTCAGATGTGTATAAGAGACA**NNNGCA** *GCCTACGATCGACGAA* 4. TCGTCGGCAGCGTCAGATGTGTATAAGAGACA**NNNTGCA** *GCCTACGATCGACGAA*
Reverse primer mix	5. GTCTCGTGGGCTCGGAGATGTGTATAAGAGACAG**ANNN** *CTCAGCTATCTAGGATC* 6. GTCTCGTGGGCTCGGAGATGTGTATAAGAGACAG**ACNNN** *CTCAGCTATCTAGGATC* 7. GTCTCGTGGGCTCGGAGATGTGTATAAGAGACAG**ACGNNN** *CTCAGCTATCTAGGATC* 8. GTCTCGTGGGCTCGGAGATGTGTATAAGAGACAG**ACGTNNN** *CTCAGCTATCTAGGATC*

7. Standard N-series Illumina barcoded adapters.
8. DNA TapeStation.
9. AMPure XP beads.

3 Methods

3.1 Construction of URA3-Peptide (URA3-P) Fusion Library

The backbone plasmid contains three necessary parts: (1) an inducible *CUP1* promoter, to enable controlled expression of the reporter protein, (2) the open reading frame (ORF) of the *URA3* gene as a reporter, and (3) the *LEU2* gene as a metabolic marker for plasmid selection and maintenance (*see* **Note 2**). Nonessential elements that were also added to the backbone are (1) the human influenza hemagglutinin (HA) tag for the purpose of immunoblotting and (2) green fluorescent protein (GFP), to determine the cellular localization of the reporter protein (*see* **Note 3**).

To ensure protein expression in their original ORF and to increase library complexity, three backbone plasmids were made, each in a different reading frame with respect to the GFP reporter. Each ORF was followed by three sequential STOP codons at different reading frames, to ensure translation termination of the fusion proteins immediately after the cloning site. An illustration of the backbone plasmid *CUP1p*-*URA3*$_{ORF}$-HA-GFP is shown in Fig. 2a. The relevant sequences that surround the cDNA cloning site are shown in Fig. 2b.

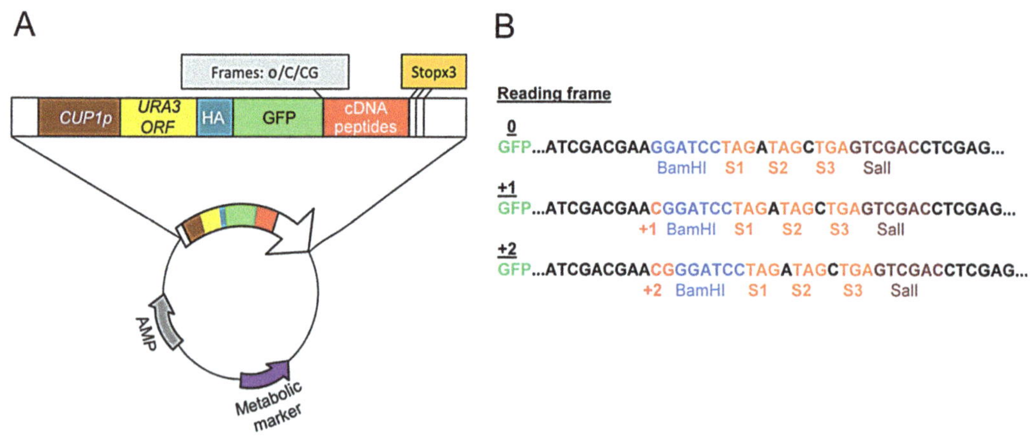

Fig. 2 Illustration of the plasmids used for the screen. (**a**) The backbone plasmid contains two selection markers: beta-lactamase that confers ampicillin resistance for bacterial selection (gray arrow) and the *LEU2* gene as a metabolic marker for plasmid selection and maintenance in yeast (purple arrow). The second part (colorful arrow) shows the reporter CUP1p-URA3orf-HA-GFP, three frames; cDNA peptides, three tandem stop codons. (**b**) Sequence representation of the three different frame shifts – GFP (green) is followed by a homogeneous plasmid region (black), different frame (red), a BamHI site for library insertion (blue), three tandem stop codons for each frame (assigned S1–3 represented in orange), and a SalI restriction site on the plasmid (brown)

1. Extract RNA from the appropriate yeast strain (*see* **Note 4**). This step can be obtained by any commonly used RNA extraction kit (*see* **Note 5**).
2. Convert mRNA to cDNA by a standard library construction kit according to the manual (*see* **Note 6**).
3. Take the entire cDNA volume (about 40 μL), split the volume equally into two tubes, and fragment the cDNA by incubation with a 4mer restriction enzyme, such as Sau3AI, for either 5 min (to receive partial digestion) or 30 min (for full digestion of the library). Two time points were chosen for digestion in order to increase library complexity. Mix the digested cDNA from the two groups.
4. Digest the three backbone plasmids (each in a different frame) with BamHI, which is complementary to the Sau3AI restriction site (*see* **Note 7**).
5. Ligate 75–100 ng of backbone plasmids with the fragmented cDNA library at a 1:3 molar ratio. Transform each ligation reaction separately to electrocompetent bacterial cells (preferentially from commercial source for maximal yield), to yield ~10^8 clones for each library.
6. Apply 2–5 mL of LB-AMP solution (according to plate size) on each plate, and gently scrape the bacteria colonies using a cell scraper. Collect the media from all plates into one 50 mL tube

for each library, and wash plates with an additional 1 mL of LB-AMP solution.

7. Spin down the cells at medium speed (4500–6000 × *g*) for 15 min. Remove the supernatant, and resuspend the cells in 5 mL of LB-AMP solution containing 25% glycerol. Aliquot 100 μL bacteria cells into 50 freezing tubes, and store samples at −80 °C for subsequent extraction.
8. For library extraction, take one tube (100 μL) from each library of the different reading frames, and grow in 100 mL LB-AMP solution at 37 °C overnight. Continue with a midi-prep plasmid extraction according to standard protocols.

3.2 Expressing URA3-P Yeast Libraries

1. Use a high efficiency protocol [15] to transform the appropriate *URA3*-P libraries into yeast cells auxotrophic for uracil and leucine, such as TRy481 (*see* Subheading 2.1), to yield ~10^7 colonies for each library (Fig. 3a). This step can be achieved by transforming 1 μg of plasmid from each library into yeast cells grown at logarithmic phase at total amount of 5 OD_{600} and plating the transformed cells on either 10 (90 × 15 mm standard) plates or on 3 (150 × 20 mm) plates.
2. Apply 2–5 mL of suitable selective SD medium for plasmid maintenance (according to plate size) on each plate, and gently scrape the yeast colonies with a cell scraper. Collect the media from all plates of the same library into one 50 mL tube, and wash with 1 mL of media for each plate.
3. Spin down the cells at 3000 × *g* for 5 min. Remove the supernatant, and resuspend the cells in 5 mL of suitable selective SD medium containing 25% glycerol. Aliquot 100 μL yeast cells into 50 freezing tubes and store samples at −80 °C.

3.3 URA3-P Library Competition Assay for Degron Selection

1. Grow the yeast cells from each frame library (three libraries total) overnight at 30 °C in a suitable selective SD medium for plasmid maintenance until a stationary phase is reached.
2. Combine 5 OD_{600} of cells from each of the overnight cultures in 15 mL tube.
3. Spin down the cells at 3000 × *g* for 5 min. Remove the supernatant, and resuspend the cells in 60 mL of the appropriate selective medium to obtain a concentration of 0.25 OD_{600}. Incubate for 2–3 h until logarithmic phase is reached (Fig. 3b).
4. Add 100 μM 5FOA and take immediately cells at an amount equivalent to 10 OD_{600}. This is the reference library at time 0 h that shows the distribution of peptides in the library without selection.

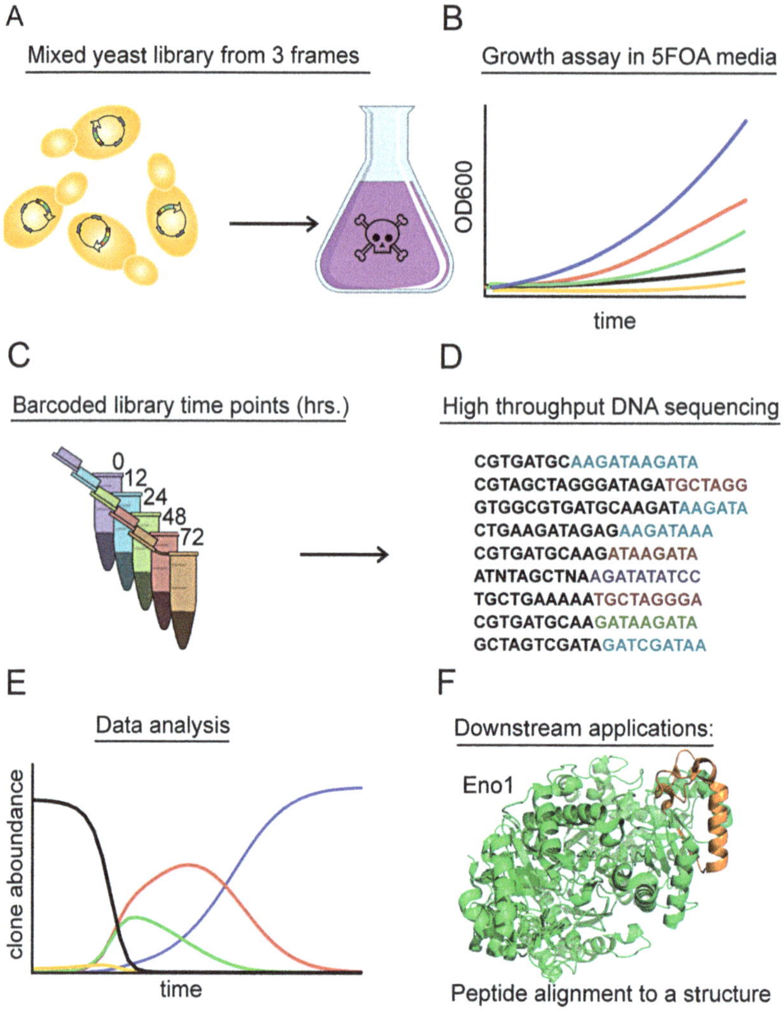

Fig. 3 Method overview. (**a**) Yeast cells are transformed with three different frame libraries. (**b**) Libraries are combined to one flask for the duration of a GiLS competition assay under plasmid-selective conditions and in the presence of 100 μM 5FOA. (**c**) Samples are collected at the indicated time points, and plasmid DNA is extracted and amplified using specific primers. (**d**) Barcoded libraries are combined and subjected to NGS. (**e**) An illustration of the expected competition and how it shapes the abundance profiles of the different clones. These profiles can be used to estimate the growth rate of each individual clone. (**f**) Optional data analysis – native degrons are assigned to known PDB structures. Shown is the structural assignment of a degron (orange) emerged from the yeast enzyme Eno1 (green) to the deposited PDB structure 1EBG

5. Dilute manually every 4–8 h (depending on the yeast strain doubling time) to maintain exponential growth (*see* **Note 8**).
6. Extract samples at desired time points (usually 12–24 h) at a quantity equivalent to 10 OD_{600}. Spin down the cells at 3000 × *g* for 5 min. Remove the supernatant, and store the pellet samples at −80 °C.

3.4 Preparation of URA3-P Library for Deep Sequencing Analysis

1. Plasmid extraction: Thaw the frozen pellets at room temperature and purify plasmid DNA from the pelleted yeast cells. This is done using any plasmid miniprep DNA extraction kit with a few modifications when using yeast as an extraction source: (1) after adding NaOH-SDS lysis buffer and mixing the solution, add 250 μL glass beads and break the cells by conducting three shaking cycles, each including 2 min of vortexing, followed by incubation for 3 min on ice; and (2) add prechilled neutralization buffer and incubate on ice for 5 min. Follow the user manual to obtain plasmid DNA concentration of 30–50 ng/μL.
2. DNA library preparation for deep sequencing:

   Amplification: Sequencing library preparation is consisted of two amplification steps. The first step (18 cycles) is performed using primers flanking the peptide (plasmid sequences), with overhangs complementary to Illumina adapter (Fig. 3c).

   A primer mix is used to increase library complexity for downstream NGS step.

   Each primer contains an Illumina read primers (underlined, RD1 for forward primers and RD2 for reverse primers), a heterogeneity added region (bold) and a plasmid region (italics) (*see* **Note 9**) (Table 2).

   For PCR, use a 10 μM total primer stocks composed of forward (1–4) or reverse (5–8) primers mixed at equal amounts using the parameters listed in Table 3.

   PCR products from the first PCR were isolated by 2× SPRI (solid-phase reversible immobilization) cleanup according to manufacturer instructions (AMPure beads).

   The second step is performed with the standard N-series Illumina barcoded adapters directly from the SPRI isolated DNA in a 50 μL reaction using the parameters listed in Table 4. Pool all barcoded libraries in equivalent amounts (*see* **Note 10**).

Table 3
Parameters of PCR 1st cycle

Cycles #	1	18			1	1
Temp	95 °C	95 °C	48 °C	68 °C	68 °C	10 °C
Time	3 min	30 s	30 s	1 min	5 min	Hold

Table 4
Parameters of PCR 2nd cycle

Cycles #	1	12			1	1
Temp	95 °C	98 °C	56 °C	72 °C	72 °C	10 °C
Time	3 min	20 s	30 s	1 min	5 min	Hold

3.5 Library Size Selection

Analyze the libraries concentration and size distribution using a DNA TapeStation or similar equipment. Size select fragments between 200 and 800 bp using AMPure XP beads for NGS (*see* **Note 11**). Submit for sequencing on an Illumina machine (Fig. 3d).

3.6 Preprocessing of NGS Data

The following steps can be performed in bash/csh or any other scripting language (perl/R/python/MATLAB) assuming the libraries are in FASTQ format and were already split by their barcodes (e.g., using bcl2fastq).

1. Identify and store the variable sequence added in library construction as a unique molecular identifier (UMI). This step is done to avoid double counting of the same molecule (*see* **Note 12**).
2. Strip each read pair of the constant flanking sequences (*see* **Note 13**).
3. Identify and store the frame by a search for the first occurrence of the restriction site – GATC (0/1/2).
4. Align every read pair that successfully passed all of the above steps to the yeast genome to assign genomic coordinates (*see* **Notes 14** and **15**).
5. Alignment correction: Due to sequencing (and other) errors, fragments may align to the genome with slight offsets. In most cases, reads should fall within a 4 bp window around bonafide restriction sites (>90%). It is advisable to either ignore misalignments (i.e., discard every read not mapped to a restriction site) or assign reads to the nearest restriction site. In cases where sequences did not appear within a 4 bp distance from a restriction site, they can be analyzed with their mapped genomic position without alteration (*see* **Note 16**).
6. At this point, an entry per read pair should have been obtained that includes the chromosomal location, the start and end coordinates, a UMI, and a relative frame. Next, collapse all reads with the identical parameters mentioned above, to be considered as only one molecule. Count the number of molecules that share the same start, end, strand orientation, and relative frame in each sample. Now each entry should have the following fields:

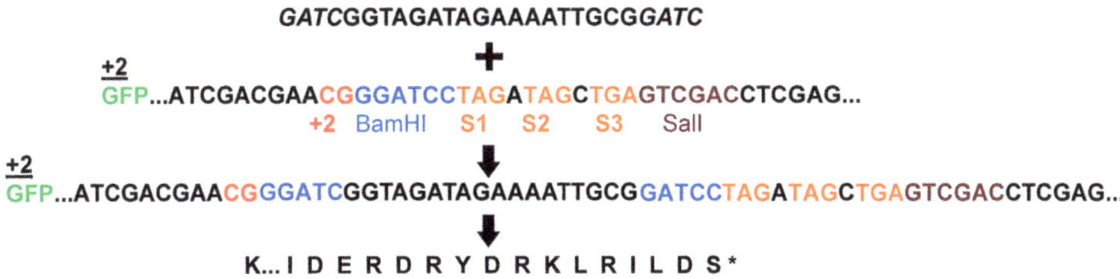

Fig. 4 Example of a sequence outcome. A given cDNA sequence, flanked by Sau3AI restriction sites (*in italics*), that was inserted to the backbone plasmid at a +2 position (*see* Fig. 2b), is then translated according to the relative frame, shown in the figure

<chr>, <start>, <end>, <str>, <fr>, #@sample1, #@sample2, #@sample3, . . .

where #@sampleX is the number of reads from sample X that mapped to chromosome <chr> between <start> and <end> on strand <str> in relative frame <fr>.

7. Frame assignment: Use the genomic coordinates, the reference genome, and the relative frame to construct the original DNA coding sequence found on the expression vector. For example, *see* Fig. 4.
8. Translate: Once the relative frame is assigned, sequences can be translated to a peptide until reaching a stop codon. For the above example, the correct translation is RDRYDRKLRILDS.
9. Filter: A final filtration of fragments is possible on different criteria. We remove peptides that are shorter than four amino acids, and remove identical peptides (that originate from duplicated sequences in the genome). In this filtration step, we keep the peptide entry with the most reads, and the other entries are discarded.

3.7 Data Analysis

3.7.1 Growth Rate Estimation

The main premise of the method described here is that under the experimental conditions, the reporter levels are linked to the degradation potency of the inserted peptide and correlated to the growth rate of the clones harboring it. Hence, the growth rate of each peptide within the population needs to be estimated during the competition assay (Fig. 3e).

1. Growth parameter: Assuming exponential growth of all yeast with different rates, a simple model for the data is:

$$X_{ij} \sim \text{Poisson}\left(C_j S_i 2^{r_i t_j}\right)$$

where X_{ij} is the measured read count for fragment i at time sample j, S_i is the initial amount of fragment i, r_i is the growth rate of yeast expressing fragment i, t_j is the elapsed time in sample j, and C_j is a scaling factor for sample j. Importantly, the scaling factor also consists of the sum of overall fragments at time t_j and another unknown scaling factor introduced in library preparation and sequencing.

To find the maximum likelihood parameters for the data, we iterate between optimizing the values of the C_j parameters and the values of fragment-specific parameters (S_i, r_i) until parameter convergence:

(a) Given the fragment parameters, the maximum likelihood estimator for C_j is the ratio between total observed and total predicted fragments:

$$C_j^* = \frac{\sum_i S_i 2^{r_i t_j}}{\sum_i X_{ij}}$$

(b) Given the scaling factors, each fragment is independent of the others, and the maximum likelihood parameters for each fragment can be obtained separately by solving a generalized linear model with Poisson noise and a log link function:

$$(r_i, S_i)^* = \underset{r,S}{\operatorname{argmin}} \sum\nolimits_j \left[X_{ij}(S + r_i t_j) - e^{S + r_i t_j}\right]$$

This is a convex optimization problem and can hence be solved efficiently and accurately by any statistical analysis program (MATLAB/python/R); we used glmfit in MATLAB.

2. Growth parameters' confidence estimation: To provide a confidence interval for the estimation of the parameters, we used two sampling techniques. The nonparametric sampling approach is described here, and an alternative approach is in the notes.

We resample the data with replacement, according to a multinomial distribution at each experimental sample. That is, we sample:

$$\overline{X_{:j}} \sim \mathrm{MN}\left(n, \frac{X_{:j} + \varepsilon}{n}\right)$$

where the ":" is the MATLAB notation for a column vector, $n = \Sigma_i(X_{ij} + \varepsilon)$, and is a small number (we've used 0.05) that avoids probability 0 for sampling any specific fragment in any specific experimental sample.

We repeat this procedure 1000 times to obtain 1000 datasets, and the parameter estimation procedure described above is performed on each of these datasets independently. Thus, the parameter estimation variance is approximated from the 1000 different estimations obtained per parameter per fragment (*see* **Note 17**).

3.7.2 Advanced Feature Extraction

Given the peptide sequences, one can extract many different features. We discuss below simple methods to extract nontrivial protein features.

1. Structure prediction: In cases where the peptides map to proteins and are in frame, it is possible to use available structure databases to obtain the secondary structure of the peptide in the solved protein structure. For yeast proteins, the collection of pre-aligned homologous structures at https://www.yeastgenome.org/cgi-bin/protein/get3d can be used. For other peptides, predicted secondary structures can be obtained from various tools. We used JPred3 [16] that has a web interface with an option of submitting queries of 800 peptides at a time (*see* Fig. 3f for an example).
2. Aggregation propensity prediction: A potentially important feature of degrons is their aggregation propensity. We used an established tool, AGGRESCAN [17], which extracts various measures for aggregation, based mostly on residue hydrophobicity. This algorithm is also available through a web interface that allows simultaneous multi-sequence query in FASTA format.
3. Motif search: Having a large cohort of peptide sequences enables the search for motif enrichment. A motif can be enriched in a set of peptides compared to a background set, e.g., a set of degrons related to a specific ubiquitin-protein ligase vs. all degrons, or in the top of an ordered list, e.g., looking for motifs within the list of degrons sorted by their potency. These and other conditions can be analyzed by using DRIMust [18], which is a web tool designed specifically for this purpose.

4 Notes

1. We switched the mating type from “a” to “alpha,” in order to enable mating with yeast strains from commercially available libraries [10], most of which are “a” type. Alternatively, the yeast strain BY4742 (*MATα his3Δ1 leu2Δ0 lys2Δ0 ura3Δ0*) can be used.
2. The backbone plasmid may vary as needed. Using other common promoters such as that of *MET25*, *GAL1*, or *GPD1* will lead to variations in the expression of Ura3. In principle, the stronger the promoter, the higher Ura3 levels; therefore the selection for degrons is more stringent. For this reason, using *CUP1*p gives a good range of expression: low expression from uninduced promotor to strong expression when induced. Different metabolic markers can also be employed, according to the auxotrophic markers on the yeast strain to be used.

3. Localization signals, such as *NLS* or *KDEL* for targeting the *URA3-P* library to the nucleus or the endoplasmic reticulum, respectively, can be added upstream to the *URA3* ORF, to identify compartment-specific degrons and compare degron potency in different cellular compartments. A termination sequence, such as $CYC1_t$, may also be considered. The terminator was found to be beneficial in stabilizing transcripts, and it mildly increases protein levels, which may be important for the identification of strong degrons.
4. Libraries can be obtained from different yeast strains, various stress conditions (heat, DNA damage, etc.), and from different organisms.
5. It is crucial to test the RNA integrity by a bioanalyzer or TapeStation, designed to test total RNA degradation, before continuing to the cDNA library construction. Low RNA integrity will cause a reduced library yield.
6. If both C- and N-terminal fragments are required, custom-made primers can be used to introduce Sau3AI restriction sites during first-strand synthesis. An example for modified primers sequence:

   cDNA Lib Primer F

   AAGCAGTGGTATCAACGCAGAGTGATCGAGGCG-GCCTTTTTTTTTTTTTTTTTTTTCC
   cDNA Lib Primer R
   AAGCAGTGGTATCAACGCAGAGTGATCATTACG-GCCGGGGG-P.

7. Make sure to get a full digest of the plasmid by two sequential additions of the restriction enzyme followed by a step of dephosphorylation to prevent religation of the linearized plasmid DNA.
8. Alternatively a chemostat can be used for the duration of the experiment.
9. When sequencing from a plasmid, adding a heterogeneity region is crucial for creating nucleotide diversity, which is required for effective cluster calling on Illumina sequencing systems and is important for the generation of high-quality data. In particular, nucleotide diversity is a key feature for the first four to seven cycles of Read 1 for all Illumina sequencing systems. The sequencing software uses images from these early cycles to identify the location of each cluster. Nucleotide diversity is also important for the first 25 cycles for evaluating the percentage of clusters passing filter (%PF). These factors are used in base calling and quality score calculations.
10. Barcoded adaptors are used to tag different experiments that may include different time points, strains, organisms, etc.

Henceforth, barcoded libraries can be mixed and handled in downstream applications as a single library.

11. Size selection is critical for obtaining a high yield sequencing data – either too long or too short sequences disrupt the sequencing procedure.
12. This can be done by searching for the prefix up to the constant priming sequence *CTCAG*.
13. This can be done using the following regular expressions: [NACGT]+GACGAA for R1 and [NACGT]+ATCTAG for R2
14. We use Bowtie2 [19], but any standard short read aligner should work just as well.
15. Optional:

    At this point you can assess restriction efficiency by counting the number of inserts that when mapped to the genome extend over a restriction site. If restriction was efficient overall (>99%), in the cases where only R1 was mapped to the genome, the sequence can be extended to the proximal genomic restriction site with reasonable certainty that this is the correct extension, especially in cases where the next stop codon occurs before the next restriction site.
16. In our experience, alignment distance from restriction sites is exponentially decaying, and it is reasonable to assume that deviations arise from sequencing (PCR, and other) errors. This can be corrected by reassigning the alignment position to the restriction site.
17. Alternative parametric sampling:

$$X_{ij} \sim \text{Poisson}\left(C_j S_i 2^{r_i t_j}\right)$$

where the parameters used are the ones estimated from the data or possibly slightly noised parameters.

References

1. Harper JW, Bennett EJ (2016) Proteome complexity and the forces that drive proteome imbalance. Nature 537(7620):328–338
2. Balchin D, Hayer-Hartl M, Hartl FU (2016) In vivo aspects of protein folding and quality control. Science 353(6294):aac4354
3. Shiber A, Ravid T (2014) Chaperoning proteins for destruction: diverse roles of Hsp70 chaperones and their co-chaperones in targeting misfolded proteins to the proteasome. Biomol Ther 4(3):704–724
4. Ji CH, Kwon YT (2017) Crosstalk and interplay between the ubiquitin-proteasome system and autophagy. Mol Cells 40(7):441–449
5. Kevei E, Pokrzywa W, Hoppe T (2017) Repair or destruction-an intimate liaison between ubiquitin ligases and molecular chaperones in proteostasis. FEBS Lett 591:2616
6. Hartl FU, Andreas B, Hayer-Hartl M (2011) Molecular chaperones in protein folding and proteostasis. Nature 475(7356):324
7. Shabek N, Zheng N (2014) Plant ubiquitin ligases as signaling hubs. Nat Struct Mol Biol 21(4):293–296
8. Collins GA, Goldberg AL (2017) The logic of the 26S proteasome. Cell 169(5):792–806
9. Ravid T, Hochstrasser M (2008) Diversity of degradation signals in the ubiquitin-

proteasome system. Nat Rev Mol Cell Biol 9 (9):679–689

10. Geffen Y, Appleboim A, Gardner RG, Friedman N, Sadeh R, Ravid T (2016) Mapping the landscape of a eukaryotic degronome. Mol Cell 63(6):1055–1065
11. Cohen I, Geffen Y, Ravid G, Ravid T (2014) Reporter-based growth assay for systematic analysis of protein degradation. J Vis Exp 93: e52021
12. Boeke JD, LaCroute F, Fink GR (1984) A positive selection for mutants lacking orotidine-5′-phosphate decarboxylase activity in yeast: 5 fluoro-orotic acid resistance. Mol Gen Genet 197:345–346
13. Gilon T, Chomsky O, Kulka RG (1998) Degradation signals for ubiquitin system proteolysis in Saccharomyces cerevisiae. EMBO J 17 (10):2759–2766
14. Boeke JD, Trueheart J, Natsoulis G, Fink GR (1987) [10] 5-Fluoroorotic acid as a selective agent in yeast molecular genetics. Methods Enzymol 154:164–175
15. Gietz RD (2014) Yeast transformation by the LiAc/SS carrier DNA/PEG method. In: Xiao W (ed) Yeast protocols. Springer, New York, NY, pp 33–44
16. Drozdetskiy A, Cole C, Procter J, Barton GJ (2015) JPred4: a protein secondary structure prediction server. Nucleic Acids Res 43(W1): W389–W394
17. Conchillo-Sole O, de Groot NS, Aviles FX, Vendrell J, Daura X, Ventura S (2007) AGGRESCAN: a server for the prediction and evaluation of "hot spots" of aggregation in polypeptides. BMC Bioinformatics 8:65
18. Leibovich L, Paz I, Yakhini Z, Mandel-Gutfreund Y (2013) DRIMust: a web server for discovering rank imbalanced motifs using suffix trees. Nucleic Acids Res 41(Web Server issue):W174–W179
19. Langmead B, Salzberg SL (2012) Fast gapped-read alignment with Bowtie 2. Nat Methods 9 (4):357–359

Chapter 10

A Method to Monitor Protein Turnover by Flow Cytometry and to Screen for Factors that Control Degradation by Fluorescence-Activated Cell Sorting

Sophie A. Comyn and Thibault Mayor

Abstract

The protein quality control network consists of multiple proteins or protein complexes that monitor proteome integrity by mediating protein folding and the removal of proteins that cannot be folded. An integral component of this network is the ubiquitin-proteasome system, which controls the degradation of thousands of cellular proteins. A number of questions remain unanswered regarding the degradation of misfolded proteins. For example, how are substrates recognized and triaged? What are the identities of the components involved? And finally, what substrates are targeted by any given component of the quality control network? Finding answers to these questions is what inspires our work in protein quality control. Further characterization of protein quality control mechanisms requires methods that can reliably quantify turnover rates of model substrates. One such method is based on flow cytometry. Here, we present protocols detailing how to assess protein stability with flow cytometry and how fluorescence-activated cell sorting (FACS) can be used to screen for factors important for protein quality control and protein turnover.

Key words Green fluorescent protein (GFP), Flow cytometry, Fluorescence-activated cell sorting (FACS), High-throughput screen, Protein quality control, Protein misfolding, Ubiquitin-proteasome system

1 Introduction

Protein homeostasis encompasses multiple protein assemblies and pathways that influence the fate of proteins from synthesis to degradation for the purpose of maintaining and regulating proteome integrity, thereby promoting viability and functionality at both the cellular and organismal levels [1]. The ubiquitin-proteasome system (UPS) plays a central role in proteolysis [2]. A major fraction of proteasome substrates consist of misfolded proteins that challenge the cell's capacity to maintain homeostatic balance. If left unchecked, the accumulation of misfolded proteins may divert resources away from essential cellular processes or result in the

Thibault Mayor and Gary Kleiger (eds.), *The Ubiquitin Proteasome System: Methods and Protocols*, Methods in Molecular Biology, vol. 1844, https://doi.org/10.1007/978-1-4939-8706-1_10,

production of potentially toxic protein aggregates [3]. Consequently, cells have adopted numerous protein quality control pathways to prevent aberrant protein aggregation. Examples include promoting protein folding and targeting terminally misfolded proteins for degradation [4–11].

Classically, protein degradation has been assayed by pulse-chase metabolic labeling or by using protein synthesis inhibitors coupled with downstream biochemical analysis [12]. These steps are labor intensive, require cell lysis, and severely limit the potential throughput of screening activities. This bottleneck was significantly overcome by the advent of fluorescently tagged proteins that allow for protein stabilities to be monitored by flow cytometry. For example, GFP-tagged Hmg2g was used as a model substrate to study degradation in the endoplasmic reticulum [13]. More recently, a CYP†--GFP substrate was used to screen a collection of yeast deletion strains missing single genes from the ubiquitin pathway (the model cytoplasmic substrate CYP† was created by removing the signal sequence from the misfolded vacuolar protein CYP*) [7]. This screen identified the parallel requirement for Ubr1 and San1 E3 ubiquitin ligases in the degradation of select cytosolic misfolded substrates [7]. The great potential for using fluorescence-activated cell sorting (FACS) to screen and enrich for degradation mutants was appreciated years ago by Dr. Randolph Hampton and colleagues [14]. However, it would be years later before the first studies to use a FACS-based high-throughput screen to look for protein homeostasis factors was performed by Yen et al. who developed global protein stability (GPS) analysis [15, 16]. GPS analysis is a method for assessing protein turnover at the proteome level in mammalian cells and was used to successfully identify substrates of the SCF ubiquitin ligase.

Flow cytometry confers a number of advantages compared to biochemical methods. First, measurements can be performed in vivo, and, for most applications, no additional processing or cell lysis is required. Second, the data are collected relatively quickly, and measurements are precise and often have low variability. Finally, the method is quantitative and sensitive enough to measure partial effects, an important advantage since as noted above; multiple ubiquitin ligases may collaborate to fully degrade the pool of any given misfolded protein.

Model substrates fused to GFP have been developed by our research group in order to characterize degradative protein quality control pathways. Previous work from our lab identified a panel of temperature-sensitive alleles of essential genes encoding for cytosolic proteins in *Saccharomyces cerevisiae* that are degraded in a proteasome-dependent manner at the nonpermissive temperature [8, 11]. The protein quality control (PQC) pathways responsible for the degradation of a number of these alleles, which contain potentially destabilizing missense mutations, remained mostly unknown. Therefore, we were interested in identifying and characterizing the cytosolic PQC factors that induce proteasome-mediated degradation of these model substrates. To monitor turnover, we fused a GFP tag C-terminally to a subset of model substrates. We then used a flow cytometry assay to assess protein stability and performed a FACS screen to isolate factors important for cytosolic PQC, which we describe here in detail. The methods below can be modified to measure the stabilities of additional proteasome substrates in yeast and can be adapted to perform similar analyses in mammalian cells grown in tissue culture.

2 Materials

Dissolve all reagents in deionized ultrapure water and store at room temperature unless specified otherwise.

2.1 Solutions, Media, and Reagents for Flow Cytometry Assay

1. 1000× Cycloheximide: Dissolve 100 mg/mL in DMSO and filter sterilize. Store in small aliquots at −20 °C, and limit the number of freeze thaw cycles. For growth media, use at 100 μg/mL.
2. ≥99.7% pure DMSO.
3. Yeast strain(s) expressing a fluorescently tagged protein of interest (*see* Subheading 3.3 for an example).
4. 50% (w/w) Dextrose: Mix equal weights of dextrose and water until dissolved and then autoclave (*see* **Note 1**).
5. 100× Synthetic Dropout: Dissolve 0.2 g adenine, 0.2 g uracil, 0.2 g L-tryptophan, 0.2 g L-histidine monohydrochloride monohydrate, 0.2 g L-methionine, 1 g L-leucine, and 0.3 g L-lysine monohydrochloride in 80 mL water. Fill to a final volume of 100 mL before filter sterilizing. Leave out the corresponding component for the dropout mix of choice.
6. 10× YNB: Dissolve 6.7 g yeast nitrogen base without amino acids in water up to 100 mL. Filter sterilize and then store at 4 °C wrapped in foil to limit exposure to light.
7. Synthetic Dropout Media: For 1 mL culture mix 100 μL 10× YNB, 40 μL 50% dextrose, 10 μL 100× synthetic dropout media, and 850 μL water.

2.2 Equipment for Flow Cytometry Assay

1. Temperature regulated shaking incubator for 3 mL to 50 mL yeast cultures.
2. Standard spectrophotometer to measure the optical density at 600 nm (OD_{600}).
3. Small vortexer.

4. Glass culture tubes.
5. Plastic 2 mL tubes with screw caps.
6. Thermomixer for 1.5 mL tubes.
7. Shaking water bath that can be heated to 37 °C.
8. Polystyrene culture tubes.
9. FACSCalibur flow cytometer.
10. CellQuest software.
11. FlowJo software.

2.3 Additional Solutions, Media, and Reagents for FACS Screen

1. Yeast knockout collection for nonessential genes with kanMX marker (haploid).
2. 500× G418: Dissolve G418 powder in water to a final concentration of 100 mg/mL and filter sterilize. Distribute into 1 mL aliquots and store at −20 °C. For plates and media, use at 200 μg/mL.
3. YPD plates: Mix 4 g yeast extract, 8 g peptone, 8 g agar, and 384 mL water and autoclave. Cool to 65 °C in a water bath before adding 16 mL 50% sterilized dextrose and an autoclaved stir bar. Mix and then pour 25 mL each into 100 × 15 mm petri dishes.
4. YPD + G418 plates. As YPD plates but also add 800 μL 500× G418 with sterile dextrose and stir bar before mixing when media is at 65 °C. Pour 25 mL each into 100 × 15 mm petri dishes.
5. 1× YP: Dissolve 10 g yeast extract (1% w/v) and 20 g peptone (2% w/v) in 800 mL water, and then adjust the volume to 1 L. Distribute into 400 mL aliquots and autoclave.
6. YPD: For 1 mL culture mix 40 μL 50% sterilized dextrose and 960 μL YP media.
7. YPD + G418: For 1 mL culture, mix 40 μL 50% sterilized dextrose with 958 μL YP, and then add 2 μL of 500× G418 immediately prior to using.
8. 75% Glycerol: Mix 75 mL 100% glycerol and 25 mL water until dissolved and then autoclave to sterilize.
9. Synthetic Dropout plates: Mix 8 g agar with 420 mL water and autoclave. Cool to 65 °C in a water bath prior to adding 50 mL 10× YNB, 20 mL 50% dextrose, 5 mL 100× synthetic dropout mix (all sterilized), and an autoclaved stir bar. Mix and then pour 25 mL each into 100 × 15 mm petri dishes.
10. 10× PBS: Dissolve 40 g NaCl (137 mM), 1 g KCl (2.7 mM), 7.2 g Na_2HPO_4 (10 mM), and 1.2 g KH_2PO_4 (1.4 mM) in 400 mL water, and bring to pH 7.4. Adjust to a final volume of

500 mL and then autoclave. Prepare a working solution from a 10× stock by diluting 10 mL into 90 mL water.

11. 1× lysis buffer: 2% Triton X-100 (v/v), 1% SDS (w/v), 100 mM NaCl, 10 mM Tris–HCl (pH 8.0), 1 mM EDTA.
12. ≥99.8% pure chloroform.
13. 100% ethanol (prechilled at −20 °C).
14. 70% ethanol (prechilled on ice).
15. TE: 10 mM Tris–HCl (pH 8.0), 1 mM EDTA.
16. 50 μL Phusion PCR Reaction Mix: 38.5 μL of ultrapure water, 10 μL 5× high-fidelity buffer, 5 μL 2 mM dNTPs, 2 μL each of 5 mM forward and reverse primers, 2 μL genomic DNA template (100 ng/μL), and 0.5 μL Phusion high-fidelity DNA polymerase. Leave reagents on ice when preparing mix.
17. Primers: Forward 5′-GAT GTC CAC GAG CTC TCT-3′; reverse 5′-CGG TGT CGG TCT CGT AG-3′; and sequencing 5′-TCG GGC TTC CCA TAC AAT CG-3′.

2.4 Additional Equipment for FACS Screen

1. Singer ROTOR robot or 96-solid-pin replicator.
2. Tissue culture cell scrapers.
3. −80 °C freezer.
4. 15 mL plastic conical tubes.
5. Benchtop centrifuge with a capacity for 15–50 mL conical tubes.
6. Benchtop microfuge.
7. Sonicator.
8. Polypropylene culture tubes.
9. BD Influx cell sorter.
10. 400 μm acid-washed glass beads.
11. Cell lysis bead beater or vortexer.
12. 1.5 mL tubes.
13. PCR machine.
14. PCR purification kit.
15. Sanger sequencing service.

3 Methods

3.1 Flow Cytometry Assay to Assess the Degradation of Proteasome Substrates in Yeast

The assay below can be used to measure the half-lives of proteasome substrates, as well as to assess how the absence or mutation of a given element of the UPS affects the degradation of a fluorescently labeled substrate in yeast. The protocol detailed below is performed in the presence of cycloheximide. In order to examine

Table 1
Plasmids used in this study

Name	Plasmid ID	Auxotrophic marker	Genotype	Lab source
pRS316	BPM42	Ura	CEN/ARS	P. Hieter
pRS313	BPM45	His	CEN/ARS	P. Hieter
P_{YDJ1}-YDJ1	BPM390	Ura	CEN/ARS	E. Craig
P_{GPD}-Guk1-7-GFP	BPM458	His	CEN/ARS	T. Mayor
P_{SSA1}-SSA1	BPM559	Ura	CEN/ARS	T. Mayor
P_{RSP5}-RSP5	BPM573	Ura	CEN/ARS	T. Mayor
P_{RSP5}-RSP5(C777A)	BPM575	Ura	CEN/ARS	T. Mayor

steady-state protein levels, we strongly recommend performing the same experimental procedure in the absence of cycloheximide (*see* **Note 2**).

1. Grow the desired yeast strains that express the fluorescent fusion proteins of interest overnight at 25 °C in 3–5 mL of the appropriate synthetic dropout media (*see* **Notes 3** and **4** and Table 1).
2. Dilute saturated overnight cultures to an $OD_{600} = 0.2$ in 3 mL of the appropriate synthetic dropout media. Incubate at 25 °C with shaking until cultures reach $OD_{600} = 0.8$–1.0 (*see* **Note 5**).
3. For each culture, distribute 1 mL into two tubes and add 1× cycloheximide to each. Vortex for a few seconds to mix (*see* **Notes 2**, **6**, and 7).
4. Incubate one tube at 25 °C and the other at 37 °C in thermomixers, and collect samples at desired time points (*see* **Note 8**).
5. Prior to running the samples on a flow cytometer, briefly vortex the culture for a few seconds, and transfer to an appropriate flow cytometry tube (*see* **Notes 9** and **10**).
6. Data are gathered using CellQuest software and a FACSCalibur flow cytometer with a 488 laser and 530/30 filter for GFP. 50,000 events are collected for each sample (*see* **Note 11**).

3.2 Data Processing

1. FCS files are exported from CellQuest acquisition software and opened with FlowJo for analysis (Fig. 1a).
2. Single, viable cells are selected for analysis by manual gating using FSC (forward scatter) versus SSC (side scatter) dot plots. This step eliminates small debris (located in the lower left portion of the graph) and clumps or doublets (higher FSC and SSC values than single yeast cells) from being included in

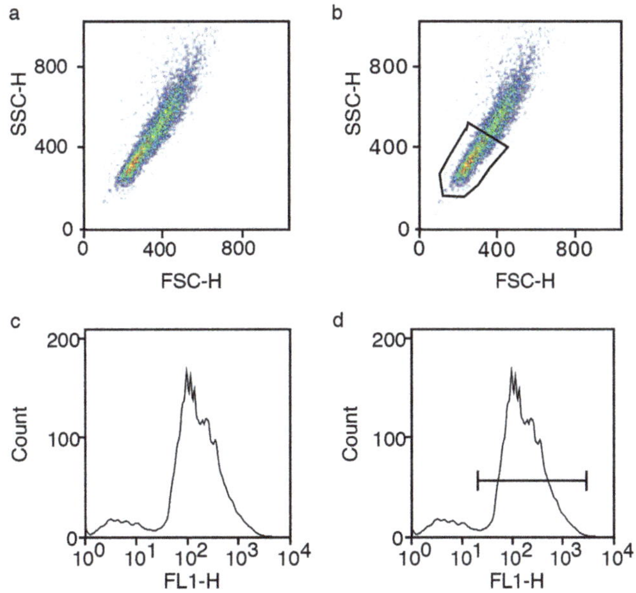

Fig. 1 Flow cytometry data processing. (**a**) Flow cytometry data is transferred from CellQuest acquisition software to FlowJo for analysis. (**b**) Single, viable cells are gated for subsequent analysis. (**c**) Histogram plot of FL1 channel. (**d**) Median fluorescence intensity is calculated by first selecting a region of interest with the ranged gate tool

subsequent analysis steps. Only cells within the area demarcated by the black line will be analyzed (Fig. 1b).

3. To calculate the median fluorescence intensity (MFI) of each sample, first create a histogram plot for the FL1 channel by selecting the histogram option from the left-hand menu. Click on the ranged gate tool and drag the mouse to set the width of the gate to equal the width of the histogram measured from the lowest point on each side. Then, from the statistics menu, select the median fluorescence intensity option and specify the FL1 channel (Fig. 1c, d).
4. The loss of fluorescence (LoF) intensity resulting from shifting cells to a higher temperature is expressed as a percentage. This value is estimated by calculating the difference between median fluorescence intensities of cultures incubated at 25 °C and 37 °C and normalized to that of the 25 °C sample ($LoF = (MFI_{25\ °C} - MFI_{37\ °C})/MFI_{25\ °C}$). For multiple strain comparisons, this LoF value is then normalized to that of a wild-type sample resulting in values ranging from 1.0 (no difference in loss of fluorescence compared to wild type) to 0 (*see* **Note 12**).
5. For publication and statistical purposes, we recommend that all experiments are performed independently at least three times

with averages and standard deviations calculated where appropriate. Statistical significance is tested using either a two-tailed unpaired Student's *t*-test for pairwise comparisons or a one-way ANOVA and post hoc Tukey HSD for multiple comparisons.

3.3 Application of the Flow Cytometry Assay to Mutant and Deletion Yeast Strains

To demonstrate the power of our flow cytometry assay to uncover key UPS participants in protein quality control, we selected the Guk1-7 allele fused to GFP as a model substrate representing a misfolded cytosolic protein. Guk1-7 is a thermally unstable mutant allele of the guanylate kinase Guk1, which contains four missense mutations generated by random PCR mutagenesis [11, 17, 18]. Previously, we demonstrated that Guk1-7-GFP degradation requires the ubiquitin ligase Ubr1 [8, 19]. Ubr1 activity alone, however, was not sufficient to account for the bulk of Guk1-7-GFP degradation. Therefore, we were interested in identifying other E3 ligases responsible for Guk1-7-GFP proteasome-mediated degradation. We noted that the ubiquitin ligase *RSP5* is responsible for an increase in the ubiquitination of misfolded proteins following acute heat stress to yeast cells [9]. To determine whether Rsp5 was also involved in the degradation of Guk1-7-GFP, we performed cycloheximide chase assays using the temperature-sensitive mutant allele *rsp5-1*. Guk1-7-GFP was expressed from a plasmid, and fluorescence levels were measured as indicted in Subheadings 3.1 and 3.2 (until **step 3**). Levels of the model substrate remained significantly higher in *rsp5-1* cells compared to wild-type cells after the temperature shift to 37 °C (between 1 and 4 h; Fig. 2a). To confirm that this enhanced stabilization was the direct consequence of a loss of *RSP5* function, we performed addback experiments, whereby wild-type *RSP5* or a catalytically inactive form (C777A) was expressed from a plasmid in *rsp5-1* cells (*see* Table 1 for all plasmids used in this study). During the 4 h incubation at 37 °C, degradation of Guk1-7-GFP was restored when wild-type Rsp5 was expressed in *rsp5-1* cells. In contrast, Guk1-7-GFP levels in *rsp5-1* cells containing an empty vector control or expressing the catalytically inactive Rsp5 (C777A) mutant had nearly equivalent Guk1-7-GFP levels that were significantly higher than those observed in the *RSP5* wild-type strain (Fig. 2b). For these experiments we analyzed three biological replicates. Protein levels were expressed based on the measured fluorescence values using the MFI as described in **step 3** of Subheading 3.2. For each replicate, values were normalized to the level in time 0 (percent of initial). P values were calculated with a one-way ANOVA and post hoc Tukey HSD to assess significance.

Chaperones are indispensable for maintaining proteome integrity as, in addition to their principal role in assisting nascent protein folding, they also prevent aberrant interactions with misfolded domains. In some cases, chaperones also facilitate targeting

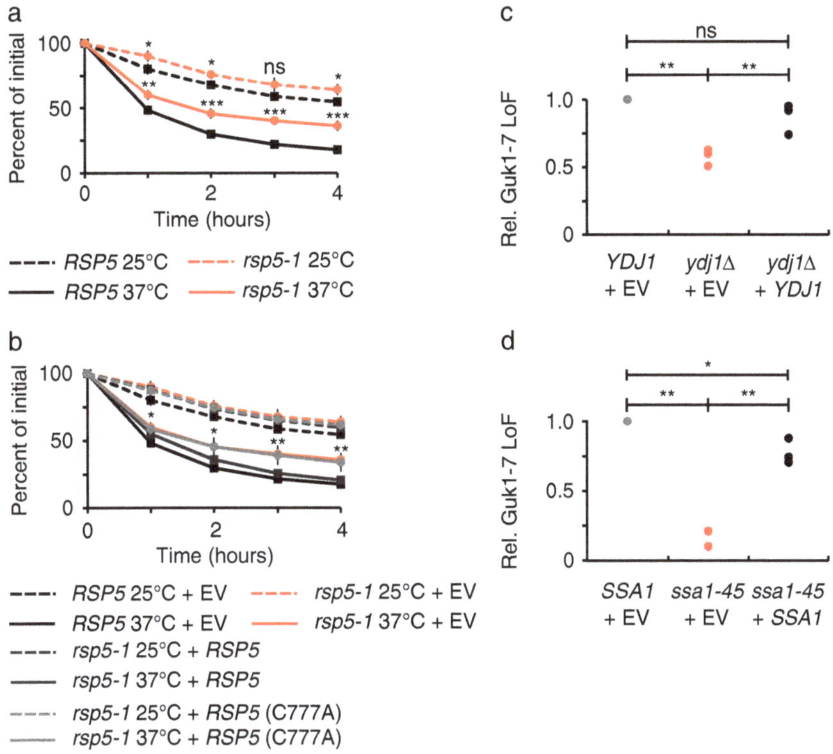

Fig. 2 Flow cytometry assays used to assess protein stability in cycloheximide chase and addback experiments. (**a**) *RSP5* and *rsp5-1* cells expressing Guk1-7-GFP were incubated with cycloheximide for a total of 4 h with samples collected at the indicated time points. *P* values were calculated using a two-tailed unpaired Student's *t*-test (*, **, ***, and ns denotes $P < 0.05$, 0.01, 0.005, and not significant, respectively). The results represent the average and standard deviation of three independent experiments. (**b**) *RSP5* cells expressing Guk1-7-GFP and a control empty vector (EV) as well as *rsp5-1* cells expressing Guk1-7-GFP and either a control empty vector, *RSP5*, or *RSP5* (C777A) were incubated with cycloheximide and grown for 4 h at 25 °C and 37 °C. Samples were collected at the indicated time points. The results represent the average and standard deviation of three independent experiments. (**c**) *YDJ1* and *ydj1Δ* cells coexpressing Guk1-7-GFP and an empty vector control or a vector expressing wild-type *YDJ1* were incubated at 37 °C with cycloheximide for 2 h. The results represent three independent experiments, for which relative LoF was normalized to 1.0 in *YDJ1* control cells. (**d**) *SSA1* and *ssa1-45* cells (in the *ssa2Δ*, *ssa3Δ*, *ssa4Δ* background) coexpressing Guk1-7-GFP and an empty vector control or a vector expressing wild-type *SSA1* were incubated with cycloheximide for 2 h. The results represent three independent experiments, for which relative LoF was normalized to 1.0 in *SSA1* control cells. *P* values (in **b**, **c**, and **d**) were calculated with a one-way ANOVA and post hoc Tukey HSD to assess significance (*, **, and ns denote $P < 0.05$, 0.01, and not significant, respectively)

proteins for degradation by recognizing exposed hydrophobic stretches that are common to misfolded proteins. Therefore, we were interested to see whether Hsp40 and Hsp70 chaperones had a role in Guk1-7-GFP degradation. Ydj1 is an Hsp40 chaperone known to associate with Rsp5 to ubiquitinate misfolded proteins following heat shock [9]. An absence of *YDJ1* resulted in Guk1-7-GFP levels 42% higher than wild type (Fig. 2c). In contrast, the LoF in *ydj1Δ* cells was restored in the presence of *YDJ1* expression from a plasmid. We next asked whether *SSA1*, a member of the Hsp70 family of molecular chaperones shown to be required for the degradation of other PQC model substrates, was also required for Guk1-7-GFP degradation [20]. For these experiments, we used *ssa1-45* cells that have a temperature-sensitive allele of *SSA1* in which the other *SSA1* paralogs (*SSA2, 3,* and *4*) were also deleted. In comparison to wild-type cells, loss of Guk1-7-GFP fluorescence was strikingly low in *ssa1-45* cells after a 2 h incubation at 37 °C (Fig. 2d). In contrast, expression of a wild-type copy of *SSA1* from a plasmid in *ssa1-45* cells recovered Guk1-7-GFP LoF levels close to those seen in wild-type cells. These results suggest that Ssa1 has a key role in promoting Guk1-7-GFP degradation. LoF in these experiments was calculated as described in **step 4** in Subheading 3.2. Three independent experiments were performed, and for each replicate, the LoF was normalized to the LoF in wild-type control cells. *P* values were calculated with a one-way ANOVA and post hoc Tukey HSD to assess significance. Together, these data suggest that Hsp70 and Hsp40 chaperones and the E3 Rsp5 play a part in promoting Guk1-7-GFP degradation.

3.4 FACS Screen for Yeast Mutants with Impaired Proteolysis of PQC Substrates

The method below describes a FACS screen using the haploid yeast deletion collection in which single nonessential genes are deleted and replaced using the kanMX module [21]. The deletion strains (~5000) from the library are first pooled together and then sorted to identify cells that display impaired turnover of the model substrate. Following cell sorting, single colonies are isolated on solid media plates for further validation.

1. Pin out the nonessential haploid yeast deletion collection onto YPD + G418 plates using a Singer robot or, manually, using a 96-pinning tool, and incubate at 30 °C for 48 h to allow colonies to grow.
2. Pipette 1 mL of YPD + G418 media onto each plate and swirl to distribute evenly over the surface.
3. Using a tissue culture cell scraper, scrape the colonies off of the agar surface, and add an appropriate volume of 75% glycerol to make a final concentration of 15%. Store at −80 °C.
4. Grow an aliquot of the −80 °C samples in YPD + G418 overnight, and transform with your GFP fusion containing plasmid of choice using the standard high-efficiency LiAc transformation method [22]. Distribute onto the appropriate selective synthetic dropout plate, and incubate at 25 °C for 48–72 h, or until colonies appear (*see* **Note 13**; Fig. 3i).
5. Transformants are collected by scraping colonies from the agar plates using 1 mL of selective media per plate. An aliquot from each is combined and then grown at 25 °C overnight in 5 mL

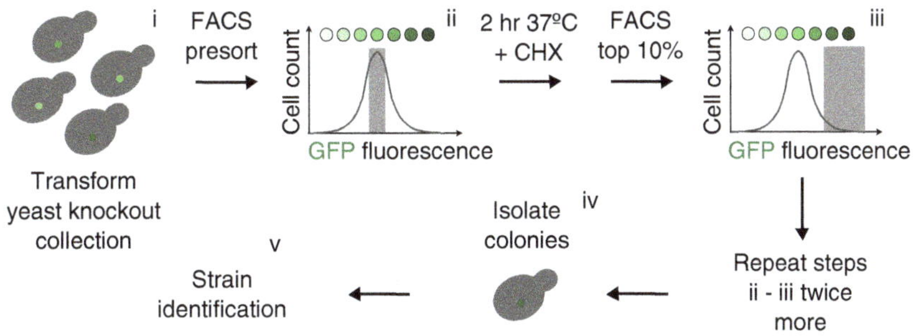

Fig. 3 Schematic representation of the FACS screen. (i) The pooled library is grown in the presence of the model substrate tagged to GFP. (ii) Cells are presorted by FACS in order to reduce the fluorescence range of the model substrate. The light gray area represents the region selected by FACS. (iii) Following a 2 h cycloheximide treatment, cells with higher fluorescence (light gray area) were sorted. The previous two steps are repeated twice. (iv) Single colonies are grown for hit validation (alternatively, cells can be maintained in pool for microarray or next-generation sequencing). (v) Validated hits in which compromised LoF was confirmed were sequenced

of selective dropout media. The overnight culture is set up such that the following morning an $OD_{600} = 0.8–1.0$ is reached.

6. The overnight culture is then transferred to a 15 mL conical tube and spun at 1800 rcf in a benchtop centrifuge for 3 min at 4 °C. The resulting pellet is resuspended in 2 mL 1× PBS.
7. Samples are sonicated (15 s total with 1 s on, 1 s off, intermittent pulses at low 20% amplitude on a Qsonica) to remove any cell clumps and placed into appropriate flow cytometry tubes (*see* **Note 14**).
8. Samples are then sorted by FACS using a BD Influx. We perform an initial presort to obtain a narrow fluorescence range. To do so, we collect approximately two million cells from the median fluorescence intensity region of the FL1 (GFP) channel (Fig. 3ii; *see* **Notes 15–17**).
9. The culture volume is raised to 1 mL with synthetic dropout media and incubated in the presence of 1× cycloheximide for 2 h at 37 °C.
10. Samples are sonicated in the synthetic dropout media as in **step 7** and then sorted again, this time selecting only for cells with GFP fluorescence in the top 10% range. Cells were sorted for 15 min such that 50,000–150,000 cells are collected (Fig. 3iii; *see* **Note 18**).
11. Sorted cells are incubated in 5 mL synthetic dropout media at 25 °C for approximately 36 h, and then the screen is repeated two more times (**steps 6–10**) for a total of three rounds of enrichment (*see* **Note 19**).

12. Following the final sorting, cells are distributed onto selective media plates and incubated at 25 °C for 48 h or until colonies appear (Fig. 3iv).
13. Isolated colonies can then be processed for downstream validation (Fig. 3v; *see* **Notes 20** and **21**).
14. Optional: Sanger sequencing can be performed for small-scale identification between 50 and 150 colonies. For each single colony to be sequenced, grow cells to saturation overnight in 5 mL YPD at 30 °C.
15. The following day, collect the cell pellet in a 15 mL conical tube by centrifuging at 1800 × *g* for 3 min at 4 °C.
16. Resuspend the cell pellet in 300 μL 1× lysis buffer, and transfer into a 2 mL screw cap tube with 300 μL glass beads for bead beating (*see* **Note 22**).
17. Collect clear lysates after a centrifugation step, and transfer the supernatant to a new 1.5 mL tube (*see* **Note 23**).
18. Add 300 μL chloroform to the lysate and vortex at 1400 rpm for 2 min at room temperature in a thermomixer. Centrifuge at 16,000 × *g* for 3 min at room temperature, and then transfer the aqueous layer into a new 1.5 mL tube.
19. Add 600 μL cold 100% ethanol and precipitate at −20 °C for a minimum of 5 min. Centrifuge at 16,000 × *g* for 5 min at room temperature, and wash the resulting pellet with 500 μL ice-cold 70% ethanol.
20. Centrifuge at 16,000 × *g* for 5 min at room temperature, and remove any remaining liquid before air drying for 5 min at room temperature.
21. Resuspend the pellet in 150 μL 1× TE. Genomic DNA can be stored at −20 °C (*see* **Note 24**).
22. A PCR reaction is required before the unique barcode can be identified by Sanger sequencing. First, dilute the genomic DNA isolated above to a 100 ng/μL aliquot in 1× TE buffer.
23. Run a 50 μL Phusion polymerase PCR reaction mix with forward (5′-GAT GTC CAC GAG CTC TCT-3′) and reverse (5′-CGG TGT CGG TCT CGT AG-3′) primers using the following PCR program: 94 °C 3:00, 30× (94 °C 0:30, 52 °C 0:30, 72 °C 2:00), 72 °C 8:00 (*see* **Notes 25** and **26**).
24. Purify the PCR reaction with a PCR purification kit according to the manufacturer's instructions, and elute with 20 μL of water (*see* **Note 27**).
25. Send ~50 ng of purified DNA from the PCR reaction for Sanger sequencing using the sequencing primer.
26. The barcode can be identified from the Sanger sequencing results by alignment with the sequence: 5′-GTC GAC CTG

CAG CGT ACG-3′. The reverse complement of the 20 base pairs immediately following the sequence comprises the barcode. Barcodes are then matched to systematic gene names by searching files available at http://www-sequence.stanford.edu/group/yeast_deletion_project/strain_a_mating_type.txt.

4 Notes

1. Dextrose can take a long time to get into solution using room temperature water. To speed up the process, we preheat water either in a microwave or in a water bath set at 65 °C.
2. Using cycloheximide to inhibit protein translation allows us to measure the rate of protein degradation. In some conditions, we found that the temperature induced degradation occurred at a slower rate in the presence of cycloheximide than without the treatment, presumably due to the higher folding capacity of the cell in the absence of protein synthesis. An alternative to cycloheximide would be to use an inducible promoter such as *GAL1*, which is rapidly inhibited in the presence of dextrose. However, some mRNA can remain stable, and in this case, translation of the assessed substrate would not be inhibited in a timely manner.
3. In our experiments, the GFP-tagged thermosensitive model substrate (e.g., Guk1-7) is ectopically expressed under a constitutive *GPD1* promoter (with a *PGK1* terminator sequence) and is supplied on a CEN/ARS plasmid. The protocols described herein can easily be modified for use with a genomically integrated fluorescent fusion protein or with a proteasome substrate other than a temperature-sensitive allele. Some substrates expressed at low levels and/or with a short half-life may not be adequately detected above the background signal in wild-type cells. It must be cautioned that GFP and other methods of tagging substrates can affect substrate stability and solubility. All substrates should be verified using an alternative tagging or detection mechanism as a control.
4. To measure the level of background autofluoresence, analyze strains of interest in the absence of a fluorescent fusion protein (e.g., using an empty plasmid as a control).
5. Growth rates will vary depending on the yeast strains selected. Generally, achieving an $OD_{600} = 0.8–1.0$ for the cultures may take between 4 and 6 h in synthetic dropout media at 25 °C.
6. In our particular case, we compare the loss of fluorescence due to the increased misfolding of the thermosensitive model substrate with the signal from cells grown at the permissive temperature. Another application of this protocol would be to

compare steady-state levels of different substrates or steady-state levels of one substrate in several yeast strains. If measuring steady-state levels, only one sample per condition would be required, and Subheading 3.1, **steps 3** and **4**, should be omitted. If the goal is to determine whether a substrate is degraded by the proteasome, the 37 °C incubation would be replaced with an incubation at 25 °C in the presence of 20 μM MG132 (dissolved in DMSO). For MG132 treatment to be effective, one should use either yeast strains containing a *pdr5Δ* background or wild-type cells should be grown in minimal media without ammonium sulfate in the presence of 0.1% proline and 0.003% SDS [23].

7. For single time point experiments, we use 2 mL plastic tubes with screw caps. For multiple time point cycloheximide chase experiments, we use 25–50 mL glass flasks.
8. Depending on the number of samples and volumes involved, we use either a thermomixer or a shaking water bath for the 37 °C incubations.
9. For flow cytometry on a FACSCalibur machine, we use 12 × 75 mm polystyrene tubes from VWR (60818-292).
10. We do not find it necessary to resuspend cultures in 1× PBS and instead run our flow cytometry samples directly from synthetic dropout media.
11. We use the following settings with a FACSCalibur machine to assess GFP fluorescence levels (voltage/gain): FSC (EOO/1.62), SSC (443/1.27), FL1 (575), FL2 (550), and FL3 (551).
12. The relative loss of fluorescence is a function of the difference between median fluorescence intensities of cultures incubated at 25 °C and 37 °C and normalized to that of the 25 °C sample. To perform multiple strain comparisons, the relative loss of fluorescence for each strain is normalized to that of the wild type, which is set to 1 (*see* Fig. 2c, d). For cycloheximide chase experiments, percent of initial values are calculated by normalizing the median GFP fluorescence intensity values for each time point to that of the initial time zero measurement (*see* Fig. 2a, b). One can then calculate half-life values ($T_{1/2}$) from the chase experimental results. For each sample, first take the natural log of the median fluorescence intensity values (MFI) for each time point. Second, create a plot of time versus the log-transformed MFI values and calculate the slope. The half-life is then calculated using the formula, $T_{1/2} = \ln(2)/k$, where k is the slope value calculated above.
13. As reported previously, we found that mutations in *WHI2* occur quite frequently in the yeast deletion collection (in addition to the strain deletion) and that these *WHI2*

mutations led to impaired turnover of misfolded cytosolic proteins. This *WHI2* phenotype was dependent on the presence of leucine in the media and was abrogated by the addition of rapamycin [19]. To avoid false-positive hits in screening for genes involved in protein quality control, we advise that the GFP reporter is cloned into a plasmid containing the *LEU2* marker (and cells are grown in SD-LEU media) or that the cells are incubated with 200 nM rapamycin for 2 h prior to the screen in Subheading 3.4, **step 4**.

14. We sort cells using a BD Influx instrument and use BD Falcon (14-959-11A) 12 × 75 mm polypropylene tubes.
15. The initial presort is used to limit the range of fluorescence intensities sampled and acts to reduce the compounding effects of cell-to-cell variability in GFP fusion expression or plasmid copy number.
16. Viability following FACS is significantly reduced. Therefore, we sort cells into tubes containing 200 μL synthetic dropout media. Ideally the final volume of collected cells should be no more than 500 μL. The duration of the presorting step will depend on the number of samples collected and can take anywhere from 30 min to an hour.
17. An aliquot of the presorted cells (one fifth to one third) should be set aside before the addition of cycloheximide for use as a reference sample. This sample acts as a baseline reference when next-generation sequencing is used to identify the deletion strains using the unique barcodes and will act to indicate which strains are under- or overrepresented prior to screening. Aliquoted reference cells are grown in 1 mL media until an OD_{600} of 0.6–1.0 has been reached and are then stored in 15% glycerol at −80 °C.
18. Sample collection for the second round of sorting is limited to 15 min to reduce the effect of a population shift toward lower fluorescence values as protein degradation continues.
19. Following each screening round, histograms of cells pre- and post-sorting should demonstrate an enrichment for cells with higher GFP fluorescence values.
20. Substrate stability in colonies isolated from the screen can be validated using the flow cytometry assay described in Subheading 3.1. We found that over two thirds of the isolated colonies displayed a clear impairment in degradation of the assessed model substrate.
21. For next-generation sequencing or microarray analysis, cells can be maintained and grown in 1 mL of liquid media until an OD_{600} of 0.6–1.0 has been reached. Cells can then be stored at −80 °C in 15% glycerol until ready to be processed (*see* also **Note 17**).

22. For yeast lysis, we use a Precellys 24 placed in a 4 °C chamber. Lysis is performed in 2 mL screw cap tubes using the following program: 30 s at 6800 rpm followed by a 90 s pause, for a total of three cycles. Alternatively, bead beading can be performed by vortexing the tube containing beads and cells for 10 min.
23. Lysate can be transferred into a new 1.5 mL tube using a 200 μL pipette tip in which the tip was cut, followed by a 16,000 × *g* centrifugation for 2 min at 4 °C. Alternatively, the 2 mL tube containing the beads can be pierced using a hot 27G needle (heated using a Bunsen burner) and placed on top of a new 1.5 mL tube that has had its cap removed. The pair of stacked tubes are then placed into a 15 mL conical tube, and the lysate is collected by centrifugation (1800 × *g* for 3 min at 4 °C). If needed the lysate can then be further cleared by centrifugation of the recovered 1.5 mL tube in a microfuge (16,000 × *g* for 2 min at 4 °C).
24. DNA concentration can be measured using a NanoDrop device using standard techniques. Approximately 200 μg of DNA can be recovered using this method.
25. When analyzing multiple samples, prepare a master mix without the DNA, and then transfer 48 μL of the master mix into each PCR tube before adding 2 μL of the DNA.
26. When the PCR is completed, 2 μL of the PCR reaction should be assessed on a 1% agarose gel to confirm the presence of a band approximately 1500 base pairs long.
27. About 800–1000 ng of DNA should be recovered.

Acknowledgments

We appreciate the insightful discussions and comments by all members, former and current, of the Mayor lab. We thank Justin Wong and Andy Johnson of the UBC Flow Cytometry Facility for their assistance with cell sorting, analysis, and training. Finally, we are grateful to the Hieter lab for access to their FACSCalibur flow cytometer. This research was funded by a Canadian Institutes of Health Research (CIHR) grant, and TM is a MSFHR new investigator.

References

1. Balch WE, Morimoto RI, Dillin A, Kelly JW (2008) Adapting proteostasis for disease intervention. Science 319(5865):916–919
2. Kleiger G, Mayor T (2014) Perilous journey: a tour of the ubiquitin-proteasome system. Trends Cell Biol 24(6):352–359
3. Geiler-Samerotte KA, Dion MF, Budnik BA, Wang SM, Hartl DL, Drummond DA (2011) Misfolded proteins impose a dosage-dependent fitness cost and trigger a cytosolic unfolded protein response in yeast. Proc Natl Acad Sci U S A 108(2):680–685

4. Finley D, Ulrich HD, Sommer T, Kaiser P (2012) The ubiquitin-proteasome system of Saccharomyces cerevisiae. Genetics 192 (2):319–360
5. Vembar SS, Brodsky JL (2008) One step at a time: endoplasmic reticulum-associated degradation. Nat Rev Mol Cell Biol 9(12):944–957
6. Gardner RG, Nelson ZW, Gottschling DE (2005) Degradation-mediated protein quality control in the nucleus. Cell 120(6):803–815
7. Heck JW, Cheusng SK, Hampton RY (2010) Cytoplasmic protein quality control degradation mediated by parallel actions of the E3 ubiquitin ligases Ubr1 and San1. Proc Natl Acad Sci U S A 107(3):1106–1111
8. Comyn SA, Young BP, Loewen CJ, Mayor T (2016) Prefoldin promotes proteasomal degradation of cytosolic proteins with missense mutations by maintaining substrate solubility. PLoS Genet 12(7):e1006184
9. Fang NN, Chan GT, Zhu M, Comyn SA, Persaud A, Deshaies RJ, Rotin D, Gsponer J, Mayor T (2014) Rsp5/Nedd4 is the main ubiquitin ligase that targets cytosolic misfolded proteins following heat stress. Nat Cell Biol 16(12):1227–1237
10. Fang NN, Ng AH, Measday V, Mayor T (2011) Hul5 HECT ubiquitin ligase plays a major role in the ubiquitylation and turnover of cytosolic misfolded proteins. Nat Cell Biol 13(11):1344–1352
11. Khosrow-Khavar F, Fang NN, Ng AH, Winget JM, Comyn SA, Mayor T (2012) The yeast ubr1 ubiquitin ligase participates in a prominent pathway that targets cytosolic thermosensitive mutants for degradation. G3 (Bethesda) 2(5):619–628
12. Yewdell JW, Lacsina JR, Rechsteiner MC, Nicchitta CV (2011) Out with the old, in with the new? Comparing methods for measuring protein degradation. Cell Biol Int 35(5):457–462
13. Gardner RG, Hampton RY (1999) A highly conserved signal controls degradation of 3-hydroxy-3-methylglutaryl-coenzyme a (HMG-CoA) reductase in eukaryotes. J Biol Chem 274(44):31671–31678
14. Cronin SR, Hampton RY (1999) Measuring protein degradation with green fluorescent protein. Methods Enzymol 302:58–73
15. Yen HC, Elledge SJ (2008) Identification of SCF ubiquitin ligase substrates by global protein stability profiling. Science 322 (5903):923–929
16. Yen HC, Xu Q, Chou DM, Zhao Z, Elledge SJ (2008) Global protein stability profiling in mammalian cells. Science 322(5903):918–923
17. Ben-Aroya S, Coombes C, Kwok T, O'Donnell KA, Boeke JD, Hieter P (2008) Toward a comprehensive temperature-sensitive mutant repository of the essential genes of Saccharomyces cerevisiae. Mol Cell 30(2):248–258
18. Ben-Aroya S, Pan X, Boeke JD, Hieter P (2010) Making temperature-sensitive mutants. Methods Enzymol 470:181–204
19. Comyn SA, Flibotte S, Mayor T (2017) Recurrent background mutations in WHI2 impair proteostasis and degradation of misfolded cytosolic proteins in Saccharomyces cerevisiae. Sci Rep 7(1):4183
20. Lee do H, Sherman MY, Goldberg AL (2016) The requirements of yeast Hsp70 of SSA family for the ubiquitin-dependent degradation of short-lived and abnormal proteins. Biochem Biophys Res Commun 475(1):100–106
21. Winzeler EA, Shoemaker DD, Astromoff A, Liang H, Anderson K, Andre B, Bangham R, Benito R, Boeke JD, Bussey H, Chu AM, Connelly C, Davis K, Dietrich F, Dow SW, El Bakkoury M, Foury F, Friend SH, Gentalen E, Giaever G, Hegemann JH, Jones T, Laub M, Liao H, Liebundguth N, Lockhart DJ, Lucau-Danila A, Lussier M, M'Rabet N, Menard P, Mittmann M, Pai C, Rebischung C, Revuelta JL, Riles L, Roberts CJ, Ross-MacDonald P, Scherens B, Snyder M, Sookhai-Mahadeo S, Storms RK, Veronneau S, Voet M, Volckaert G, Ward TR, Wysocki R, Yen GS, Yu K, Zimmermann K, Philippsen P, Johnston M, Davis RW (1999) Functional characterization of the S. cerevisiae genome by gene deletion and parallel analysis. Science 285(5429):901–906
22. Gietz RD, Schiestl RH (2007) Quick and easy yeast transformation using the LiAc/SS carrier DNA/PEG method. Nat Protoc 2(1):35–37
23. Liu C, Apodaca J, Davis LE, Rao H (2007) Proteasome inhibition in wild-type yeast Saccharomyces cerevisiae cells. BioTechniques 42 (2):158, 160, 162

Chapter 11

E. coli-Based Selection and Expression Systems for Discovery, Characterization, and Purification of Ubiquitylated Proteins

Olga Levin-Kravets, Tal Keren-Kaplan, Ilan Attali, Itai Sharon, Neta Tanner, Dar Shapira, Ritu Rathi, Avinash Persaud, Noa Shohat, Anna Shusterman, and Gali Prag

Abstract

Ubiquitylation is an eukaryotic signal that regulates most cellular pathways. However, four major hurdles pose challenges to study ubiquitylation: (1) high redundancy between ubiquitin (Ub) cascades, (2) ubiquitylation is tightly regulated in the cell, (3) the transient nature of the Ub signal, and (4) difficulties to purify functional ubiquitylation apparatus for in vitro assay. Here, we present systems that express functional Ub cascades in *E. coli*, which lacks deubiquitylases, Ub-dependent degradations, and control mechanisms for ubiquitylation. Therefore, expression of an ubiquitylation cascade results in the accumulation of stable ubiquitylated protein that can be genetically selected or purified, thus circumventing the above challenges. Co-expression of split antibiotic resistance protein fragments tethered to Ub and ubiquitylation targets along with ubiquitylation enzymes (E1, E2, and E3) gives rise to bacterial growth on selective media. We show that ubiquitylation rate is highly correlated with growth efficiency. Hence, genetic libraries and simple manipulations in the selection system facilitate the identification and characterization of components and interfaces along Ub cascades. The bacterial expression system also facilitates the detection of ubiquitylated proteins. Furthermore, the expression system allows affinity chromatography-based purification of milligram quantities of ubiquitylated proteins for downstream biochemical, biophysical, and structural studies.

Key words Bacterial genetics, Selection, Identification, Expression and purification of ubiquitylated proteins

1 Introduction

Ubiquitylation cascades form a huge connectivity map (Fig. 1). A large set of enzymes including few E1s, dozens of E2s, and hundreds of E3s *write* the code by attaching ubiquitin (Ub) to thousands of targeted proteins [1, 2]. The Ub code is then *read* by a few hundreds of Ub receptors that decode signals into cellular responses [3, 4]. Moreover, Ub signals are *edited* and efficiently erased by about hundred deubiquitylases (DUBs) [5] and

Thibault Mayor and Gary Kleiger (eds.), *The Ubiquitin Proteasome System: Methods and Protocols*, Methods in Molecular Biology, vol. 1844, https://doi.org/10.1007/978-1-4939-8706-1_11,

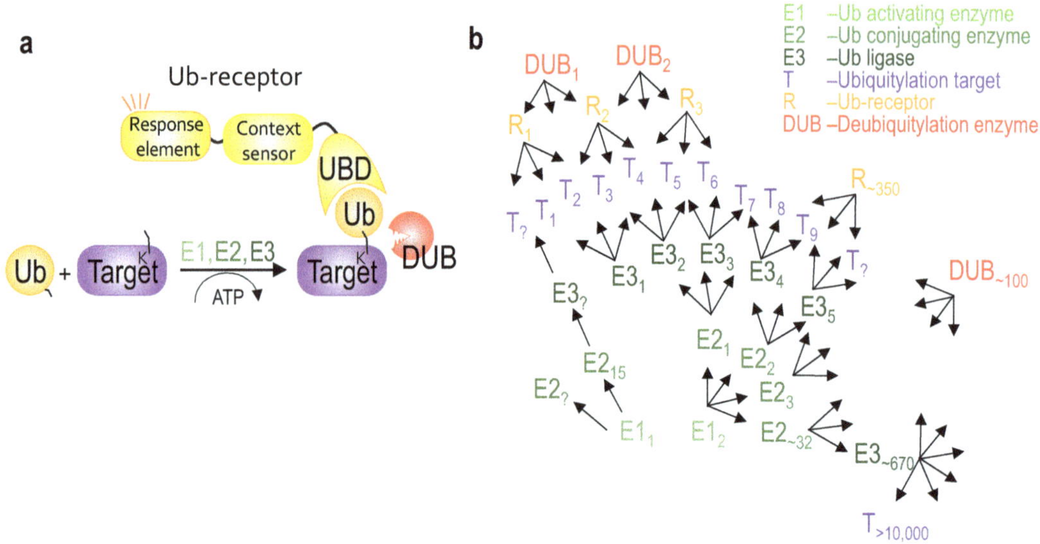

Fig. 1 Scheme of the ubiquitylation connectivity map. (**a**) Shows a simple scheme of ubiquitylation cascade including Ub and its target, the ubiquitylation enzymes (E1, E2, and E3), deubiquitylase (DUB), and Ub receptors. (**b**) Shows the size and the redundancy (multiplex) of the Ub cascades in human proteome

Ub-dependent degradation processes mediated by the proteasomes and lysosomes [6–8].

Despite all advancements in research, characterizing these cascades has long been hampered by the following major challenges in the field: (1) the multiplex structure of the network (i.e., redundancy between E2/E3s and their cognate ubiquitylation targets) poses difficulties to assign the connectivity between cognate components; (2) ubiquitylation is rapidly removed by deubiquitylases and often leads to proteasomal/lysosomal degradations, posing difficulties to detect and purify ubiquitylated proteins; (3) Ub cascades are tightly regulated in the cell, making it difficult to study them without proper manipulations (e.g., induction); and (4) many E3s and their targets are unstable and difficult to purify in their full-length forms, posing challenges to set up in vitro ubiquitylation experiments.

To circumvent these hurdles, we developed *E. coli*-based methods that facilitate the characterization of ubiquitylation and the purification of ubiquitylated proteins for downstream biochemical and biophysical studies. In addition, the system facilitates structural characterization of the Ub cascade as it provides a simple readout without the need of protein purification.

Co-expression of split antibiotic resistance protein fragments tethered to Ub and ubiquitylation target together with a functional ubiquitylation apparatus results in a covalent assembly of the resistance protein, giving rise to bacterial growth on selective media (Fig. 2) [9]. Novel components can be identified by expression of libraries of E2s, E3s, or potential targets.

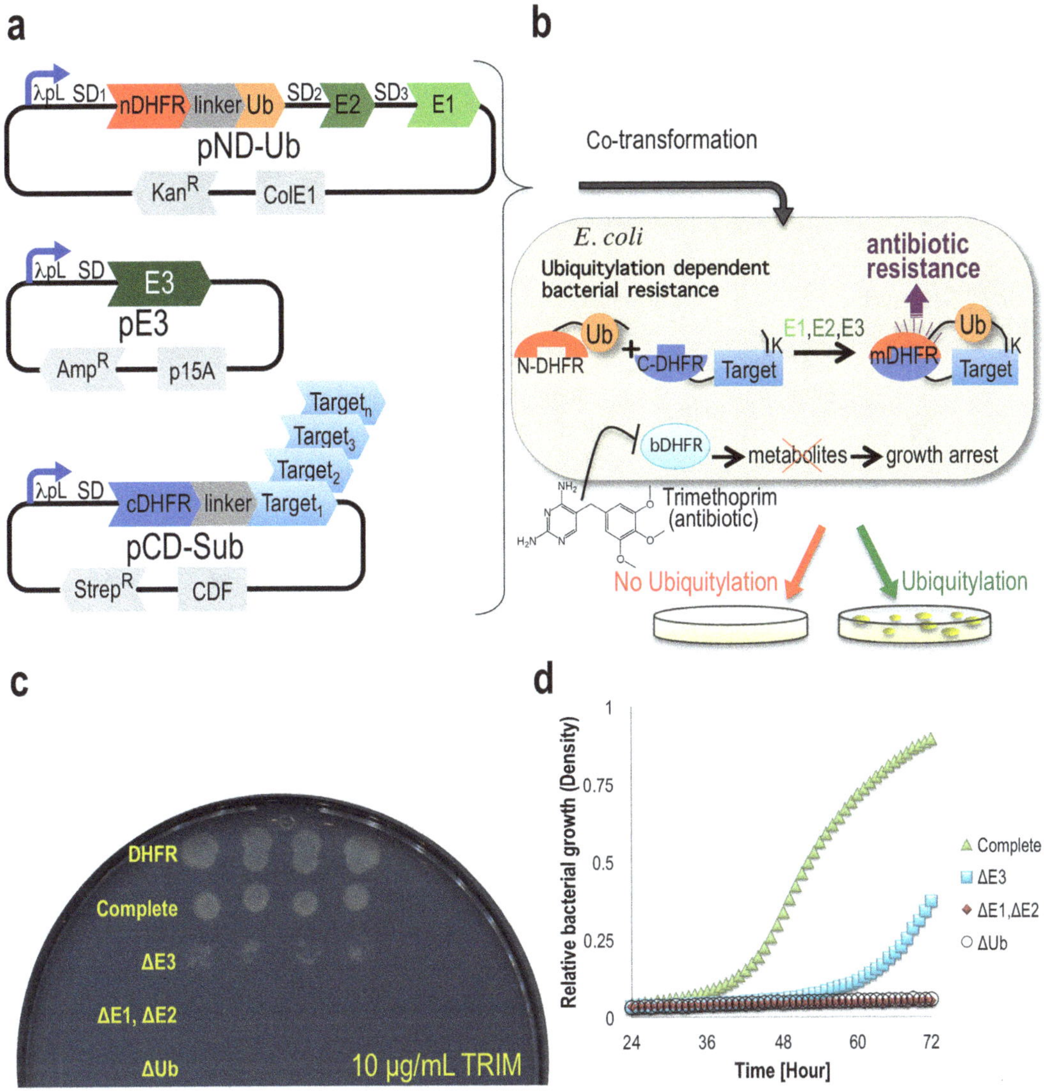

Fig. 2 A bacterial genetic selection system for ubiquitylation discovery. (**a**) Shows the structure of the plasmids that expresses elements of the ubiquitin system. The ubiquitylation apparatus is expressed from two or three compatible plasmids. Each plasmid harbors a constitutive λ-phage pL promoter, a different *ori* (origin of replication), and different antibiotic resistance marker. pND-Ub co-expresses the N-terminal fragment of DHFR fused to Ub, E1, and E2 from an operon-like cassette, whereas each ORF (open reading frame) is preceded by a Shine-Dalgarno (SD) sequence. pE3 expresses the E3 ligase. pCD-Sub expresses the C-terminal fragment of DHFR fused to a ubiquitylation target. (**b**) Shows the concept of the *E. coli*-based genetic selection system for ubiquitylation. Ubiquitylation-dependent assembly of DHFR and bacterial growth under restrictive conditions is shown (reproduced with permission from ref. 12). (**c**) Substrate discovery by the selection system. A screen of yeast library in the pCD-Sub against Rsp5 as an E3 ligase uncovered Sem1 as novel target. The last image of time-laps scanning experiment (72 h) is shown. Complete a system that contains E1, E2, E3, Ub, and Sem1 as ubiquitylation targets, DHFR, and full-length DHFR (positive control). (**d**) Shows quantitation of the bacterial growth assay ($N = 4$) in the presence of 10 mg/mL TRIM (**c** and **d** were reproduced with permission from ref. 9)

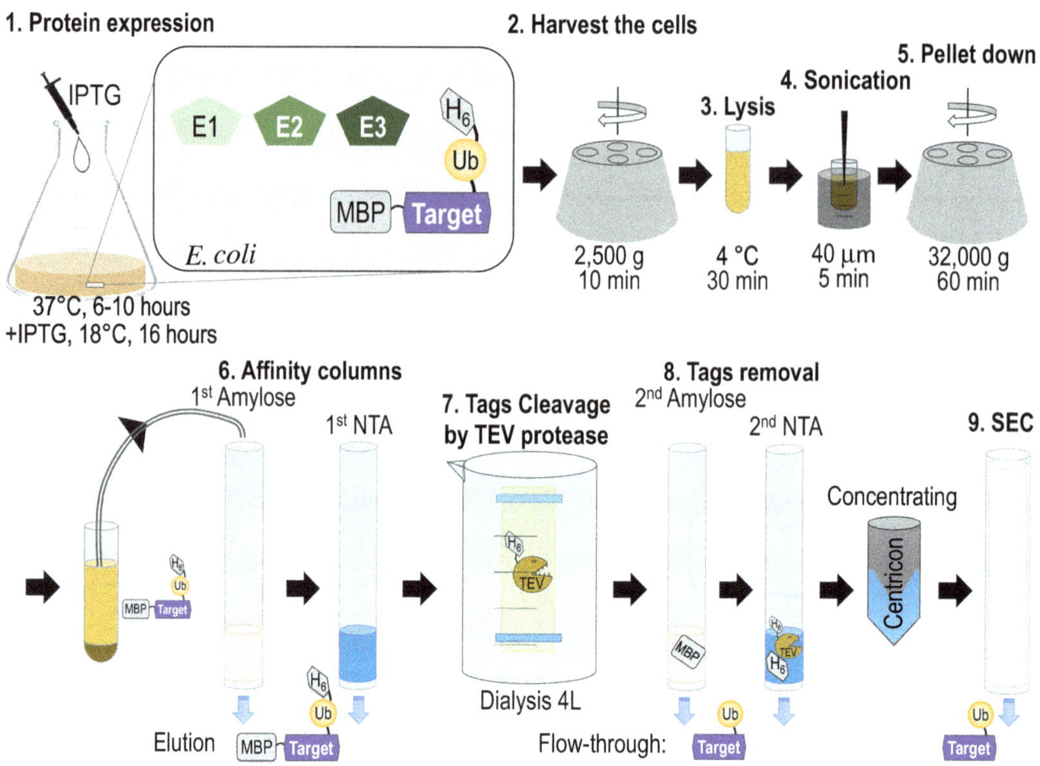

Fig. 3 Purification strategy for the ubiquitylated proteins. The main steps along the purification protocol are shown. See some of the purification products of Ub-Rpn10 along the protocol in the SDS-PAGE in Fig. 4a

To directly detect the ubiquitylation and to purify ubiquitylated proteins, we co-express affinity-tagged Ub and affinity-tagged ubiquitylation target along with its cognate ubiquitylation cascade [9–12]. Double affinity tag purification steps produce milligrams of highly pure ubiquitylated protein for downstream biochemistry, biophysicals, and structural studies (Fig. 3).

2 Materials

2.1 Selection System for Ubiquitylation Discovery

1. pNDUb, pCDSUb, and pE3 (optional) vectors (*see* **Note 1** and Fig. 2a).
2. Competent W3110 *E. coli* strain: F- λ- rph-1 INV(rrnD, rrnE) (*see* **Note 2**).
3. 10× Davis minimal salts: 70 g K_2HPO_4, 30 g KH_2PO, 20 g $(NH_4)_2SO_4$, 5 g $C_6H_5Na_3O_7$ in 1 L with dH_2O (*see* **Note 3**).
4. Davis medium [13]: 7 g K_2HPO_4, 3 g KH_2PO, 2 g $(NH_4)_2SO_4$, 0.5 g $C_6H_5Na_3O_7$, 2 g glucose, 100 mg $MgSO_4$, 10 mg vitamin B1 in 1 L with dH_2O (*see* **Note 3**).
5. Davis plates: Davis medium with 1.5% Bacto agar (*see* **Note 3**).

6. 5 mg/mL trimethoprim (TRIM) in methanol (*see* **Note 4**).
7. 30 mg/mL kanamycin in dH_2O (filter sterilized).
8. 50 mg/mL streptomycin in dH_2O (filter sterilized).
9. 100 mg/mL ampicillin in dH_2O (filter sterilized; optional).
10. Luria-Bertani (LB) medium [14].
11. LB plates with 34 μg/mL kanamycin and 25 μg/mL streptomycin.
12. LB plates with 23 μg/mL kanamycin, 16 μg/mL streptomycin and 33 μg/mL ampicillin (optional: *see* **step 2** in Subheading 3.1).
13. Office scanner or UV camera such as DNA gel doc system.
14. Scanning manager application: http://www.phys.huji.ac.il/bio_physics/nathalie/publications.html [15].
15. Image analysis software: Fiji or ImageJ [16].

2.2 Direct Detection and Purification of Ubiquitylated Proteins from E. coli

We provide here an example protocol for purification of Ub-Rpn10. Adjustments will be needed for purification of other ubiquitylated proteins.

1. pGEN6, pCOG30, and pCOG31 plasmids (*see* **Note 1**) [10].
2. Competent Rosetta 2 BL21 (λDE3) *E. coli* cells.
3. Bacterial growth media: Terrific broth (TB) [14], Luria-Bertani (LB) [14], and Super Optimal broth with catabolite repression (SOC).
4. LB plates with 12.5 μg/mL streptomycin, 17 μg/mL chloramphenicol, 12 μg/mL tetracycline, and 50 μg/mL ampicillin.
5. 1 M Isopropyl β-D-1-thiogalactopyranoside (IPTG).
6. Lysis buffer: 20 mM Tris–HCl (pH 7.0), 200 mM NaCl, 1 mM EDTA, 1 mM DTT, 0.2 mg/mL Lysozyme, 10 μg/mL DNaseI, and 0.1 mM protease inhibitor AEBSF.
7. Sonicator.
8. Purification reagents: Amylose, NTA (Ni^{2+}), Q-Sepharose Fast Flow resins and Superdex 75 size exclusion chromatography (SEC) column.
9. Amylose-binding buffer: 20 mM Tris–HCl (pH 7.0), 200 mM NaCl, 1 mM EDTA, 1 mM DTT.
10. Amylose elution buffer: 150 mM Tris–HCl (pH 8.0), 50 mM NaCl, 10 mM maltose, 20 mM imidazole.
11. Wash buffer 1: 50 mM Tris–HCl (pH 7.5), 150 mM NaCl, 20 mM imidazole, 2% Triton X-100.
12. Wash buffer 2: 50 mM Tris–HCl (pH 7.5), 150 mM NaCl, 40 mM imidazole, 2% Triton X-100.

13. Wash buffer 3: 50 mM Tris–HCl (pH 7.5), 150 mM NaCl, 40 mM imidazole.
14. Elution buffer: 50 mM Tris–HCl (pH 7.5), 150 mM NaCl, 500 mM imidazole.
15. Purified recombinant His_6-TEV (*Tobacco etch virus*) protease.
16. Dialysis buffer: 50 mM Tris–HCl (pH 8.0), 150 mM NaCl, 0.5 mM EDTA, and 1 mM DTT.
17. A dialysis bag (10 kDa cutoff).
18. Amicon centrifugal filter device for protein concentration (10 kDa cutoff).
19. FPLC chromatography system.

3 Methods

3.1 Selection of Ubiquitylated Proteins

1. The selection system requires transformation of typically two or three plasmids with different antibiotic resistance markers and different origins of replication. We recommend using large quantities of DNA and competent cells to ensure successful transformation. Co-transfrom 400 ng of pNDUb and pCDSUb vectors (Fig. 2a); in some cases add pE3 a 3rd vector that expresses cognate E3 ligase (*see* **Note 1**) into 200 μL W3110 competent cells (*see* **Note 2**).
2. Spread the entire content of the tube on pre-warmed LB plates supplemented with 34 μg/mL kanamycin and 25 μg/mL streptomycin (or 23 μg/mL kanamycin, 16 μg/mL streptomycin, and 33 μg/mL ampicillin, for selection of three plasmids).
3. Isolate a single colony and grow in 5 mL LB media supplemented with same antibiotic concentrations at 37°C with shaking overnight.
4. Harvest cells by gentle centrifugation at ~ 900 g for about 5 min and resuspend thoroughly in 1 mL 1× Davis salts and then add another 4 mL 1× Davis salts.
5. Dilute samples 1:6, and measure the optical density (OD_{600}).
6. Adjust the culture density to 0.3 in a new sterile 1.5 mL Eppendorf tube using 1× Davis salts solution. Vortex samples briefly, and spot 2.5 μL of each transformant (*see* **Note 1**) on Davis agar plate supplemented with 10 or 0 (control) μg/mL of TRIM. Spot in a row at least three technical repetitions for standard errors as demonstrated in Fig. 2c.
7. Incubate plates for 3–4 days at 26–28 °C and scan or photograph (we use a UV camera such as DNA gel doc system).
8. To quantitate the growth efficiency (time-lapse scanning), place spotted plates on an A4/US-letter office scanner

(Epson Perfection V37) situated in an incubator [15]. Incubate plates for 3–4 days at 26–28 °C and scan every hour (*see* **Note 5**).

9. Analyze images using Fiji or ImageJ [16]. Download and install the "Time Series Analyzer V3" plugin algorithm into the ImageJ/Fiji from http://rsb.info.nih.gov/ij/plugins/time-series.html [17].

   To read the image into Fiji, on the "File" menu, select "Import," then "Image Sequence," and choose selected images. Check the "Use virtual stack" and click OK. Crop import images as desired and save to a new directory (*see* **Note 6**). To quantify the growth, invoke "Time Series Analyzer V3" from the "Plugin" menu. Check the "Add on click," then select "Auto ROI (Region Of Interest) Properties," and adjust to "10 × 10" (pixels). Measure the optical density of each set of spots by clicking on the desired spot. Circle showing the ROI will appear. For each culture type, randomly select several statistical-repeat circles in each spot. In the "Time Series V3" window, hit the "Get average" button. "Time trace Average" and "Plot Values" windows including a plot and the numeric data will be opened. Copy the raw numeric data to a spreadsheet (such as MS Excel) to assemble data of different culture types. We recommend repeating the spotting assay in at least three independent cell cultures (Fig. 2c) (*see* **Note 7**).

10. For analyses in 96-well plates, adjust optical density (OD_{600}) of culture to 0.15 using Davis medium supplemented with 0 or 10 μg/mL TRIM. Transfer the 0.2 mL cells to a 96-well plate and grow at 30 °C with shaking. Monitor the bacterial growth by measuring the optical density (OD_{600}) hourly (30 or 60 min) using a microplate spectrophotometer (*see* **Note 8**).

3.2 Purification of Ubiquitylated Proteins

A protocol for the purification of ubiquitylated Rpn10 is presented below (Figs. 3 and 4a). For the purification of different ubiquitylated targets, modifications such as pH and osmotic values of the buffers may be crucial according to physiochemical parameters of the protein of interest.

1. Similar to the transformation step for the selection system (above), we recommend using large quantities of DNA and competent cells. Co-transform 250–400 ng of each plasmid (pGEN6, pCOG30, and pCOG31 plasmids, *see* **Note 9**) in 200 μL Rosetta 2 (λDE3) competent cells [10].
2. Prior to the recovery stage, add 250 μL of SOC media to the tube and incubate for 1 h at 37 °C in a shaking incubator. Spread the entire content of the tube on pre-warmed LB plates supplemented with 12.5 μg/mL streptomycin, 17 μg/mL

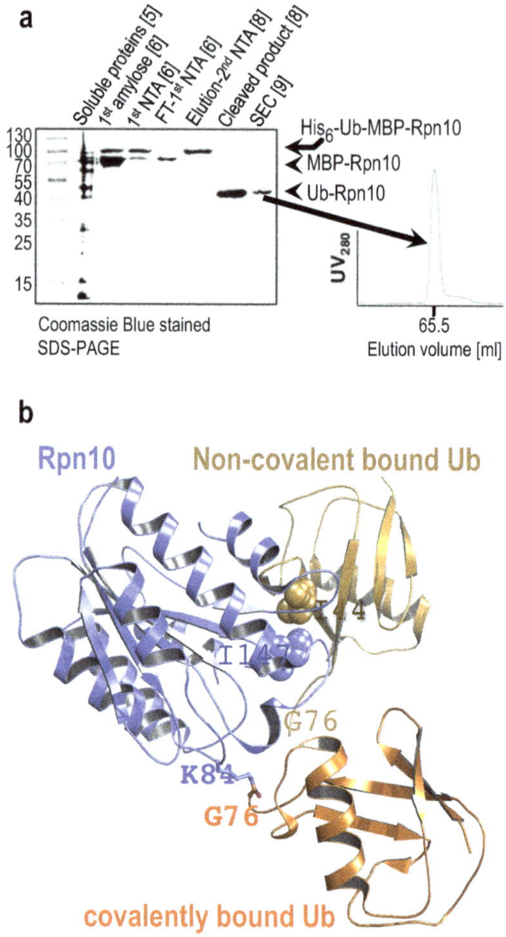

Fig. 4 Purification of ubiquitylated target facilitates structure determination of the ubiquitylated state. (**a**) Coomassie blue-stained SDS-PAGE showing the purification steps Ub-Rpn10. The numbers in the brackets indicate the associated purification step schemed in Fig. 3 (the figure was reproduced with permission from ref. 12). (**b**) Structure of ubiquitylated Rpn10. The protocol yielded 12 mg of purified Ub-Rpn10 from 9 L of *E. coli* culture. Modified Rpn10 was crystallized, and the structure was determined; PDB code 5LN1 [12]

chloramphenicol, 12 μg/mL tetracycline, and 50 μg/mL ampicillin.

3. Next day, inoculate colonies from the plate to the 10–100 mL TB media supplemented with antibiotics and grow overnight at 37 °C in an incubator with shaking.

4. Next morning, transfer 10–30 mL of overnight culture to the 1.5 L of TB media supplemented with antibiotics and grow for 6–10 h at 37 °C in a shaking incubator at 180 rpm. Once the culture appears dense and "cloudy" ($OD_{600} \cong 2.0$), lower the temperature down to ~18 °C and induce it with IPTG to a final

concentration of 1 mM. Grow for additional 16–20 h (Step 1 in Fig. 3).

5. Harvest the cells by 2500 × *g* centrifugation for 10 min, and resuspend the pellet in lysis buffer (Steps 2 and 3 in Fig. 3).
6. Stir the sample gently at 4 °C on a magnetic stirring plate for 30 min for better lysis (Step 3 in Fig. 3).
7. Disrupt cells using sonication (time, 5 min; pulse on, 5 s; pulse off, 25 s; amplitude, 40 μm (peak-to-peak)) (Step 4 in Fig. 3).
8. Centrifuge the lysate at 32,000 × *g* (Sorvall SS-34 rotor) for 45 min at 4 °C. Following first spin, transfer supernatant to a new tube and centrifuge under the same conditions for another 15 min (Step 5 in Fig. 3).
9. Load sample (supernatant) on an amylose resin column equilibrated with amylose-binding buffer, which allows binding of the *apo* and the ubiquitylated forms of Rpn10 (Step 6 in Fig. 3). Wash the column with 1–2 L of amylose-binding buffer. Elute the sample with amylose elution buffer.
10. Load the eluant onto a pre-equilibrated NTA (Ni^{2+}) affinity column with wash buffer 1 (Step 6 in Fig. 3). Wash the column extensively to remove nonspecific binding of the *apo* protein (non-ubiquitylated) with the following buffers: wash buffer 1, wash buffer 2, and wash buffer 3. Elute the protein from the column with gradient of imidazole elution buffer (Step 6 in Fig. 3).

The below steps are optional: pending on your purpose, additional purification steps may be included to cleave and remove the affinity tags (*see* refs. 11, 12).

11. Digest the eluted sample with His_6-TEV proteases (0.5–1 mg) overnight in a dialysis bag against 4 L of dialysis buffer at 4 °C (Step 7 in Fig. 3).
12. To remove the MBP-tagged proteins, pass the sample from the dialysis bag on a second amylose column, pre-equilibrated with amylose-binding buffer (Step 8 in Fig. 3). Collect the flow-through (unbound fraction). To remove His_6-tagged proteins, pass the second amylose column flow-through on a second NTA column, pre-equilibrated with wash buffer 1. Collect the flow-through (unbound fraction) (Step 8 in Fig. 3).
13. Concentrate the sample to a final volume of 2–5 mL (depending on the size of sample loop) using an Amicon centrifugal filter and load on appropriate SEC column (Step 9 in Fig. 3). For crystallization of ubiquitylated Rpn10, an additional purification step of ion exchange was performed prior the SEC column [11].

14. Concentrate all combined fractions with the protein of interest to a desired concentration, aliquot, and flash freeze in the liquid nitrogen. Store in −80 °C.

4 Notes

1. A list of available pNDUb and pCDSub vectors can be found in Supplementary Table 1 in Levin-Kravets et al. (2016) [9]. pND-Ub co-expresses the N-terminal fragment of DHFR fused to Ub, E1, and E2 from an operon-like cassette with a λ-pL promoter. Each of the ORFs is preceded by a Shine-Dalgarno (SD) sequences. pE3 expresses the E3 ligase. pCD-Sub expresses the C-terminal fragment of DHFR fused to a ubiquitylation target (this plasmid sometimes co-expresses the E3 ligase). We suggest to add positive and negative control cultures such as full-length DHFR; known complete ubiquitylation cascade expresses E1, E2, E3, and substrate ("complete"), ΔE1, ΔE2, ΔE3, ΔUb, etc. [9].
2. Co-transformation of two or three plasmids requires high-efficiency competent cells. We use the *Inoue* method for preparation of high-quality competent cells [18].
3. Davis minimal medium and plates are prepared as follows. Prepare 10× Davis salts by weighing listed components, and add dH_2O up to a volume of 1 L, then mix, and autoclave. Prepare a 100× Davis supplements by weighing 8 g glucose, 400 mg $MgSO_4$, and 40 mg vitamin B1, and add dH_2O up to a volume of 40 mL and sterilize it by filtration through a 0.22 μm filter (do *not* autoclave). For plates, weigh 6 g of Bacto agar, add dH_2O up to a volume of 348 mL, and then mix and autoclave. Once the agar is autoclaved, swirl to mix evenly and "temper" at room temperature. Add 40 mL of the 10× Davis salts, 4 mL of the 100× Davis supplements, and complete with sterile dH_2O to a volume of 400 mL. Pour plates with and without TRIM at final concentration of 10 μg/mL.
4. Preparation of TRIM stock solution: weigh 5 mg TRIM and add methanol up to a volume of 1 mL, and sterilize by filtration through a 0.22 μm filter.
5. We recommend scanning plates from the bottom (i.e., do *not* scan the plates up-side-down). To gain good contrast between the growing colonies and the plate, cover the plate with a sterile black lid. Clear lids can be sprayed in advance with black paint. We reuse the black lids by washing/sterilizing them with 50% bleach followed by 96% denatured ethanol.
6. Save cropped images in a new directory (file).

7. We recommend preparing new transformants for each culture type.
8. Growth in liquid Davis medium for 3–4 days using 96-well plate requires special humidity conditions to eliminate evaporation and accumulation of dew on the lead which interferes with OD_{600} measurements.
9. A list of available pGEN and pCOG vectors can be found in Supplementary Table SIV in Keren-Kaplan et al. (2012) [10]. The generic plasmid (pGEN) expresses His_6-Ub, E1-activating enzyme, and E2-conjugating enzyme. The cognate plasmid (pCOG) encodes a selected substrate for ubiquitylation and its cognate E3 ligase. In some cases, we express the E3 ligase in a third plasmid as described in refs. 11, 12.

Acknowledgments

We are very thankful to A Weissman, A Chitnis, R Vierstra, I Dikic, D Katzmann and S Misra, R Klevit, and Scott Emr for the fruitful discussions and for providing reagents.

We also thank S Ben-Aroya and D Rotin for validation experiments in cells and D Wolf and O Kleifeld for mass spectrometry analyses. This research was supported by grants from the Israeli Science Foundation (grant no. 1695/08, 464/11, and 651/16), from the EC FP7 Marie Curie International Reintegration Grant (PIRG03-GA-2008-231079), and from the Israeli Ministry of Health (5108) to GP.

References

1. Schulman BA, Wade Harper J (2009) Ubiquitin-like protein activation by E1 enzymes: the apex for downstream signalling pathways. Nat Rev Mol Cell Biol 10:319–331. https://doi.org/10.1038/nrm2673
2. Metzger MB, Hristova VA, Weissman AM (2012) HECT and RING finger families of E3 ubiquitin ligases at a glance. J Cell Sci 125:531–537. https://doi.org/10.1242/jcs.091777
3. Hurley JH, Lee S, Prag G (2006) Ubiquitin-binding domains. Biochem J 399:361–372. https://doi.org/10.1042/BJ20061138
4. Komander D, Rape M (2012) The ubiquitin code. Annu Rev Biochem 81:203–229. https://doi.org/10.1146/annurev-biochem-060310-170328
5. Mevissen TET, Komander D (2017) Mechanisms of deubiquitinase specificity and regulation. Annu Rev Biochem 86:159–192. https://doi.org/10.1146/annurev-biochem-061516-044916
6. Hurley JH, Schulman BA (2014) Atomistic autophagy: the structures of cellular self-digestion. Cell 157:300–311. https://doi.org/10.1016/j.cell.2014.01.070
7. Henne WM, Buchkovich NJ, Emr SD (2011) The ESCRT pathway. Dev Cell 21:77–91. https://doi.org/10.1016/j.devcel.2011.05.015
8. Collins GA, Goldberg AL (2017) The logic of the 26S proteasome. Cell 169:792–806. https://doi.org/10.1016/j.cell.2017.04.023
9. Levin-Kravets O, Tanner N, Shohat N et al (2016) A bacterial genetic selection system for ubiquitylation cascade discovery. Nat Methods 13:945–952. https://doi.org/10.1038/nmeth.4003
10. Keren-Kaplan T, Attali I, Motamedchaboki K et al (2012) Synthetic biology approach to

reconstituting the ubiquitylation cascade in bacteria. EMBO J 31:378–390. https://doi.org/10.1038/emboj.2011.397

11. Keren-Kaplan T, Prag G (2012) Purification and crystallization of mono-ubiquitylated ubiquitin receptor Rpn10. Acta Crystallogr Sect F Struct Biol Cryst Commun 68:1120–1123. https://doi.org/10.1107/S1744309112034331
12. Keren-Kaplan T, Peters LZ, Levin-Kravets O et al (2016) Structure of ubiquitylated-Rpn10 provides insight into its autoregulation mechanism. Nat Commun. https://doi.org/10.1038/ncomms12960
13. Davis BD (1948) Isolation of biochemically deficient mutants of bacteria by penicillin. J Am Chem Soc 70:4267. https://doi.org/10.1021/ja01192a520
14. Sambrook J, Fritsch EF, Maniatis T (1989) Molecular cloning: a laboratory manual. Cold Spring Harb. Lab. Press, Cold Spring Harbor, NY
15. Levin-Reisman I, Gefen O, Fridman O et al (2010) Automated imaging with ScanLag reveals previously undetectable bacterial growth phenotypes. Nat Methods 7:737–741. https://doi.org/10.1038/nmeth.1485
16. Schindelin J, Arganda-Carreras I, Frise E et al (2012) Fiji: an open-source platform for biological-image analysis. Nat Methods 9:676–682. https://doi.org/10.1038/nmeth.2019
17. Balaji J (2014) Time Series Analyzer. https://imagej.nih.gov/ij/plugins/time-series.html
18. Inoue H, Nojima H, Okayama H (1990) High efficiency transformation of Escherichia coli with plasmids. Gene 96:23–28. https://doi.org/10.1016/0378-1119(90)90336-P

Part III

Structural Approaches as Applied to Enzymes that Participate in the Ubiquitin Proteasome System

Chapter 12

Strategies to Trap Enzyme-Substrate Complexes that Mimic Michaelis Intermediates During E3-Mediated Ubiquitin-Like Protein Ligation

Frederick C. Streich Jr. and Christopher D. Lima

Abstract

Most cellular functions rely on pathways that catalyze posttranslational modification of cellular proteins by ubiquitin (Ub) and ubiquitin-like (Ubl) proteins. Like other posttranslational modifications that require distinct writers, readers, and erasers during signaling, Ub/Ubl pathways employ distinct enzymes that catalyze Ub/Ubl attachment, Ub/Ubl recognition, and Ub/Ubl removal. Ubl protein conjugation typically relies on parallel but distinct enzymatic cascades catalyzed by an E1-activating enzyme, an E2 carrier protein, and an E3 ubiquitin-like protein ligase. One major class of E3, with ca. 600 members, harbors RING or the RING-like SP-RING or Ubox domains. These RING/RING-like domains bind and activate the E2-Ubl thioester by stabilizing a conformation that is optimal for nucleophilic attack by the side chain residue (typically lysine) on the substrate. These RING/RING-like domains typically function together with other domains or protein complexes that often serve to recruit particular substrates. How these RING/RING-like E3 domains function to activate the E2-Ubl thioester while engaged with substrate remains poorly understood. We describe a strategy to generate and purify a unique $E2_{Ubc9}$-Ubl_{SUMO} thioester mimetic that can be cross-linked to the $Substrate_{PCNA}$ at Lys164, a conjugation site that is only observed in the presence of $E3_{Siz1}$. We describe two techniques to cross-link the $E2_{Ubc9}$-Ubl_{SUMO} thioester mimetic active site to the site of modification on PCNA and the subsequent purification of these complexes. Finally, we describe the reconstitution and purification of the $E2_{Ubc9}$-Ubl_{SUMO}-PCNA complex with the $E3_{Siz1}$ and purification that enabled its crystallization and structure determination. We think this technique can be extended to other E2-Ubl-substrate/E3 complexes to better probe the function and specificity of RING-based E3 Ubl ligases.

Key words Ubiquitin, Ubiquitin-like protein, SUMO, Smt3, E2-conjugating enzyme, Ubc9, E3, Ligase, Siz/PIAS, PCNA, Complex reconstitution, Structure

1 Introduction

Posttranslational modification of cellular proteins with ubiquitin and ubiquitin-like (Ubl) proteins (such as Nedd8, SUMO (Smt3 in yeast), ISG15, etc.) represents parallel signaling pathways that control and regulate nearly every aspect of cellular homeostasis [1–3]. Most Ubl proteins are expressed as pro-proteins that are

Thibault Mayor and Gary Kleiger (eds.), *The Ubiquitin Proteasome System: Methods and Protocols*, Methods in Molecular Biology, vol. 1844, https://doi.org/10.1007/978-1-4939-8706-1_12, © Springer Science+Business Media, LLC, part of Springer Nature 2018

proteolytically processed to generate their mature forms, each by a Ubl-specific protease, to generate the C-terminal Gly-Gly motif that is subsequently activated for conjugation [4–8]. Each member of the canonical Ubl family can be conjugated to cellular targets, typically through the concerted activities of at least three enzymes that include an E1 Ubl-activating enzyme, an E2 Ubl carrier protein, and an E3 Ubl ligase [9–13]. These conjugates can be reversed by proteases similar to those that catalyze Ubl protein maturation, effectively reversing the signal and recycling the Ubl protein, the substrate, or both.

The E1-activating enzyme catalyzes ATP-coupled activation of the Ubl C-terminus and E1-thioester bond formation [14–16] and subsequent formation of the E2-Ubl thioester between the E2 active site cysteine and Ubl C-terminus (Fig. 1a) [17, 18]. Finally, E3 ligases help catalyze formation of the isopeptide bond between the Ubl and the target substrate (Fig. 1b) [9, 19, 20]. A large number of these conjugation events utilize RING-type Ubl E3 ligases [21, 22]. These ligases harbor RING domains [23, 24] (or the highly related SP-RING [25–29] or Ubox [30] domains) that bind the E2-Ubl thioester complex to promote a closed and activated conformation of the E2-Ubl thioester, amenable to nucleophilic attack (usually by a primary amine, typically a lysine residue) present on the target substrate [31–48]. The family of proteins containing RING-like domains is large (~600 members) and diverse. The RING-like domains are found appended to (or complexed with) a diversity of additional domains, many of which recruit the target substrate [22]. A major dearth of knowledge in the field of Ubl conjugation is how this diverse repertoire of substrate recruitment domains functions in concert with the RING-like domain/E2-Ubl thioester complexes to execute targeted modification of particular substrates.

To better understand how this catalytic machinery functions during conjugation to a specific position on a particular substrate, it is often useful to determine the three-dimensional structure of the reconstituted complex. Historically, reconstituting these complexes in a manner amenable to structural studies proved challenging. First, the E2-Ubl thioester complex is labile in the presence of the E3, which can activate it for nondiscriminate discharge to water or buffer components, particularly in the conditions and protein concentrations necessary for structural studies. Other groups developed E2-Ubl thioester mimetics with stabilized covalent linkages between the E2 active site (Cys to Ser or Cys to Lys) and the Ubl C-terminus, making them recalcitrant to discharge of the Ubl. These mimetics have proved useful in obtaining RING/E2-Ubl complex structures (Fig. 1c, d) [33–50]. Second, interactions between the E2-Ubl thioester and the E3 and the E2-Ubl/E3 complex and the target substrate are often too transient with rapid association, turnover, and dissociation [51]. This makes

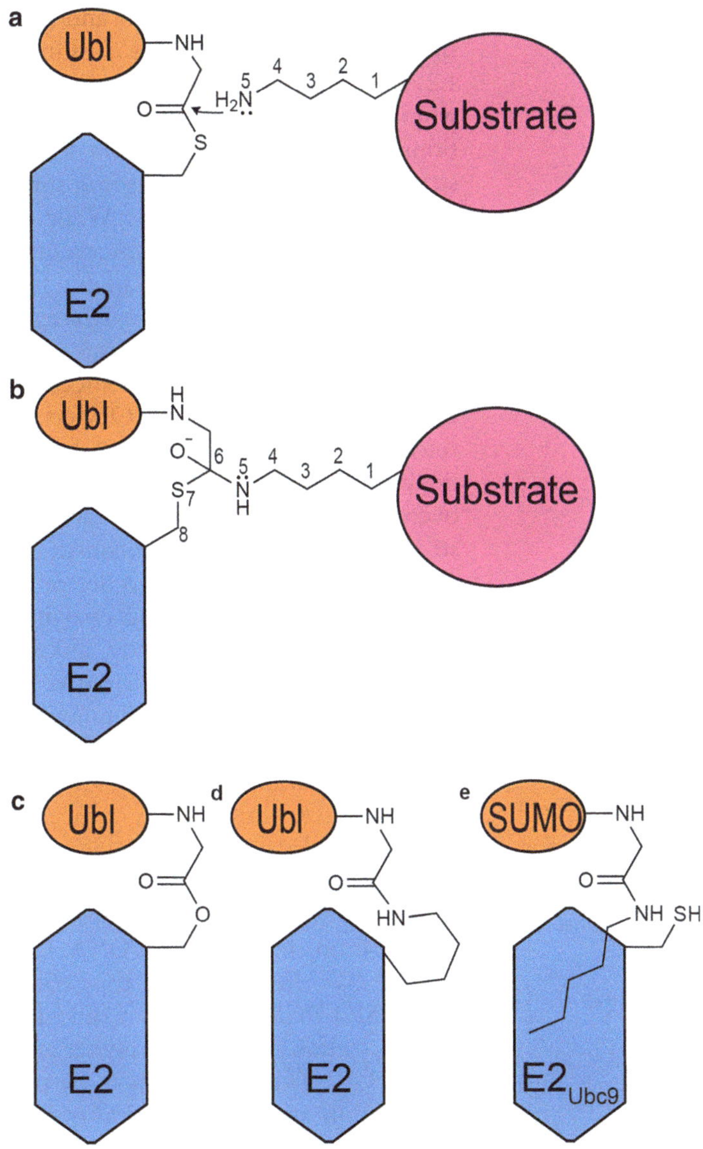

Fig. 1 Chemical diagrams for Ubl transfer from the E2-Ubl thioester complex to the target substrate and chemical diagrams of thioester mimetics used in structural studies. (**a**) Chemical diagram of a lysine residue undergoing nucleophilic attack of a native E2-Ubl thioester. (**b**) Chemical diagram of the predicted tetrahedral intermediate transition state formed after nucleophilic attack on the way to isopeptide bond formation. Chemical diagram of an E2-Ubl thioester mimetic formed using (**c**) an E2 catalytic Cys to Ser mutant, which forms an oxyester linkage or (**d**) an E2 catalytic Cys to Lys mutant, which forms an isopeptide linkage. (**e**) Chemical diagram of the thioester mimetic utilized in the current protocol. Alanine 129, neighboring the catalytic cysteine, is mutated to Lys. The E1 charges the E2 catalytic Cys, followed by nucleophilic attack by the installed Lys129 to form and E2-Ubl thioester mimetic with a stable isopeptide linkage and an intact active site cysteine residue available for chemical cross-linking

stoichiometric reconstitution and purification challenging. Given these issues, it is perhaps unsurprising that the majority of E2/RING E3 structures resulted from co-crystallization strategies, where crystallization serves to promote protein-protein interactions. Similar co-crystallization strategies have proven useful in structure determination of some substrates in complex with substrate recognition domains. While useful, these strategies have largely been unsuccessful for reconstituting complete E2-Ubl/E3/substrate complexes. Notably, the Schulman laboratory utilized alternative cross-linking strategies to those discussed here to trap other E2-Ubl/E3/Substrate complexes [52–55].

We pursued numerous unsuccessful attempts to reconstitute a complex between the SUMO-specific $E3_{Siz1}$, the $E2_{Ubc9}$-SUMO thioester, and the proliferating cell nuclear antigen (PCNA), a well-established target of SUMO modification in yeast [56–58]. To overcome this challenge, we adopted a chemical cross-linking strategy to aid in complex reconstitution [59]. Other groups have utilized E2 active site Cys to Ser or Lys substitutions that can be used with the E1-activating enzyme to catalyze formation of a stable oxyester or isopeptide linkage, respectively, to the Ubl C-terminus (instead of the native thioester linkage) to create more stable E2-Ubl thioester mimetics (Fig. 1c, d). We created an alternative thioester mimetic with the $E2_{Ubc9}$ A129K substitution (adjacent to the active site Cys93) to move the E2-Ubl linkage to an alternative position just adjacent to the active site cysteine (Fig. 1e). This has the advantage of leaving the active site cysteine available for chemical cross-linking to the target substrate. With respect to the substrate, we substituted the E3-dependent site of modification, in this case PCNA Lys164, to cysteine for cross-linking. We hypothesized this covalently linked multi-protein E2-SUMO-PCNA complex would present multiple binding sites to $E3_{Siz1}$, thus generating a higher affinity, stoichiometric complex with $E3_{Siz1}$. This turned out to be partially true; however further stabilization and reconstitution of the complex were achieved by fusing a second molecule of SUMO to the C-terminus of the E3 in a location that enabled formation of a stable backside interaction with the $E2_{Ubc9}$, as observed previously (Fig. 2) [60–62]. This strategy resulted in the successful reconstitution of an enzyme-substrate complex that crystallized readily.

Presented here are two chemical cross-linking strategies, both of which employ straightforward and technically accessible methods that utilize standard protein purification equipment and commercially available reagents. The first technique utilizes bismaleimidoethane (BMOE) that creates a cross-link that is stable to reducing agents post reaction between the respective cysteine residues. Utilizing BMOE, we reconstituted the E2-SUMO/E3/substrate complex and solved the crystal structure. The structure revealed that the BMOE linkage was suboptimal, perhaps due to

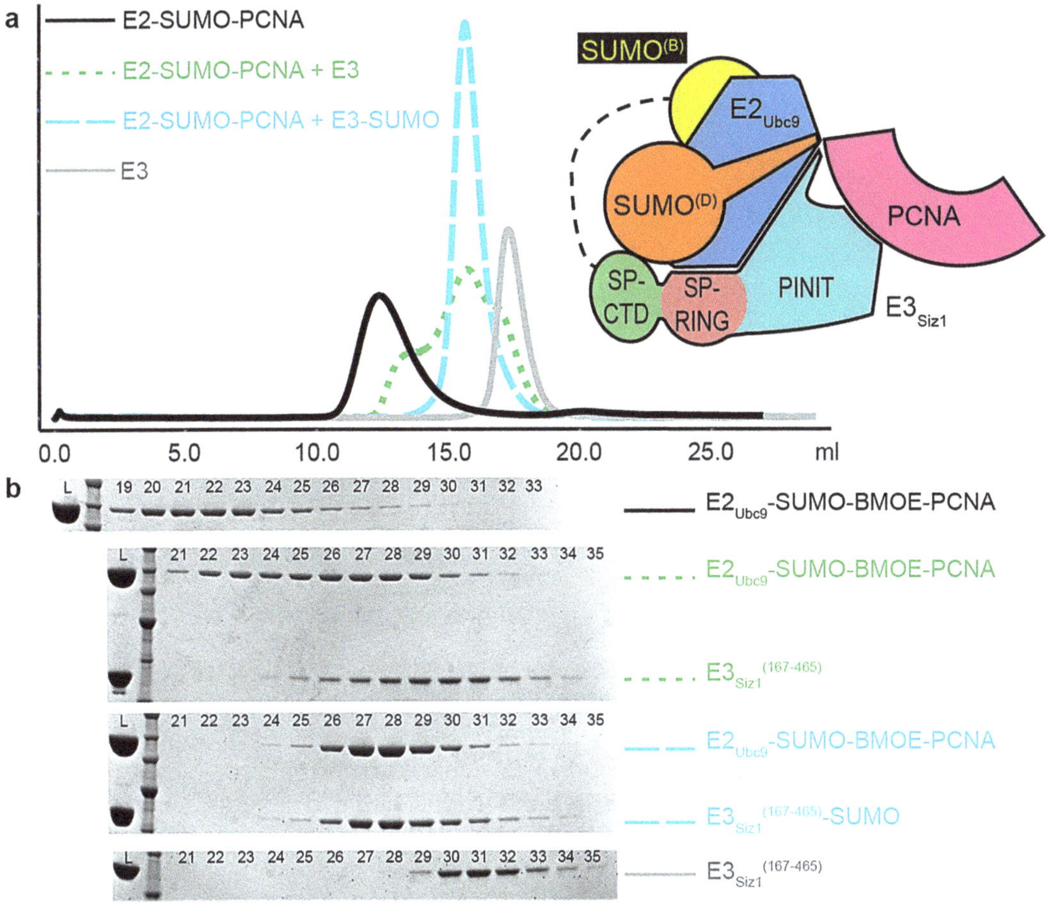

Fig. 2 $E2_{Ubc9}$-Ubl_{SUMO}/$E3_{Siz1}$/$substrate_{PCNA}$ complex reconstitution. (**a**) UV trace of Superose 6 analytical gel filtration runs to investigate complex formation between individual components. An E3-SUMO fusion along with chemical cross-linking of the E2 active site to the site of modification on PCNA were both required for stoichiometric complex reconstitution, depicted in the cartoon complex. (**b**) SDS-PAGE analysis of the eluted fractions from the indicated gel filtration runs, Coomassie stained

its bulky maleimide ring structures and linker that creates a suboptimal distance between the E2 and substrate after cross-linking (Fig. 3). This led us to employ 1,2-ethanedithiol (EDT) as an alternative cysteine-to-cysteine cross-linker, creating an atomic bridge with the same number of atoms as the predicted tetrahedral intermediate (Fig. 1b and Fig. 4). Although less stable to reducing agents, in comparison to a BMOE cross-link, EDT is a better mimic of conformation and distance for isopeptide bond formation despite having unique dihedral angles at the two disulfide bonds. The EDT strategy also led to successful enzyme-substrate complex reconstitution, crystallization, and structure determination.

The strategy of E2-Ubl thioester mimetic formation, cross-linking of the E2 active site to the site of modification on the substrate, and reconstitution with the E3 may prove useful in

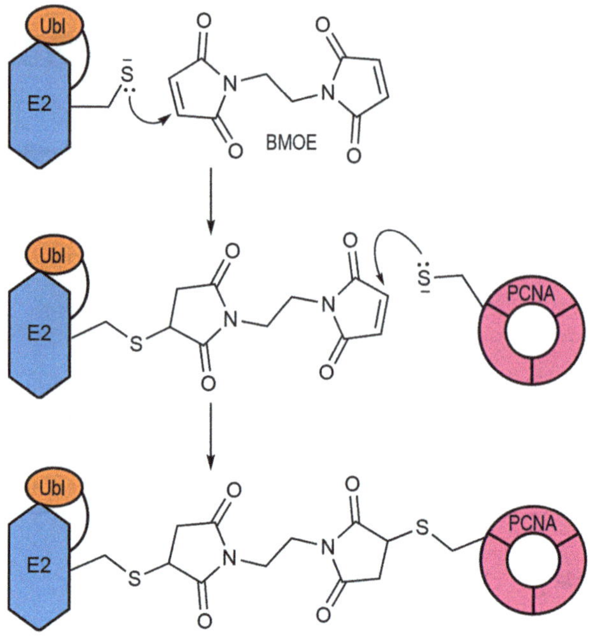

Fig. 3 Chemical diagram of BMOE based cross-linking of the E2 and target substrate PCNA. BMOE, bismaleimidoethane

reconstituting other E2-Ubl/E3/substrate complexes for structural study (*see* **Note 1**). Elucidating structural details for other such complexes will shed light on the manner in which the E2-Ubl/RING association works in concert with the immense variety and diversity of substrate recruitment domains/substrate interactions to affect specific conjugation across the variety of extant RING E3 ligases.

2 Materials

2.1 Cloning, Protein Expression, and Purification

Stock solutions should be prepared in water unless otherwise noted and 0.2 μM filtered.

1. 18.2 MΩ-cm ultrapure water for all solutions.
2. Plasmids: pET15b, pET21a, pET28b, pSMT3, and TOPO-SMT3 (*see* **Note 2**).
3. Oligonucleotide primers.
4. Thermostable proofreading fusion polymerase (Herculase II, Agilent).
5. dNTP mixture.
6. Thermocycler.
7. Luria-Bertani (LB) medium: 10 g tryptone, 10 g NaCl, 5 g yeast extract, 1.5 g Tris-HCl in 1 L water, autoclaved.

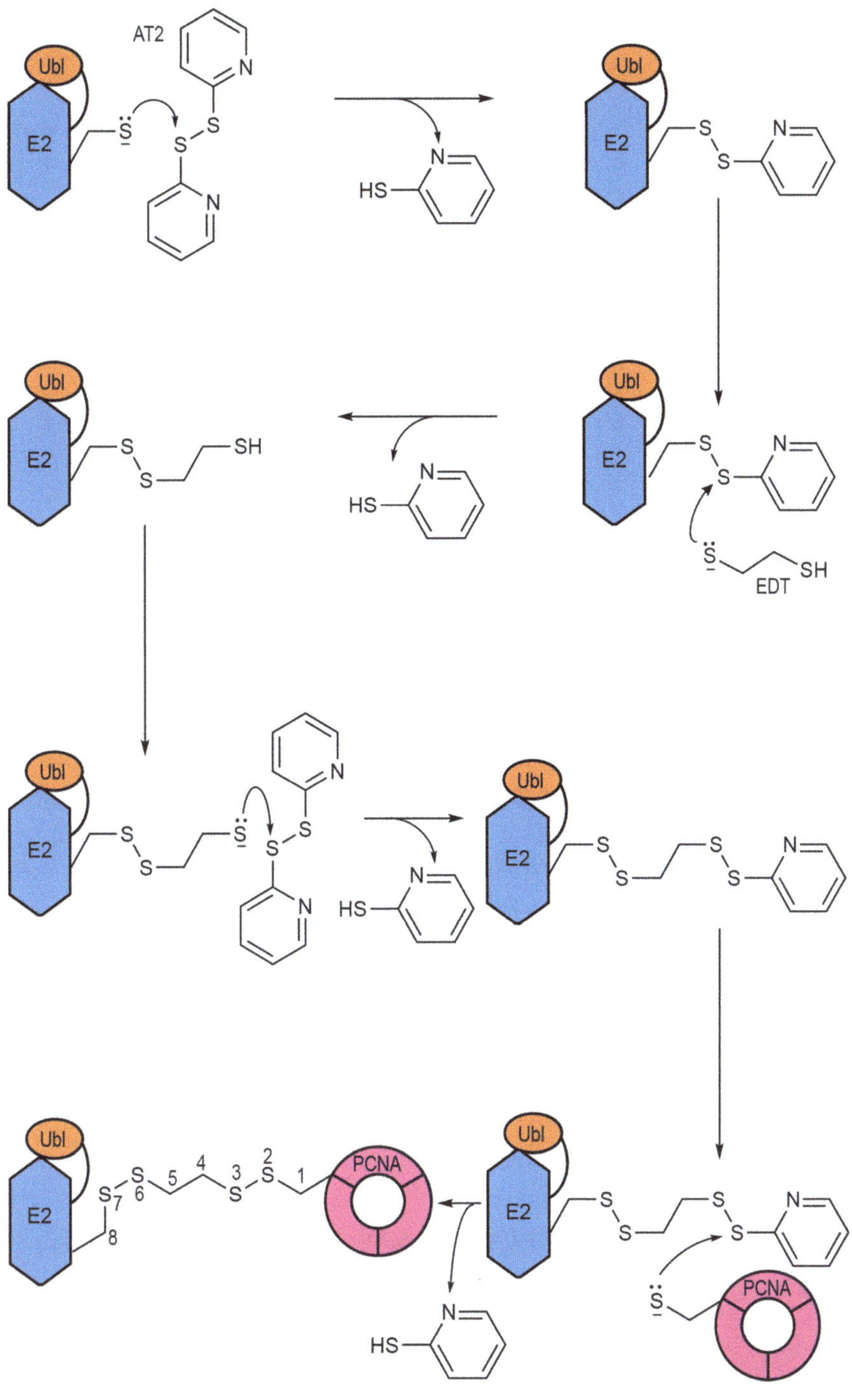

Fig. 4 Chemical diagram of the progressive steps to cross-link the E2 and target substrate with EDT. The number of atoms present in the EDT bridging cross-link equals that in the predicted tetrahedral intermediate (Fig. 1b). *AT2,* Aldrithiol-2; *EDT,* 1,2-ethanedithiol

8. Luria-Bertani (LB) agar medium: 10 g tryptone, 10 g NaCl, 5 g yeast extract, 1.5 g Tris-HCl, 15 g agar in 1 L water, autoclaved and poured in plates.
9. Super Broth (SB) medium: 32 g tryptone, 20 g yeast extract, 5 g NaCl, 9.87 g potassium phosphate dibasic in 1 L water, autoclaved.
10. Antibiotics: 100 mg/mL ampicillin in water; 50 mg/mL kanamycin in water; 30 mg/mL chloramphenicol in ethanol.
11. Bacterial strains: *Escherichia coli* (*E. coli*) Top10 (Thermo Fisher) or *E. cloni* (Lucigen) (for cloning), *E. coli* BL21 (DE3) RIL Codon Plus (Agilent) (for expression).
12. Shaking incubators capable of a temperature range of 18–37 °C.
13. 2 L baffled shaking flasks.
14. 1 M isopropyl β-D-thiogalactopyranoside (IPTG) in water.
15. Bovine thrombin (Sigma): 1 U/μL resuspended in 10 mM Na_2HPO_4, 1.8 mM KH_2PO_4, 137 mM NaCl, 2.7 mM KCl (pH 7.4) (PBS). Aliquot, flash freeze, and store at −20 °C.
16. Ulp1 protease catalytic domain (amino acids 403–621) [63]: 3 mg/mL in 20 mM Tris-HCl (pH 8.0), 350 mM NaCl, 1 mM β-mercaptoethanol (BME), 10% glycerol (*see* **Note 3**).
17. HIS-GST-Ulp1 protease catalytic domain (amino acids 403–621): 4.8 mg/mL in 20 mM Tris-HCl (pH 8.0), 350 mM NaCl, 1 mM BME (*see* **Note 3**).
18. 1 M N-Ethylmaleimide (NEM) in ethanol stock solution, diluted to 20 mM in water.
19. 1 M Tris-HCl (pH 8.0 at 4 °C).
20. 14.3 M BME.
21. 100 mM phenylmethylsulfonyl fluoride (PMSF) in ethanol.
22. 1 M $MgCl_2$.
23. 2 M $CaCl_2$.
24. 4 M imidazole.
25. 10% IGEPAL CA-630.
26. 5 M NaCl.
27. 10 mg/mL lysozyme in 20 mM Tris-HCl (pH 8.0), 50 mM NaCl. Aliquot, flash freeze, and store at −20 °C.
28. 10 mg/mL DNAse I in 20 mM Tris-HCl (pH 8.0), 50 mM NaCl. Aliquot, flash freeze, and store at −20 °C.
29. Suspension buffer: 50 mM Tris-HCl (pH 8.0), 20% sucrose.
30. Lysis Buffer A: 50 mM Tris-HCl (pH 8.0), 20% sucrose, 20 μg/mL lysozyme, 5 mM BME, 2.5 mM $MgCl_2$, 0.5 mM $CaCl_2$, 20 mM imidazole, 0.05% IGEPAL CA-630, 100 μg/mL DNAse I, 350 mM NaCl (*see* **Note 4**).

31. Lysis Buffer B: 50 mM Tris-HCl (pH 8.0), 20% sucrose, 20 μg/mL lysozyme, 5 mM BME, 1 mM PMSF, 2.5 mM $MgCl_2$, 0.5 mM $CaCl_2$, 20 mM imidazole, 0.05% IGEPAL CA-630, 100 μg/mL DNAse I, 350 mM NaCl.
32. Lysis Buffer C: 50 mM Tris-HCl (pH 8.0), 20% sucrose, 20 μg/mL lysozyme, 5 mM BME, 1 mM PMSF, 2.5 mM $MgCl_2$, 0.5 mM $CaCl_2$, 0.05% IGEPAL CA-630, 100 μg/mL DNAse I, 350 mM NaCl.
33. Lysis Buffer D: 50 mM Tris-HCl (pH 8.0), 20% sucrose, 20 μg/mL lysozyme, 5 mM BME, 2.5 mM $MgCl_2$, 0.5 mM $CaCl_2$, 0.05% IGEPAL CA-630, 100 μg/mL DNAse I, 150 mM NaCl.
34. Branson Ultrasonics Digital Sonifier with 1/2-in. horn or equivalent.
35. NiNTA Superflow agarose resin (Qiagen).
36. Buffer A: 20 mM Tris-HCl (pH 8.0), 350 mM NaCl, 20 mM imidazole, 1 mM BME.
37. Buffer B: 20 mM Tris-HCl (pH 8.0), 350 mM NaCl, 250 mM imidazole, 1 mM BME.
38. Buffer C: 20 mM Tris-HCl (pH 8.0), 350 mM NaCl, 1 mM BME.
39. Buffer D: 20 mM Tris-HCl (pH 8.0), 100 mM NaCl, 1 mM BME.
40. Buffer E: 20 mM Tris-HCl (pH 8.0), 1 M NaCl, 1 mM BME.
41. Buffer F: 20 mM Tris-HCl (pH 8.0), 50 mM NaCl, 1 mM BME.
42. Buffer G: 20 mM Tris-HCl (pH 8.0), 200 mM NaCl, 1 mM BME.
43. AKTA-FPLC (GE Healthcare) or an equivalent workstation equipped with gel filtration columns (HiLoad 26/60 Superdex 75 and Superdex 200) and ion exchange columns (Mono-Q 10/100 GL and Mono-S 10/100 GL) (*see* **Note 5**).
44. Amicon Ultra centrifugal filter devices (Millipore) with 3K, 10K, and 30K molecular weight cutoffs.
45. Bradford reagent.
46. NanoDrop 2000 Spectrophotometer (Thermo Scientific) or equivalent.
47. Glutathione Sepharose 4B (GE Healthcare).
48. SDS-polyacrylamide gel electrophoresis (SDS-PAGE):
    (a) NuPAGE system (Invitrogen) for SDS-PAGE utilizing either MES or MOPS running buffer.
    (b) 12% and 4–12% gradient polyacrylamide BIS-Tris gels (Invitrogen).

49. Ammonium sulfate, $(NH_4)_2SO_4$ (powder).
50. 2K and 8K molecular weight cutoff dialysis tubing.

2.2 E2-SUMO Formation Reaction

1. 1 M BIS-Tris Propane (pH 9.5).
2. 1 M $MgCl_2$.
3. 5 M NaCl.
4. 10% Triton X-100.
5. 1 M DTT.
6. 100 mM ATP in 10 mM HEPES (pH 7.5).

2.3 E2-SUMO-PCNA Cross-Linking Reaction

1. 1 M TCEP in 10 mM HEPES (pH 7.5).
2. PD-10 (GE Healthcare) and/or Zeba (Thermo Fisher) desalting columns, or equivalent.
3. Cross-linking buffer: 20 mM HEPES, 200 mM NaCl, 5 mM EDTA.
4. Bismaleimidoethane (BMOE, Pierce) solution: 100 mM BMOE in dimethyl sulfoxide (DMSO).
5. Aldrithiol-2 (AT2) solution: 300 mM AT2 in DMSO.
6. 1,2-Ethanedithiol (EDT) solution: 300 mM EDT in DMSO.
7. Buffer H: 20 mM Tris-HCl (pH 8.0), 350 mM NaCl.
8. Buffer I: 20 mM Tris-HCl (pH 8.0), 200 mM NaCl.
9. Buffer J: 20 mM Tris-HCl (pH 8.0), 1000 mM NaCl.

2.4 Final Reconstitution with the E3

1. Buffer K: 20 mM Tris-HCl (pH 8.0), 50 mM NaCl, 5 mM EDTA.

3 Methods

3.1 Protein Expression and Purification (Described Previously [29, 32, 58, 64–66])

3.1.1 Cloning and Purification of ΔN18Smt3

1. Residues 19-98 of yeast Smt3, representing mature C-terminally processed Smt3 lacking the disordered N-terminal tail, are amplified by PCR from yeast genomic DNA and subsequently cloned into the 5′ NdeI and 3′ XhoI sites of pET28b. The expressed protein will have an N-terminal, thrombin cleavable, His_6 fusion tag.
2. Transform the plasmid into *E. coli* BL21 (DE3) RIL Codon Plus cells.
3. Inoculate 50 mL of LB or SB with the Codon Plus cells harboring the plasmid and let grow at 37 °C overnight for ca. 16 h with shaking. Inoculate 4 × 1 L of Super Broth in 2 L baffled shaking flasks, from the overnight culture to a final optical density at 600 nM (OD_{600}) = 0.1, and let grow at

37 °C. When the culture $OD_{600} = 1.0–2.0$, induce with 1 mM IPTG, and let grow 3–4 h at 30 °C.

4. Collect cells by centrifugation (4500 × *g*). Suspend cell pellets in suspension buffer (2 mL of buffer per g of wet cell mass). Flash freeze in liquid nitrogen and store at −80 °C for future use.
5. Rapidly thaw suspended cells in warm water. On ice, adjust composition of the cell solution to that of Lysis Buffer A by adding each of the components from stock solutions, with rapid mixing (assuming a volume increase of ca. 1.2-fold).
6. On ice/water slurry with rapid stirring (ideally in a cold room), sonicate at 60% output for ca. 1 min for every 100 mL of lysate with 5 s on, 8 s off to disrupt cells.
7. Pellet cell debris for 30 min at 40,000 × *g* at 4 °C and collect the cleared supernatant (*see* **Note 6**).
8. Apply the supernatant to 10 mL of NiNTA resin in a chromatography column, equilibrated with Buffer A. Wash column with 10 column volumes of Buffer A followed by elution with Buffer B. Locate the peak of the eluate with Bradford reagent and/or analyze by SDS-PAGE.
9. Pool the eluate peak and incubate with 100 units of thrombin and dialyze overnight against 2 L Buffer C in 2K molecular weight cutoff dialysis tubing at 4 °C. Confirm cleavage of the His_6 fusion tag by SDS-PAGE.
10. Resolve the dialysate over Superdex 75 gel filtration column equilibrated with Buffer C. Analyze the fractions by SDS-PAGE.
11. Collect the peak fractions and concentrate to ca. 30–45 mg/mL. Measure the OD_{280}, and determine the final protein concentration with the extinction coefficient 1490/M/cm or 1 $OD_{280} = 6.22$ mg/mL. Extinction coefficient determined by Protean software within Lasergene 8 suite (DNASTAR). Aliquot, flash freeze, and store at −80 °C for later use.

3.1.2 Cloning and Purification of the Yeast SUMO/Smt3 E1-Activating Enzyme

1. The yeast SUMO E1-activating enzyme is a heterodimer complex of Aos1 and Uba2. The yeast genes for full-length Aos1 and C-terminally truncated Uba2 (ΔC-term, amino acids 1–554) are amplified by PCR from yeast genomic DNA.
2. Yeast Aos1 is cloned into pET15b using the 5′ NcoI and 3′ BamHI sites, and Uba2ΔC-term is ligated into pET28b using the 5′ NdeI and 3′ XhoI sites. Uba2ΔC-term is expressed with an N-terminal, thrombin cleavable, His_6 fusion tag.
3. Transform both plasmids, together, into *E. coli* BL21 (DE3) RIL Codon Plus cells for coexpression.

4. The Codon Plus cells with the plasmids are inoculated into 200 mL of LB or SB and let grow at 37 °C overnight for ca. 16 h with shaking. Inoculate 12 × 1 L of Super Broth in 2 L baffled shaking flasks, from the overnight culture to a final $OD_{600} = 0.1$, and let grow at 37 °C. When the culture reaches $OD_{600} = 1.0–2.0$, cold shock in ice water for 30 min, induce with 2% ethanol and 1 mM IPTG. The temperature is shifted to 18 °C and cells are induced overnight for ca. 16 h.
5. Cells are harvested, and the pellets are processed for lysis by sonication as described in Subheading 3.1.1, with the exception of using Lysis Buffer B composition.
6. Apply the supernatant to 10 mL of NiNTA resin in a chromatography column, equilibrated with Buffer A. Wash column with 10 column volumes of Buffer A followed by elution with Buffer B. Locate the peak of the eluate with Bradford reagent and/or analyze by SDS-PAGE.
7. Pool the eluate peak and incubate with 100 units of thrombin and dialyze overnight against 2 L Buffer C in 8K molecular weight cutoff dialysis tubing at 4 °C. Confirm cleavage of the His_6 fusion tag by SDS-PAGE.
8. Resolve the dialysate over Superdex 200 gel filtration column equilibrated with Buffer C. Analyze the fractions by SDS-PAGE.
9. Pool peak fractions containing the Aos1/Uba2ΔC-term heterodimer complex. Dilute with 20 mM Tris-HCl (pH 8.0), 1 mM BME to final [NaCl] = 100 mM or desalt to Buffer D.
10. Load this mixture onto anion-exchange resin (Mono-Q). The protein is eluted with a linear gradient from Buffer D to 35% Buffer E over 20 column volumes. The Aos1/Uba2ΔC-term complex elutes between ca. 200 and 300 mM NaCl.
11. Collect the peak fractions and concentrate to ca. 10 mg/mL. Measure the OD_{280}, and determine the final protein concentration with the extinction coefficient 65,710/M/cm or 1 OD_{280} = 1.55 mg/mL. Aliquot, flash freeze, and store at −80 °C for later use.

3.1.3 Cloning and Purification of the Yeast E2 $Ubc9_{C5S/A129K/K153R}$ Mutant

1. The SUMO-specific yeast E2, Ubc9, is amplified by PCR from yeast genomic DNA and subsequently cloned into the 5′ NdeI and 3′ HindIII sites of pET28b. The expressed protein will have an N-terminal, thrombin cleavable, His_6 fusion tag.
2. Three successive site-directed mutagenesis reactions generate the K153R mutation, the A129K/K153R mutation combination, and the C5S/A129K/K153R mutation combination yielding the pET28b-$E2_{Ubc9}^{C5S/A129K/K153R}$ construct (*see* **Note 7**).

3. The plasmid is transformed into *E. coli* BL21 (DE3) RIL Codon Plus cells.
4. Inoculate 50 mL of LB or SB with the Codon Plus cells harboring the plasmid, and let grow at 37 °C overnight for ca. 16 h with shaking. Inoculate 4 × 1 L of Super Broth in 2 L baffled shaking flasks from the overnight culture to a final $OD_{600} = 0.1$, and let grow at 37 °C. When the culture reaches $OD_{600} = 1.0$–2.0, induce with 1 mM IPTG, and let grow 3–4 h at 30 °C.
5. Cells are harvested, and the pellets are processed for lysis by sonication as described in Subheading 3.1.1, again using Lysis Buffer A composition.
6. Apply the supernatant to 10 mL of NiNTA resin in a chromatography column, equilibrated with Buffer A. Wash column with 10 column volumes of Buffer A followed by elution with Buffer B. Locate the peak of the eluate with Bradford reagent and/or analyze by SDS-PAGE.
7. Pool the eluate peak and incubate with 100 units of thrombin and dialyze overnight against 2 L Buffer C in 8K molecular weight cutoff dialysis tubing at 4 °C. Confirm cleavage of the His_6 fusion tag by SDS-PAGE.
8. Resolve the dialysate over Superdex 75 gel filtration column equilibrated with Buffer C. Analyze the fractions by SDS-PAGE.
9. Typically, the protein is sufficiently pure at this point; however, if further purity is required, pool peak fractions containing $Ubc9^{C5S/A129K/K153R}$. Dilute with 20 mM Tris-HCl (pH 8.0), 1 mM BME to final [NaCl] = 100 mM or desalt to Buffer D.
10. Load this mixture onto cation-exchange resin (Mono-S). The protein is eluted with a linear gradient from Buffer D to 35% Buffer E over 20 column volumes. The $Ubc9^{C5S/A129K/K153R}$ protein elutes at approximately 150–230 mM NaCl.
11. Collect the peak fractions and concentrate to between 15 and 45 mg/mL. Measure the OD_{280} and determine the final protein concentration with the extinction coefficient 39,545/M/cm or 1 $OD_{280} = 0.45$ mg/mL. Aliquot, flash freeze, and store at −80 °C for later use.

3.1.4 Cloning and Purification of the Yeast E3 $Siz1_{(167\text{-}449)}^{C361D}$-Δ N18Smt3 Fusion Protein

1. The SUMO-specific yeast E3, Siz1, catalytic fragment (residues 167–449) is amplified by PCR from yeast genomic DNA and subsequently cloned into the 5′ BamHI and 3′ HindIII sites of pSmt3, creating pSmt3-$Siz1_{(167\text{-}449)}$. Subsequently, ΔN18Smt3 is amplified from the plasmid in Subheading 3.1.1 and ligated into the 5′ HindIII and 3′ XhoI of the pSmt3-$Siz1_{(167\text{-}449)}$ plasmid, creating pSmt3-$Siz1_{(167\text{-}449)}$-Δ

N18Smt3. The expressed Siz1$_{(167\text{-}449)}$-ΔN18Smt3 fusion protein will have an N-terminal, Ulp1 cleavable, His$_6$-Smt3 fusion tag.

2. The Siz1 C361D mutation is introduced into pSmt3-Siz1$_{(167\text{-}449)}$-ΔN18Smt3 by site-directed mutagenesis. Cysteine 361 occurs within the SP-RING domain but is not involved in Zn^{2+} coordination [29] and may be prone to oxidation and Siz1 inactivation. In *Candida albicans* and *Caenorhabditis elegans*, the homologous residue is an aspartate, thus our choice for mutation.
3. The plasmid is transformed into *E. coli* BL21 (DE3) RIL Codon Plus cells.
4. Inoculate 200 mL of LB or SB with the Codon Plus cells harboring the plasmid, and let grow at 37 °C overnight for ca. 16 h with shaking. Inoculate 8 × 1 L of Super Broth in 2 L baffled shaking flasks from the overnight culture to a final $OD_{600} = 0.1$, and let grow at 37 °C. When the culture reaches $OD_{600} = 1.0–2.0$, cold shock in ice water for 30 min, then induce with 2% ethanol and 1 mM IPTG, and let grow ca. 16 h at 18 °C.
5. Cells are harvested, and the pellets are processed for lysis by sonication as described in Subheading 3.1.1, with the exception of using Lysis Buffer C composition.
6. Apply the supernatant to 10 mL of NiNTA resin in a chromatography column, equilibrated with Buffer A. Wash column with 10 column volumes of Buffer A followed by elution with Buffer B. Locate the peak of the eluate with Bradford reagent and/or analyze by SDS-PAGE.
7. Pool the eluate peak and incubate with 1:10,000 His$_6$-Ulp1 or His$_6$-GST-Ulp1 (mg Ulp1:mg Smt3-Siz1-ΔN18Smt3) (*see* **Note 3**) and dialyze overnight against 2 L Buffer C in 8K molecular weight cutoff dialysis tubing at 4 °C. Confirm cleavage of the His$_6$-Smt3 fusion tag by SDS-PAGE.
8. If using His$_6$-GST-Ulp1, pass dialysate cleavage product over 3 mL glutathione beads equilibrated in Buffer C collecting the unadsorbed protein flow-through.
9. Resolve the flow-through over Superdex 200 gel filtration column equilibrated with Buffer C. Analyze the fractions by SDS-PAGE.
10. Pool peak fractions and dilute with 20 mM Tris-HCl (pH 8.0), 1 mM BME to final [NaCl] = 100 mM or desalt to Buffer D.
11. Load this mixture onto anion-exchange resin (Mono-Q). The protein is eluted with a linear gradient from Buffer D to 35%

Buffer E over 20 column volumes. The $Siz1_{(167\text{-}449)}^{C361D}$-Δ N18Smt3 protein elutes between ca. 220 and 280 mM NaCl.

12. Collect the peak fractions and concentrate to ca. 10–12 mg/mL. Measure the OD_{280} and determine the final protein concentration with the extinction coefficient 33,890/M/cm or 1 OD_{280} = 1.23 mg/mL. Aliquot, flash freeze and store at −80 °C for later use.

3.1.5 Cloning and Purification of Yeast $PCNA_{K127G/K164C}$ and $PCNA_{K77D/C81E/R110D/K127G/K164C}$

1. The current chemical cross-linking protocol was performed with both trimeric ($PCNA^{K127G/K164C}$) and monomeric ($PCNA^{K77D/C81E/R110D/K127G/K164C}$). Both monomeric and trimeric versions were reconstituted into final E2-SUMO/E3/substrate complexes and crystallized; however the crystals for trimeric did not diffract to sufficient resolution to yield a structure.
2. Full-length PCNA is amplified by PCR from yeast genomic DNA and subsequently cloned into the 5′ NdeI and 3′ XhoI sites of pET21a in order to express a tag-less protein with native termini.
3. Serial site-directed mutagenesis generates the K127G mutation and then the K127G/K164C mutation combination to utilize a thiol-based cross-linking strategy at the E3-dependent site of modification (K164). Subsequent mutagenesis created the K77D/C81E/R110D/K127G/K164C combination of mutations to create monomeric PCNA with the same ability to chemically cross-link.
4. The plasmids are transformed, separately, into *E. coli* BL21 (DE3) RIL Codon Plus cells.
5. Inoculate 50 mL of LB or SB with the Codon Plus cells harboring the plasmid, and let grow at 37 °C overnight for ca. 16 h with shaking. Inoculate 4 × 1 L of Super Broth in 2 L baffled shaking flasks from the overnight culture to a final OD_{600} = 0.1, and let grow at 37 °C. When the culture reaches OD_{600} = 1.0–2.0, induce with 1 mM IPTG and let grow ca. 16 h at 18 °C.
6. Cells are harvested, and the pellets are processed for lysis by sonication as described in Subheading 3.1.1, with the exception of utilizing Lysis Buffer D composition.
7. Collect the supernatant and, with rapid stirring at 4 °C, slowly add dry $(NH_4)_2SO_4$ to 40% (for trimeric $PCNA^{K127G/K164C}$) or 50% (for monomeric $PCNA^{K77D/C81E/R110D/K127G/K164C}$). Let this mix at 4 °C for 4 h to overnight. Pellet the precipitate at 40,000 × *g* for 30 min at 4 °C. Collect the supernatant and, with rapid stirring at 4 °C, slowly add dry $(NH_4)_2SO_4$ to 70%. Let mix 4 h at 4 °C. Pellet the precipitate at 40,000 × *g* for 30 min at 4 °C. Dissolve the pellet in 4 mL 20 mM Tris-HCl

(pH 8.0), 1 mM BME. Dialyze the dissolved pellet(s) against 2 L Buffer C for 4 h to overnight at 4 °C in 8K MWCO dialysis tubing.

8. Resolve the dialysate over Superdex 200 gel filtration column equilibrated with Buffer C (may require two runs). Analyze the fractions by SDS-PAGE.
9. Pool peak fractions and dilute with 20 mM Tris-HCl (pH 8.0), 1 mM BME to final [NaCl] = 200 mM or desalt to Buffer G.
10. Load one fourth of this mixture (to avoid overloading) onto anion-exchange resin (Mono-Q). The protein is eluted with a linear gradient from Buffer G to 30% Buffer E over 10 column volumes. The trimeric $PCNA^{K127G/K164C}$ protein elutes between ca. 330 and 440 mM NaCl, and monomeric $PCNA^{K77D/C81E/R110D/K127G/K164C}$ elutes between ca. 200 and 380 mM NaCl. Repeat for three more column runs. Analyze the fractions by SDS-PAGE.
11. Collect the peak fractions and concentrate $PCNA_{K127G/K164C}$ to ca. 70 mg/mL and measure the OD_{280} and determine the final protein concentration with the extinction coefficient 6585/M/cm or 1 OD_{280} = 4.38 mg/mL. Concentrate monomeric $PCNA^{K77D/C81E/R110D/K127G/K164C}$ to 40–100 mg/mL and measure the OD_{280} and determine the final protein concentration with the extinction coefficient 6460/M/cm or 1 OD_{280} = 4.46 mg/mL. Aliquot, flash freeze and store at −80 °C for later use.

3.2 E2-SUMO Formation Reaction

1. The E2-SUMO thioester mimetic is formed in a reaction containing the $E1^{Aos1/Uba2}$, $E2_{Ubc9}$, and $Ubl_{\Delta N18SUMO}$, followed by purification of the mimetic from the E1, the unreacted E2 and SUMO, and in some instances any E2-2(SUMO) or E2-$(SUMO)_2$ side products (*see* **Note 8**). Our early explorations with the formation of the E2-SUMO mimetic, prior to employing the chemical cross-linking strategy, were performed with $E2_{Ubc9}{}^{A129K/K153R}$ (Fig. 5). The $E2_{Ubc9}{}^{C5S/A129K/K153R}$ mutant, necessary for specificity of cross-linking, behaves similarly in these types of reactions. In our explorations, we found that the presence of the $E3_{Siz1}$ increased the rate of E2-SUMO formation, in the case of the $E2_{Ubc9}{}^{A129K/K153R}$ mutant but not the $E2_{Ubc9}{}^{C93K}$ mutant. We interpret this result to indicate that, in the case of the $E2_{Ubc9}{}^{C5S/A129K/K153R}$ mutant, SUMO is transferred from the E1 onto the E2 active site cysteine to form an E2-SUMO thioester. This thioester is then attacked by the lysine installed at position 129 to form the isopeptide linkage adjacent to the active site cysteine. These data suggest binding of the E3 to the charged E2 stimulates conjugation to the installed lysine residue by orienting SUMO in the closed

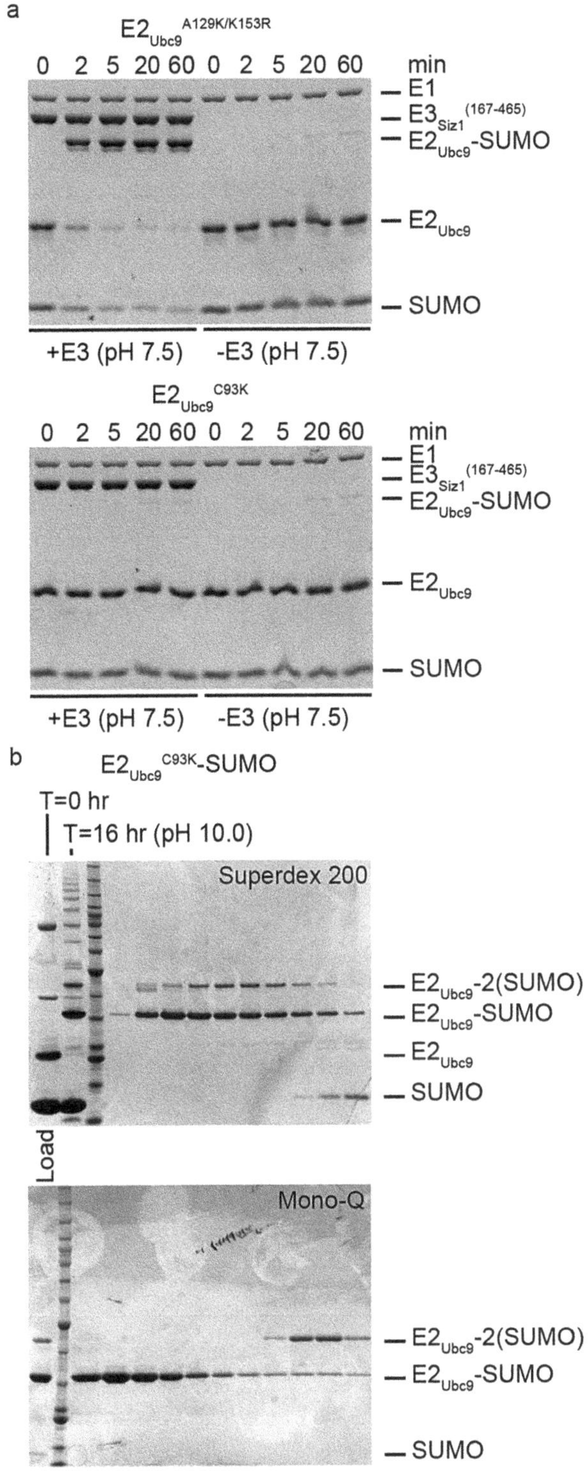

Fig. 5 $E2_{Ubc9}$-SUMO thioester mimetic formation and purification. (**a**) SDS-PAGE analysis of a time course for reactions forming the $E2_{Ubc9}{}^{C93K}$-SUMO versus $E2_{Ubc9}{}^{A129K}$ -SUMO mimetics in the presence and absence of the E3, Coomassie stained. Time points are 0, 2, 5, 20, and 60 min. The presence of the E3 speeds the

and most productive conformation for nucleophilic attack. Our final preparative protocol did not employ E3 to form the E2-SUMO mimetic as we opted to bypass the requirement of the E3 by performing the reactions at higher pH to ensure lysine deprotonation with fewer protein components.

2. Set up a reaction with 50 mM BIS-Tris propane (pH 9.5), 50 mM NaCl, 10 mM $MgCl_2$, 0.1% Tween-20, 1 mM DTT, 0.5 μM $E1_{Aos1/Uba2}$, 100 μM $E2_{Ubc9}^{C5S/A129K/K153R}$, and 200 μM ΔN18Smt3. Save aliquot for analysis by SDS-PAGE. Initiate the reaction with the addition of 2 mM buffered ATP. Let reaction incubate 1 h at 30 °C (Fig. 5) (*see* **Note 9**).
3. Filter the reaction, and resolve over Superdex 75 gel filtration column equilibrated with Buffer C. Analyze the fractions by SDS-PAGE.
4. Collect the peak fractions and concentrate to ca. 30–40 mg/mL. Measure the OD_{280}, and determine the final protein concentration with the extinction coefficient 41,035/M/cm or 1 OD_{280} = 0.66 mg/mL. Aliquot, flash freeze, and store at −80 °C for later use.

3.3 E2-SUMO-PCNA Cross-Linking Reaction

3.3.1 Cross-Linking with BMOE

1. BMOE is a robust and highly reactive reagent to generate stable cross-links between two cysteine residues (Fig. 3).
2. Rapidly thaw the E2-Smt3 thioester mimetic and PCNA proteins to be cross-linked. Add TCEP to 1 mM and incubate 15 min at 22 °C. Desalt to cross-linking buffer (*see* **Note 10**).
3. To 800 μL of 493 μM and desalted $E2_{Ubc9}^{C5S/A129K/K153R}$-ΔN18Smt3, add 12 μL of 100 mM BMOE (ca. threefold excess BMOE over E2-SUMO mimetic), and incubate 3 min at 22 °C. Desalt to cross-linking buffer to remove excess BMOE.
4. Mix the desalted E2-Smt3-BMOE with the 480 μL of 3290 μM desalted $PCNA^{K77D/C81E/R110D/K127G/K164C}$ (ca. fourfold excess PCNA) (*see* **Note 11**), and let react 15 min at 22 °C (Fig. 6).
5. Quench with 1 μL BME.

Fig. 5 (continued) reaction in the case of A129K, but not C93K. This unexpected observation suggests E3 binding activates the E2-SUMO thioester for nucleophilic attack by the lysine exogenously installed proximal to its own active site cysteine. This observation supports the model that the E1 catalyzes transfer of the SUMO to Cys93 and not directly to the installed Lys129. (**b**) SDS-PAGE analysis of the purification of the $E2_{Ubc9}^{C93K}$-SUMO thioester mimetic, Coomassie stained. Gel filtration can isolate the $E2_{Ubc9}$-SUMO mimetic from the E1, free E2, and free SUMO. The larger than expected shift in stokes radius and trailing peak results, in part, from the non-covalent $E2_{Ubc9}$/SUMO backside interaction creating a daisy chain effect. (**c**) The purification of the $E2_{Ubc9}^{A129K}$-Ubl_{SUMO} thioester mimetic proceeds similar to the C93K mutant

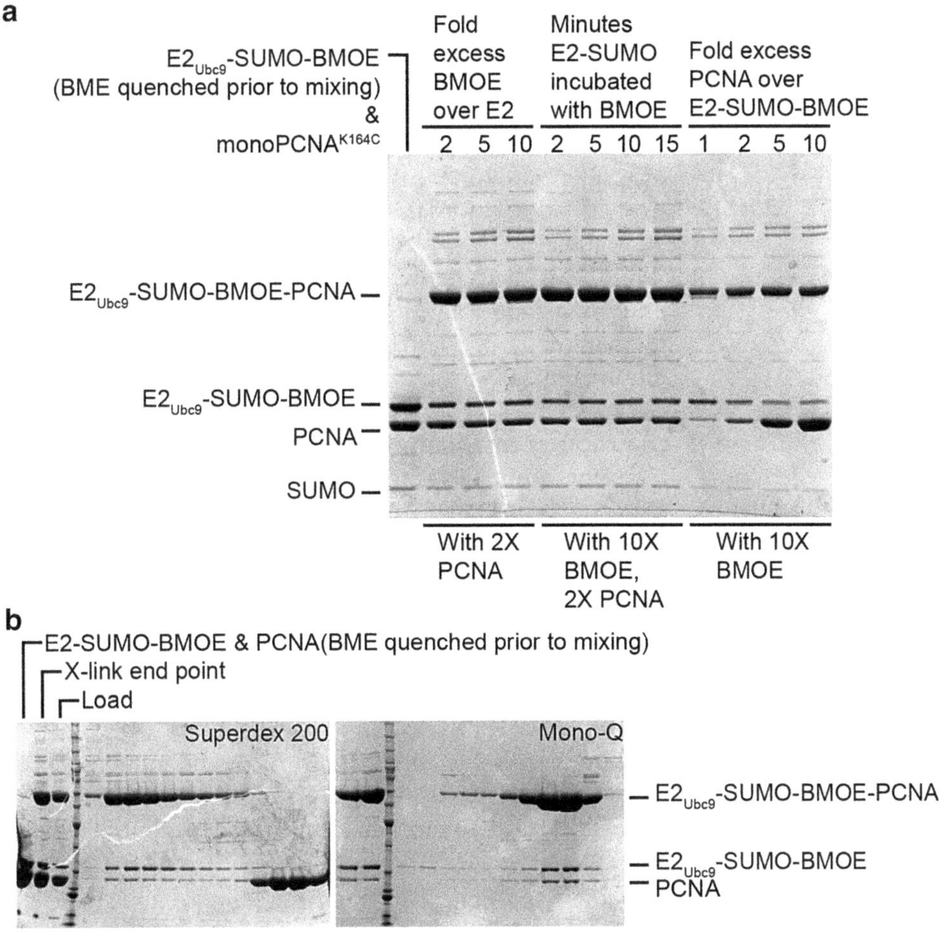

Fig. 6 BMOE cross-linking reaction and complex purification. (**a**) SDS-PAGE analysis of reactions exploring optimal BMOE concentrations, time of reaction, and ratio of PCNA to E2-SUMO-BMOE for maximizing yield of final product, Coomassie stained. (**b**) SDS-PAGE analysis of the purification of the E2-SUMO-BMOE-PCNA cross-linked complex, Coomassie stained

6. Resolve the quenched reaction over Superdex 200 gel filtration column equilibrated with Buffer C. Analyze the fractions by SDS-PAGE.
7. Pool peak fractions and dilute with 20 mM Tris-HCl (pH 8.0), 1 mM BME to final [NaCl] = 200 mM or desalt to Buffer G.
8. Load this mixture onto anion-exchange resin (Mono-Q). The protein is eluted with a linear gradient from Buffer G to 30% Buffer E over 10 column volumes. The Ubc9-Smt3-BMOE-PCNA complex elutes approximately at the [NaCl] that the source PCNA species eluted originally. Analyze the fractions by SDS-PAGE.
9. Collect the peak fractions and measure the OD_{280} and determine the final protein concentration with the extinction

coefficient 81,510/M/cm or 1 OD_{280} = 1.20 mg/mL (for E2-SUMO-BMOE-PCNA$^{K77D/C81E/R110D/K127G/K164C}$) or with the extinction coefficient 81,635/M/cm or 1 OD_{280} = 1.20 mg/mL (for E2-SUMO-BMOE-PCNA$^{K127G/K164C}$). Proceed to final reconstitution with E3 or concentrate to at least 10 mg/mL, aliquot, flash freeze, and store at −80 °C for later use.

3.3.2 Cross-Linking with EDT

1. While BMOE is a robust cross-linker, with two bulky maleimide rings and long distance between reactive centers, it is a poor mimetic of the distance and geometry expected for the tetrahedral intermediate (Figs. 1b and 3). Cross-linking via EDT provides a bridge between the E2 active site cysteine and the Lys to Cys mutation on the target substrate that has the same number of atoms as the predicted tetrahedral intermediate (Fig. 4). Thus, EDT is a better structural mimetic of the reaction intermediate, despite the two disulfide bonds that impart unique dihedral angles.
2. Flash thaw the E2-Smt3 thioester mimetic and PCNA proteins to be cross-linked. Add TCEP to 1 mM and incubate 15 min at 22 °C. Desalt to cross-linking buffer (*see* **Note 12**).
3. To 400 μL of 1238 μM and desalted $E2_{Ubc9}^{C5S/A129K/K153R}$-Δ N18Smt3, add 4 μL of 300 mM AT2, and incubate 15 min at 22 °C. The AT2 forms a disulfide adduct to the active site cysteine. Desalt to cross-linking buffer to remove excess AT2 (Fig. 4).
4. Add 5 μL of 300 mM EDT to the desalted E2-Smt3-AT2 and incubate 15 min at 22 °C. The AT2 acts as a good leaving group and helps catalyze the formation of the E2-Smt3-EDT disulfide adduct on the active site cysteine. Desalt to cross-linking buffer to remove excess EDT (Fig. 4).
5. Add 5 μL of 300 mM AT2 to the desalted E2-Smt3-EDT and incubate 15 min at 22 °C. This second round of AT2 forms a disulfide adduct to the distal end of the EDT molecules attached to the E2 active site cysteines (Fig. 4). This second AT2 will act as a leaving group to catalyze the disulfide bond formation between the distal end of the EDT and the PCNA cysteine. Desalt to cross-linking buffer.
6. Mix the desalted E2-Smt3-EDT-AT2 with the 825 μL of 1575 μM and desalted PCNA$^{K77D/C81E/R110D/K127G/K164C}$ (ca. threefold excess PCNA), and let react 20 min at 22 °C (Fig. 7).
7. Resolve the cross-linked reaction over Superdex 200 gel filtration column equilibrated with Buffer H. Analyze the fractions by SDS-PAGE.

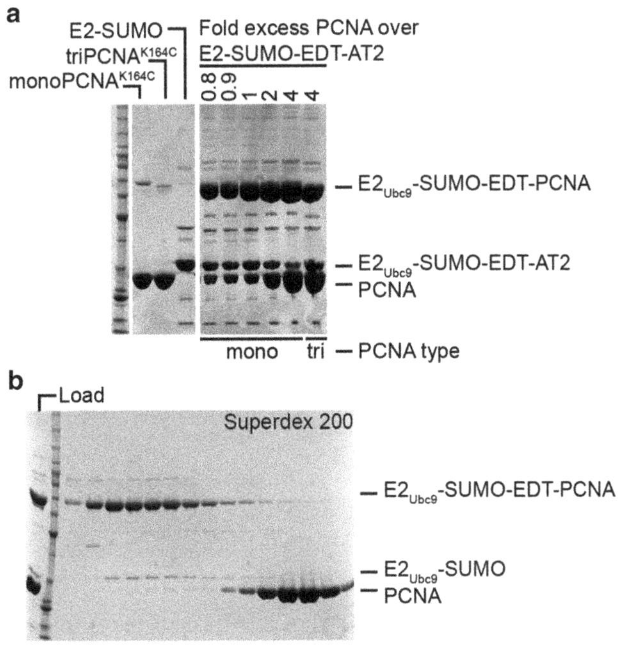

Fig. 7 EDT cross-linking reaction and complex purification. (**a**) SDS-PAGE analysis of reactions exploring the optimal PCNA ratio to E2-SUMO-EDT-AT2 for maximizing yield of final product, Coomassie stained. (**b**) SDS-PAGE analysis of the gel filtration purification of the E2-SUMO-EDT-PCNA complex, Coomassie stained

8. Pool peak fractions and dilute with 20 mM Tris-HCl (pH 8.0), 1 mM BME to final [NaCl] = 200 mM or desalt to Buffer I.
9. Load this mixture onto anion-exchange resin (Mono-Q). The protein is eluted with a linear gradient from Buffer I to 30% Buffer J over 10 column volumes. The Ubc9-Smt3-EDT-PCNA complex elutes approximately at the [NaCl] that the source PCNA species eluted originally. Analyze the fractions by SDS-PAGE.
10. Collect the peak fractions and measure the OD_{280} and determine the final protein concentration as in Subheading 3.3.1. Because of the potential instability of the EDT linkage, proceed to final reconstitution with E3.

3.4 Reconstitution with E3

1. For the final reconstitution, mix the $E2_{Ubc9}$-Ubl_{Smt3}-PCNA cross-linked complex with the $E3_{Siz1}$-ΔN18Smt3 in a 1:1 stoichiometry. Dialyze to Buffer F (if using BMOE) or Buffer K (if using EDT) (*see* **Note 13**).
2. Resolve the reconstituted complex over Superdex 200 gel filtration column equilibrated with Buffer F or K, depending on BMOE versus EDT as the cross-linker, respectively. Analyze the fractions by SDS-PAGE (Fig. 8).

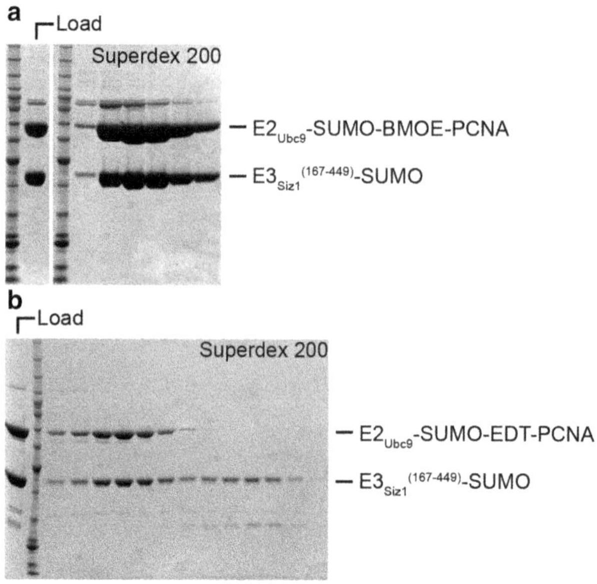

Fig. 8 Final reconstitution of the E2-SUMO-PCNA cross-linked complex with the E3-SUMO fusion. SDS-PAGE analysis of the gel filtration purification of the final reconstituted complex between (**a**) the E2-SUMO-BMOE-PCNA complex or (**b**) the E2-SUMO-EDT-PCNA complex with E3-SUMO fusion, Coomassie stained

3. Collect the peak fractions and concentrate to 10–15 mg/mL. Proceed to downstream structure determination technique of choice or concentrate to at least 10 mg/mL, aliquot, flash freeze, and store at −80 °C for later use.

4 Notes

1. The generality of this strategy will depend primarily on the particular E2 and target substrate and the number and surface accessible cysteine residues to which cross-linking is not desired. In the case of the E2, it may be necessary to mutate any cysteine, other than the active site, to a compatible residue (perhaps serine, threonine, alanine, or an alternative side chain based on sequence analysis and alignment with orthologous proteins). In the case of the target substrate, whether any surface exposed cysteine residues need to be removed by mutation should be determined empirically with pilot cross-linking assays. For instance, Cys22, Cys30, and Cys62 remained in our engineered PCNA K164C monomeric substrate, while Cys22, Cys30, Cys62, and Cys81 remained in the trimeric version. While not employed here, it is possible that the E3 could provide specificity during the cross-linking reaction.

2. The pSMT3 vectors derive from pET-28b and include an N-terminal His_6 fusion to *S. cerevisiae* SUMO/Smt3 [63]. pSMT3 enables cloning of the protein of interest (POI) in the MCS, via the BamHI and XhoI sites, in frame with the N-terminal His_6-Smt3 to generate the His_6-Smt3-Gly-Gly-Ser-POI fusion polypeptide. The Ulp1 protease cleaves the His_6-Smt3 from the POI, C-terminal to the di-Gly motif, liberating the POI with an N-terminal serine residue artifact. The TOPO-SMT3 vector is modified pSMT3, with custom TOPO adaptation (Invitrogen) for downstream flap ligation. The N-terminal Smt3 fusion can enhance expression, solubility, and yield for challenging proteins.
3. This protocol describes formation of a Smt3/SUMO conjugate. Even trace amounts of contaminating Ulp1 can quickly reverse the conjugation reaction. A pilot cleavage reaction should be performed to identify the minimal amount of Ulp1 needed to cleave the protein of interest in the time frame necessary. Additionally, using His_6-GST-Ulp1 (purified via the His_6 tag) allows for cleavage of the His_6-Smt3 fusion tag and the ability to subsequently deplete the Ulp1 with immobilized glutathione resin. Additionally, trace Ulp1 can deposit on columns, become solubilized, and can contaminate protein preparations previously free of Ulp1. To prevent this, any remaining Ulp1 can be inactivated by applying 10 mL of 20 mM NEM to columns prior to equilibration and use. NEM can be inactivated with excess BME.
4. PMSF is only used for E1 and E3 lysis where there is an observed benefit. Imidazole is omitted from the E3 lysis buffer to minimize the exposure of the RING domain to imidazole where it might compete for Zn^{2+} coordination.
5. All buffers and solutions for use during FPLC are prepared with ultrapure water, 0.2 μM filtered and degassed for at least 1 h to maximize reproducibility and column integrity.
6. All steps of protein purification were performed at 4 °C or on ice. Each sample is passed through a 0.2 μM filter prior to application on chromatography columns. All proteins are concentrated (ideally to at least 10 mg/mL or greater), aliquoted, and flash frozen prior to storage at −80 °C.
7. The C5S mutation in Ubc9 results in a single cysteine in the E2, the active site cysteine, which is critical for specific cross-linking in later steps. The K153R mutation is introduced to prevent a known E2 SUMO conjugate from forming [67]. The A129K mutation is introduced to install a lysine residue proximal to the active site cysteine in a position that enables nucleophilic attack of the E2-SUMO thioester and formation of a stable isopeptide linkage. This position for the lysine works well

for $E2_{Ubc9}$ and $E2_{Rad6}$; however, other E2s may differ in their active site configurations. The extension of this technique to more divergent E2s may require exploration of alternative positions for installation of a suitable lysine.

8. Separation of the $E2_{Ubc9}$-Ubl from unreacted $E2_{Ubc9}$ or any $E2_{Ubc9}$-2(Ubl) or $E2_{Ubc9}$-$(Ubl)_2$ represents one of the more challenging aspects of this strategy. While $E2_{Ubc9}$ binds to Mono-S resin, full resolution of $E2_{Ubc9}$ from $E2_{Ubc9}$-Smt3 is not readily achieved. Full-length Smt3 binds to Mono-Q resin which enables resolution of free $E2_{Ubc9}$ from $E2_{Ubc9}$-Smt3. Additionally, $E2_{Ubc9}$ with two full-length Smt3 proteins (i.e., $E2_{Ubc9}$-2(Smt3) or $E2_{Ubc9}$-$(Smt3)_2$) elutes from the Mono-Q at higher [NaCl] than $E2_{Ubc9}$ with a single Smt3 ($E2_{Ubc9}$-Smt3). However, this purification advantage is lost with the ΔN18Smt3 construct since lowering the [NaCl] sufficiently to bind the Mono-Q causes the $E2_{Ubc9}$-ΔN18Smt3 complex to precipitate. This advantage is also not observed for the few ubiquitin E2s we have tested. In these instances, the E2, E2-Ubl, E2-2(Ubl), and E2-$(Ubl)_2$ species do not fully resolve from one another on gel filtration or ion exchange. In these cases, it is paramount to optimize the reaction conditions to maximize yield of desired product and minimize components that are challenging to remove (*see* **Note 9**). In cases such as these, two additional strategies may be employed. First, leaving the N-terminal His_6 tag on the Ubl can allow for immobilization of the Ubl and E2-Ubl mimetic on NiNTA resin while washing away unreacted E2 [35]. Second, a His_6-Smt3-Ubl (or perhaps His_6-GST-Ubl, His_6-MBP-Ubl, etc.) construct could be used to form the E2-His_6-Smt3-Ubl mimetic. This strategy provides the advantage of a sufficiently large difference in the Stokes radius for E2-2(Ubl) or E2-$(Ubl)_2$ to better separate them from E2-Ubl by gel filtration. The His_6-Smt3 (or His_6-GST, His_6-MBP, etc.) can be subsequently removed with Ulp1 (or Thrombin, TEV, etc.).

9. When extending this technique to other E2-Ubl mimetics, it will be advantageous to screen a range of E1, E2, and Ubl concentrations in addition to a range of reaction times and pH to identify conditions that maximize E2-Ubl conjugate yield while minimizing formation of undesirable E2-2(Ubl) or E2-$(Ubl)_2$ species.

10. An alternate technique would be to purify the E2-Ubl thioester mimetic and substrate to be cross-linked in the presence of TCEP instead of BME. TCEP is compatible with BMOE cross-linking; however, in the present protocol, we removed it by desalting at the same time as removing BME. Because BME will react and interfere with BMOE based cross-linking, it is crucial to remove it from all proteins prior to cross-linking.

Once proteins are cross-linked with BMOE, they are stable and no longer susceptible to reducing agents.

11. For monomeric PCNA, it was desirable to attempt to react all of the E2-Ubl-BMOE species and maximize yield, since the monomeric unmodified PCNA could be easily separated from the E2-Ubl-BMOE-PCNA species. In the case of trimeric PCNA, the reverse was true, and the incubation was maintained at a 1.2- to 1.5-fold excess of E2-Ubl-BMOE over PCNA in an attempt to modify all three Cys164 positions around the PCNA ring to achieve near homogeneity and to then purify away any unreacted E2-Ubl-BMOE by gel filtration.
12. The EDT cross-link is susceptible to reducing agents. It is critical to remove all BME or other reducing agents from the input proteins, the columns, and FPLC system prior to purification of the EDT cross-linked proteins. For this reason, it may be advantageous to use the more robust BMOE technique to optimize reconstitution conditions and structure determination before attempting EDT preparation and reconstitution for structural analysis.
13. The stoichiometry of this complex is due in part to the E3-SUMO fusion, where the SUMO binds the backside of $E2_{Ubc9}$ in addition to the other E2/E3 and E3/PCNA binding interactions. Because the backside affinities between E2s and Ubls differ, it is possible that other complexes may not reconstitute with similar stoichiometry to enable purification by gel filtration. In such cases, structural analysis may require a co-crystallization strategy.

Acknowledgments

This work was supported in part by GM065872 and GM118080 (NIH/NIGMS, C.D.L) and P30CA008748 (NIH/National Cancer Institute). The content is the authors' responsibility and does not represent the official views of the NIH. C.D.L is a Howard Hughes Medical Institute Investigator.

References

1. Kerscher O, Felberbaum R, Hochstrasser M (2006) Modification of proteins by ubiquitin and ubiquitin-like proteins. Annu Rev Cell Dev Biol 22:159–180
2. Hochstrasser M (2009) Origin and function of ubiquitin-like proteins. Nature 458:422–429
3. Gareau JR, Lima CD (2010) The SUMO pathway: emerging mechanisms that shape specificity, conjugation and recognition. Nat Rev Mol Cell Biol 11:861–871
4. Rose IA, Warms JV (1983) An enzyme with ubiquitin carboxy-terminal esterase activity from reticulocytes. Biochemistry 22:4234–4237
5. Pickart CM, Rose IA (1985) Ubiquitin carboxyl-terminal hydrolase acts on ubiquitin

carboxyl-terminal amides. J Biol Chem 260:7903–7910

6. Mayer AN, Wilkinson KD (1989) Detection, resolution, and nomenclature of multiple ubiquitin carboxyl-terminal esterases from bovine calf thymus. Biochemistry 28:166–172
7. Amerik AY, Hochstrasser M (2004) Mechanism and function of deubiquitinating enzymes. Biochim Biophys Acta 1695:189–207
8. Mevissen TET, Komander D (2017) Mechanisms of deubiquitinase specificity and regulation. Annu Rev Biochem 86:159–192
9. Hershko A, Ciechanover A, Heller H, Haas AL, Rose IA (1980) Proposed role of ATP in protein breakdown: conjugation of protein with multiple chains of the polypeptide of ATP-dependent proteolysis. Proc Natl Acad Sci U S A 77:1783–1786
10. Schulman BA, Harper JW (2009) Ubiquitin-like protein activation by E1 enzymes: the apex for downstream signalling pathways. Nat Rev Mol Cell Biol 10:319–331
11. Streich FC Jr, Lima CD (2014) Structural and functional insights to ubiquitin-like protein conjugation. Annu Rev Biophys 43:357–379
12. Streich FC, Haas AL (2010) Activation of ubiquitin and ubiquitin-like proteins. In: Groettrup M (ed) Conjugation and deconjugation of ubiquitin family modifiers. Landes Bioscience, Austin, pp 1–16
13. Cappadocia L, Lima CD (2018) Ubiquitin-like protein conjugation: structures, chemistry, and mechanism. Chem Rev 118(3):889–918
14. Haas AL, Rose IA (1982) The mechanism of ubiquitin activating enzyme. A kinetic and equilibrium analysis. J Biol Chem 257:10329–10337
15. Haas AL, Warms JV, Hershko A, Rose IA (1982) Ubiquitin-activating enzyme. Mechanism and role in protein-ubiquitin conjugation. J Biol Chem 257:2543–2548
16. Haas AL, Warms JV, Rose IA (1983) Ubiquitin adenylate: structure and role in ubiquitin activation. Biochemistry 22:4388–4394
17. Pickart CM, Rose IA (1985) Functional-heterogeneity of ubiquitin carrier proteins. J Biol Chem 260:1573–1581
18. Haas AL, Bright PM (1988) The resolution and characterization of putative ubiquitin carrier protein isozymes from rabbit reticulocytes. J Biol Chem 263:13258–13267
19. Hershko A, Heller H, Elias S, Ciechanover A (1983) Components of ubiquitin-protein ligase system. Resolution, affinity purification, and role in protein breakdown. J Biol Chem 258:8206–8214
20. Hershko A, Heller H, Eytan E, Reiss Y (1986) The protein substrate binding site of the ubiquitin-protein ligase system. J Biol Chem 261:11992–11999
21. Lorick KL, Jensen JP, Fang S, Ong AM, Hatakeyama S, Weissman AM (1999) RING fingers mediate ubiquitin-conjugating enzyme (E2)-dependent ubiquitination. Proc Natl Acad Sci U S A 96:11364–11369
22. Deshaies RJ, Joazeiro CA (2009) RING domain E3 ubiquitin ligases. Annu Rev Biochem 78:399–434
23. Freemont PS, Hanson IM, Trowsdale J (1991) A novel cysteine-rich sequence motif. Cell 64:483–484
24. Barlow PN, Luisi B, Milner A, Elliott M, Everett R (1994) Structure of the C3HC4 domain by 1H-nuclear magnetic resonance spectroscopy. A new structural class of zinc-finger. J Mol Biol 237:201–211
25. Johnson ES, Gupta AA (2001) An E3-like factor that promotes SUMO conjugation to the yeast septins. Cell 106:735–744
26. Kahyo T, Nishida T, Yasuda H (2001) Involvement of PIAS1 in the sumoylation of tumor suppressor p53. Mol Cell 8:713–718
27. Takahashi Y, Toh-e A, Kikuchi Y (2001) A novel factor required for the SUMO1/Smt3 conjugation of yeast septins. Gene 275:223–231
28. Hochstrasser M (2001) SP-RING for SUMO: new functions bloom for a ubiquitin-like protein. Cell 107:5–8
29. Yunus AA, Lima CD (2009) Structure of the Siz/PIAS SUMO E3 ligase Siz1 and determinants required for SUMO modification of PCNA. Mol Cell 35:669–682
30. Aravind L, Koonin EV (2000) The U box is a modified RING finger - a common domain in ubiquitination. Curr Biol 10:R132–R134
31. Hamilton KS, Ellison MJ, Barber KR, Williams RS, Huzil JT, McKenna S, Ptak C, Glover M, Shaw GS (2001) Structure of a conjugating enzyme-ubiquitin thiolester intermediate reveals a novel role for the ubiquitin tail. Structure 9:897–904
32. Reverter D, Lima CD (2005) Insights into E3 ligase activity revealed by a SUMO-RanGAP1-Ubc9-Nup358 complex. Nature 435:687–692
33. Dou H, Buetow L, Sibbet GJ, Cameron K, Huang DT (2012) BIRC7-E2 ubiquitin conjugate structure reveals the mechanism of ubiquitin transfer by a RING dimer. Nat Struct Mol Biol 19:876–883
34. Page RC, Pruneda JN, Amick J, Klevit RE, Misra S (2012) Structural insights into the conformation and oligomerization of

E2~ubiquitin conjugates. Biochemistry 51:4175–4187

35. Plechanovová A, Jaffray EG, Tatham MH, Naismith JH, Hay RT (2012) Structure of a RING E3 ligase and ubiquitin-loaded E2 primed for catalysis. Nature 489:115–120
36. Pruneda JN, Littlefield PJ, Soss SE, Nordquist KA, Chazin WJ, Brzovic PS, Klevit RE (2012) Structure of an E3:E2~Ub complex reveals an allosteric mechanism shared among RING/U-box ligases. Mol Cell 47:933–942
37. Dou H, Buetow L, Sibbet GJ, Cameron K, Huang DT (2013) Essentiality of a non-RING element in priming donor ubiquitin for catalysis by a monomeric E3. Nat Struct Mol Biol 20:982–986
38. Branigan E, Plechanovova A, Jaffray EG, Naismith JH, Hay RT (2015) Structural basis for the RING-catalyzed synthesis of K63-linked ubiquitin chains. Nat Struct Mol Biol 22:597–602
39. Hodge CD, Ismail IH, Edwards RA, Hura GL, Xiao AT, Tainer JA, Hendzel MJ, Glover JN (2016) RNF8 E3 ubiquitin ligase stimulates Ubc13 E2 conjugating activity that is essential for DNA double strand break signaling and BRCA1 tumor suppressor recruitment. J Biol Chem 291:9396–9410
40. Koliopoulos MG, Esposito D, Christodoulou E, Taylor IA, Rittinger K (2016) Functional role of TRIM E3 ligase oligomerization and regulation of catalytic activity. EMBO J 35:1204–1218
41. Lechtenberg BC, Rajput A, Sanishvili R, Dobaczewska MK, Ware CF, Mace PD, Riedl SJ (2016) Structure of a HOIP/E2~ubiquitin complex reveals RBR E3 ligase mechanism and regulation. Nature 529:546–550
42. Sanchez JG, Chiang JJ, Sparrer KM, Alam SL, Chi M, Roganowicz MD, Sankaran B, Gack MU, Pornillos O (2016) Mechanism of TRIM25 catalytic activation in the antiviral RIG-I pathway. Cell Rep 16:1315–1325
43. Wagner JM, Roganowicz MD, Skorupka K, Alam SL, Christensen D, Doss G, Wan Y, Frank GA, Ganser-Pornillos BK, Sundquist WI, Pornillos O (2016) Mechanism of B-box 2 domain-mediated higher-order assembly of the retroviral restriction factor TRIM5alpha. elife 5:e16309
44. Wright JD, Mace PD, Day CL (2016) Secondary ubiquitin-RING docking enhances Arkadia and Ark2C E3 ligase activity. Nat Struct Mol Biol 23:45–52
45. Dove KK, Olszewski JL, Martino L, Duda DM, Wu XS, Miller DJ, Reiter KH, Rittinger K, Schulman BA, Klevit RE (2017) Structural studies of HHARI/UbcH7 approximately Ub reveal unique E2 approximately Ub conformational restriction by RBR RING1. Structure 25:890–900.e5
46. Yuan L, Lv Z, Atkison JH, Olsen SK (2017) Structural insights into the mechanism and E2 specificity of the RBR E3 ubiquitin ligase HHARI. Nat Commun 8:211
47. Dawidziak DM, Sanchez JG, Wagner JM, Ganser-Pornillos BK, Pornillos O (2017) Structure and catalytic activation of the TRIM23 RING E3 ubiquitin ligase. Proteins 85:1957–1961
48. Nomura K, Klejnot M, Kowalczyk D, Hock AK, Sibbet GJ, Vousden KH, Huang DT (2017) Structural analysis of MDM2 RING separates degradation from regulation of p53 transcription activity. Nat Struct Mol Biol 24:578–587
49. Eddins MJ, Carlile CM, Gomez KM, Pickart CM, Wolberger C (2006) Mms2-Ubc13 covalently bound to ubiquitin reveals the structural basis of linkage-specific polyubiquitin chain formation. Nat Struct Mol Biol 13:915–920
50. Kamadurai HB, Souphron J, Scott DC, Duda DM, Miller DJ, Stringer D, Piper RC, Schulman BA (2009) Insights into ubiquitin transfer cascades from a structure of a UbcH5B~ubiquitin-HECT(NEDD4L) complex. Mol Cell 36:1095–1102
51. Kleiger G, Saha A, Lewis S, Kuhlman B, Deshaies RJ (2009) Rapid E2-E3 assembly and disassembly enable processive ubiquitylation of cullin-RING ubiquitin ligase substrates. Cell 139:957–968
52. Kamadurai HB, Qiu Y, Deng A, Harrison JS, Macdonald C, Actis M, Rodrigues P, Miller DJ, Souphron J, Lewis SM, Kurinov I, Fujii N, Hammel M, Piper R, Kuhlman B, Schulman BA (2013) Mechanism of ubiquitin ligation and lysine prioritization by a HECT E3. elife 2:e00828
53. Brown NG, VanderLinden R, Watson ER, Qiao R, Grace CR, Yamaguchi M, Weissmann F, Frye JJ, Dube P, Ei Cho S, Actis ML, Rodrigues P, Fujii N, Peters JM, Stark H, Schulman BA (2015) RING E3 mechanism for ubiquitin ligation to a disordered substrate visualized for human anaphase-promoting complex. Proc Natl Acad Sci U S A 112:5272–5279
54. Brown NG, VanderLinden R, Watson ER, Weissmann F, Ordureau A, Wu KP, Zhang W, Yu S, Mercredi PY, Harrison JS, Davidson IF, Qiao R, Lu Y, Dube P, Brunner MR, Grace CR, Miller DJ, Haselbach D, Jarvis MA, Yamaguchi M, Yanishevski D, Petzold G, Sidhu SS, Kuhlman B, Kirschner MW, Harper

JW, Peters JM, Stark H, Schulman BA (2016) Dual RING E3 architectures regulate multiubiquitination and ubiquitin chain elongation by APC/C. Cell 165:1440–1453

55. Qiao R, Weissmann F, Yamaguchi M, Brown NG, VanderLinden R, Imre R, Jarvis MA, Brunner MR, Davidson IF, Litos G, Haselbach D, Mechtler K, Stark H, Schulman BA, Peters JM (2016) Mechanism of APC/CCDC20 activation by mitotic phosphorylation. Proc Natl Acad Sci U S A 113: E2570–E2578
56. Pfander B, Moldovan GL, Sacher M, Hoege C, Jentsch S (2005) SUMO-modified PCNA recruits Srs2 to prevent recombination during S phase. Nature 436:428–433
57. Papouli E, Chen S, Davies AA, Huttner D, Krejci L, Sung P, Ulrich HD (2005) Crosstalk between SUMO and ubiquitin on PCNA is mediated by recruitment of the helicase Srs2p. Mol Cell 19:123–133
58. Armstrong AA, Mohideen F, Lima CD (2012) Recognition of SUMO-modified PCNA requires tandem receptor motifs in Srs2. Nature 483:59–63
59. Streich FC Jr, Lima CD (2016) Capturing a substrate in an activated RING E3/E2-SUMO complex. Nature 536:304–308
60. Capili AD, Lima CD (2007) Structure and analysis of a complex between SUMO and Ubc9 illustrates features of a conserved E2-Ubl interaction. J Mol Biol 369:608–618
61. Duda DM, van Waardenburg RC, Borg LA, McGarity S, Nourse A, Waddell MB, Bjornsti MA, Schulman BA (2007) Structure of a SUMO-binding-motif mimic bound to Smt3p-Ubc9p: conservation of a non-covalent ubiquitin-like protein-E2 complex as a platform for selective interactions within a SUMO pathway. J Mol Biol 369:619–630
62. Knipscheer P, van Dijk WJ, Olsen JV, Mann M, Sixma TK (2007) Noncovalent interaction between Ubc9 and SUMO promotes SUMO chain formation. EMBO J 26:2797–2807
63. Mossessova E, Lima CD (2000) Ulp1-SUMO crystal structure and genetic analysis reveal conserved interactions and a regulatory element essential for cell growth in yeast. Mol Cell 5:865–876
64. Bernier-Villamor V, Sampson DA, Matunis MJ, Lima CD (2002) Structural basis for E2-mediated SUMO conjugation revealed by a complex between ubiquitin-conjugating enzyme Ubc9 and RanGAP1. Cell 108:345–356
65. Lois LM, Lima CD (2005) Structures of the SUMO E1 provide mechanistic insights into SUMO activation and E2 recruitment to E1. EMBO J 24:439–451
66. Yunus AA, Lima CD (2009) Purification of SUMO conjugating enzymes and kinetic analysis of substrate conjugation. Methods Mol Biol 497:167–186
67. Knipscheer P, Flotho A, Klug H, Olsen JV, van Dijk WJ, Fish A, Johnson ES, Mann M, Sixma TK, Pichler A (2008) Ubc9 sumoylation regulates SUMO target discrimination. Mol Cell 31:371–382

Chapter 13

Small-Angle X-Ray Scattering for the Study of Proteins in the Ubiquitin Pathway

Jean-François Trempe and Kalle Gehring

Abstract

Small-angle X-ray scattering (SAXS) is an invaluable complement to other biophysical methods used to interrogate the structure and dynamics of proteins. Here, we describe the standard experimental protocol used in our laboratory to analyze proteins in the ubiquitin pathway. The method addresses buffer selection, data collection using an in-house X-ray source, diagnostic tests to assess data quality, and computational approaches to interpret SAXS data.

Key words X-ray scattering, SAXS, Ubiquitin, Kratky plot, Guinier analysis, Pair-distance distribution function

1 Introduction

Small-angle X-ray scattering (SAXS) is an extremely useful technique for studying the structure and dynamics of proteins in the ubiquitin pathway. Applications range from the straightforward confirmation of crystal structures to sophisticated modeling of protein dynamics and ensembles. In this protocol, we cover the basics of sample preparation, data acquisition, and analysis.

SAXS measurements are performed on an ensemble of molecules, and the interpretation of the data critically depends on the homogeneity of the protein solution. Sample purity and buffer choice are paramount to the success of a SAXS experiment. After purification of the protein of interest using affinity and/or ion-exchange chromatography, it is highly recommended to perform size-exclusion chromatography (SEC). This achieves two goals: (1) it removes protein aggregates and helps ensure a *monodisperse* sample and (2) it exchanges the protein into the buffer to be used for SAXS measurements. Because the scattering profile of the protein is obtained by subtracting the scattering of a buffer, it is imperative that the composition of the reference solution be identical to that of the protein solution.

Thibault Mayor and Gary Kleiger (eds.), *The Ubiquitin Proteasome System: Methods and Protocols*, Methods in Molecular Biology, vol. 1844, https://doi.org/10.1007/978-1-4939-8706-1_13,

The method we introduce specifically refers to an in-house X-ray source for data acquisition, but it may be easily adapted to synchrotron radiation facilities. Given the high intensity of X-ray beams available at synchrotrons, exposure times of a few seconds are usually sufficient to detect protein scattering. Those facilities are typically equipped with robotic liquid handling systems to automatically load samples into the X-ray beam. In addition, some synchrotron radiation facilities (CHESS, SSRL, ALS) now offer size-exclusion chromatography coupled to SAXS data acquisition, where images are acquired every second during the chromatographic run. This method has several advantages, such as the ability to resolve oligomeric assemblies that are in slow exchange, improved baseline subtraction, and avoiding radiation damage as each fraction of the sample is exposed for only a short period of time. Moreover, overlapping components can be resolved computationally through evolving factor analysis, as recently implemented with the software *BioXTAS Raw* [1–3].

Computational approaches for using SAXS data to probe protein structures are an active area of research. We briefly describe two approaches: ab initio modeling and back-calculating scattering from a known protein structure. Both have been widely used to study the ubiquitin system. As a check against crystallization artifacts, SAXS confirmed the solution structure of the ubiquitin ligase parkin [4, 5]. In the absence of protein crystals, different types of modeling have been used to generate structures for complexes of Atg7 and Atg3 [6], Ddi2 dimers [7, 8], and parkin with phospho-ubiquitin [5]. SAXS can also be used to characterize protein mobility, for example, in E2~Ub conjugates [9].

2 Materials

2.1 Size-Exclusion Chromatography

1. Filtered and deionized water, with resistivity 18.2 MΩ cm at 25 °C.
2. SAXS buffer, typically 50 mM Tris–HCl, pH 7.5, 100 mM NaCl, 5 mM DTT, 3% glycerol.
3. Concentrated protein solution above 10 mg/mL (*see* **Note 1**).
4. Fast protein liquid chromatography (FPLC) system, capable of flow rates between 0.05 and 10 mL/min, capable of handling back pressure above 20 bar.
5. Injection loop, 100 μL, with needle injection port.
6. High-resolution gel filtration column, with a total volume less than 3 mL and a column height equal or greater than 150 mm.
7. Centrifugal ultrafiltration device (500 μL), with a molecular weight cutoff (MWCO) matched to retain the protein of interest. For example, use MWCO of 10 kDa for proteins above 20 kDa.

8. Hamilton syringe, 100 μL with thin metal needle.
9. 0.25 μm filter device, connected to a standard laboratory vacuum.

2.2 Data Collection

1. In-house small-angle X-ray diffraction system, equipped with a temperature-controlled sample holder.
2. Sealed sample cell with capillary that holds 10–40 μL of protein solution, compatible with the sample holder of the diffraction system.
3. 2D X-ray detector for intensity counting.

2.3 Primary Data Processing and Data Analysis

1. Computer with ≥2 GHz processor and ≥8 GB RAM, running either Mac OSX, Windows, or Linux operating systems.

3 Methods

3.1 Buffer Selection

Selection of an appropriate buffer solution for data acquisition is an important step guided by considerations of protein stability, solubility, and aggregation. The following aspects should be considered when choosing its components:

1. *pH*: The pH of a solution can affect protein stability, solubility, and oligomerization. It is wise to pick a pH that is at least one unit away from the isoelectric point of the protein, as this tends to favor solubility. This in turn will dictate the choice of the buffer. For example, the protein parkin has a pI of 6.6, so measurements were performed in Tris buffer at pH 7.5–8 [4, 5].
2. *Salts*: Studies in pure water should be avoided since this can lead to long-range electric interactions between proteins, which require greater sample dilution to remove interparticle interactions. Salts such as NaCl or KCl often improve protein solubility. High salt concentrations (above 0.5 M) should be avoided since they reduce the contrast in electron density between the bulk solvent and protein.
3. *Glycerol*: This is a very common additive and used principally because it improves the thermal stability of proteins and reduces protein-protein interactions improving monodispersity [10]. Other polyols such as saccharides (sucrose, glucose, trehalose) may be used, and ideal concentrations are between 2 and 5%. Higher concentrations are not recommended as they increase the buffer density and reduce the contrast with the protein.
4. *Reducing agents*: Chemicals such as dithiothreitol (DTT), β-mercaptoethanol (BME), and tris(2-carboxyethyl)phosphine

(TCEP) are frequently used to prevent formation of disulfide bonds between cysteine side chains. Thiol-based agents such as DTT and BME have the added advantage of scavenging free radicals generated by the X-ray beam and extend the stability of the protein solution during exposure. As much as 10 mM of these agents can be used. Ascorbic acid has also been suggested to scavenge effectively free radicals [11].

5. *Detergents*: Some proteins require detergents to remain in solution. Detergents can be used in SAXS, but because micelles scatter strongly, they must be used below the critical micellar concentration (CMC). Thus, it is recommended to use detergents such as CHAPS, CHAPSO, or octyl β-D-glucopyranoside, which have high CMCs (>8 mM) and can be used in the range of 0.1–0.4%.

3.2 Size-Exclusion Chromatography

All steps should be performed at 4 °C unless indicated otherwise.

1. Filter buffer under a vacuum to remove small particles and debris.
2. Purge FPLC pumps with deionized water, and equilibrate the SEC column for 1 column volume (CV) at 0.15 mL/min. Monitor pressure and ensure it does not exceed the pressure limit indicated by the manufacturer.
3. Purge FPLC pumps with the SAXS buffer, and equilibrate the SEC column for at least 1.5 CV at 0.15 mL/min.
4. Spin protein solution for 5 min at 13,000 × *g* in a microcentrifuge. Transfer the supernatant (~80 μL) to a clean tube.
5. Set the injection valve in "load" mode, rinse injection loop with buffer (3 × 100 μL), and load ~75 μL of the protein solution supernatant in the loop.
6. Resume flow at 0.15 mL/min and set the injection valve to "inject" for elution.
7. Collect 100 μL fractions, starting before the void volume (*see* column specifications). Use UV absorbance at 280 nm to monitor protein elution (Fig. 1).
8. For each fraction of the peak, measure protein concentration from the theoretical absorption coefficient. Ensure that the protein concentration is at least 4 mg/mL for the most concentrated fraction. Fractions may be pooled.
9. If necessary, use ultrafiltration to concentrate the protein fractions to achieve concentrations above 4 mg/mL. Keep 50 μL of the protein solution aside in case the ultrafiltration procedure induces aggregation.
10. Keep buffer fractions eluted before the void volume for reference measurements.

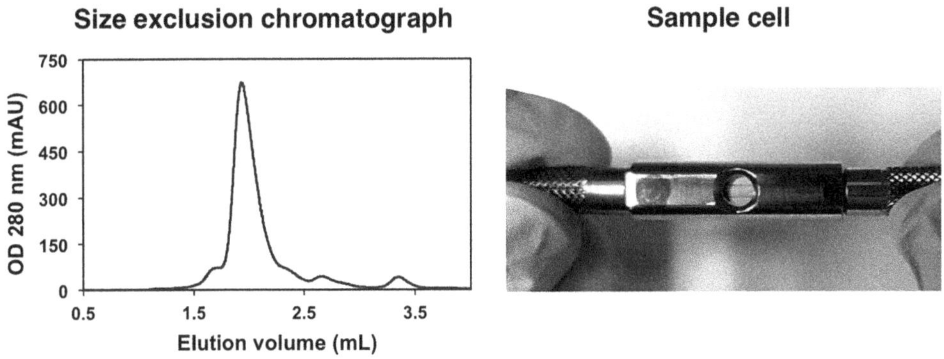

Fig. 1 A SEC profile and picture of a sample cell

3.3 Data Collection

1. Unscrew caps from sample cell and fill with 40 μL of buffer using a regular narrow-end pipette tip. Ensure that the liquid does not contain bubbles in the central window that would interfere with measurements (Fig. 1).
2. Screw caps on both ends of the cell, and check that the sample stays in place.
3. Adjust the temperature of the sample holder to 4 °C. Mount the cell in the sample holder, close the access lid or door, and evacuate the beam path until the pressure reaches less than 2 mbar (*see* **Note 2**).
4. Perform a short exposure (10 s) to ensure that the beam stop and sample cell are aligned correctly. Then expose the buffer with X-rays for 10 min and collect images.
5. Open the sample enclosure, empty the sample cell, and fill with the most concentrated protein solution (above 4 mg/mL). Expose with X-rays for 10 min and collect images.
6. Process the data and subtract the protein and buffer data sets to obtain the scattering of the protein (*see* Subheading 3.4). If the data quality is good, the sample may be exposed further with X-rays for another 20–60 min in 10 min steps (*see* **Note 3**).
7. The sample cell is washed extensively with buffer and filled with a solution of lower protein concentration (typically half as concentrated) and data collected for at least as long as for the most concentrated sample.
8. **Step 7** may be repeated with additional dilutions.
9. Finally, the sample cell should be washed and filled again with buffer in order to collect a long exposure (at least as long as the longest protein sample exposure) with the buffer as the reference measurement.

3.4 Primary Data Processing and Quality Assessment

Because different X-ray systems and detectors may be used, we will assume that the user has a set of 1D scattering files (.dat extension) corresponding to the buffer and protein samples at different concentrations. The ASCII formatted files should have three columns: the scattering vector ($s = 4\pi \sin \theta/\lambda$), the scattering intensity $I(s)$, and the standard error on I. The scattering curve is typically generated from radial averaging of 2D profiles, under the assumption of isotropic scattering. For a Kratky camera with line collimation, line averaging is performed, followed by desmearing using the beam profile.

1. Load the files in the software for primary data processing, such as *Primus* or *BioXTAS Raw* [2, 12].
2. For each protein measurement, subtract the reference buffer. The resulting file corresponds to the scattering from the protein alone + the ordered solvent shell (Fig. 2a) (*see* **Note 4**).
3. For a single concentration, data acquired at different times should be compared to ensure that there are no time-dependent changes. Those can arise from aggregation, radiation damage, or even loss of sample if the cell was not properly sealed. The data sets that show no significant variation over time can be averaged to improve the signal-to-noise ratio.
4. A Guinier plot (Fig. 2) is produced to evaluate the monodispersity of the sample and determine the radius of gyration R_g and the forward scattering I_0 [13] (*see* **Note 5**).
5. Concentration-dependent interactions can be assessed by comparison of the scattering profiles acquired at different concentrations. The profiles should normally be identical after scaling to account for concentration differences. In particular, the Guinier plots should yield the same R_g and I_0 values. If not, then the assumption of infinite dilution does not hold true. When samples show aggregation at high concentrations, the scattering curves obtained at different concentrations can be merged with the small angle data obtained from diluted samples and the wide angle data obtained from concentrated samples. The assumption is that concentration-dependent artifacts only affect the scattering at very small angles where the signal is strongest and not the scattering at larger angles where the signal is weak.
6. The invariant Q_r ratio is calculated to estimate the molecular weight of the particle in solution in a shape-independent manner [14]. The ratio is independent of the concentration and is calculated from a single data set (*see* **Note 6**).
7. The Kratky plot ($s^2 I(s)$ as a function of s) should be computed to determine whether the protein is folded or not (Fig. 2). The resulting plot will have a bell-shaped curve if the protein is

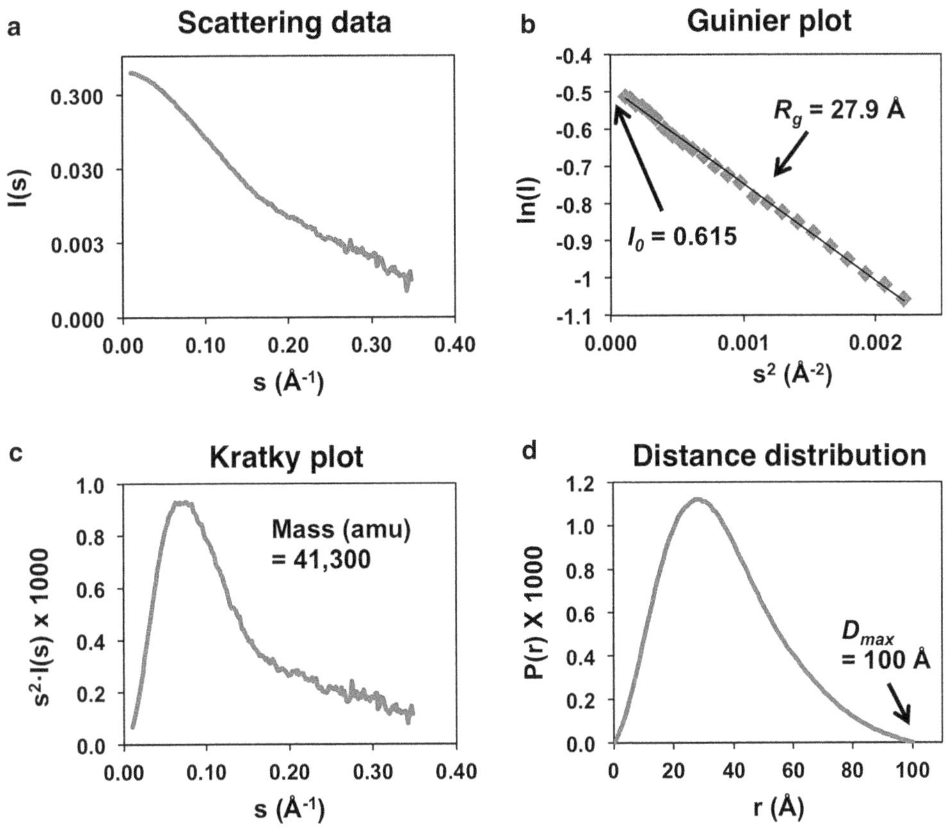

Fig. 2 Principal plots for quality assessment and interpretation of SAXS data (**a**) Scattering data for parkin 141-465, plotted on a log scale on the y-axis. (**b**) Guinier plot for the scattering data. The I0 parameter is derived from the y-intercept, while the Rg value is derived from the slope. (**c**) Kratky plot with a bell-shaped curve characteristic of folded proteins. The s value at the peak can be used to estimate the molecular weight. (**d**) Pair-distance distribution P(r) derived by Fourier transformation of the scattering data. The Dmax value is given by the x-intercept when P(r)=0

folded, a plateau if unfolded, and a partial bell followed by a plateau for a folded protein with disordered segments [15] (*see* **Note** 7).

3.5 Ab Initio Modeling

The procedure described here is for modeling data using a series of dummy particles to define the particle shape, which is very commonly used and applicable to any system. For other types of analysis, such as rigid-body docking or ensemble of flexible proteins, we refer the reader to specialized texts [16, 17].

1. Using the program *GNOM* [12], calculate the pair-distance distribution function $P(r)$ by performing an indirect transform of the scattering profile. The function is a bounded integral that can be thought as a histogram of every interatomic distance r in a protein (Fig. 2d). For a monodisperse particle, the $P(r)$ function is exactly zero at $r = 0$ and $r \geq D_{\max}$, which is the

distance between the two furthest atoms. The parameter D_{max} can be optimized such that the $P(r)$ falls smoothly to zero at $r = D_{max}$, while the back-calculated scattering profile still fits the data.

2. The $P(r)$ function is then used as input to reconstruct the protein shape using the program *GASBOR* [12] (*see* **Note 8**). The critical inputs are:
   - The portion of the curve to be fitted (normally the entirety, i.e., 1.0)
   - The symmetry of the particle (P1 for a monomer, P2 for a dimer, etc.)
   - The number of residues in the monomer
   - The expected overall particle shape (oblate, prolate, or unknown).
3. The reconstruction yields a single PDB file with multiple dummy residues and solvent atoms. The procedure should be repeated at least 20 times and may be run in parallel on multiple CPUs to accelerate the procedure. Back-prediction of the scattering from each file should yield a χ^2 value below 2.5 for a good fit (*see* **Note 9**).
4. The multiple PDB files can then be averaged using the program *DAMAVER*, which superposes all coordinates with each other, rejects outliers, and selects the most typical for averaging [12]. The averaged coordinate file is then filtered to a cutoff volume corresponding to the average volume of individual models.
5. The averaged filtered coordinates can be converted into a pseudo density volume using the *Situs* program *pdb2vol* [18]. The density file can be opened in the molecular visualization software *Chimera* [19] and rendered as a contoured surface with a volume that corresponds to the Porod volume of the particle (in Å^3, ~1.7× the molecular weight in Da). Molecular models obtained from crystallography or NMR can be fitted to the density, as shown previously by our group [5, 7, 20].

3.6 Validation of Crystal Structure Assembly

Another important use of SAXS data is in the validation of complexes or assemblies observed in crystal structures. Here, we describe a simple procedure to assess the agreement between a structure and an experimental scattering profile. In the example given here, the crystal structure of parkin with a linker deletion contained two molecules per asymmetric unit [5]. Because no electron density was visible between the ubiquitin-like (Ubl) and C-terminal domains, the packing gave rise to two topologically feasible configurations. The crystal structure alone could not unambiguously assign the position of the Ubl domain (Fig. 3, *left panels*).

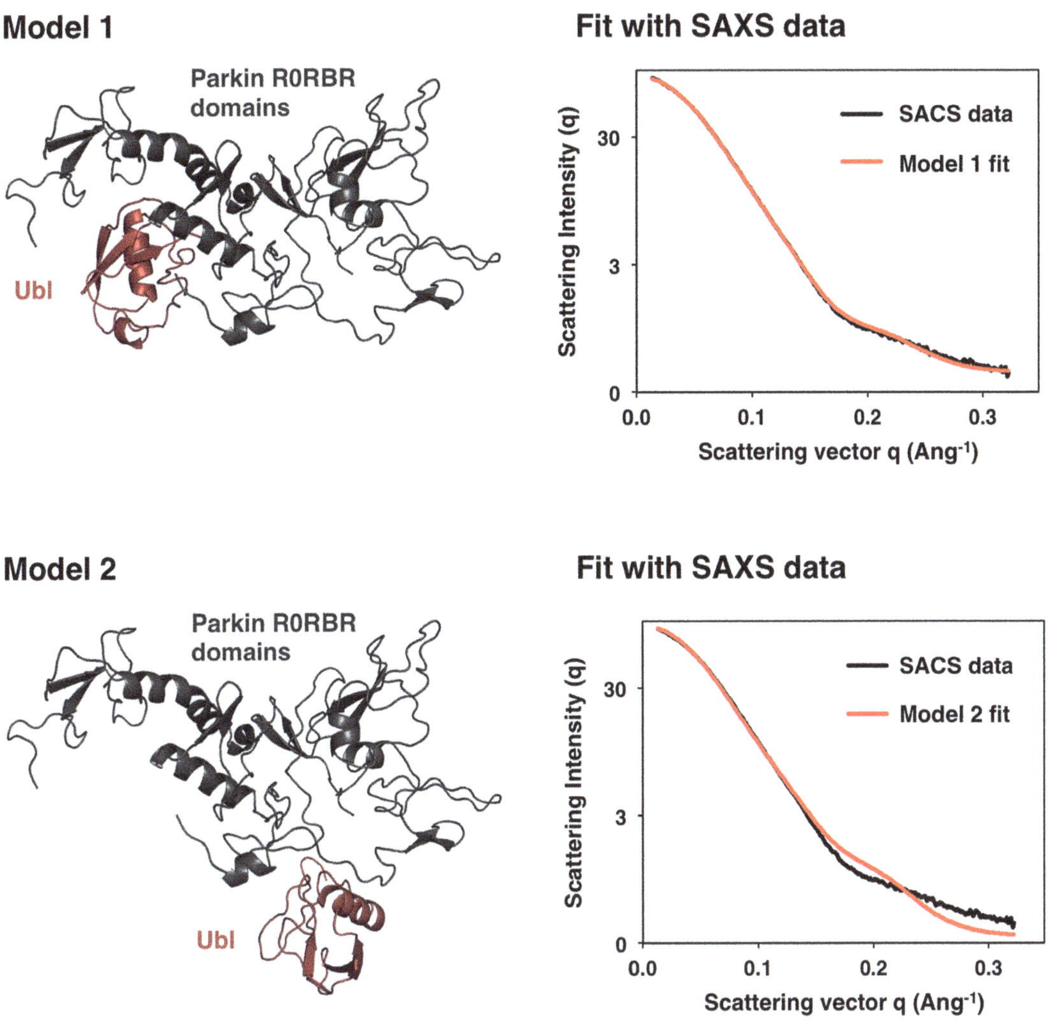

Fig. 3 An application of SAXS analysis. Uncertainty in the domain connectivity in the crystal structure of parkin was resolved by comparison of the experimental and back-calculated scattering curves [5]

1. Edit the PDB files to generate the assembly to be investigated.
2. Run the program *CRYSOL* to back-calculate the scattering from the PDB file [21].
3. The critical input parameters for the computation are:
   - The PDB file
   - The scattering profile (.dat format)
   - The maximum *s* value (adjust to data set, should be >0.25 Å^{-1})
4. The output of the computation is a calculated scattering profile, as well as a χ^2 value to assess agreement. From the graphs, it is quite clear that model #1 is the correct one (Fig. 3, *right panels*). This is confirmed by the χ^2 value of 1.6 for model #1, compared to a value of 7.6 for model #2.

4 Notes

1. Five- to tenfold dilution will take place on SEC depending on sample loss, dead volumes, etc., and therefore protein concentrations above 20 mg/mL need to be injected to achieve an eluted concentration of 2–4 mg/mL. Lower concentrations (10–20 mg/mL) can be used for inline SEC-SAXS at synchrotron radiation facilities.
2. If the sample cell is not properly sealed, the vacuum of the beam path will cause the sample to escape the cell. It is a good practice to start the data acquisition with the buffer sample. After evacuating the camera, remove the sample and visually inspect the cell to ensure that all buffer remains inside. It's much better to find out the cell was not properly assembled with buffer rather than your precious protein preparation.
3. The signal-to-noise ratio is proportional to the square root of the acquisition time. To obtain the same signal-to-noise ratio with half the protein concentration, the exposure time must be increased by four. Longer exposure times, however, lead to more radiation damage. In general, the acquisition time should be kept to less than 2 h for a modern liquid gallium jet or rotating copper anode source.
4. The subtracted scattering should drop to close to zero at large scattering angles ($s > 0.3$ Å^{-1}), with a background value about 100-fold smaller than the intensity at zero angle (I_0) (Fig. 2). If the subtracted intensity is below zero, this implies that the reference buffer scattered X-rays more than the protein, which often indicates an air bubble in the protein sample. Conversely, a high background value implies the reference sample was incorrectly measured and needs to be repeated.
5. A Guinier plot is a graph of $I(s)$ as a function of s^2. In the low-angle region ($s < 0.05$ Å^{-1}), the plot should be linear for a monodisperse system of globular particles, such as folded proteins. The y-intercept corresponds to $\ln(I_0)$, and the slope m can be used to calculate the radius of gyration using the equation $R_g = \sqrt{-3m}$. The range of linearity should extend to the limit $R_g s_{max} \geq 1.25$.
6. The Q_r is derived from the invariant volume-of-correlation V_c, which is equal to I_0 divided by the integral $\int_0^\infty sI(s)ds$. In practice, the latter is estimated by the sum of the function $sI(s)$ over the entire s range multiplied by the spacing between data points. This enables calculation of the ratio $Q_r = V_c^2/R_g$, which can be used to determine the molecular weight of the particle, depending on its density. For proteins, the molecular weight is given by $Q_r/0.1231$. This value should be within ±10% of the theoretical oligomeric molecular weight.

7. For globular proteins, the s value where $s^2 I(s)$ is a maximum ($s_{\max}$) is inversely proportional to the molecular weight of the particle. This property can be used to estimate the molecular weight via the empirical equation MW = exp $(-2.67 \times \ln(s_{\max}) - 3.25)$, derived from a number of molecular weight standards.
8. The scattering profile can also be used to calculate the shape by fitting the data in reciprocal space. The procedure requires more computing time, and thus should only be used for systems with less than 500–600 residues, since the computing time increases with the square of the residue number. As an example, for a 500-residue protein, the computation takes 7.5 min in real space and 11 min in reciprocal space on a 2.5 GHz Intel Core i7 processor.
9. The χ^2 value for the discrepancy between experimental and calculated scattering intensities is given by the formula $\chi^2 = \frac{1}{N_{p-1}} \sum_i \left[\frac{I_{\exp} - I_{\text{calc}}}{\sigma}\right]^2$, where σ is the experimental error on every data point, and N_{p} is the number of points. This value is highly dependent on how the error is actually measured. We found that many X-ray detectors tend to underestimate the error, which leads to an overestimate of the χ^2 value. The error can be more accurately measured by computing the standard error for consecutive points in a sliding window of about 6 points, or 0.003 Å^{-1}. In the high-angle region, this value corresponds to the noise level of the detector.

References

1. Meisburger SP, Taylor AB, Khan CA, Zhang S, Fitzpatrick PF, Ando N (2016) Domain movements upon activation of phenylalanine hydroxylase characterized by crystallography and chromatography-coupled small-angle X-ray scattering. J Am Chem Soc 138 (20):6506–6516. https://doi.org/10.1021/jacs.6b01563
2. Nielsen SS, Toft KN, Snakenborg D, Jeppesen MG, Jacobsen JK, Vestergaard B, Kutter JP, Arleth L (2009) BioXTAS RAW, a software program for high-throughput automated small-angle X-ray scattering data reduction and preliminary analysis. J Appl Crystallogr 42:965–974
3. Rasool S, Soya N, Truong L, Croteau N, Lukacs GL, Trempe JF (2018) PINK1 autophosphorylation is required for ubiquitin recognition. EMBO reports 19(4):e44981
4. Trempe JF, Sauve V, Grenier K, Seirafi M, Tang MY, Menade M, Al-Abdul-Wahid S, Krett J, Wong K, Kozlov G, Nagar B, Fon EA, Gehring K (2013) Structure of parkin reveals mechanisms for ubiquitin ligase activation. Science 340:1451–1455. science.1237908 [pii]. https://doi.org/10.1126/science.1237908
5. Sauvé V, Lilov A, Seirafi M, Vranas M, Rasool S, Kozlov G, Sprules T, Wang J, Trempe JF, Gehring K (2015) A Ubl/ubiquitin switch in the activation of Parkin. EMBO J 34(20):2492–2505. https://doi.org/10.15252/embj.201592237
6. Taherbhoy AM, Tait SW, Kaiser SE, Williams AH, Deng A, Nourse A, Hammel M, Kurinov I, Rock CO, Green DR, Schulman BA (2011) Atg8 transfer from Atg7 to Atg3: a distinctive E1-E2 architecture and mechanism in the autophagy pathway. Mol Cell 44 (3):451–461 S1097-2765(11)00767-2 [pii]. https://doi.org/10.1016/j.molcel.2011.08.034
7. Trempe JF, Saskova KG, Siva M, Ratcliffe CD, Veverka V, Hoegl A, Menade M, Feng X, Shenker S, Svoboda M, Kozisek M,

Konvalinka J, Gehring K (2016) Structural studies of the yeast DNA damage-inducible protein Ddi1 reveal domain architecture of this eukaryotic protein family. Sci Rep 6:33671. https://doi.org/10.1038/srep33671

8. Siva M, Svoboda M, Veverka V, Trempe JF, Hofmann K, Kozisek M, Hexnerova R, Sedlak F, Belza J, Brynda J, Sacha P, Hubalek M, Starkova J, Flaisigova I, Konvalinka J, Saskova KG (2016) Human DNA-damage-inducible 2 protein is structurally and functionally distinct from its yeast Ortholog. Sci Rep 6:30443. https://doi.org/10.1038/srep30443
9. Pruneda JN, Stoll KE, Bolton LJ, Brzovic PS, Klevit RE (2011) Ubiquitin in motion: structural studies of the ubiquitin-conjugating enzyme approximately ubiquitin conjugate. Biochemistry 50(10):1624–1633. https://doi.org/10.1021/bi101913m
10. Davis-Searles PR, Saunders AJ, Erie DA, Winzor DJ, Pielak GJ (2001) Interpreting the effects of small uncharged solutes on protein-folding equilibria. Annu Rev Biophys Biomol Struct 30:271–306. https://doi.org/10.1146/annurev.biophys.30.1.271
11. Grishaev A (2012) Sample preparation, data collection, and preliminary data analysis in biomolecular solution X-ray scattering. Curr Protoc Protein Sci Chapter 17:Unit17.14. https://doi.org/10.1002/0471140864.ps1714s70
12. Petoukhov MV, Franke D, Shkumatov AV, Tria G, Kikhney AG, Gajda M, Gorba C, Haydyn MDT, Konarev PV, Svergun DI (2012) New developments in the ATSAS program package for small-angle scattering data analysis. J Appl Crystallogr 45:342–350
13. Jacques DA, Trewhella J (2010) Small-angle scattering for structural biology--expanding the frontier while avoiding the pitfalls. Protein Sci 19(4):642–657. https://doi.org/10.1002/pro.351
14. Rambo RP, Tainer JA (2013) Accurate assessment of mass, models and resolution by small-angle scattering. Nature 496(7446):477–481. https://doi.org/10.1038/nature12070
15. Semisotnov GV, Kihara H, Kotova NV, Kimura K, Amemiya Y, Wakabayashi K, Serdyuk IN, Timchenko AA, Chiba K, Nikaido K, Ikura T, Kuwajima K (1996) Protein globularization during folding. A study by synchrotron small-angle X-ray scattering. J Mol Biol 262(4):559–574. https://doi.org/10.1006/jmbi.1996.0535
16. Tsutakawa SE, Hura GL, Frankel KA, Cooper PK, Tainer JA (2007) Structural analysis of flexible proteins in solution by small angle X-ray scattering combined with crystallography. J Struct Biol 158(2):214–223. https://doi.org/10.1016/j.jsb.2006.09.008
17. Bernado P, Mylonas E, Petoukhov MV, Blackledge M, Svergun DI (2007) Structural characterization of flexible proteins using small-angle X-ray scattering. J Am Chem Soc 129(17):5656–5664. https://doi.org/10.1021/ja069124n
18. Wriggers W, Chacon P (2001) Using Situs for the registration of protein structures with low-resolution bead models from X-ray solution scattering. J Appl Crystallogr 34:773–776
19. Pettersen EF, Goddard TD, Huang CC, Couch GS, Greenblatt DM, Meng EC, Ferrin TE (2004) UCSF chimera – a visualization system for exploratory research and analysis. J Comput Chem 25(13):1605–1612. https://doi.org/10.1002/jcc.20084
20. Trempe JF, Shenker S, Kozlov G, Gehring K (2011) Self-association studies of the bifunctional N-acetylglucosamine-1-phosphate uridyltransferase from Escherichia coli. Protein Sci 20(4):745–752. https://doi.org/10.1002/pro.608
21. Svergun DI, Barberato C, Koch MHJ (1995) CRYSOL - a program to evaluate X-ray solution scattering of biological macromolecules from atomic coordinates. J Appl Crystallogr 28:768–773

Chapter 14

Methods for Preparing Cryo-EM Grids of Large Macromolecular Complexes

Leifu Chang and David Barford

Abstract

The recent resolution revolution in cryo-electron microscopy has generated a huge interest in the technique for determining atomic resolution structures of large and dynamic macromolecular complexes that are intractable to crystallography and NMR. A key element of success in cryo-EM is the quality of the specimen vitrified on the cryo-EM grid. In this chapter we outline methods for cryo-EM grid sample preparation.

Key words Cryo-electron microscopy (cryo-EM), Cryo-EM grids, APC/C, Multi-subunit complexes

1 Introduction

The recent resolution revolution in cryo-electron microscopy (cryo-EM) has transformed all fields of biological research by enabling the determination to near-atomic resolution of macromolecules ranging from proteins as small as 64 kDa to large MDa multi-subunit molecular machines responsible for orchestrating complex biological processes [1–11]. These advances have revolutionized structural biology by hugely expanding both the range of macromolecules whose structures can be determined and by providing a description of macromolecular dynamics. One of the advantages of cryo-EM is that compared with protein crystallography and NMR, relative little sample at low concentration is required to generate cryo-EM grids, and moreover, the method tolerates compositional and conformational heterogeneity. However, even these modest requirements often surpass the quantities that can be easily isolated given the low natural abundance of many complexes. Because this limits the opportunities for structural and biophysical studies, the development of heterologous overexpression methods to reconstitute recombinant complexes in vitro has played a significant role in exploiting the opportunities provided by the new cryo-EM methods.

Thibault Mayor and Gary Kleiger (eds.), *The Ubiquitin Proteasome System: Methods and Protocols*, Methods in Molecular Biology, vol. 1844, https://doi.org/10.1007/978-1-4939-8706-1_14, © Springer Science+Business Media, LLC, part of Springer Nature 2018

In this chapter we present protocols for preparing cryo-EM grids suitable for determining atomic resolution structures of a representative MDa multi-subunit complex, the anaphase-promoting complex/cyclosome (APC/C), by cryo-EM. In cryo-EM the biological specimen is imaged in a thin layer of amorphous vitreous ice. The specimen is embedded in ice suspended within holes of an EM grid. Typically the grid is 3 mm in diameter constructed from either copper or gold with grid bars spaced ~100 μm apart to generate a mesh of 80 × 80 μm squares. The grid bars support a thin foil, conventionally carbon, but now also gold (UltrAuFoil) [12], approximately 200–400 Å thick. This foil is perforated with 1–2 μm circular holes that support the vitreous ice with the embedded protein particles.

In cryo-EM, biological samples in a native solution are flash frozen to create a thin layer of vitreous ice. The ideal ice in holes should just cover the embedded protein particles so that the protein is in a hydrated state and the background noise due to the ice is minimized. However, the extended air-water interface is problematic. Denaturation of proteins is frequently observed caused by the high surface tension at the air-water interface, a particular problem for large fragile protein complexes. To circumvent this problem, a layer of continuous thin carbon film is commonly used to cover the holes [13]. There are several advantages of using thin carbon film. First, because particles are absorbed by the carbon, it controls one side of the air-water interface and has proven very useful to maintain the stability of large protein complexes. Second, by absorbing protein particles, it will concentrate macromolecules beneficial for rare samples. Third, the signal from carbon is useful for more accurate contrast transfer function (CTF) parameter estimation, an important step for achieving high-resolution reconstructions by cryo-EM. Lastly, thin carbon film can sometimes be used to overcome preferred orientations of the macromolecule.

Traditionally the film (also referred to as the substrate) is composed of an approximately 50 Å thick layer of amorphous carbon. However, the carbon substrate contributes to the background of the image, reducing contrast and image signal, significantly degrading the alignment of small proteins (<200 kDa). To alleviate these problems, recently extremely thin and electron-transparent pristine graphene has been developed as a support layer [14]. Graphene oxide is an alternative [15] that has also been successfully applied [16] (e.g., *see* [17]).

In this chapter we outline methods for preparing cryo-EM grids of the APC/C in complex with the coactivator Cdh1 and inhibitor Emi1 (this complex is an assembly of 21 subunits from a total of 16 different proteins). Expression and purification of this complex were described in [18, 19]. The methods explained here are applicable to other multi-subunit complexes. For these studies, EM grids covered with a thin amorphous carbon film were used in preference to both EM grids without a carbon support and EM grids with a graphene oxide support. Using the latter two EM grid types, the APC/C was observed to be disassembled.

2 Materials

1. Carbon rod (Agar Scientific shaped carbon rods 3.05 mm × 55 mm).
2. Mica (mica sheets 75 mm × 25 mm in size, 0.15 mm thick).
3. Tweezers (Dumont tweezers 5).
4. Glass petri dishes.
5. Filter paper (Whatman qualitative filter paper, Grade 1, 90 mm).
6. Carbon evaporator (Edward Auto 306 Turbo Evaporator).
7. Glow discharge unit (Edwards sputter coater S150B)
8. Plasma cleaner (Fischione MODEL 1070 NanoClean).
9. Vitrobot (Thermo Fisher Scientific Vitrobot Mark III or IV).
10. Electron microscope (e.g., Thermo Fisher Scientific Titan Krios).
11. Quantifoil holey carbon grids.
12. Ultrapure water (resistivity of 18.2 MΩ cm at 25 °C).
13. Liquid nitrogen.
14. Ethane.

3 Methods

3.1 Carbon Film Preparation

1. Mount the carbon rods in the carbon evaporator. Prepare two carbon rods, one with a sharp tip and the other with a flattened end. Assemble them according to the evaporator manual so that the sharp tip of one rod is connected to the flat end of another rod. Ensure a good connection is maintained (usually through a spring mechanism).
2. Cleave a piece of mica (75 mm × 25 mm). Insert a tweezer tip into a mica sheet from one edge. Then tear the mica with two pairs of tweezers. This creates two fresh and clean mica surfaces for carbon deposition. Place them on a piece of filter paper in a glass petri dish (fresh surface on top). Put a spiral wire next to the mica to be used for monitoring carbon thickness while coating. Place the mica in the chamber of the evaporator, just under the tip of carbon rod.
3. Pump down the chamber to ~3×10^{-7} mbar. This typically takes 4 h. The vacuum can be further improved by filling the cold-trap with liquid nitrogen.

4. Preheat the carbon rod. This allows outgassing and removal of surface contamination. Use the metal disk guard to prevent carbon deposition onto the mica at this stage.
5. Coating. Gradually increase the current to allow carbon evaporation from the tip. Carefully monitor the vacuum. When the vacuum reaches 2×10^{-6} mbar, reduce the current to zero and allow the vacuum to recover to 3×10^{-7} mbar. Start coating again and repeat this process several times until the desired carbon thickness is achieved on the mica. Check the shadow under the spiral wire to visually judge the thickness.
6. Allow the system to cool down for 1 h or longer before removing the mica.
7. Label and store the mica along with the filter paper for later use (*see* **Note 1**).

3.2 Transferring Carbon Film on EM Grids

1. Clean Quantifoil holey carbon grids using a glow discharge unit. Place the grids onto a glass slide, and place them into the chamber. Pump down the chamber to 0.2 mbar. Turn on the tension and glow discharge for 30–60 s.
2. Fill a glass petri dish with ultrapure water to 80% full. Cut a piece of filter paper and place it into the water. Place cleaned Quantifoil grids on the filter paper.
3. Cut a small piece of mica coated with carbon film. It is important to note the top side of the carbon. Hold the mica with a pair of tweezers (carbon side on top), and dip the mica gradually into water with a tilt angle of roughly 30°. This allows the carbon film to separate from the mica and to float on water (Fig. 1a; *see* **Note 2**).
4. Use one hand to drag the filter paper slowly out of water from the edge of the petri dish. Do this with care and keep the grids static on the filter paper. Simultaneously control the position of the floated carbon film with tweezers on the other hand so that the carbon film covers the grids as they are dragged out of the petri dish (Fig. 1b).

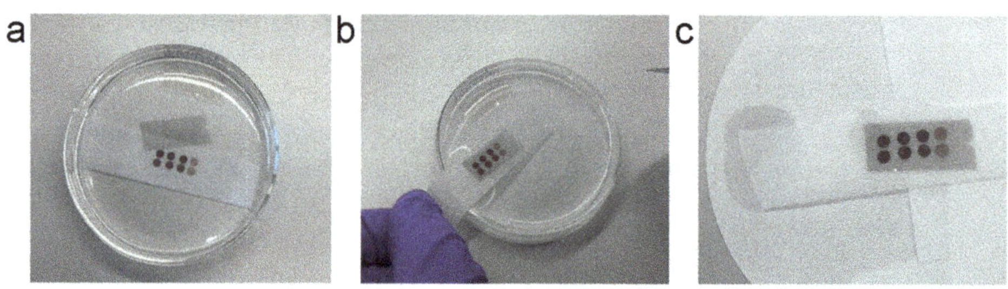

Fig. 1 Method for transferring thin carbon film onto EM grids. (**a**) Float carbon film on water in a glass dish. (**b**) Drag out filter paper and allow carbon film to cover all grids. (**c**) Place grids on a piece of clean filter paper to allow them to dry at room temperature

5. Place the filter paper on a second clean piece of filter paper, and let it dry for at least 1 h before use (Fig. 1c).

3.3 Plasma Cleaning

1. Place the grids prepared in the previous step into slots of the holder, and insert the holder into the chamber of the plasma cleaner (Fischione Model 1070 NanoClean).
2. Pump the vacuum down to high vacuum ($<10^{-5}$ mbar).
3. Set the gas mixture of argon (Ar) and O_2 to a ratio of 9:1 and the timer to 15–50 s depending on the carbon thickness. Thicker carbon can be treated for a longer period to increase hydrophilicity without breaking the carbon (*see* **Note 3**).
4. Start plasma cleaning. The mixed gas (Ar:O_2 at a ratio of 9:1) will fill the chamber to a final vacuum of ~0.02 mbar. The plasma cleaning starts automatically.
5. Use the treated grids for preparing cryo-EM grids within 1 h.

3.4 Freezing Grids

1. Vitrobot settings: 4 °C, 100% humidity, waiting time 30 s, blotting offset 0, blotting time 6 s, blot only one time (*see* **Note 4**).
2. Apply 3 μL of protein sample (typically 0.12 mg/mL for the APC/C [19], equivalent to ~0.1 μM) to the EM grid (*see* **Note 5**).
3. Freeze grids using the Vitrobot protocol (*see* **Note 6**).

3.5 Checking Grids with an Electron Microscope

1. Using a cryo-electron microscope, check the particle concentration on cryo-electron micrographs (*see* **Note 7**). This should be determined before making many more grids, thereby saving grids and time (*see* **Note 8**).
2. Collect high-resolution data. Figure 2 shows an image of a representative cryo-EM grid for APC/C using a Titan Krios microscope with the Volta phase plate.

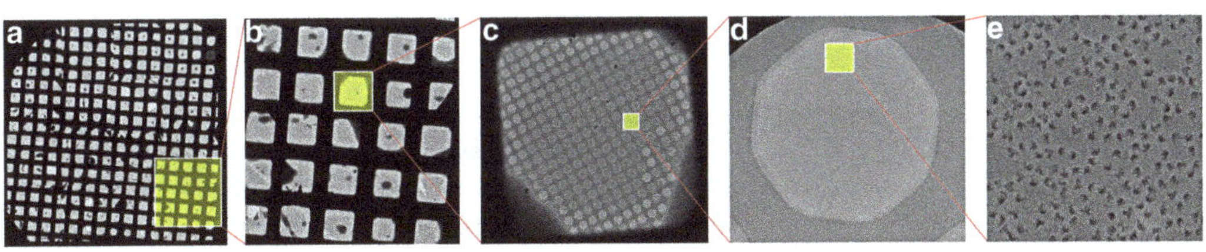

Fig. 2 An example of a good cryo-EM grid. The grid used here is a Quantifoil R3.5/1 grid. The figure shows from left to right views of the grid at increasing magnification. (**a**) Overall 3 mm diameter grid showing metal grid bars that define a field of square holes and which supports the carbon foil. (**b**) Zoom over an array of 5 × 5 squares; each square measures 80 × 80 μm. (**c**) View of a single grid square showing the holes in the carbon foil. The holes are covered by a thin carbon support. (**d**) View of an individual hole (3.5 μm). (**e**) A cryo-electron micrograph of the APC/C, nominal magnification of × 105,000, taken on a Titan Krios microscope with Volta phase plate

4 Notes

1. Make several batches of carbon film and select good ones for use. Too thin (<3 nm) carbon is unstable and tends to break during plasma cleaning or blotting. Carbon that is too thick (>10 nm) generates too much background in cryo-EM images. A 5–10-nm-thick carbon works well for the APC/C (mol mass, 1.2 MDa).
2. If a batch of carbon film tends to break easily when floating on water, it is not good in general. Good carbon film should be strong.
3. The plasma cleaning process removes surface carbon and generates thinner carbon. It is important to take this into consideration when determining the treatment time. Thicker carbon (8–10 nm) with longer treatment yields good ice distribution on grids with relatively low background noise.
4. Before freezing grids, determine which parameters to use on a specific blotting machine. This is usually based on experience and can be determined systematically by experimental variation of parameters such as blotting force, blotting time, etc.
5. To determine the correct macromolecule concentration for making cryo-EM grids, perform a dilution series and analyze them on the microscope. The correct concentration can be estimated based on negative stain EM. To obtain a similar particle density in cryo-EM images, typically a threefold higher concentration is used.
6. Grid type makes a difference, although the sample only contacts the thin continuous carbon. It is beneficial to try different types of grids (e.g., R1.2/1.3, R2/2, R3.5/1 in 200/300/400 mesh). For APC/C sample, R3.5/1 in 200 mesh tends to work better than other types.
7. Make a note of every grid. This is very useful for optimizing the conditions and saving time screening. Expectations of grids based on experience will facilitate the screening process.
8. Other factors affect the quality of cryo-EM grids. In practice, even if the same conditions are used, the grids can look completely different when viewed in an electron microscope. This problem seems to be more serious for thin carbon-covered grids. Possible causes include (a) the exact force applied on grids when blotting is not the same and (b) plasma cleaning is variable (i.e., different grids may receive different treatment as their positions are different in the chamber). So a decision has to be made so that enough grids are made without waste. Fortunately only a few good grids are required for collecting enough data (in principle one is enough).

References

1. Bai XC, McMullan G, Scheres SH (2015) How cryo-EM is revolutionizing structural biology. Trends Biochem Sci 40:49–57
2. Vinothkumar KR, Henderson R (2016) Single particle electron cryomicroscopy: trends, issues and future perspective. Q Rev Biophys 49:e13
3. Henderson R (2015) Overview and future of single particle electron cryomicroscopy. Arch Biochem Biophys 581:19–24
4. Cheng Y (2015) Single-particle cryo-EM at crystallographic resolution. Cell 161:450–457
5. Fernandez-Leiro R, Scheres SH (2016) Unravelling biological macromolecules with cryo-electron microscopy. Nature 537:339–346
6. Nogales E, Scheres SH (2015) Cryo-EM: a unique tool for the visualization of macromolecular complexity. Mol Cell 58:677–689
7. Merino F, Raunser S (2017) Electron cryo-microscopy as a tool for structure-based drug development. Angew Chem Int Ed Engl 56:2846–2860
8. Glaeser RM (2016) How good can cryo-EM become? Nat Methods 13:28–32
9. Nogales E (2016) The development of cryo-EM into a mainstream structural biology technique. Nat Methods 13:24–27
10. Boland A, Chang L, Barford D (2017) The potential of cryo-electron microscopy for structure-based drug design. Essays Biochem 61:543–560
11. Passmore LA, Russo CJ (2016) Specimen preparation for high-resolution cryo-EM. Methods Enzymol 579:51–86
12. Russo CJ, Passmore LA (2014) Electron microscopy: ultrastable gold substrates for electron cryomicroscopy. Science 346:1377–1380
13. Russo CJ, Passmore LA (2016) Progress towards an optimal specimen support for electron cryomicroscopy. Curr Opin Struct Biol 37:81–89
14. Russo CJ, Passmore LA (2014) Controlling protein adsorption on graphene for cryo-EM using low-energy hydrogen plasmas. Nat Methods 11:649–652
15. Pantelic RS, Meyer JC, Kaiser U, Baumeister W, Plitzko JM (2010) Graphene oxide: a substrate for optimizing preparations of frozen-hydrated samples. J Struct Biol 170:152–156
16. Boland A, Martin TG, Zhang Z, Yang J, Bai XC, Chang L, Scheres SH, Barford D (2017) Cryo-EM structure of a metazoan separase-securin complex at near-atomic resolution. Nat Struct Mol Biol 24:414–418
17. Martin TG, Boland A, Fitzpatrick A, Scheres S (2016) Graphene Oxide Grid Preparation. Available via figshare. https://figshare.com/articles/Graphene_Oxide_Grid_Preparation/3178669
18. Zhang Z, Yang J, Barford D (2016) Recombinant expression and reconstitution of multi-protein complexes by the USER cloning method in the insect cell-baculovirus expression system. Methods 95:13–25
19. Chang L, Zhang Z, Yang J, McLaughlin SH, Barford D (2015) Atomic structure of the APC/C and its mechanism of protein ubiquitination. Nature 522:450–454

Part IV

Methods to Study 26S Proteasome Function

Chapter 15

Recombinant Expression, Unnatural Amino Acid Incorporation, and Site-Specific Labeling of 26S Proteasomal Subcomplexes

Jared A. M. Bard and Andreas Martin

Abstract

The 26S proteasome is the major regulated protease in eukaryotes and is responsible for degrading ubiquitinated substrates. It consists of a barrel-shaped 20S core peptidase and one or two 19S regulatory particles, which recognize, unfold, and translocate substrates into the core. The regulatory particle can be further divided into two multi-subunit complexes: the base and the lid. Here we present protocols for expressing the *Saccharomyces cerevisiae* base and lid recombinantly in *Escherichia coli* and purifying the assembled subcomplexes using a tandem affinity purification method. The purified complexes can then be reconstituted with 20S core to form fully functional proteasomes. Furthermore, we describe a method for incorporating the unnatural amino acid *p*-azido-L-phenylalanine into the recombinant complexes at any residue position, allowing for non-disruptive site-specific modifications of these large assemblies. The use of recombinant proteins allows for complete mutational control over the proteasome regulatory particle, enabling detailed studies of the mechanism by which the proteasome processes its substrates. The ability to then specifically modify residues in the regulatory particle opens the door to a wide range of previously impossible biochemical and biophysical studies. The techniques described below for incorporating unnatural amino acids into the proteasomal subcomplexes should be widely transferable to other recombinant proteins, whether individually purified or in larger multi-subunit assemblies.

Key words 26S proteasome, Recombinant expression, Macromolecular complex, Tandem affinity purification, Unnatural amino acid incorporation, Click chemistry, *p*-Azido-L-phenylalanine

1 Introduction

The 26S proteasome is the major protease in eukaryotic cells, where it is not only responsible for protein quality control and homeostasis through the degradation of misfolded, damaged, and obsolete polypeptides but also controls a myriad of vital cellular processes by the specific turnover of regulatory proteins [1–3]. Due to this role as one of the most critical proteome regulators, elucidating the proteasome's structure, mechanisms of action, and regulation has been of prime interest in the ubiquitin-proteasome field. However,

Thibault Mayor and Gary Kleiger (eds.), *The Ubiquitin Proteasome System: Methods and Protocols*, Methods in Molecular Biology, vol. 1844, https://doi.org/10.1007/978-1-4939-8706-1_15, © Springer Science+Business Media, LLC, part of Springer Nature 2018

the high complexity of this 2.5 MDa molecular machine with more than 34 different subunits, its compositional heterogeneity, and the critical dependence of eukaryotic cell viability on fully functional proteasomes have previously limited in vitro studies and mutational analyses. To circumvent these limitations of working with endogenous proteasomes, we developed systems for the heterologous expression of two proteasomal subcomplexes, the lid and the base, from *Saccharomyces cerevisiae* in *Escherichia coli* and the incorporation of unnatural amino acids to allow specific fluorescence labeling and advanced spectroscopic studies of proteasome function.

At the center of the 26S proteasome is the 20S core peptidase that is capped on one or both ends by the 19S regulatory particle, which recognizes appropriate substrates by their covalently attached ubiquitin modifications, mechanically unfolds them, and translocates the unfolded polypeptides through a narrow axial pore into the 20S core for proteolytic cleavage [4–6]. This regulatory particle can be further subdivided into the base and lid subcomplexes [7, 8], which also represent the intermediates of proteasome assembly in vivo [9] and thus can be used as stable building blocks together with the 20S core peptidase to reconstitute the 26S holoenzyme in vitro. The base contains six distinct AAA+ ATPases, Rpt1-Rpt6, that form a heterohexameric ring and constitute the molecular motor of the proteasome. It also includes three non-ATPases, in yeast called Rpn1, Rpn2, and Rpn13, with Rpn1 and Rpn13 functioning as ubiquitin receptors [10, 11]. The assembly of the base proceeds through the initial formation of modules containing pairs of Rpt subunits [9], which then associate to form the ATPase hexamer in the order Rpt1-Rpt2-Rpt6-Rpt3-Rpt4-Rpt5 [12], with Rpn1 bound to Rpt1 and Rpt2 and Rpn2 plus Rpn13 associated with Rpt6 and Rpt3. Correct pairing of Rpt subunits in the three modules as well as their proper assembly into the base subcomplex is controlled by four assembly factors, Rpn14, Hsm3, Nas2, and Nas6 [13–15]. While Nas2 is released during this process [12], Nas6, Rpn14, and Hsm3 remain bound to the C-terminal tails of Rpt subunits until they are displaced by the base docking with the 20S core peptidase [16].

The lid subcomplex includes six PCI (proteasome-CSN-initiation factor 3) domain-containing subunits, Rpn3, Rpn5, Rpn6, Rpn7, Rpn9, and Rpn12, the small peptide Sem1, and two MPN (Mpr1-Pad1 N-terminal) domain-containing subunits, Rpn8 and Rpn11, with the latter representing the essential Zn^{2+}-dependent deubiquitinase of the proteasome [17–19]. A helical bundle formed by the C-terminal helices of the eight globular lid subunits functions as a hub to control the ordered lid assembly [20], which is independent of any additional factors [9]. Another ubiquitin receptor subunit, Rpn10, is not stably associated with the isolated lid or base but interacts with both subcomplexes in the assembled regulatory particle.

Given this knowledge about the in vivo formation of lid and base, we created recombinant expression systems in *E. coli* with three compatible plasmids coding for either the nine lid subunits or the nine subunits of the base plus the four essential assembly factors, as described below (Subheadings 3.1 and 3.2). While the lid subunits associate without additional chaperones, the base subunits assemble efficiently only in the presence of the assembly factors. Subcomplexes are isolated by tandem affinity purification using tags on two subunits of the lid and the base followed by size-exclusion chromatography. The use of affinity tags on two separate subunits ensures that only fully assembled complexes are purified and thus corrects for differences in the expression levels or stability of individual subunits.

One of the advantages of these heterologous expression systems is that they can be easily combined with techniques for the incorporation of unnatural amino acids (UAAs) [21]. By co-expressing a synthetase/tRNA pair that has been evolved to recognize one of the amber/ochre/opal codons, an UAA can be placed anywhere within the expressed proteins. These UAAs allow for the incorporation of a variety of useful chemical properties into proteins, including reactive groups for cross-linking [22], photo-caged amino acids such as lysine [23], and handles for orthogonal labeling [21]. The protocol below (Subheading 3.3) describes how to incorporate the UAA *p*-azido-L-phenylalanine (AzF) at a UAG codon that has been introduced at a specific position in the base or lid subcomplex. The azide group of the AzF can then be conjugated to an alkyne through a cycloaddition reaction, often referred to as "click" chemistry, allowing for the site-specific modification of the complex. Incorporation of AzF into the base will be used as an example below, though the protocol can be easily adapted to the lid or any other desired recombinant protein.

The vector used for UAA incorporation, pAM87, was constructed based on a synthetase variant (pAzFRS.2.t1) evolved by the Isaacs lab and a vector (pUltra) designed by the Schultz lab that also encodes the UAG-recognizing tRNA [24, 25]. Proteins with the UAG codon inserted in the desired position are expressed in *E. coli* strains carrying pAM87 with AzF added to the growth media. Once the protein is purified, the azide group of the incorporated AzF can be modified with an alkyne-containing molecule using Cu-free click chemistry [26]. In the protocol below, the base is modified with a fluorophore linked to dibenzocyclooctyne (DBCO), a strained cyclooctyne that allows for mild, copper-free conjugation [27]. Though designed to be bio-orthogonal, it is known that DBCO reacts with exposed cysteines [28]. To suppress this reaction, free thiols are protected with Ellman's reagent before the labeling. Once the labeling reaction is complete, the cysteines are restored to their native state through reduction with dithiothreitol (DTT).

2 Materials

2.1 Reagents and Equipment for Base and Lid Expression and Purification

1. Electrocompetent *E. coli* BL21 Star (DE3) cells [29].
2. Electroporation apparatus.
3. Expression plasmids (*see* Table 1).
4. 1000× ampicillin: 300 mg/mL ampicillin sodium in 50% EtOH.
5. 1000× kanamycin: 50 mg/mL kanamycin monosulfate in water.
6. 1000× chloramphenicol: 25 mg/mL chloramphenicol in 100% EtOH.
7. LB-antibiotic plates: 1% (w/v) tryptone, 0.5% (w/v) yeast extract, 0.5% (w/v) NaCl, 1.5% (w/v) agar. Autoclave, cool to 60 °C, or lower in a water bath before adding appropriate antibiotics to 1× from 1000× stocks.
8. dYT liquid media: 1.6% (w/v) tryptone, 1% (w/v) yeast extract, 0.5% (w/v) NaCl. Autoclave.
9. Terrific broth (TB): 47 g of Novagen terrific broth (Millipore) in 1 L of water +4 mL of glycerol. Autoclave.
10. 2.5 L Ultra Yield baffled shaking flask (Thomson).
11. 0.5 M IPTG: 0.5 M isopropyl-beta-D-thiogalactoside in water.

Table 1
Expression plasmids used in this protocol

Plasmid #	Description	Vector backbone	Antibiotic marker
pAM81 [32]	Rpn1, Rpn2, Rpn13	pETDuet	Ampicillin
pAM82 [32]	FLAG-Rpt1, Rpt2, 6xHis-Rpt3, Rpt5, Rpt6, Rpt4	pCOLADuet	Kanamycin
pAM83 [32]	RIL rare tRNAs, Nas6, Nas2, Hsm3, Rpn14	pACYCDuet	Chloramphenicol
pAM85 (modified from [33])	Rpn9, Rpn11, Rpn8, MBP-HRV-Rpn6, Rpn5	pETDuet	Ampicillin
pAM86 (modified from [33])	6xHis-HRV-Rpn12, Rpn7, Rpn3	pCOLADuet	Kanamycin
pAM80 [33]	Sem1, Hsp90	pACYCDuet	Chloramphenicol
pAM87 (constructed from [24, 25])	pAzFRS.2.t1, UAG tRNA	pCDF	Spectinomycin
pAM82 (modified from [32])	pAM82 with the UAG codon inserted at the desired location	pCOLADuet	Kanamycin

12. 100× PMSF: 0.2613 g phenylmethylsulfonyl fluoride in 15 mL ethanol (safety note: protease inhibitors are hazardous and should only be handled after taking appropriate safety precautions).
13. 1000× aprotinin: 1 mg/mL aprotinin in 25% methanol/75% H_20 (safety note: protease inhibitors are hazardous and should only be handled after taking appropriate safety precautions).
14. 1000× pepstatin = 1 mg/mL pepstatin A in 90% methanol/ 10% acetic acid (safety note: protease inhibitors are hazardous and should only be handled after taking appropriate safety precautions).
15. 1000× leupeptin = 1 mg/mL leupeptin hemisulfate in H_20 (safety note: protease inhibitors are hazardous and should only be handled after taking appropriate safety precautions).
16. Benzonase nuclease.
17. NiA buffer: 60 mM HEPES, pH 7.6, 100 mM NaCl, 100 mM KCl, 10 mM $MgCl_2$, 20 mM imidazole, 5% glycerol. Cool to 4 °C, adjust the pH, and then filter using a 0.22 μm filter.
18. NiB buffer: NiA buffer + 250 mM imidazole. Add dry imidazole to NiA buffer, and then readjust pH to 7.6.
19. Ultrasonic homogenizer with horn attachment.
20. 125 mL metal beaker (Polar Ware).
21. 5 mL HisTrap FF crude (GE).
22. 5 M NaOH.
23. 0.5 M ATP: Dissolve 9.08 g ATP disodium trihydrate in 15 mL of water. Adjust pH to 7.0 with 5 M NaOH, and then add water to a final volume of 30 mL. Aliquot and store at −20 °C.
24. Bradford reagent.
25. 2× SDS loading buffer: 100 mM Tris–HCL, pH 6.8, 4% (w/v) sodium dodecyl sulfate (SDS), 0.2% (w/v) bromophenol blue, 20% (w/v) glycerol. Add 10% (v/v) dithiothreitol (DTT) just before using.
26. Anti-FLAG M2 agarose (Sigma).
27. Gravity drip column.
28. TBS: 50 mM Tris–HCl, pH 7.6, 150 mM NaCl.
29. 3× FLAG peptide: Dissolve peptide (MDYKDHDGDYKDH-DIDYKDDDDK) in TBS. Adjust pH to 7.6 using 5 M NaOH. Then adjust final concentration of FLAG peptide to 20 mg/ mL using TBS.
30. 100 kDa cutoff Ultra-15 and Ultra-0.5 centrifugal filter concentrator (Amicon).

31. Spin-X 0.22 μm 0.5 mL cellulose acetate centrifuge filter (Corning).
32. Superose 6 Increase 10/300 GL size-exclusion column (GE).
33. GF Buffer: 30 mM HEPES, pH 7.6, 50 mM NaCl, 50 mM KCl, 10 mM $MgCl_2$, 5% glycerol. Cool to 4 °C, adjust the pH, and then filter with a 0.22 μm filter.
34. 0.5 M TCEP: Add 5.73 g tris(2-carboxyethyl)phosphine-HCl to 30 mL of water. Adjust pH to 7.0 with 5 M NaOH, and then adjust final volume to 40 mL with water. Aliquot and store at −20 °C.
35. 1 mg/mL BSA: Add 10 mg of bovine serum albumin (fatty acid free) to 10 mL of GF buffer. Adjust the concentration to 15 μM by measuring the A_{280} and using an extinction coefficient of 43,824 M^{-1} cm^{-1}.
36. Amylose resin (NEB).
37. 0.5 M maltose: Add 18 g maltose monohydrate to water, adjust to 100 mL final volume, and then filter sterilize with a 0.22 μm filter into a sterile bottle.
38. Human rhinovirus 3C protease (Thermo Fisher).
39. 8% NuPAGE SDS-PAGE gels (Thermo Fisher).
40. 10% NuPAGE SDS-PAGE gels (Thermo Fisher).
41. 20× NuPAGE MOPS running buffer (Thermo Fisher).
42. Coomassie blue stain: 0.25% Coomassie blue R-250, 45% methanol, 45% water, 10% glacial acetic acid. Filter before using.
43. Coomassie destain: 20% methanol, 10% glacial acetic acid.
44. UV spectrophotometer.
45. Quartz cuvette, 20 μL sample volume (Hellma).

2.2 Reagents for Expression and Purification of Multi-subunit Complexes Incorporating p-Azido-L-Phenylalanine

1. 1000× spectinomycin: 100 mg/mL spectinomycin dihydrochloride pentahydrate in water.
2. Unnatural amino acid incorporation plasmid (*see* Table 1).
3. 10× phosphate buffer: 0.17 M KH_2PO_4, 0.72 M K_2HPO_4. Autoclave.
4. UAA media: Add 24 g yeast extract, 20 g of tryptone, and 10 mL of glycerol to 880 mL of water. Autoclave. Prepare 1 L of UAA media by adding 100 mL of 10× phosphate buffer to 900 mL of media.
5. *p*-Azido-L-phenylalanine (Amatek Chemical).
6. 5 mM DTNB: Add 5 mg of 5,5′-dithiobis-(2-nitrobenzoic acid) to 5 mL of GF buffer. Make fresh.

7. 1 M DTT: Add 1.54 g dithiothreitol to 10 mL of water. Aliquot and store at −20 °C.
8. 40 mM fluorophore-DBCO: Dissolve fluorophore conjugated to dibenzocyclooctyne (DBCO) in DMSO to 40 mM (Click Chemistry Tools).

3 Methods

3.1 Expression and Purification of the Base

3.1.1 *E. coli* Transformation and Expression of the Base

1. Prepare electrocompetent *E. coli* BL21 Star (DE3) cells, and then transform them with 50 ng each of pAM81, pAM82, and pAM83 (*see* Fig. 1 and Table 1 for more information on the plasmids). Plate on LB + ampicillin/kanamycin/chloramphenicol. Grow at 37 °C overnight (*see* **Note 1**).
2. Select a single colony from the plate to inoculate 50 mL of dYT medium containing ampicillin (300 μg/mL), chloramphenicol (25 μg/mL), and kanamycin (50 μg/mL). Grow overnight at 30 °C, shaking at 180 rpm.
3. Pellet the overnight culture in a conical tube by centrifuging at 3000 rcf for 4 min, pour off the supernatant, and resuspend the pellet of cells in 50 mL of fresh dYT. Repeat this wash step

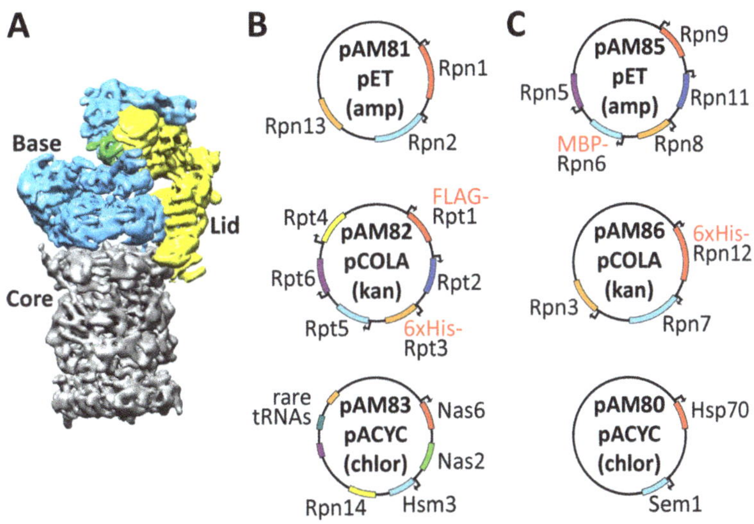

Fig. 1 (**a**) The 26S proteasome consists of three subcomplexes, the 20S core (gray), the base (blue, containing the AAA+ motor), and the lid (yellow, containing the deubiquitinating enzyme Rpn11 shown in green). (**b**, **c**) Maps of the plasmids used for expression of the base (**b**) and lid (**c**). Each plasmid was constructed by insertion of the respective proteasome-subunit genes from *Saccharomyces cerevisiae* S288C into a duet vector (pETDuet, pCOLADuet, pACYCDuet). The majority of genes are expressed with their own T7 promoter (indicated by an arrow at the beginning of the ORF), but there is only one T7 terminator per plasmid

once. Resuspend the pellet in 25 mL of dYT and divide evenly between three 2.5 L baffled flasks, each containing 1 L of TB + ampicillin (300 μg/mL), chloramphenicol (25 μg/mL), and kanamycin (50 μg/mL) (*see* **Notes 2** and **3**).

4. Grow cells at 37 °C in a shaker at 180 rpm for 3–5 h, until OD_{600} reaches 0.7. Turn down the temperature to 30 °C, wait 30 min, and then induce protein expression by adding 1 mL of 0.5 M IPTG (0.5 mM final concentration) to each flask.
5. Incubate cells for 5 h at 30 °C, shaking at 180 rpm, and then change temperature to 16 °C and let induce overnight.
6. Spin down the culture in three 1 L centrifuge bottles for 15 min at 3500 rcf, pour off the supernatant, and resuspend the cell pellets in cold NiA buffer + protease inhibitors and Benzonase (1 μg/mL aprotinin, 1 μg/mL pepstatin, 1 μg/mL leupeptin, 174 μg/mL PMSF, 50 units of Benzonase). The final volume of the combined, resuspended pellets should be 75 mL. The pellets can be transferred to 50 mL conical tubes and stored at −80 °C or carried on directly to the purification.

3.1.2 Purification of the Base

1. If using frozen pellets, thaw them in a water bath. **All remaining purification steps should be performed at 4 °C.** Combine the lysate into one 125 mL metal beaker, and use an ultrasonic homogenizer to sonicate the lysate on ice for 3 min at 75% amplitude, using a cycle of 15 s on and 60 s off (*see* **Note 4**).
2. Centrifuge the lysate for 30 min at 26,000 rcf in a pre-chilled rotor at 4 °C.
3. Prepare a 5 mL HisTrap charged with Ni^{2+} by washing the resin with 15 mL of NiA buffer (*see* **Note 5**).
4. Flow the lysate supernatant at 1 mL/min over the HisTrap. Take 1 μL of lysate supernatant, and then take another 1 μL of HisTrap flow-through for gel samples.
5. Wash the HisTrap with 50 mL of NiA + 1 mM ATP. **All following buffers should contain 1 mM ATP to ensure the stability of the complex** (*see* **Note 6**).
6. Elute the protein from the HisTrap with 20 mL of NiB buffer +1 mM ATP. Take 2 μL of the eluate for a gel sample.
7. Prepare 5 mL of anti-FLAG M2 agarose in a gravity drip column by washing with NiA + 1 mM ATP (*see* **Notes 7** and **8**).
8. Flow the nickel eluate over the anti-FLAG resin 4–5 times, exposing the protein to the resin for at least 30 min. Take 2 μL of the flow-through for a gel sample to ensure good depletion of FLAG-Rpt1.

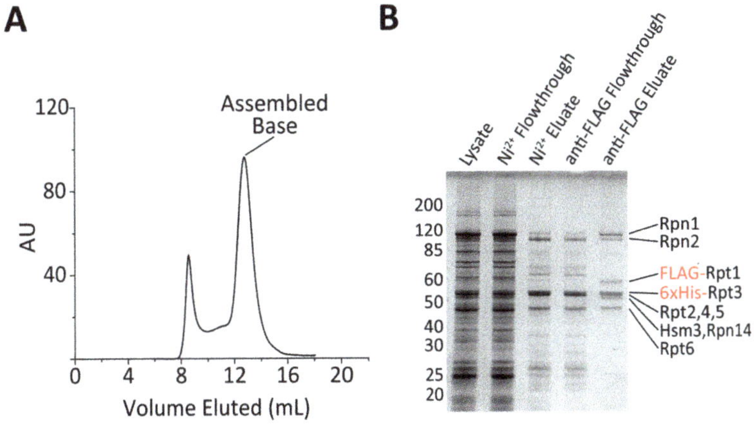

Fig. 2 (**a**) The trace from a typical size-exclusion run of purified base is shown. The peak between the void and the assembled base contains misassembled base and on a gel looks similar to properly assembled base. (**b**) Gel samples from a typical base purification run on a 10% NuPAGE gel in MOPS buffer and stained using Coomassie blue

9. Wash the anti-FLAG column 2× with 25 mL of NiA + 1 mM ATP.
10. Elute with 12 mL of NiA + 1 mM ATP + 0.15 mg/mL 3× FLAG peptide. Take 5 μL of FLAG eluate for a gel sample (*see* **Note 9**).
11. Concentrate FLAG elution to 400 μL in a 15 mL 100 kDa cutoff spin concentrator by spinning at 2000 rcf at 4 °C (*see* **Note 10**).
12. Spin filter the concentrated protein using a 0.22 μm filter (*see* **Note 11**).
13. Load and run the protein on a pre-equilibrated Superose 6 Increase 10/300 GL size-exclusion column using GF buffer + 0.5 mM TCEP and 0.5 mM ATP (*see* Fig. 2 for an example size-exclusion trace and a typical purification gel).
14. Collect the peak indicated in Fig. 2, typically from 12 mL to 14 mL when run at a flow rate of 0.4 mL/min at 4 °C. Concentrate the protein by successive spins in a 0.5 mL 100 kDa cutoff spin concentrator to a final volume of 100 μL. Take 1 μL of concentrated protein for a gel sample.
15. As the presence of ATP masks the A_{280} signal for the protein, the Bradford assay is used to measure concentration of the complex with bovine serum albumin (BSA) as a standard. Typical concentrations are 5–15 μM base (*see* **Note 12**).
16. Mix gel samples with 2× SDS loading buffer, and confirm the purity of the final complex by SDS-PAGE.

17. Aliquot the protein into 8 μL aliquots, flash freeze in liquid nitrogen, and store at −80 °C.

3.2 Expression and Purification of the Lid

3.2.1 E. coli Transformation and Expression of the Lid

1. Prepare electrocompetent *E. coli* BL21 Star (DE3) cells, and then transform with 50 ng of pAM85, pAM86, and pAM80 (*see* Fig. 1 and Table 1 for more information on the plasmids). Plate on LB + ampicillin/kanamycin/chloramphenicol. Grow at 37 °C overnight (*see* **Note 1**).
2. Select a single colony from the plate to inoculate 50 mL of dYT medium containing ampicillin (300 μg/mL), chloramphenicol (25 μg/mL), and kanamycin (50 μg/mL). Grow overnight at 30 °C, shaking at 180 rpm.
3. Pellet the 50 mL of culture in a conical tube by centrifuging at 3000 rcf for 4 min, pour off the supernatant, and resuspend the pellet of cells in 50 mL of fresh dYT. Repeat this wash step once. Resuspend the pellet in 25 mL of dYT, and divide evenly between three 2.5 L baffled flasks, each containing 1 L of TB + ampicillin (300 μg/mL), chloramphenicol (25 μg/mL), and kanamycin (50 μg/mL) (*see* **Notes 2** and **3**).
4. Grow cells at 37 °C in a shaker at 180 rpm for 3–5 h, until OD_{600} reaches 0.7. Turn down the temperature to 18 °C, wait 30 min, and then induce protein expression by adding 1 mL of 0.5 M IPTG (0.5 mM final) to each flask. Let the cells induce overnight.
5. Spin down the culture in three 1 L centrifuge bottles for 15 min at 3500 rcf, pour off the supernatant, and resuspend the pellets in cold NiA buffer + protease inhibitors and Benzonase (1 μg/mL aprotinin, 1 μg/mL pepstatin, 1 μg/mL leupeptin, 174 μg/mL PMSF, 50 units of Benzonase). The final volume of the combined, resuspended pellets should be 75 mL. The pellets can be transferred to 50 mL conical tubes and stored at −80 °C or carried on directly to the purification.

3.2.2 Purification of the Lid

1. If using frozen pellets, thaw them in a water bath. **All remaining purification steps should be performed at 4 °C.** Combine the lysate into one 125 mL metal beaker, and use an ultrasonic homogenizer to sonicate the lysate on ice for 3 min at 75% amplitude, using a cycle of 15 s on and 60 s off (*see* **Note 4**).
2. Centrifuge the lysate for 30 min at 26,000 rcf in a pre-chilled rotor at 4 °C.
3. Prepare a 5 mL HisTrap charged with Ni^{2+} by washing the resin with 15 mL of NiA buffer (*see* **Note 5**).
4. Flow the lysate supernatant at 1 mL/min over the HisTrap. Take 1 μL of lysate supernatant, and then take another 1 μL of HisTrap flow-through for gel samples.

5. Wash the HisTrap with 100 mL of NiA.
6. Elute the protein from the HisTrap with 20 mL of NiB buffer. Take 2 μL of the eluate for a gel sample.
7. Prepare 7.5 mL of amylose resin in a gravity drip column by washing with 30 mL of NiA.
8. Flow the nickel eluate over the amylose resin 4–5 times, exposing the protein to the resin for at least 30 min. Take 2 μL of the flow-through for a gel sample to ensure good depletion of MBP-Rpn6.
9. Wash the amylose column twice with 25 mL of NiA.
10. Elute with 15 mL of NiA + 10 mM maltose. Take 5 μL of amylose eluate for a gel sample.
11. To cleave off the MBP fusion from Rpn6 and the 6xHis tag from Rpn12, add 20 μL of 2 units/μL human rhinovirus 3C protease to the eluate. Incubate overnight at 4 °C or for 2–3 h at room temperature. After cleavage, take a 5 μL sample. Confirm complete cleavage through SDS-PAGE analysis.
12. Concentrate protein to 400 μL in a 15 mL 100 kDa cutoff spin concentrator by spinning at 2000 rcf (*see* **Note 10**).
13. Spin filter the concentrated protein using a 0.22 μm filter.
14. Load and run the protein on a pre-equilibrated Superose 6 Increase 10/300 GL size-exclusion column using GF buffer + 0.5 mM TCEP (*see* Fig. 3 for an example of size-exclusion trace and a typical purification gel).

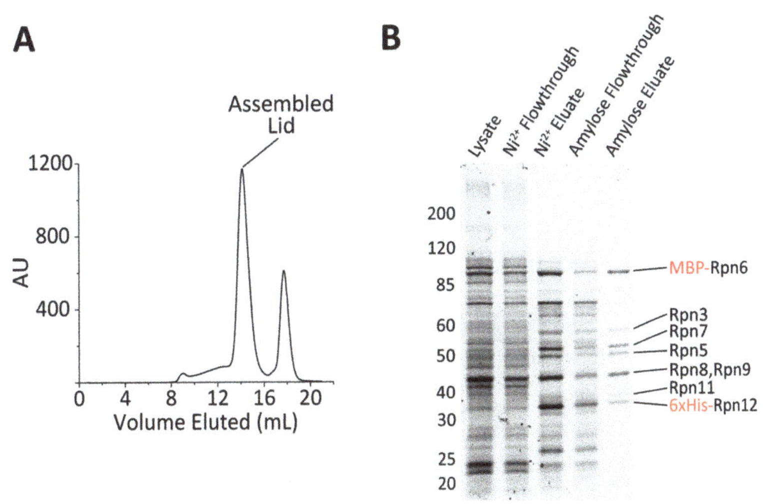

Fig. 3 (**a**) The trace from a typical size-exclusion run of purified lid is shown. The peak after the assembled lid contains mostly cleaved MBP. (**b**) Gel samples from a typical lid purification run on a 10% NuPAGE gel in MOPS buffer and stained using Coomassie blue

15. Collect the peak indicated in Fig. 3, typically from 13.5 mL to 15 mL when run at 0.4 mL/min at 4 °C. Concentrate the protein by successive spins in a 0.5 mL 100 kDa cutoff spin concentrator to a final volume of 100–200 μL. Take 1 μL of concentrated protein for a gel sample.
16. Quantify the concentration by measuring the A_{280}. The extinction coefficient calculated from the sequence is 349,000 M^{-1} cm^{-1}, and the expected concentration is 10–50 μM (*see* **Note 13**).
17. Mix gel samples with 2× SDS loading buffer, and confirm the purity of the final complex by SDS-PAGE.
18. Aliquot the protein into 8 μL aliquots, flash freeze in liquid nitrogen, and store at −80 °C.

3.3 Expression and Purification of Multi-subunit Complexes Incorporating Unnatural Amino Acids

This protocol will be presented as an abbreviated version of the base protocol presented above, focusing on the unnatural amino acid specific instructions. For more detail on the procedures, see protocol Subheading 3.1. The modifications should be transferable to the purification of other complexes.

3.3.1 *E. coli* Transformation and Protein Expression

1. Prepare electrocompetent *E. coli* BL21 Star (DE3) cells carrying pAM87, which contains the synthetase and tRNA for incorporation of AzF at the UAG amber codon. pAM87 confers resistance to spectinomycin. Please *see* **Note 1** for a protocol.
2. Use standard molecular cloning techniques to insert the UAG amber codon at the desired location in the base subcomplex (*see* **Notes 14–16**).
3. Transform pAM81, pAM82*, and pAM83 into the pAM87-containing electrocompetent *E. coli.*
4. Inoculate 2 × 50 mL cultures of dYT + ampicillin (300 μg/mL), chloramphenicol (25 μg/mL), kanamycin (50 μg/mL), and spectinomycin (100 μg/mL); grow at 37 °C overnight (*see* **Note 17**).
5. Wash the overnight cultures with fresh dYT twice, resuspend in 25 mL of dYT, and then divide the cells evenly between 6 L of dYT + 0.5× antibiotic (ampicillin (150 μg/mL), chloramphenicol (12.5 μg/mL), kanamycin (12.5 μg/mL), and spectinomycin (50 μg/mL)), in 2.5 L baffled flasks (*see* **Notes 2** and **3**).
6. Grow the cultures at 37 °C until OD_{600} 0.7 is reached. Spin down the cultures in clean 1 L centrifuge bottles for 15 min at 3500 rcf at room temperature.

7. Gently (without vortexing) resuspend the cultures in 1 L total of UAA media, leading to a sixfold concentration of cells. Divide into two 2.5 L baffled flasks to maximize aeration (*see* **Note 18**).
8. Add 0.24 g of AzF to each 500 mL culture (2 mM final), and then shake for 30 min at 30 °C at 180 rpm (*see* **Note 19**).
9. Add 1 mL of 0.5 M IPTG (1 mM final) to each flask to induce expression of both the proteins of interest and the synthetase. Let induce at 30 °C for 5 h and then 16 °C overnight.
10. Spin down the culture in two 1 L centrifuge bottles for 15 min at 3500 rcf at 4 °C, pour off the supernatant, and resuspend the pellets in cold NiA buffer + protease inhibitors and Benzonase (1 μg/mL aprotinin, 1 μg/mL pepstatin, 1 μg/mL leupeptin, 174 μg/mL PMSF, 50 units of Benzonase). The final volume of the combined, resuspended pellets should be 75 mL. The pellets can be transferred to 50 mL conical tubes and stored at −80 °C or carried on directly to the purification.

3.3.2 Purification of the Base

1. Follow **steps 1–11** of Subheading 3.1.2. In brief, the protein should be purified using Ni^{2+} affinity and anti-FLAG affinity columns and then concentrated to 400 μL (*see* **Note 20**).
2. To protect surface-exposed cysteines, add DTNB to 150 μM final concentration from a freshly made 5 mM stock, incubate at 22 °C for 10 min (solution will turn yellow), and then cool the protein back down to 4 °C (*see* **Note 21**).
3. Add the DBCO-fluorophore to a final concentration of 300 μM and incubate at 4 °C overnight (*see* **Note 22**).
4. Quench the reaction with 1 mM AzF from a 100 mM stock, then add DTT to a final concentration of 5 mM from a 1 M stock to de-protect the cysteines and resolve any inter-protein disulfide bonds.
5. Follow **steps 12–16** of protocol Subheading 3.1.2 to run the protein over the size-exclusion column. *See* Fig. 4 for examples of labeling specificity. Typical yields are about half of the yield for unlabeled complexes.

4 Notes

1. It is important to use properly prepared electrocompetent cells. The protocol from Seidman et al. (2001) can be followed as a reference [29]. Chemically competent cells are not efficient at taking up the large plasmids used here. It is possible to make electrocompetent cells from BL21 Star (DE3) *E. coli* already transformed with pAM83. This is advantageous because the efficiency of a double transformation is much higher than that

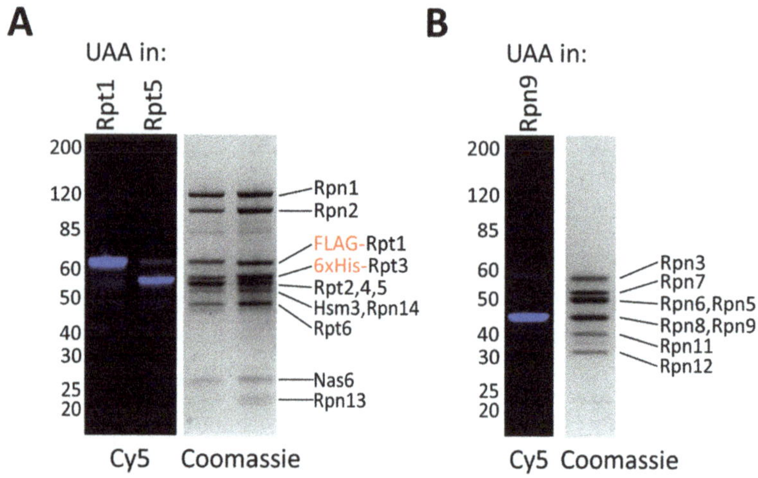

Fig. 4 (**a**) Base with AzF incorporated at either Rpt1 or Rpt5 was labeled with Cy5-DBCO. Shown here are the Cy5 fluorescence channel and Coomassie staining of those samples run on an 8% NuPAGE gel. (**b**) Lid with AzF incorporated into Rpn9 was labeled with Cy5-DBCO. As in A, the Cy5 fluorescence channel and Coomassie staining of those samples run on an 8% NuPAGE gel are shown

of the triple transformation. For protocol Subheading 3.3.1 it is necessary to make electrocompetent cells containing pAM87 or another plasmid because quadruple transformations are not feasible.

2. The wash step is important to remove any secreted β-lactamase in the media, which would diminish selection for the ampicillin resistance containing plasmid pAM81.
3. The minimum recommended volume to prep is three liters, though the protocol can be scaled up.
4. The proper sonication conditions will vary depending on the instrument used and should be optimized to maximize lysis without heating the sample. Other methods of cell lysis such as using a French press will also work.
5. The purification can also be performed using Ni-NTA agarose. Modify the procedure by batch binding 5 mL of Ni-NTA beads with the cleared lysate for 45 min and then transferring to a gravity drip column for the wash and elution.
6. The purification can be monitored using Bradford reagent. Aliquot 40 μL of Bradford reagent into the wells of a 96-well plate. Then take 5 μL of column eluate and mix thoroughly with the Bradford reagent. The presence of protein will be indicated by a color change from brown to blue. After 50 mL of washing, the eluate from the HisTrap may still react with

Bradford, but you should nevertheless continue to the elution step. Excessive washing of the nickel column reduces yield of the final complex.

7. FLAG agarose must be regenerated before the first use by following the manufacturer's recommended protocol. The resin can then be reused by regenerating after each purification.
8. The amount of FLAG agarose used is often the limiting factor for the final yield of base. Fresh FLAG agarose has higher binding capacity than used and regenerated resin. It is recommended to confirm good depletion of FLAG-Rpt1. If FLAG-Rpt1 is still visible in the flow-through of the FLAG column, the FLAG binding, washing, and eluting steps (Subheading 3.1.2, **steps 7–10**) can be repeated with freshly regenerated FLAG agarose.
9. Be sure to use 3 × FLAG peptide (MDYKDHDGDYKDHDIDYKDDDDK), as 1 × FLAG peptide does not effectively compete off the FLAG-Rpt1.
10. To prevent aggregation at the bottom of the concentrator, mix the sample by inverting the concentrator every 5–10 min.
11. Occasionally not all of the protein solution will go through the spin filter. If the filter has clogged, it may be necessary to switch to a fresh filter. Pool the filtered protein before running on the size-exclusion column.
12. Using BSA as a standard may result in concentrations inaccurate by as much as twofold due to the different reactivity of BSA with the Bradford reagent compared to the base proteins. For a more accurate concentration, the sample can be submitted for amino acid analysis. The molecular weight of the complex is 652,000 kDa, including Rpt1–6, Rpn1, Rpn2, Rpn13, Nas6, Hsm3, and Rpn14. Nas2 does not stay bound.
13. It is recommended not to use a NanoDrop to measure the protein concentration as the high viscosity of the solutions interferes with accurate readings. Instead, measurements are typically taken in a 20 μL cuvette using a UV-Vis spectrophotometer.
14. The efficiency of incorporation seems to be dependent on the location of the amber codon within the protein, but this effect is difficult to predict [30]. Therefore, it is recommended to test out multiple locations within the protein to identify the ones that give the highest protein yields.
15. Another challenge is that some locations also yield truncated products, so it is also recommended to have a C-terminal tag or other means of selecting for full-length products. In the case of the base, truncated products do not incorporate well into the complex and are thus not purified with fully assembled base.

16. To ensure proper protein production, it is also necessary to replace any UAG codons that are being used as stop codons with either the ochre or opal stop codons.
17. The cells grow slowly under quadruple antibiotic selection, so it may be necessary to first inoculate a 5 mL culture, let that grow overnight, and then use it to inoculate the two 50 mL cultures at 1:500.
18. Concentrating the cells allows for a much higher cell density per liter of culture, allowing for more protein production for each gram of unnatural amino acid used.
19. This step gives the cells time to take up the AzF.
20. As AzF is sensitive to reducing agents, they should be avoided if possible until after the labeling step described in Subheading 3.3.2 **step 3**. It has been reported that the azide group is somewhat compatible with βME if reducing agent is required [31].
21. Enough DTNB should be added to be in excess of the estimated number of exposed cysteines in solution, which varies between protein complexes.
22. The optimal concentration of fluorophores varies between proteins due to differences in the accessibility of the AzF. It is recommended to optimize the concentration of DBCO-fluorophore. The efficiency of labeling can be calculated by the ratio of the concentration of fluorophore in the final protein solution (as measured by absorbance) to the concentration of protein. Specificity can be estimated by subjecting wild-type protein with no AzF incorporated to the same labeling conditions. The labeling was found to be significantly more specific at 4 °C than at room temperature.

Acknowledgments

We thank the members of the Martin Lab for helpful discussions and the Prof. P. Schulz Lab at the Scripps Research Institute as well as the Prof. F. Isaacs Lab at Yale University for providing plasmid constructs for unnatural amino acid incorporation. J.A.M.B. acknowledges support from the NSF Graduate Research Fellowship. This research was funded in part by the US National Institutes of Health (R01-GM094497 to A.M.), the US National Science Foundation CAREER Program (NSF-MCB-1150288 to A.M.), and the Howard Hughes Medical Institute (A.M.).

References

1. Finley D (2009) Recognition and processing of ubiquitin-protein conjugates by the proteasome. Annu Rev Biochem 78(1):477–513. https://doi.org/10.1146/annurev.biochem.78.081507.101607
2. Goldberg AL (2007) Functions of the proteasome: from protein degradation and immune surveillance to cancer therapy. Biochem Soc Trans 35(Pt 1):12–17. https://doi.org/10.1042/BST0350012
3. Suraweera A, Munch C, Hanssum A, Bertolotti A (2012) Failure of amino acid homeostasis causes cell death following proteasome inhibition. Mol Cell 48(2):242–253. https://doi.org/10.1016/j.molcel.2012.08.003
4. Glickman MH, Rubin DM, Fried VA, Finley D (1998) The regulatory particle of the Saccharomyces cerevisiae proteasome. Mol Cell Biol 18(6):3149–3162
5. Groll M, Bajorek M, Kohler A, Moroder L, Rubin DM, Huber R, Glickman MH, Finley D (2000) A gated channel into the proteasome core particle. Nat Struct Biol 7 (11):1062–1067. https://doi.org/10.1038/80992
6. Rabl J, Smith DM, Yu Y, Chang SC, Goldberg AL, Cheng Y (2008) Mechanism of gate opening in the 20S proteasome by the proteasomal ATPases. Mol Cell 30(3):360–368. https://doi.org/10.1016/j.molcel.2008.03.004
7. Glickman MH, Rubin DM, Coux O, Wefes I, Pfeifer G, Cjeka Z, Baumeister W, Fried VA, Finley D (1998) A subcomplex of the proteasome regulatory particle required for ubiquitin-conjugate degradation and related to the COP9-signalosome and eIF3. Cell 94 (5):615–623
8. Saeki Y, Tanaka K (2012) Assembly and function of the proteasome. Methods Mol Biol 832:315–337. https://doi.org/10.1007/978-1-61779-474-2_22
9. Budenholzer L, Cheng CL, Li Y, Hochstrasser M (2017) Proteasome structure and assembly. J Mol Biol 429(22):3500–3524. https://doi.org/10.1016/j.jmb.2017.05.027
10. Shi Y, Chen X, Elsasser S, Stocks BB, Tian G, Lee BH, Shi Y, Zhang N, de Poot SA, Tuebing F, Sun S, Vannoy J, Tarasov SG, Engen JR, Finley D, Walters KJ (2016) Rpn1 provides adjacent receptor sites for substrate binding and deubiquitination by the proteasome. Science 351(6275):aad9421. https://doi.org/10.1126/science.aad9421
11. Husnjak K, Elsasser S, Zhang N, Chen X, Randles L, Shi Y, Hofmann K, Walters KJ, Finley D, Dikic I (2008) Proteasome subunit Rpn13 is a novel ubiquitin receptor. Nature 453(7194):481–488. https://doi.org/10.1038/nature06926
12. Tomko RJ Jr, Funakoshi M, Schneider K, Wang J, Hochstrasser M (2010) Heterohexameric ring arrangement of the eukaryotic proteasomal ATPases: implications for proteasome structure and assembly. Mol Cell 38 (3):393–403. https://doi.org/10.1016/j.molcel.2010.02.035
13. Saeki Y, Toh EA, Kudo T, Kawamura H, Tanaka K (2009) Multiple proteasome-interacting proteins assist the assembly of the yeast 19S regulatory particle. Cell 137 (5):900–913. https://doi.org/10.1016/j.cell.2009.05.005
14. Funakoshi M, Tomko RJ Jr, Kobayashi H, Hochstrasser M (2009) Multiple assembly chaperones govern biogenesis of the proteasome regulatory particle base. Cell 137 (5):887–899. https://doi.org/10.1016/j.cell.2009.04.061
15. Roelofs J, Park S, Haas W, Tian G, McAllister FE, Huo Y, Lee BH, Zhang F, Shi Y, Gygi SP, Finley D (2009) Chaperone-mediated pathway of proteasome regulatory particle assembly. Nature 459(7248):861–865. https://doi.org/10.1038/nature08063
16. Park S, Li X, Kim HM, Singh CR, Tian G, Hoyt MA, Lovell S, Battaile KP, Zolkiewski M, Coffino P, Roelofs J, Cheng Y, Finley D (2013) Reconfiguration of the proteasome during chaperone-mediated assembly. Nature 497(7450):512–516. https://doi.org/10.1038/nature12123
17. Verma R, Aravind L, Oania R, McDonald WH, Yates JR 3rd, Koonin EV, Deshaies RJ (2002) Role of Rpn11 metalloprotease in deubiquitination and degradation by the 26S proteasome. Science 298(5593):611–615. https://doi.org/10.1126/science.1075898
18. Yao T, Cohen RE (2002) A cryptic protease couples deubiquitination and degradation by the proteasome. Nature 419(6905):403–407. https://doi.org/10.1038/nature01071
19. Tomko RJ Jr, Hochstrasser M (2011) Incorporation of the Rpn12 subunit couples completion of proteasome regulatory particle lid assembly to lid-base joining. Mol Cell 44 (6):907–917. https://doi.org/10.1016/j.molcel.2011.11.020
20. Estrin E, Lopez-Blanco JR, Chacon P, Martin A (2013) Formation of an intricate helical bundle dictates the assembly of the 26S proteasome

lid. Structure 21(9):1624–1635. https://doi.org/10.1016/j.str.2013.06.023

21. Chin JW, Santoro SW, Martin AB, King DS, Wang L, Schultz PG (2002) Addition of p-azido-L-phenylalanine to the genetic code of Escherichia coli. J Am Chem Soc 124 (31):9026–9027
22. Chin JW, Martin AB, King DS, Wang L, Schultz PG (2002) Addition of a photocrosslinking amino acid to the genetic code of Escherichiacoli. Proc Natl Acad Sci U S A 99 (17):11020–11024. https://doi.org/10.1073/pnas.172226299
23. Chen PR, Groff D, Guo J, Ou W, Cellitti S, Geierstanger BH, Schultz PG (2009) A facile system for encoding unnatural amino acids in mammalian cells. Angew Chem 48 (22):4052–4055. https://doi.org/10.1002/anie.200900683
24. Amiram M, Haimovich AD, Fan C, Wang YS, Aerni HR, Ntai I, Moonan DW, Ma NJ, Rovner AJ, Hong SH, Kelleher NL, Goodman AL, Jewett MC, Soll D, Rinehart J, Isaacs FJ (2015) Evolution of translation machinery in recoded bacteria enables multi-site incorporation of nonstandard amino acids. Nat Biotechnol 33(12):1272–1279. https://doi.org/10.1038/nbt.3372
25. Chatterjee A, Sun SB, Furman JL, Xiao H, Schultz PG (2013) A versatile platform for single- and multiple-unnatural amino acid mutagenesis in Escherichia coli. Biochemistry 52(10):1828–1837. https://doi.org/10.1021/bi4000244
26. Agard NJ, Prescher JA, Bertozzi CR (2004) A strain-promoted [3 + 2] azide-alkyne cycloaddition for covalent modification of biomolecules in living systems. J Am Chem Soc 126 (46):15046–15047. https://doi.org/10.1021/ja044996f
27. Ning X, Guo J, Wolfert MA, Boons GJ (2008) Visualizing metabolically labeled glycoconjugates of living cells by copper-free and fast huisgen cycloadditions. Angew Chem 47 (12):2253–2255. https://doi.org/10.1002/anie.200705456
28. van Geel R, Pruijn GJ, van Delft FL, Boelens WC (2012) Preventing thiol-yne addition improves the specificity of strain-promoted azide-alkyne cycloaddition. Bioconjug Chem 23(3):392–398. https://doi.org/10.1021/bc200365k
29. Seidman CE, Struhl K, Sheen J, Jessen T (2001) Introduction of plasmid DNA into cells. Curr Protoc Mol Biol Chapter 1:Unit1 8. https://doi.org/10.1002/0471142727.mb0108s37
30. Schinn SM, Bradley W, Groesbeck A, Wu JC, Broadbent A, Bundy BC (2017) Rapid in vitro screening for the location-dependent effects of unnatural amino acids on protein expression and activity. Biotechnol Bioeng 114 (10):2412–2417. https://doi.org/10.1002/bit.26305
31. Tian H, Sakmar TP, Huber T (2016) A simple method for enhancing the bioorthogonality of cyclooctyne reagent. Chem Commun 52 (31):5451–5454. https://doi.org/10.1039/c6cc01321j
32. Beckwith R, Estrin E, Worden EJ, Martin A (2013) Reconstitution of the 26S proteasome reveals functional asymmetries in its AAA+ unfoldase. Nat Struct Mol Biol 20(10):1164–1172. https://doi.org/10.1038/nsmb.2659
33. Lander GC, Estrin E, Matyskiela ME, Bashore C, Nogales E, Martin A (2012) Complete subunit architecture of the proteasome regulatory particle. Nature 482 (7384):186–191. https://doi.org/10.1038/nature10774

Chapter 16

Native Gel Approaches in Studying Proteasome Assembly and Chaperones

Jeroen Roelofs, Anjana Suppahia, Kenrick A. Waite, and Soyeon Park

Abstract

Proteasomes are complex molecular machines that consist of 66 subunits. The assembly of these complexes is highly coordinated in a process that requires at least ten proteasome-specific molecular chaperones. One of the challenges in studying assembly intermediates is their relatively low abundance as compared to the proteasome holoenzyme. Therefore, superior separating techniques are crucial for analyses of proteasomal complexes in general and studies in the assembly in particular. For this reason, native gel analyses have been abundantly used in studying proteasomes, as they provide a high resolution. Native gels are very versatile and can be used in various combinatorial approaches. In this chapter, we outline two approaches to prepare samples for native gels. The first is a yeast cryogrinding method and the second a core particle (CP)-base reconstitution approach. We describe the native gel electrophoresis, as well as various downstream analyses, including 2D native-SDS-PAGE. These techniques and approaches can all be used, often in parallel, to gain a variety of information about activity and composition of the complexes separated by native gel. The potential combined approaches are discussed in this review.

Key words Native gel, Proteasome, Molecular chaperone, Assembly, AAA-ATPase, Activity assay, LLVY-AMC, Fluorescent label, 2D gel

1 Introduction

Cells contain many large protein complexes that execute a variety of important functions. One well-known example is the ribosome, which undergoes a complex process of assembly, maturation, and quality control prior to becoming a functional holoenzyme [1]. While ribosomes are rather unique due to their tight associations of RNA with protein components, they are not alone in their dependence on molecular chaperones to ensure accurate and efficient assembly. Other examples are spliceosomes, nucleosomes, and the proteasomes [2]. Proteasomes are large protease complexes found in the cytosol as well as nucleus of eukaryotic cells. Studying proteasome assembly has provided a wealth of biochemical, cellular, and structural information on the assembly process [3]. Besides key

Thibault Mayor and Gary Kleiger (eds.), *The Ubiquitin Proteasome System: Methods and Protocols*, Methods in Molecular Biology, vol. 1844, https://doi.org/10.1007/978-1-4939-8706-1_16,

insights into the regulation of this important complex in the cell, it has also revealed several principles that seem to be shared by assembly chaperones in general. Proteasomes consist of a core particle (CP) that is cylindrically shaped as a result of four stacked heptameric rings ($\alpha_{1-7}\beta_{1-7}\beta_{1-7}\alpha_{1-7}$). The proteolytic active sites are localized on the inner surface of this structure. Its formation is guided by proteasome-specific chaperones. Half-CP ($\alpha_{1-7}\beta_{1-6}$) lacking the β7 subunit is a readily detectable intermediate. The incorporation of β7 coincides with the dimerization of two half-CPs, autocatalytic processing of pro-peptides, and degradation of the chaperone Ump1. This process induces structural changes that lower the affinity for the chaperones Pba1-Pba2 [4–6]. The ends of the CP cylinder serve as docking sites for various regulators of the proteasome, the binding of which are mutually exclusive. These regulators include PA200 (Blm10 in yeast); PI31 (Fub1 in yeast); PA28/REGα, β, and γ (all absent from yeast); and the regulatory particle (19S or RP). While the RP is most abundant, the other regulators can be seen on native gels and in purified samples of proteasomal complexes (*see* **Note 1**); for example, Blm10-CP and Blm10-CP-RP can be detected readily in yeast total lysates. The main focus in proteasome research and assembly is on the RP and RP-CP complex for several reasons. First, the RP-CP complexes are the major degradative proteasome complex in the cell. Second, RP contains six AAA-ATPases and is the only known proteasome component to hydrolyze ATP. As such, this complex is responsible for all energy-dependent degradation by the proteasome. Third, among the regulators, only the RP is known to contain a number of ubiquitin receptors and bind ubiquitinated substrates [7]. Biochemical studies have shown that RP can be divided into two subassemblies, base and lid. These complexes have also been identified using native gel analyses of various strains with defects in proteasome assembly. For example, many mutants defective in base assembly accumulate the lid sub-complex [8–11].

One of the challenges in studying assembly is that the immature complexes of interest are a minor species in the cell. The majority of each specific subunit is present in the mature functional complex. Furthermore, it is challenging to distinguish between subcomplexes that are assembly intermediates versus the complexes that result from intracellular or post-lysis dissociation. Nevertheless, there are several approaches available to study this. One approach utilizes unique components that are only present on assembly intermediates. For the proteasome, ten specific assembly chaperones have been identified [12, 13]. Five are specific for the RP, known by their yeast names Nas6, Rpn14, Hsm3, Nas2, and Adc17. The other five, Pba1, Pba2, Pba3, Pba4, and Ump1, function in CP assembly. All ten chaperones have orthologues in humans and are specific for proteasome assembly. They have almost exclusively been identified as components of assembly intermediates instead of the holoenzyme. The human orthologue of Nas6, gankyrin, is a notable exception as it is reported to have roles in regulating the degradation of specific factors by the proteasome and is implicated as an oncogene in liver cancer [14]. In general, however, the presence of these chaperones in complexes is considered to be an indicator of the presence of assembly intermediates (*see* **Note 2**). Thus, affinity tags on these chaperones can be used to purify assembly intermediates [4, 9, 15]. A second approach is to separate lysates or purified samples based on size or specific biochemical properties, using conditions where mature complexes will behave differently from assembly intermediates. Here, native gels, FPLC, and ultracentrifugation have been commonly used to separate subcomplexes from mature forms [8, 9, 16]. A third approach to purify intermediates consists of two steps. First, all complexes are purified using an affinity tag. Next, a second tag on a subunit that is only present on the mature complex is utilized to deplete mature complexes from the sample. This approach has been less utilized, at least in proteasome research, due to an initial focus on the chaperone-assisted proteasome assembly and the potential complications resulting from some instability of certain complexes.

Both base and CP rely on assembly chaperones for efficient complex formation, whereas lid does not. Lid can assemble without specific chaperones when the subunits are heterologously expressed in bacteria and no lid-specific assembly factors have been identified in yeast or mammals. Recent studies indicate that lid assembly is driven by the subunits themselves [17, 18]. These papers also highlight powerful new in vitro approaches to study proteasome assembly, as now the heterologous co-expression of numerous proteasome subunits in *E. coli* allows for the purification of assembly intermediates that can be used to reconstitute proteasomes.

As described in the preceding sections, the low abundance of assembly intermediates relative to the mature complex can make their analyses challenging. To increase the levels and odds of detecting these assembly intermediates, research in proteasome assembly has heavily relied on genetic manipulations that cause an enrichment in, or disappearance of, assembly intermediates. In mammalian cells, this has been done using a siRNA approach [19], while in yeast, knockouts and combinations of mutants have been commonly used to manipulate the system [8, 9, 20]. Consider a conceptually simple example where two subcomplexes A and B combine with other subcomplexes to form a mature complex. Both intermediates can be stable in the cell and accumulate in some small amount. Reducing the abundance of complex B, e.g., by deleting a chaperone associated with complex B, could yield an increase in complex A. This has been observed for the proteasome where the absence of chaperones required for base results in increased levels of lid [8–11, 21]. Lysates or complexes purified

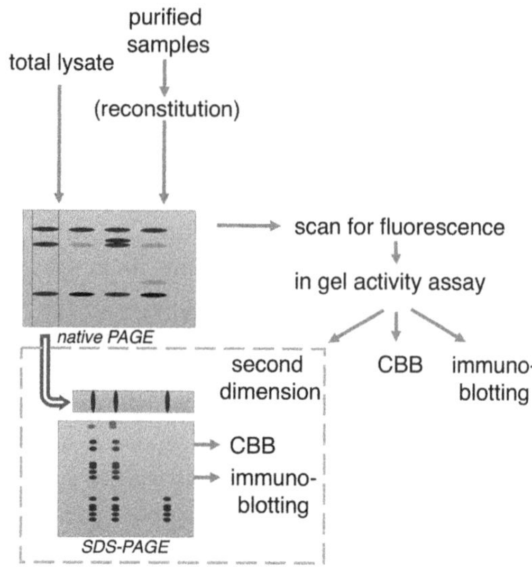

Fig. 1 Flowchart of procedures to analyze proteasome assembly by native PAGE. Samples for native gel are generally total lysates or purified proteasome complexes. Following separation, native gels can be analyzed sequentially by scanning for fluorescent proteins (if a subunit was fluorescently tagged, *see* Fig. 2b left three panels), followed by in-gel activity assay (Fig. 2b fourth panel). Next gel can be used in one of the three ways. Either the gel is stained with Coomassie dye, transferred to PVDF for immunoblotting, or lanes or bands are excised for a second dimension of SDS-PAGE, after which the 2D gel can be stained with Coomassie dye or used for immunoblotting

from wild-type and genetically manipulated strains can be analyzed with techniques mentioned above and described below to reveal differences and deduce mechanisms.

Whichever approach has been utilized in studying proteasome assembly, they almost universally utilize native gels as part of the analyses. In the method described here, we show a potential flow of how various analyses using native gels at the center can be utilized in your research (Fig. 1).

2 Materials

Prepare all solutions using Milli-Q water (18 mΩ at room temperature, referred to as ddH_2O onward). Procedures are executed at room temperature unless stated otherwise. Standard equipment for protein gel electrophoresis was used for native gels as well as second-dimension SDS-PAGE. The same equipment also allows for wet transfer of proteins to PVDF membranes for immunoblotting. We used the mini-PROTEAN electrophoresis system (Bio-Rad) with 1.5-mm-thick spacers and combs.

2.1 Total Cell Lysate

1. Lysis buffer A: 50 mM Tris–HCl (pH 8.0), 5 mM $MgCl_2$, 1 mM EDTA, 1 mM ATP.
2. Cryogrinding system (OPS Diagnostics, CG 08-10). This contains porcelain/zirconium mortars, pestles suitable with drill connection, cordless drill, and cryocooler with rack.
3. Liquid nitrogen.

2.2 Purification of Base

1. Lysis buffer B: 50 mM Tris–HCl (pH 8.0), 5 mM $MgCl_2$, 1 mM EDTA, 1 mM ATP, 10% glycerol.
2. Wash buffer B: 50 mM Tris–HCl (pH 7.5), 5 mM $MgCl_2$, 1 mM EDTA, 100 mM NaCl, 1 mM ATP, 10% glycerol.
3. Dissociation buffer: 50 mM Tris–HCl (pH 7.5), 5 mM $MgCl_2$, 1 mM EDTA, 1 M NaCl, 1 mM ATP, 10% glycerol.
4. Elution buffer B: 50 mM Tris–HCl (pH 7.5), 5 mM $MgCl_2$, 0.2 mM DTT, 1 mM ATP, 10% glycerol.
5. IgG resin (MP Biomedicals, 55961).
6. Cobalt agarose beads.
7. TEV protease: AcTEV (Invitrogen, 12575-015).
8. Centrifugal devices for sample concentration (Pall, Nanosep 10 K omega).

2.3 Purification of Chaperones

1. Lysis buffer C: 20 mM Tris–HCl (pH 8.0), 100 mM NaCl, 1 mM EDTA, 1% Triton X-100 supplemented with protease inhibitors.
2. Wash buffer C: 1× PBS with 0.1% Triton X-100 and 0.02% sodium azide.
3. Elution buffer C: 50 mM Tris–HCl (pH 6.8), 150 mM NaCl, 1 mM EDTA, 1 mM DTT, 0.01% Triton X100.
4. Glutathione agarose resin.
5. PreScission Protease (aka C3 protease, GE Healthcare, 27-0843-01).
6. Protease inhibitor tablets.
7. Slide-A-Lyzer Dialysis Cassette.

2.4 Reconstitution Assays and Analysis of Chaperone Function

1. Buffer A: 50 mM Tris–HCl (pH 7.5), 5 mM $MgCl_2$, 1 mM EDTA, 10% glycerol, 1 mM ATP.
2. Affinity-purified base, lid, and CP.
3. Affinity-purified chaperones.
4. ATP regeneration system: 52 mM creatine phosphate, 0.5 mg/mL creatine phosphokinase.

2.5 Native PAGE

1. 5× native running buffer: 450 mM Tris–Borate, 25 mM $MgCl_2$, and 2.5 mM EDTA (*see* **Note 3**).

2. 5× native loading buffer: 250 mM Tris–HCl (pH 7.5), 50% glycerol (V/V), ~50 ng/mL xylene cyanol.
3. 0.5 M ATP. Dissolve in ddH_2O and adjust pH to 7.0 using NaOH. Store as aliquots in −80°C.
4. 40% acrylamide (acrylamide/Bis solution 37.5:1).
5. 10% ammonium persulfate (APS). Dissolve 1 g APS into 7 mL ddH_2O, resuspend, and adjust volume to 10 mL to obtain a 10% APS solution (*see* **Note 4**).
6. TEMED (*N*,*N*,*N*′,*N*′-tetramethylethylene-diamine).

2.6 Detection of Proteasomes in Native Gels

2.6.1 Fluorescent Screening of Native Gel

1. Typhoon 9410.

2.6.2 LLVY-AMC In-Gel Activity Assay

1. Stock solutions of 1 M Tris–HCl (pH 7.5), 1 M $MgCl_2$, and 0.5 M ATP.
2. Substrate suc-LLVY-AMC (Bachem, I1395) is suspended in dimethylformamide to a concentration of 10 mM, aliquoted, and stored at −20°C (*see* **Notes 5** and **6**).

2.6.3 CBB Staining of Gels

1. Fixing solution: 50% (V/V) methanol, 10% (V/V) acetic acid in ddH_2O.
2. Coomassie Brilliant Blue (CBB) G-250 staining solution. Add ~80 mg of Brilliant Blue G in 1 L ddH_2O, stir for 2–3 h using stir bar and plate, and add 3 mL 12 N HCl. Keep solution in dark by wrapping bottle in aluminum foil.
3. Destaining solution: water or 40% (V/V) methanol, 10% (V/V) acetic acid in ddH_2O.

2.6.4 Immunoblotting of Native Gels

1. Methanol.
2. Transfer buffer: 25 mM Tris, 190 mM glycine (optional: 20% methanol).
3. PVDF membrane.
4. Whatman paper.
5. Mini Trans-Blot module.

2.7 Two-Dimensional Native-SDS-PAGE

1. 1× SDS sample buffer: 65 mM Tris–HCl (pH 6.8), 2% SDS, 10% glycerol, 100 mM DTT 0.01% w/v bromophenol blue.
2. 1× SDS running buffer: 25 mM Tris, 192 mM glycine, 0.1% SDS.
3. APS, TEMED, and 40% acrylamide (*see* Subheading 2.5).

Table 1
Stacking gel and separating gel solution mixture

	Stacking gel (5%) (mL)	Separating gel (12%) (mL)
ddH_2O	1.48	4.3
40% acrylamide solution	0.25	3.0
1.0 M Tris–HCl (pH 6.8)	0.25	–
1.5 M Tris–HCl (pH 8.8)	–	2.5
10% SDS	0.02	0.1
10% APS (*see* native PAGE)	0.02	0.1
TEMED (*see* native PAGE)	0.002	0.004

4. 12% SDS-PAGE gel. Mix reagents as indicated in Table 1.
5. Agarose.
6. Water-saturated n-butanol (*see* **Note 7**).
7. Bio-Rad PROTEAN-Tetra 2D comb (optional).
8. Razor blade, single edge no. 9.
9. Single lane incubation or blotting tray, such as Bio-Rad small incubation tray chamber.

3 Methods

3.1 Total Cell Lysate

For native gels, cell lysis has to occur under native conditions. This can be achieved in various ways; however, some are more prone to complex dissociation or poor resolution on native gels as compared to others. One of the challenges for lysis of yeast cells is the need to disrupt the cell wall. This can be achieved enzymatically using zymolyase or lyticase; however, this is not commonly used as it requires incubation at 30°C, which can induce physiological changes. More commonly, the cell wall is mechanically disrupted. In glass bead lysis, a yeast cell pellet is resuspended in lysis buffer, combined with acid-washed glass beads, and vortexed for 1–2 min to mechanically disrupt the cell wall. Besides vortexing, specialized beading machines exist such as a bead beater. While these approaches have been used successfully, they are prone to disruption of the proteasome complexes in our hands and might require careful care to prevent warming of the sample. An alternative approach is the French press. This lysis method has been commonly used to purify affinity-tagged proteasomes and can provide a gentle lysis. A disadvantage is that generally larger volumes of cells are required. This makes this method less convenient and more

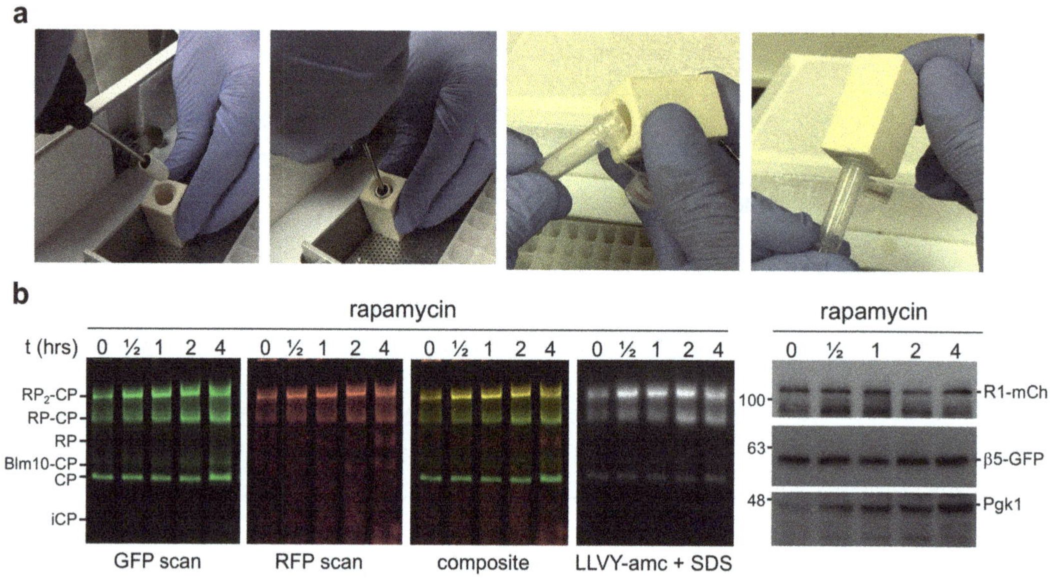

Fig. 2 Yeast total cell lysate via cryogrinding using cordless drill. (**a**) Mortar and electric drill-connected pestle are used for cryogrinding of small amount of cell droplets that have been frozen in liquid nitrogen. After lysis, powder can be transferred to 2 mL screwcap tube as shown and frozen for later resuspension in lysis buffer or used directly. (**b**) Yeast strain with GFP-tagged β5 and mCherry-tagged Rpn1 was treated with rapamycin, and cells were collected at indicated timepoints, lysed by cryogrinding, and separated on native gel. Next gel was scanned on the typhoon 9410 for GFP and mCherry fluorescence (three panels on the left). Followed by an in-gel suc-LLVY-AMC activity assay with 0.02% SDS. After lysis and resuspension of frozen powder in lysis buffer, 50 μL of lysate was mixed with 10 μL 6× SDS sample buffer and boiled for 5 min. 10 μL was loaded on SDS-PAGE gel and immunoblotted for mCherry, GFP, and the loading control Pgk1. iCP indicates the 15S immature CP complex

wasteful when the goal is to analyze a number of lysates on native gel (native gels require only a small amount of lysate). In our experience cryogrinding is the best approach. This can be done with liquid nitrogen using mortar and pestle. Alternatively, a grinding mill, e.g., from RETSCH, can be used. Here beads and jars of different volumes are available. While this approach is more reproducible and thorough as compared to grinding by hand, it is also rather cumbersome and time consuming for larger amounts of samples as it requires pair-wise processing. Our favorite approach is using small porcelain/zirconium mortars and pestles that can be connected to a cordless drill. This is done in a cryocooling box, which allows for relatively easy and reproducible creation of cryo-lysates (Fig. 2).

1. Harvest the equivalent of 100 ODs of cells (*see* **Note 8**).
2. Wash cells with cold ddH_2O.
3. Resuspend pellet in 40 μL of ddH_2O.
4. Add liquid nitrogen to a small vessel and submerge a strainer to catch pellet. Use proper eye protection and cryo-gloves when

working with liquid nitrogen and ultra-cooled materials and instruments.

5. Using a p200 pipette, slowly let droplets of the cell suspension drip into liquid nitrogen resulting in the formation of small frozen beads of yeast. Pellets will sink once freezing is complete. Transfer these frozen droplets to 1.5 mL tube, e.g., by using chilled tweezers (*see* **Note 9**).
6. Samples can be stored at −80°C or directly used in the next step.
7. To lyse pellets, pour liquid nitrogen into cryocooler.
8. Cool mortars and pestles (one for each sample) in a small vessel with a lid (small Styrofoam box) filled with liquid nitrogen. The colder everything stays the better for grinding. We found that precooling of the mortars in the cryocooler was not sufficient as it resulted in a slushy pellet.
9. Retrieve mortar and pestle from liquid nitrogen, and place on metal rack in cryocooler (Fig. 2a).
10. Add the yeast frozen droplets to the mortar (for 100 ODs equivalent, we use the half of the total pellet).
11. Attach pestle to cordless drill, and grind continuously 10 s clockwise followed by 10 s counterclockwise until pellet becomes a fine powder. If too much pellet was added, the powder can overflow the mortar here.
12. Scrape sides of mortar with liquid nitrogen-chilled spatula (*see* **Note 10**).
13. Use a tube that fits tightly on the top of the mortar to collect the material by inverting the mortar and tapping on the bottom. 2 mL screw-capped tubes fit nicely on the mortar (Fig. 2a).
14. Ground yeast cells can be stored at −80°C as powder or can be resuspended in lysis buffer for further analyses.
15. Resuspend the ground yeast cells in proteasome lysis buffer A, and clear lysates by centrifugation at 4°C at 16,000 × *g* for 5 min.
16. Collect the supernatant. Measure protein concentration of your samples for equal protein loading. Alternatively, depending on the experimental setting, it can be preferable to load equal volumes on gel as it would represent lysates from the same amount of cells.

3.2 Purification of Base

Described is the method for the purification of the proteasome base complex from yeast cells harboring a protein A-tagged Rpt1 subunit. Although the base can be efficiently separated from the proteasome holoenzyme through high-salt wash (1 M NaCl) [22], the isolated base did not readily reconstitute into the proteasome holoenzyme. A major improvement in this protocol is that the base remains functional and can be fully reconstituted into the proteasome holoenzyme [8, 23]. Key to this improvement is the addition of 10% glycerol to all buffers throughout purification. Purity of the base is further improved utilizing a strain lacking *RPN10* (*see* **Note 11**). ProA-TEV-Rpt1 precipitates endogenous base in complex with the assembly chaperones, in addition to the proteasome holoenzyme. Although high-salt wash is used during base purification, some chaperones remain bound to the base due to high binding affinities between the base and chaperones [23–25]. To ensure that any endogenous chaperone is absent in the base purification, the base can be purified from yeast strains lacking individual chaperones [25]. As compared to base purification, CP purifications are more straightforward as CP is a very stable complex and does not require nucleotide unlike RP and base [22]. While these techniques purify proteasome subcomplexes, these complexes are largely derived from dissociation of assembled proteasome holoenzyme, and thus are not true assembly intermediates. Nevertheless, they have been successfully used as model systems in reconstitution assays and various experiments to dissect the assembly process and the role of proteasome-specific assembly chaperones [23, 25].

1. Inoculate 50 mL YPD with SY36 (a strain expressing ProA-TEV-Rpt1 [22]), and grow overnight at 30°C.
2. Inoculate 10 mL of the overnight culture into three Fernbach flasks containing 1.5 L YPD each. Grow at 30°C with shaking at 200 rpm overnight. Culture should reach an OD_{600} ~10–15.
3. Harvest cells by centrifugation using GS3 rotor at 4250 × *g* for 5 min or equivalent.
4. Resuspend pellets in small amount of cold water, and combine into one bottle. Fill bottle to top with water to wash cells (*see* **Note 12**).
5. Spin down cells at 4250 × *g* for 5 min using GS3 rotor. The wet weight of pellet will be approximately 75 g. Store the pellet frozen at −80°C, or proceed to the next step.
6. Resuspend pellet in ~1.5 volume of lysis buffer B. For a 75 g pellet, use 100 mL buffer.
7. Lyse cells using the French press 1250 psi, and collect lysate in 250 mL centrifuge bottles. The pH of lysate should be 7–7.5 (*see* **Note 13**).
8. To clear lysate, centrifuge lysate in GSA rotor and tubes (16,000 × *g* for 25 min at 4°C).
9. Filter supernatant through cheese cloth, which will help to absorb some of the white lipid layer that typically forms on top.

10. Add IgG resin that has been washed in lysis buffer, to the supernatant. Use about 2.0 mL of resin for 4.5 L culture.
11. Incubate 1 h at 4°C under continuous rotation.
12. Collect sample in Bio-Rad chromatography column (*see* **Note 14**).
13. Wash with 25 bed volumes (BV) wash buffer B.
14. Close the column and add 5 BV dissociation buffer.
15. Incubate at room temperature for 1 h under constant rotation (*see* **Note 15**).
16. Wash column with 50 BV of dissociation buffer.
17. Wash column with 15 BV of elution buffer B.
18. For elution, close column, and add 0.75 BV elution buffer that contains His-tagged TEV protease. Stir gently to make sure that resin and TEV protease mix.
19. Leave column at 30°C for 1 h or overnight at 4°C.
20. Collect the first eluate.
21. Add 1/2 BV elution buffer B without TEV. Act quickly to avoid drying of resin.
22. Collect the second elution and repeat this step.
23. Combine all fractions, and add 35 μL Co^{2+} resin to remove TEV protease (Talon or Ni^{2+}-NTA resin works equally well for this purpose).
24. Incubate at 4°C for ~20 min under constant rotation.
25. Remove Co^{2+} resin by spinning down resin (2 min 500 × *g*) and collecting the supernatant.
26. Concentrate eluate at 4°C using a centrifugal concentrator with a 10 kDa cutoff (like Centricon 10 MWCO). This procedure generally yields 100 μL of base at approximately 1 μg/μL.

3.3 Purification of Chaperones

Plasmids encoding N-terminally GST-tagged chaperones are expressed in Rosetta 2 *E. coli* strains. Expression and purification follow standard procedures using glutathione resin. Purifications have been successful using batch approaches as well as column-based method. If necessary, chaperones can be further purified following elution by GST-tagged PreScission Protease cleavage using size-exclusion chromatography.

Grow Rosetta 2 cells containing plasmids of interest overnight at 37 °C in LB containing chloramphenicol and antibiotics for plasmid selection (ampicillin or kanamycin).

1. Next day, inoculate 200 mL of LB containing relevant antibiotics with 5 mL overnight culture, and grow at 37 °C until OD_{600} = ~0.6.

2. Induce expression by adding IPTG to a final concentration of 0.3 mM.
3. Incubate cells for 3 h at 25 °C or overnight at 18 °C.
4. Harvest cells by centrifugation. The pellet can be stored at −80 °C for later processing.
5. Resuspend the pellet in 10 mL lysis buffer C.
6. Lyse by French press using manufacturer's protocol.
7. Clear lysate by centrifugation at 20,000 × *g* for 30 min at 4 °C.
8. Transfer supernatant to a 15 mL tube, and add 200 μL glutathione agarose resin that has been washed in lysis buffer.
9. Incubate for 1 h at 4 °C under constant rotation.
10. Spin down resin by centrifugation (500 × *g*, 1 min, 4 °C).
11. Wash resin with 14 mL wash buffer C.
12. Repeat wash four times.
13. Wash with 14 mL elution buffer C.
14. To elute chaperones and remove the GST-tag, eluate was incubated with GST-tagged PreScission, which cleaves a PreScission cleavage site located between the GST and chaperone. Here, GST-PreScission remains bound to the beads eliminating an additional step to remove the protease from the sample.
15. Add 1 bed volume elution buffer C supplemented with 5 μL PreScission Protease.
16. Incubate overnight at 4 °C.
17. Precipitate resin (500 g, 1 min, 4 °C) and collect supernatant.
18. To retrieve maximum yield, add 1 BV of elution buffer C to the beads, centrifuge, and collect supernatant. Repeat twice, and combine all supernatants, or analyze on SDS-PAGE for fractions that contain the chaperone.
19. Purified chaperones can be dialyzed into reconstitution assay buffer (*see* Subheading 2.4) using a Slide-A-Lyzer. Float Slide-A-Lyzer in 500 mL reconstitution assay buffer with stirring at 4 °C, replace buffer after 4 h, and continue incubation overnight.
20. Purified chaperones can be aliquoted and stored at −80 °C.

3.4 Reconstitution Assays and Analysis of Chaperone Function

The proteasome holoenzyme forms via the associations of base, lid, and CP. To study the function of base-binding chaperones in this process, we have established in vitro reconstitution assays. Here, we describe two strategies for using base-CP reconstitution to assay chaperone action. These strategies overcome the challenges that chaperones act rapidly and base-CP complexes are scarce in vivo. Individual chaperones can be added in specific amounts, combinations, and at specific time points during base-CP complex formation. In addition, different nucleotides (ATP, ADP, and ATPγS) can be used to create specific nucleotide states of the base, which are linked to its maturation status during proteasome assembly [23, 25]. These strategies have also been applied to examine chaperone actions during the base-lid assembly process [8, 25].

3.4.1 Strategy I with Free Base and Chaperones

1. Equilibrate the purified base (150–300 nM) in 10–20 μL buffer A supplemented with specific nucleotide (ADP, ATP, or ATPγS) at 1–2 mM for 10 min at room temperature. If ATP is used, include an ATP regeneration system to maintain ATP concentration during reconstitution.
2. To test chaperone actions, add chaperones to the base at desirable molar ratios. To ensure that the base is constantly bound by the chaperones, we add the chaperones in molar excess (e.g., at tenfold) to base. For testing the action of all three base-binding chaperones, we combine them first, and then add them simultaneously to the base (*see* **Note 16**).
3. Incubate samples at room temperature for 10 min to allow for chaperone-base binding.
4. Add purified CP to the base at a molar ratio of CP to base of 1:2, and incubate samples at 30°C for 15, 30, and 60 min.
5. At the end of the incubation, briefly spin down the tubes, and add 5× native loading buffer. Immediately proceed to Native PAGE.

3.4.2 Strategy II with Immobilized Base and Chaperones

1. For preparation of resin-bound base, follow base purification procedure as specified in 3.2 up to the step immediately before TEV protease elution. Approximately 30 mL of cryo-powders generate sufficient amount of resin-bound base, which can be kept in the 4°C for several days.
2. Add 10 μL resin-bound base in 10–20 μL buffer A supplemented with ADP, ATP, or ATPγS at 1–2 mM. The amount of the base is estimated to be approximately 0.8 pmol. Leave the tubes for 10 min at room temperature to equilibrate the base in specific nucleotides. If ATP is used, include an ATP regeneration system to maintain ATP concentration during reconstitution.
3. To test chaperone actions, we add a tenfold molar excess of chaperones to the base.
4. Incubate tubes at room temperature for 10 min to allow for chaperone-base binding.
5. Add CP at 1:2 molar ratio to the base.
6. Rotate the tubes in the lab rotator for 45 min in the 30°C incubator.

7. Collect bead-bound material by centrifugation for 15 s at 16,000 × *g*, and wash using 500 μL of wash buffer. Repeat the washing step twice. Maintain the concentration of the nucleotides at **step 2** throughout washing steps. If ATPγS was used, we typically keep it at 0.5 mM during wash since it is a non-hydrolyzable ATP analog and is also costly.
8. Add 50 μL of TEV elution buffer to the beads to elute base-bound complexes.
9. Collect the eluates by passing the material through Bio-Rad microspin column.
10. Add Laemmli sample buffer and boil for 5 min at 95°C.
11. Analyze the samples by SDS-PAGE and immunoblotting for base subunits, chaperones, and CP (*see* **Note 17**).

3.5 Native PAGE

Proteasomes and proteasome subcomplexes are well suited for separation on native PAGE and show resolutions comparable to other commonly used separating techniques, such as size-exclusion chromatography or density gradient centrifugation. The disadvantage compared to these techniques, however, is that the sample will no longer be free in solution but embedded in the gel matrix. Nevertheless, various downstream analyses and assays are still possible (Fig. 1). For native PAGE, lab-made gels with or without stacking gel are commonly used to great effect, although costlier precast gels can be also used but have the disadvantage that they lack ATP within the gel matrix. ATP is critical to maintain the integrity of the ATPase-containing proteasomal complexes. Samples in these native gels are resolved based on the charge, the size, and the shape of the complex. As such we have seen 26 kDa proteins migrate between complexes in the MDa range.

1. Prepare your mini-PAGE system by thoroughly cleaning glass plates, drying them, and assembling the system according to the manufacturer's instruction. We use the mini-PROTEAN Tetra system from Bio-Rad, but other systems can work equally well. The native gels are very sensitive for electrophoretic anomalies. We have achieved best migration patterns for proteasomal complexes using 1.5-mm-thick spacers and 10-well combs.
2. Prepare 10 mL of gel polymerization solution. We normally use 3.5% to separate CP, and RP, and RP-CP complexes on gel. Acrylamide concentrations can be increased for resolving certain assembly intermediates. Take a 15 mL tube, and add 2 mL of 5× native running buffer, 0.9 mL 40% acrylamide, 7.1 mL ddH_2O, and 10 μL 0.5 M ATP. Mix by inverting tube three times while avoiding bubble formation. When ready to pour gel, add 90 μL 10% APS and 9 μL TEMED, which will start the

polymerization process. Mix the entire contents by inverting the tube several times (*see* **Note 18**).

3. Using a serological pipette, pipette ~9.5 mL of the solution between the glass plates in the middle of the gel assembly using a smooth continuous flow to create a homogenous gel. Stop when the top of the shorter glass plate is reached. Using a slight angle, insert the comb starting on one side. Prevent trapping of bubbles underneath the comb area to avoid electrophoretic anomalies. Close the 15 mL tube, and keep on bench to monitor gel polymerization, which can take up to 30 min depending on temperature and light levels.
4. Prepare 500 mL running buffer by combining 100 mL of 5× running buffer, 500 μL ATP, and 400 mL of cold ddH_2O.
5. Remove gel from casting stand and gently rinse wells with ddH_2O. Assemble gel in electrophoresis apparatus. Fill inner chamber to the top with running buffer, and use the remainder for outer chamber.
6. Prepare samples by adding 1/4 of the sample volume of 5× native loading buffer, mix by pipetting, and spin solution down if needed. Carefully load samples onto the gel, using a normal p2–200 tip or gel-loading tip. For the latter, be careful to avoid poking the bottom of the well. Loading empty lanes with 10 μL of 1× native sample buffer might improve overall migration behavior of neighboring lanes as well as help to track the migration of the dye. Lanes on both ends of the gel tend to generate smiling effects, similar to SDS-PAGE gels.
7. Add cap to the apparatus and transfer to a 4 °C fridge. Run gel at ~100 Volts for 3 h. Depending on your desired migrations and specific conditions, gels can run longer or shorter. However, this period normally gives a nice separation of CP, CP-Blm10, CP-RP, and CP-RP_2 from yeast when conducting in-gel activity assays (*see* **Note 19** and Fig. 2b).

3.6 Detection of Proteasomes in Native Gels

3.6.1 Fluorescent Screening of Native Gel

If proteasome subunits are genetically engineered to harbor fluorescent protein, such as GFP, mCherry, CFP, or YFP, the subcomplexes containing the tagged subunit can be monitored on gels using scanning instruments with the correct lasers and filter combinations. Furthermore, fluorescent inhibitors have been developed that bind to catalytic subunits. These can be used to monitor proteasomal complexes via native gel as well [26]. We use the Typhoon 9410 scanner to detect fluorescently tagged proteasomes in native gels. Although the gel can be scanned directly between glass plates, the plates can cause background. This can be reduced using nonluminescent glass plates (Biocraft, Japan) [10]. However, to avoid these problems, we generally disassemble the gel sandwich, and place the gel directly on the scanner plate (*see* **Note 20**). Once the gel has been placed on scanning area, use instrument settings that best match the excitation and emission spectra of the fluorescent marker. On the Typhoon, GFP can be detected using a 488 nm laser for excitation and a 526 nm short-pass emission filter. For YFP the 532 nm laser was used with a 555 nm band-pass filter. To detect mCherry excitation was achieved with a 633 nm laser and data collected using a 580 nm band-pass filter. GFP and mCherry can be successfully monitored in the same gel with little cross-bleeding [10, 27, 28] (Fig. 2b). However, the filters are not necessarily specific to a single fluorescent protein and that some combinations result in cross-bleed of signals. Furthermore, under some settings and growing conditions, a background band can be observed. Therefore, analyses should include negative control (untagged protein) and positive controls (singly tagged).

3.6.2 LLVY-AMC In-Gel Activity Assay

1. Make 20 mL of peptidase assay solution by combining 1.0 mL Tris–HCl (pH 7.5) with 100 μL 1 M $MgCl_2$, 40 μL 0.5 M ATP, and 100 μL 10 mM suc-LLVY-AMC, and add to dish slightly bigger than mini gel, such as a square petri dish.
2. Once the desired electrophoresis period has been reached, remove one glass plate from the sandwich. Dislodge the gel into the assay solution carefully, as the gel tears easily under its own weight due to its low acrylamide concentration. Keeping the gel facing down on the other glass plate closely above the petri dish can facilitate the transfer (*see* **Notes 20** and **21**).
3. Incubate gel for 15 min at 30 °C. For samples derived from human cells, we use 37 °C.
4. Transfer gel to your imaging system. We use a Syngene Gbox. AMC has an optimum emission at about 460 nm, and using a 440 nm band-pass filter gives us increased sensitivity as compared to standard ethidium bromide filter.
5. The presence of small amount of SDS (0.02%) in the assay buffer will allow for the opening of ends of the cylindrical CP, which are normally closed by the flexible N-termini of the CP α-subunits. Ideally, the assays with and without SDS are done in parallel. Nevertheless, we found it very convenient to first image our samples in the absence of SDS, then add 40 μL 10% SDS per 20 mL of assay solution, incubate the gel for an additional 15 min, and image again. The second image will show activation of free CP (Fig. 3b) as well as 26S complexes that are inhibited by Ecm29 [29, 30].

3.6.3 CBB Staining of Gels

1. Transfer gel to fixing solution and incubate for 20 min on rocker.
2. Replace fixing solution with CBB staining solution to cover the gel (~25 mL).

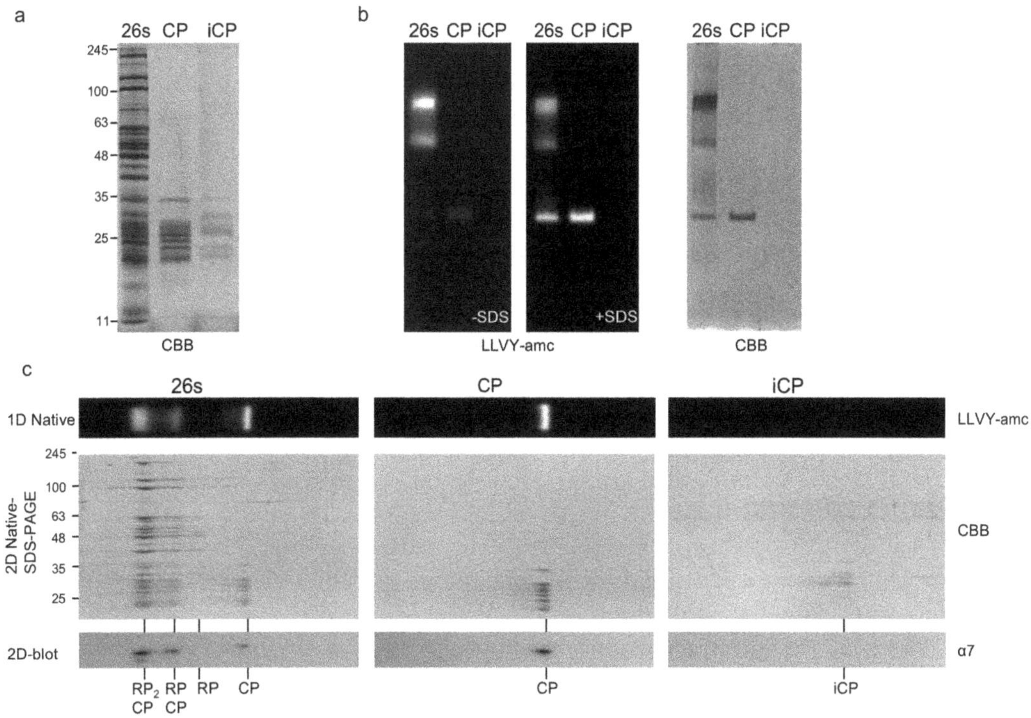

Fig. 3 2D electrophoreses native-SDS-PAGE. (**a**) Affinity purified proteasome complexes were separated on a 12% SDS-PAGE gel and stained with Coomassie dye. Protein A-tagged subunits were Rpn11, β4, and the chaperone Ump1 for 26S, CP, and iCP, respectively. (**b**) Samples were also separated on native gel, followed by suc-LLVY-AMC in-gel activity assay without and with 0.02% SDS, and stained with Coomassie dye. The SDS opens the gate of the CP allowing for entry of the substrate into free CP. RP also opens the gate thereby eliminating the need of SDS for visualization of RP-CP complexes. (**c**) Top panel shows native gel lane excised for 2D analysis. Middle panel shows Coomassie dye-stained 2D native-SDS-PAGE. Lower panel shows cropped immunoblot of 2D native-SDS-PAGE blotted for the CP subunit α-7

3. Optional: microwave for ~10 s to speed up the staining (*see* **Note 22**).
4. Incubate at room temperature while rocking until desired intensity has been reached. Transfer gel to water or destaining solution to enhance the contrast, as it will reduce gel background (*see* **Note 23** and Fig. 3a).

3.6.4 Immunoblotting of Native Gels

Immunoblotting of native gels can be precarious and not always straightforward. The protein complexes need to be negatively charged and bind SDS to allow transfer out of gel onto a PVDF membrane. To facilitate this, we first incubate our gel for 10 min in SDS-PAGE running buffer at room temperature. Next, to immunoblot the gel, standard wet transfer techniques are used (*see* **Note 24**). Please note that the assembly of the 3.5% native gel is particularly tricky. The best approach is to first place your PVDF membrane on top of the Whatman paper and then position the native gel on top of the PVDF, making sure to keep all wet. This allows for easier positioning as compared to assembling your native gel on the Whatman paper. Before use, the PVDF membrane has to be hydrated with methanol, followed by equilibration in ddH_2O and transfer buffer. Avoid applying excessive pressure on the gel, as this could stretch or distort native gels. We generally conduct transfer at 20 V (~90 mAmp) overnight in the cold room. Proceed immunoblotting following standard procedures (*see* **Note 24**).

3.7 Two-Dimensional Native-SDS-PAGE

Two-dimensional native to SDS-PAGE gels can provide a powerful approach to analyze samples. For purified samples, it can reveal the subunit composition of specific bands on the native gel and thus provide a more comprehensive analysis of the composition as compared to performing immunoblots for different subunits [31]. It can be also used to identify unknown factors that specifically purify with specific species of the sample. For total cell lysates, Coomassie Brilliant Blue or silver stain analyses are not informative due to the large amount of protein complexes; therefore we combine 2D gels with immunoblotting. While native gels can be directly used for subsequent immunoblotting, this can be challenging under some circumstances. The extra step of a second dimension prior to immunoblotting has two major advantages. First, detection sensitivity for the protein of interest increases substantially. This suggests that the SDS-PAGE step either results in better denaturing or exposure of epitopes of the proteins or the transfer efficiency is enhanced for migration from SDS-PAGE to PVDF as compared to native page to PVDF. Second, it provides molecular weight information. This can be valuable if the protein of interest is modified posttranslationally. Furthermore, antibodies might show cross-reactivity with other proteins of a different molecular weight. Since in native PAGE polypeptides migrate in complexes, the molecular weight information is absent. Thus, careful controls should be included, and caution should be taken in interpretation of immunoblot signals of native PAGE. Including a second dimension of SDS-PAGE in analyses provides the molecular weight information, thereby allowing for a more reliable interpretation of the immunoblot data (Fig. 3c).

1. Make 12% minigel according to manufacturer's instruction. After polymerization of the separating gel, rinse minigel with ddH_2O. Next, prepare and pour the stacking gel using either a 2D gel comb if available, or add stacking solution till a little below the top of the short plate (*see* **Note 25**). For the latter, use water saturated n-butanol to create a straight surface that is sealing the air for efficient polymerization. Once ready to proceed with 2D gel, remove comb from the SDS-PAGE gel and rinse with ddH_2O.

2. Excise lane (or bands) of interest from your native gel (*see* **Note 26**). Carefully transfer the gel piece to, e.g., a Bio-Rad small incubation tray chamber filled with 1× SDS-PAGE sample buffer. Incubate under gentle rocking for 15 min (*see* **Note 27**).
3. Prepare an agarose solution by dissolving 0.5% agarose in 1× SDS running buffer (*see* **Note 28**). Once fully dissolved (look carefully for the absence of any transparent clumps), transfer the solution to a couple of 1.5 mL tubes, and keep in heating block >65°C to prevent premature solidification. This solution is used to seal the native gel piece in place on top of the second-dimension SDS-PAGE gel (*see* **step 6** below).
4. Carefully take your native gel lane from the 1× SDS sample buffer and position horizontal on the separating gel. Carefully slide/push the gel in position (*see* **Note 29**).
5. Secure the gel lane into position by pipetting melted agarose solution on and around the gel piece (*see* **Note 30**).
6. Once solidified, assemble the SDS-PAGE gel into the running apparatus, fill chambers with running buffer, load protein marker, and run gel as you would run a normal SDS-PAGE gel. Then proceed to further downstream analyses like immunoblotting or CBB or silver staining.

4 Notes

1. Both ends of the CP can bind these regulators, and hybrid forms exist, e.g., where Blm10 binds to one end of the CP and RP the other end (Blm10-CP-RP). As such, even purifications with a tag to one of the regulators can yield other regulators in a purification.
2. Another convenient indicator of assembly intermediates exists for the CP, since five β subunits (β1, β2, β5, β6, β7) contain a propeptide that are proteolytically removed upon formation of mature CP. The size of these subunits is thus a reliable indicator of the assembly stage.
3. This solution is made by combining 54.5 g Tris–Base, 27.8 g boric acid, 25 mL 1 M $MgCl_2$, and 5 mL 0.5 M EDTA (pH 7.5), pH is not adjusted, and this combination of chemicals should result in a pH a bit higher than 8. The solution is filter sterilized, stored at 4°C, and can be used up to a few months.
4. Solution is stable at 4 °C for ~2–3 weeks, or aliquots can be stored in −20 °C for longer periods to avoid repeated freeze-thaw cycles.

5. The suc-LLVY-AMC (LLVY stands for the amino acids that are part of this peptide) is light sensitive, and the stock should be kept in dark. During assay period no special protection from light is generally required. Over time this solution has a tendency to precipitate. If the solution precipitates, the tube can be centrifuged to precipitate the crystalized material. Normally, sufficient amount of substrate remains in solution, and in-gel activity assays are not impeded. Using the solution in more quantitative activity assays is not advised. An alternative solvent is DMSO.
6. suc-LLVY-AMC is most commonly used and monitors mainly the chymotrypsin-like active sites (β5 residues). However, substrates are commercially available to monitor trypsin-like (β2) or caspase-like (β1) activity if desired [32].
7. Water-saturated butanol prevents drying of the gel by absorption of water from the butanol solution. Mix water and butanol thoroughly, let phases separate and used is top layer. Due to the higher difference in density as compared to simply using water to layer on top of the polymerizing solution, the butanol provides a straight and sharp edge. Rinse away butanol using ddH_2O prior to proceeding to next step.
8. Culture size depends on the physiological state of the sample you desire. Normally, logarithmically growing cells are desired. To achieve this, we generally inoculate a 100 mL culture to OD_{600} ~0.2 using a small overnight culture. These cells are grown for several hours till OD ~1 is reached; spinning down the cells in this 100 mL culture will give you the 100 OD equivalent of cells.
9. Do not yet seal the lid as it will pop open from evaporating N_2 (l) and some pellet can be lost.
10. We found one half of a flat-ended tweezer with the far end wrapped in tape works well to scrape out the material that remains in the bottom of the mortar.
11. The purification of base from Rpt1-tagged strains yields a small amount of 26S proteasomes that are resistant to treatment aimed at proteasome dissociation. The use of an Rpt1-tagged strain that has a deletion of *RPN10* eliminates the residual 26S proteasome in the base purification.
12. Omitting wash step can lower the pH of the lysate and cause premature dissociation of proteasome complexes.
13. Alternatively, yeast cells can be directly frozen as droplets and ground under cryogenic conditions using a RETSCH Mill.
14. Use appropriate column for resin volume; for 0.75 mL resin, we use the 1.5 cm diameter, which actually gives a rather small resin bed. For 2 mL resin, a bigger column can be used.

15. This step leads to the disruption of RP into the lid and base. We close the top and bottom of the column and attach it horizontal to a Labquake and keep it rotating. Just leaving the column at room temperature without movement does not appear to give good disruption. Also, we observed better dissociation at RT as compared to 4°C.
16. To test chaperone actions on preformed base-CP complexes, allow the base-CP complexes to form first, and then add the specific chaperones of interest.
17. Instead of using immobilized base, immobilized CP can be used [23] for base-CP reconstitution. In this case, free base and chaperones are added at specific points during the assays.
18. Native PAGE can be prone to leakage. This can be reduced by avoiding chipped glass plates and careful attention to lining up the bottoms of glass plates and spacers. Most leakage occurs at the ends where glass plates and spacers meet; using a very small amount of Vaseline only here can greatly reduce the occurrence of leaks.
19. Note that the dye reaches the bottom of the gel after about 2 h, but longer running times are often preferred. If gels are scanned for fluorescence, they can still be assayed for in-gel activity followed by CBB staining or immunoblotting.
20. To facilitate the handling of these low percentage native gels that have a consistency weaker than stacking gels, keeping surfaces wet is crucial. Also, gels should not be pinched between two fingers, as the weight of the hanging gel will cause tearing. Instead, roll up the gel during handling, grab it all at once with all five fingers, or fold it once and grab it as a whole using all five fingers.
21. To easily dislodge the gel from the plate, the green wedge that the Bio-Rad provides to separate the glass plates can be used to run it through the corners to detach the gel from the edges of glass plate.
22. Solution should not boil during this step; if so shorten the time in microwave.
23. CBB R-250 is less suited for this protocol. Both CBB chemicals have similar sensitivity; however, CBB R-250 will cause strong staining of gel, and bands are less prominent after destaining.
24. The transfer to PVDF can be done with or without 20% methanol. We routinely eliminate methanol to reduce organic waste; however, for some proteins, the methanol can be crucial. For example, the presence of methanol in the transfer dramatically increases the detection of free ubiquitin.
25. The percentage of the SDS-PAGE can be adjusted based on the MW of the proteins of interest. Our standard 11 or 12% allows

a reasonable separation of the lower molecular weight CP subunits and provides a good separation of the higher molecular weight proteasome subunits without the need for a gradient gel.

26. Doing the excision of the lane on a UV box after the activity assay will facilitate this process as the activity in the different lanes can be used to guide the excision. Avoid back-forward saw/cutting motion. We normally push a razorblade straight down to cut an area, raise the blade, and move to the next part of the region to be cut. Others have used a pizza cutter [33].
27. Whenever handling a lane from native gel, holding it on one edge can cause the gel to rip under its own weight. Carefully position, e.g., a wide razor blade or the green plate wedge from Bio-Rad underneath the center of the gel strip to lift the gel.
28. We generally use a 50 mL tube filled with 10 mL of 1× SDS running buffer and ~50 mg of agarose. Heat in microwave in short bursts to avoid overboiling with the SDS.
29. Align the excised lane on top of separating gel. Here, we generally position the top of the native gel strip on the side where we load the marker. Be careful to prevent trapped bubbles or gaps, as they will cause deformations when running the second dimension. Use gentle pressure to push the gel strip into position and partly between the two glass plates starting from one end. A blunt thin object, like the opposite end of the razor blade, will facilitate this process. Deformations from a straight angle of the strip with respect to the separating gel result in broadening of the dots on the final 2D gel in the x-axis direction.
30. 2D gel combs will have a lane for protein marker. If not using 2D comb, prior to adding agar, insert a single 1.5-mm-thick comb tooth next to the gel lane in the agarose for loading of the protein marker.

Acknowledgments

Many of these techniques described above have been developed or optimized and tweaked by many experts in the field, and the power of their use has been displayed in numerous excellent papers. We apologize for not being able to provide a comprehensive reference to all those contributions in this publication. This work was supported in part by grants to J.R. from NIH (R01-GM118660 and R15-GM112142).

References

1. Kressler D, Hurt E, Bassler J (2017) A puzzle of life: crafting ribosomal subunits. Trends Biochem Sci 42(8):640–654. https://doi.org/10.1016/j.tibs.2017.05.005
2. Ellis RJ (2013) Assembly chaperones: a perspective. Philos Trans R Soc Lond B Biol Sci 368(1617):20110398. https://doi.org/10.1098/rstb.2011.0398
3. Budenholzer L, Cheng CL, Li Y, Hochstrasser M (2017) Proteasome structure and assembly. J Mol Biol 429(22):3500–3524. https://doi.org/10.1016/j.jmb.2017.05.027
4. Wani PS, Rowland MA, Ondracek A, Deeds EJ, Roelofs J (2015) Maturation of the proteasome core particle induces an affinity switch that controls regulatory particle association. Nat Commun 6:6384. https://doi.org/10.1038/ncomms7384
5. Ramos PC, Hockendorff J, Johnson ES, Varshavsky A, Dohmen RJ (1998) Ump1p is required for proper maturation of the 20S proteasome and becomes its substrate upon completion of the assembly. Cell 92(4):489–499. https://doi.org/10.1016/S0092-8674(00)80942-3
6. Kock M, Nunes MM, Hemann M, Kube S, Dohmen RJ, Herzog F, Ramos PC, Wendler P (2015) Proteasome assembly from 15S precursors involves major conformational changes and recycling of the Pba1-Pba2 chaperone. Nat Commun 6:6123. https://doi.org/10.1038/ncomms7123
7. Shi Y, Chen X, Elsasser S, Stocks BB, Tian G, Lee BH, Shi Y, Zhang N, de Poot SA, Tuebing F, Sun S, Vannoy J, Tarasov SG, Engen JR, Finley D, Walters KJ (2016) Rpn1 provides adjacent receptor sites for substrate binding and deubiquitination by the proteasome. Science 351(6275):aad9421. https://doi.org/10.1126/science.aad9421
8. Park S, Roelofs J, Kim W, Robert J, Schmidt M, Gygi SP, Finley D (2009) Hexameric assembly of the proteasomal ATPases is templated through their C termini. Nature 459(7248):866–870. https://doi.org/10.1038/nature08065
9. Roelofs J, Park S, Haas W, Tian G, McAllister FE, Huo Y, Lee BH, Zhang F, Shi Y, Gygi SP, Finley D (2009) Chaperone-mediated pathway of proteasome regulatory particle assembly. Nature 459(7248):861–865. https://doi.org/10.1038/nature08063
10. Saeki Y, Toh EA, Kudo T, Kawamura H, Tanaka K (2009) Multiple proteasome-interacting proteins assist the assembly of the yeast 19S regulatory particle. Cell 137(5):900–913. https://doi.org/10.1016/j.cell.2009.05.005
11. Funakoshi M, Tomko RJ Jr, Kobayashi H, Hochstrasser M (2009) Multiple assembly chaperones govern biogenesis of the proteasome regulatory particle base. Cell 137(5):887–899. https://doi.org/10.1016/j.cell.2009.04.061
12. Bedford L, Paine S, Sheppard PW, Mayer RJ, Roelofs J (2010) Assembly, structure, and function of the 26S proteasome. Trends Cell Biol 20(7):391–401. https://doi.org/10.1016/j.tcb.2010.03.007
13. Tomko RJ Jr, Hochstrasser M (2013) Molecular architecture and assembly of the eukaryotic proteasome. Annu Rev Biochem 82:415–445. https://doi.org/10.1146/annurev-biochem-060410-150257
14. Higashitsuji H, Liu Y, Mayer RJ, Fujita J (2005) The oncoprotein gankyrin negatively regulates both p53 and RB by enhancing proteasomal degradation. Cell Cycle 4(10):1335–1337. https://doi.org/10.4161/cc.4.10.2107
15. Tomko RJ Jr, Funakoshi M, Schneider K, Wang J, Hochstrasser M (2010) Heterohexameric ring arrangement of the eukaryotic proteasomal ATPases: implications for proteasome structure and assembly. Mol Cell 38(3):393–403. https://doi.org/10.1016/j.molcel.2010.02.035
16. Kusmierczyk AR, Kunjappu MJ, Funakoshi M, Hochstrasser M (2008) A multimeric assembly factor controls the formation of alternative 20S proteasomes. Nat Struct Mol Biol 15(3):237–244. https://doi.org/10.1038/nsmb.1389
17. Tomko RJ Jr, Taylor DW, Chen ZA, Wang HW, Rappsilber J, Hochstrasser M (2015) A single alpha helix drives extensive remodeling of the proteasome lid and completion of regulatory particle assembly. Cell 163(2):432–444. https://doi.org/10.1016/j.cell.2015.09.022
18. Estrin E, Lopez-Blanco JR, Chacon P, Martin A (2013) Formation of an intricate helical bundle dictates the assembly of the 26S proteasome lid. Structure 21(9):1624–1635. https://doi.org/10.1016/j.str.2013.06.023
19. Kaneko T, Murata S (2012) Using siRNA techniques to dissect proteasome assembly pathways in mammalian cells. Methods Mol Biol 832:433–442. https://doi.org/10.1007/978-1-61779-474-2_30
20. Tallec BL, Peyroche A (2012) Using DNA damage sensitivity phenotypes to characterize

mutations affecting proteasome function. Methods Mol Biol 832:363–371. https://doi.org/10.1007/978-1-61779-474-2_25

21. Barrault MB, Richet N, Godard C, Murciano B, Le Tallec B, Rousseau E, Legrand P, Charbonnier JB, Le Du MH, Guerois R, Ochsenbein F, Peyroche A (2012) Dual functions of the Hsm3 protein in chaperoning and scaffolding regulatory particle subunits during the proteasome assembly. Proc Natl Acad Sci U S A 109(17):E1001–E1010. https://doi.org/10.1073/pnas.1116538109
22. Leggett DS, Glickman MH, Finley D (2005) Purification of proteasomes, proteasome subcomplexes, and proteasome-associated proteins from budding yeast. Methods Mol Biol 301:57–70. https://doi.org/10.1385/1-59259-895-1:057
23. Park S, Li X, Kim HM, Singh CR, Tian G, Hoyt MA, Lovell S, Battaile KP, Zolkiewski M, Coffino P, Roelofs J, Cheng Y, Finley D (2013) Reconfiguration of the proteasome during chaperone-mediated assembly. Nature 497(7450):512–516. https://doi.org/10.1038/nature12123
24. Le Tallec B, Barrault MB, Guerois R, Carre T, Peyroche A (2009) Hsm3/S5b participates in the assembly pathway of the 19S regulatory particle of the proteasome. Mol Cell 33(3):389–399. https://doi.org/10.1016/j.molcel.2009.01.010
25. Li F, Tian G, Langager D, Sokolova V, Finley D, Park S (2017) Nucleotide-dependent switch in proteasome assembly mediated by the Nas6 chaperone. Proc Natl Acad Sci U S A 114(7):1548–1553. https://doi.org/10.1073/pnas.1612922114
26. Verdoes M, Florea BI, Menendez-Benito V, Maynard CJ, Witte MD, van der Linden WA, van den Nieuwendijk AM, Hofmann T, Berkers CR, van Leeuwen FW, Groothuis TA, Leeuwenburgh MA, Ovaa H, Neefjes JJ, Filippov DV, van der Marel GA, Dantuma NP, Overkleeft HS (2006) A fluorescent broad-spectrum proteasome inhibitor for labeling proteasomes in vitro and in vivo. Chem Biol 13(11):1217–1226. https://doi.org/10.1016/j.chembiol.2006.09.013
27. Waite KA, De-La Mota-Peynado A, Vontz G, Roelofs J (2016) Starvation induces proteasome autophagy with different pathways for core and regulatory particles. J Biol Chem 291(7):3239–3253. https://doi.org/10.1074/jbc.M115.699124
28. Enenkel C (2012) Using native gel electrophoresis and phosphofluoroimaging to analyze GFP-tagged proteasomes. Methods Mol Biol 832:339–348. https://doi.org/10.1007/978-1-61779-474-2_23
29. Lee SY, De la Mota-Peynado A, Roelofs J (2011) Loss of Rpt5 protein interactions with the core particle and Nas2 protein causes the formation of faulty proteasomes that are inhibited by Ecm29 protein. J Biol Chem 286(42):36641–36651. https://doi.org/10.1074/jbc.M111.280875
30. De La Mota-Peynado A, Lee SY, Pierce BM, Wani P, Singh CR, Roelofs J (2013) The proteasome-associated protein Ecm29 inhibits proteasomal ATPase activity and in vivo protein degradation by the proteasome. J Biol Chem 288(41):29467–29481. https://doi.org/10.1074/jbc.M113.491662
31. Kleijnen MF, Roelofs J, Park S, Hathaway NA, Glickman M, King RW, Finley D (2007) Stability of the proteasome can be regulated allosterically through engagement of its proteolytic active sites. Nat Struct Mol Biol 14(12):1180–1188. https://doi.org/10.1038/nsmb1335
32. Li Y, Tomko RJ Jr, Hochstrasser M (2015) Proteasomes: Isolation and activity assays. Curr Protoc Cell Biol 67:3.43. 1–3.43.20. https://doi.org/10.1002/0471143030.cb0343s67
33. Hochstrasser M, Funakoshi M (2012) Disulfide engineering to map subunit interactions in the proteasome and other macromolecular complexes. Methods Mol Biol 832:349–362. https://doi.org/10.1007/978-1-61779-474-2_24

Chapter 17

Measuring the Overall Rate of Protein Breakdown in Cells and the Contributions of the Ubiquitin-Proteasome and Autophagy-Lysosomal Pathways

Zhe Sha, Jinghui Zhao, and Alfred L. Goldberg

Abstract

In certain physiological or pathological states (e.g., starvation, heat shock, or muscle atrophy) and upon drug treatments, the overall rate of protein degradation in cells may increase or decrease. These adaptations and pathological responses can occur through alterations in substrate flux through the ubiquitin-proteasome pathway (UPP), the autophagy-lysosomal system, or both. Therefore, it is important to precisely measure the activities of these degradation pathways in degrading cell proteins under different physiological states or upon treatment with drugs. In particular, proteasome inhibitors have become very important agents for treating multiple myeloma and very useful tools in basic research. To evaluate rigorously their efficacy and the cellular responses to other inhibitors, it is essential to know the degree of inhibition of protein breakdown. Unfortunately, commonly used assays of the activities of the UPP or autophagy rely on qualitative, indirect approaches that do not directly reflect the actual rates of protein degradation by these pathways. In this chapter, we describe isotopic pulse-chase methods to directly measure overall rates of protein degradation in cells by radiolabeling cell proteins and following their subsequent degradation to radioactive amino acids, which diffuse from cells into the medium and can be easily quantitated. While pulse-chase methods have often been used to follow degradation of specific proteins, the methods described here allow quantification of the total cellular activity in degrading either long-lived proteins (the great bulk of cell constituents) or the fraction with short half-lives. Moreover, by use of specific inhibitors of proteasomes or lysosomes, it is also possible to measure precisely the total contributions of the UPP or lysosomal proteases. These approaches have already been proven very useful in defining the effects of inhibitors, growth factors, nutrients, ubiquitination, and different proteasome activators on overall proteolysis and on substrate flux through the proteasomal and lysosomal pathways.

Key words Protein degradation, Ubiquitin-proteasome pathway (UPP), Autophagy-lysosomal system, Radiolabeling, ^{3}H-phenylalanine, Pulse-chase, Liquid scintillation counting, 26S proteasomes, Ubiquitin

1 Introduction

In mammalian cells under standard culture conditions, the ubiquitin-proteasome pathway (UPP) accounts for 70% or more of the protein degradation, while the autophagy-lysosomal system

Thibault Mayor and Gary Kleiger (eds.), *The Ubiquitin Proteasome System: Methods and Protocols*, Methods in Molecular Biology, vol. 1844, https://doi.org/10.1007/978-1-4939-8706-1_17,

degrades the remainder [1, 2]. It is now well established that the total amount of protein degradation by these two pathways is tightly regulated and can vary in different physiological and pathological conditions. For example, there is growing evidence that proteasome activity and protein degradation are diminished in several neurodegenerative diseases associated with the accumulation of misfolded aggregation-prone proteins, such as tau [3]. Conversely, a key event in muscle atrophy and cachexia is the increase in overall protein degradation through the coordinated activation of both the UPP and autophagy [4, 5]. Protein degradation can also be inhibited by many synthetic inhibitors of the proteasome, ubiquitin conjugation, or lysosomal function. The proteasome inhibitors, bortezomib, carfilzomib, and ixazomib, are now the preferred treatments for multiple myeloma, and there are many ongoing clinical trials with these inhibitors for other diseases [6–8]. In addition, proteasome inhibitors (e.g., MG132) are very widely used as research tools in many areas of biology. In such studies, it can be very important to directly measure the actual rates of cellular protein breakdown and the specific contributions of the UPP and autophagy-lysosome pathways. The approaches described here provide the most physiologically relevant measurements of the cell's degradative activities and allow us to accurately evaluate how substrate flux through these pathways changes under different physiological conditions or with drug treatments.

Unfortunately, the methods commonly used to measure the activities of the proteasome or autophagy in cells or the extent of their inhibition do not directly measure the actual rates of proteolysis in cells. Proteasome activity is most commonly evaluated simply by assaying the peptidase activities of the 20S particles' catalytic β-subunits using small fluorogenic peptides [9] or the recently developed active-site probes that bind covalently to the β-subunits' active sites [10–12]. However, in cells, proteasomes degrade ubiquitinated proteins, not short peptides, and peptidase assays often do not reflect the actual extent of changes in the cellular rate of protein breakdown by the UPP. Peptidase assays measure only the ability of peptides to enter the 20S's catalytic chamber and their subsequent hydrolysis [13], but there are many additional steps that are essential for the degradation of ubiquitinated proteins [14], and they can also change under different physiological conditions or upon drug treatment. These steps include the binding of ubiquitinated substrates to specific 26S receptor subunits, substrate deubiquitination, opening of the substrate entry channel into the 20S particle, and protein unfolding, translocation, and hydrolysis. Protocols to measure many of these steps are discussed in the Chapter 19 by Kim et al.; and methods to purify 26S proteasomes for such studies are presented in the Chapter 18 by Kuo et al. Also, approaches to investigate the regulation of proteasome function by subunit phosphorylation are covered in the Chapter 20 by VerPlank and Goldberg.

Other common approaches to evaluate the activities of the UPP and autophagy share the similar limitation of not being able to provide direct information on changes in the actual rates of protein degradation. For example, proteasome function and inhibition are also often monitored in cells and tissues by following the accumulation of ubiquitinated proteins. However, the abundance of ubiquitin conjugates is affected not only by their rates of degradation by proteasomes, but also by the actions of cellular deubiquitinating enzymes (DUBs) and the rate of protein ubiquitination. Therefore, the accumulation of ubiquitinated proteins does not necessarily indicate proteasome inhibition. In fact, inhibition of mTOR or heat shock elevates Ub conjugate levels in cells by increasing protein ubiquitination, while also increasing overall degradation by the UPP [15]. Similarly, the activity of the autophagy-lysosome pathway is generally monitored by microscopic measurements of the abundance of autophagic vacuoles or by the western blot assays of marker proteins, such as lipidated LC3 [16]. However, neither approach measures the actual amount of cell proteins that are being degraded by lysosomes at any time.

This chapter describes our methods to overcome these limitations and to directly measure the actual rates of protein degradation in cells. For this purpose, we pre-label cell proteins with a radioactive amino acid and then determine the rate of the hydrolysis of these pre-labeled proteins to amino acids. In cells, proteins are hydrolyzed by the proteasome or lysosome to small peptides, which are further cleaved by cytosolic endo- and exo-peptidases to amino acids that are transported across the plasma membrane into the extracellular medium. By using a large excess of nonradioactive amino acids in the medium as the chase, it is possible to prevent the rapid reincorporation of the labeled amino acid back into new proteins and also to promote the counter-transport of these labeled residues into the medium. The radioactive amino acids in the medium are then separated from any radiolabeled proteins by precipitating the proteins with trichloroacetic acid (TCA). Both the TCA-soluble radioactivity and the total radioactivity initially incorporated into cell proteins are measured by liquid scintillation counting. We can then calculate the mean rate of protein degradation in cells (% degradation/h) [4]. This pulse-chase approach not only measures the overall protein degradation rate, but it is also used to determine the degradation that is specifically catalyzed by the UPP or the autophagy-lysosome pathways by specifically inhibiting either proteasomal or lysosomal function. Such information is very useful in evaluating mechanisms of cell growth and the effects of disease, hormones, or drugs. For example, these methods have recently enabled the discovery of the role of protein kinase A and Dyrk2 in enhancing proteasome activity and overall protein degradation ([17, 18] *see* Chapter 20 by VerPlank and Goldberg). By contrast, the commonly used measures of proteasomal peptidases or LC3-II assays fail to provide such quantitative information, and therefore these assays can be misleading. For example, the major proteasome inhibitors can block the chymotrypsin-like activity to near completion at nanomolar concentrations, but cause only about a 50% inhibition of the proteasomal degradation of cell proteins [19] due to the continued functioning of the other proteasome sites.

By varying the length of the cell's exposure to the radioactive precursors (the pulse) and introducing a chase period, it is possible to follow specifically the degradation of long-lived cell proteins, which are the overwhelming majority of cell constituents. Under normal growth conditions, these proteins are degraded mainly by the UPP [15] and to a smaller extent by autophagy. By using a short pulse of the precursor (without a chase), it is possible also to monitor the loss of the rapidly degraded (short-lived) proteins, which include misfolded proteins and regulatory proteins (e.g., transcription factors, rate-limiting enzymes, many oncogenes, and tumor suppressors). Their degradation seems to be catalyzed solely by UPP. For most studies, the pulse-chase method is most useful to follow the degradation of long-lived proteins since their mean rates are linear for hours. Therefore, the effects of inhibitors or physiological factors on their overall degradation can be very precisely measured. By contrast, short-lived proteins may disappear in between minutes and 1–2 h, and even a short pulse also labels some long-lived stable proteins. Therefore, their mean degradation rates are continually changing and difficult to evaluate.

This method is actually an adaptation for cultured mammalian cells of approaches introduced almost 50 years ago for studying protein degradation in *E. coli* [20, 21] and the acceleration of degradation of long-lived proteins upon starvation for amino acids, carbon source, or charged tRNAs [22]. Also, similar pulse-chase studies with a brief pulse made possible the discovery of the rapid selective degradation of misfolded abnormal proteins containing amino acid analogs or puromycin in *E. coli* [20, 21], mitochondria [20], and reticulocytes [23, 24]. In fact, the existence of the non-lysosomal proteolytic degradative system, which we now call the UPP, was first demonstrated through the measurement of the degradation of radiolabeled proteins to TCA-soluble peptides in cell-free systems [23].

Analogous nonradioactive methods have also often been used to assay hydrolysis of cell proteins to specific amino acids, which are assayed by specific fluorometric assays or HPLC [21]. These nonradioactive methods have been widely used in physiological studies to evaluate overall protein degradation in isolated rodent tissues, as well as the actual contributions of lysosomes or the UPP [25]. For example, such approaches first indicated that protein degradation by the UPP is activated in skeletal muscle upon fasting, glucocorticoid treatment, cancer cachexia, or denervation [26]. A recently developed nonradioactive pulse-chase method to follow protein degradation uses azidohomoalanine (AHA) [27, 28], which is incorporated into proteins in place of methionine. This method allows quantification by mass spectrometry of labeled proteins after binding to avidin beads with a "click" reaction between the alkyne-bearing biotin and the azide moiety of AHA. However, this method is less accurate for determining the rates of protein degradation, because it measures the net decline in the cellular content of AHA-containing proteins (which is very slow for long-lived proteins). Although incorporation of AHA does not seem to cause large changes in protein stability [29], cautions in using such methods seem necessary because incorporation of amino acid analogs can markedly perturb protein conformations and promote rapid degradation [21, 23].

2 Materials

1. The appropriate complete cell growth media for your cells (*see* **Note 1**).
2. [^{3}H] Phenylalanine (Phe L-[3,4,5 ^{3}H]; American Radiolabeled Chemicals, ART0614), stock 1 mCi/mL in 0.01 M HCl.
3. Phenylalanine.
4. Scintillation vials (RPI, 125514-SP).
5. Bio-Safe II™ Complete Counting Cocktail (scintillation fluid) (RPI, 111195-CS).
6. Trichloroacetic acid.
7. PerkinElmer Tri-Carb 2910TR Liquid Scintillation Analyzer controlled with QuantaSmart™ software.
8. 4 M NaOH.
9. Bortezomib (LC Laboratories, B-1408).
10. TAK-243 (a.k.a. MLN-7243, Active Biochem, A-1384).
11. Concanamycin A.
12. Cycloheximide.

3 Methods

The duration of the exposure to radioactive precursors and of the chase period determines which types of cell proteins are preferentially labeled. Protocol Subheading 3.1 describes a method to label selectively long-lived cellular proteins by first exposing cells to the

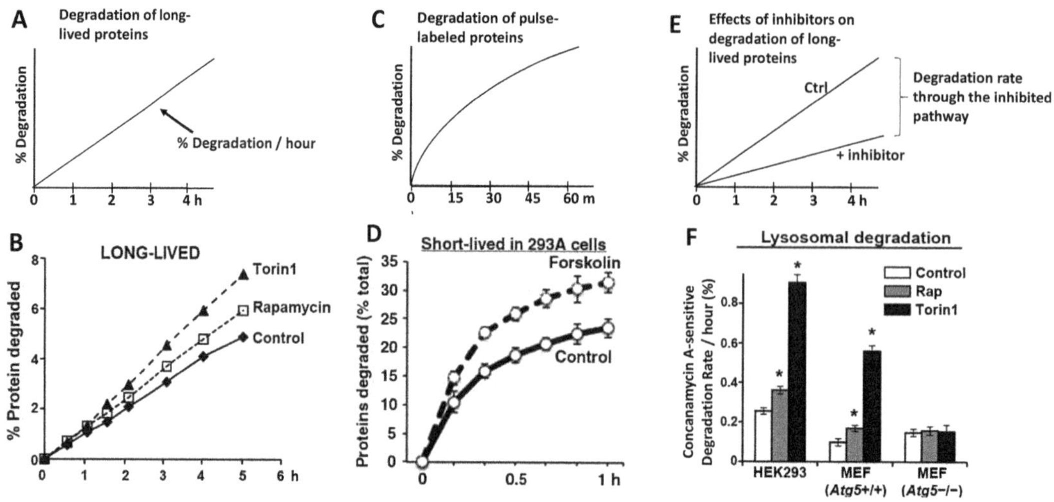

Fig. 1 Application of protein degradation assay protocols described in Subheadings 3.1–3.3. (**a**) Protocol Subheading 3.1 measures the degradation of long-lived proteins (radiolabeled for 20 h then chased for 2 h to deplete short-lived radiolabeled proteins), whose degradation rate is constant. Therefore, the precise degradation rate can be measured from the slope of the linear degradation curve. (**b**) Application of Protocol Subheading 3.1: By measuring long-lived protein degradation in HEK293A cells treated with two mTOR inhibitors (rapamycin or Torin1, which are anticipated to activate autophagy), we validated that the ability of cells to degrade protein was indeed elevated by these inhibitors. Figure adapted from [31]. (**c**) Protocol Subheading 3.2 measures the degradation of mostly short-lived proteins labeled by a short pulse (5–20 min). Their mean degradation rate decreases rapidly over time due to the depletion of short-lived proteins, so that the degradation curve is nonlinear, and it is not possible to calculate a constant degradation rate. (**d**) Application of Protocol Subheading 3.2: The degradation of proteins pulse-labeled for 10 min was measured. Treatment with a protein kinase A (PKA) activator, forskolin, greatly increased the amount of pulse-labeled proteins that were degraded over 1 h. Note that the degradation curve is nonlinear and no degradation rate could be calculated. The percentage of degraded proteins should be compared at each time point to draw conclusions. Figure adapted from [18]. (**e**) Protocol Subheading 3.3 measures the degradation of long-lived proteins by a specific degradation pathway, by using a specific inhibitor to completely block this pathway. Subtracting these two degradation rates allow us to calculate the rate of inhibitor-sensitive degradation, which represents the ability of cells to degrade long-lived proteins through this particular pathway. (**f**) Application of Protocol Subheading 3.3: To determine whether the increased degradation in cells treated with mTOR inhibitors (as in **b**) is caused by the lysosome pathway, the long-lived protein degradation rate was measured in cells whose lysosomes were completely inhibited by concanamycin A. This allows us to compare concanamycin A-sensitive (or lysosomal) degradation rates between control cells or cells treated with mTOR inhibitor. We found, using this approach, that the lysosome contributes to increased degradation in wild-type HEK293A or MEF cells, but not in cells with deficient autophagy ($Atg5^{-/-}$). Figure adapted from [31]

radioactive amino acid for 20 h and then chasing with an excess of nonradioactive amino acid for 2 h to allow breakdown of the short-lived proteins. Although the long-lived proteins are heterogeneous mixtures, their mean half-lives are similar, and therefore this approach gives a highly reproducible degradation rate that appears linear for up to 10 h. The slope of this line represents the percent of cell proteins hydrolyzed per hour (i.e., degradation rate) (Fig. 1a). On the other hand, a brief pulse labeling (for 5–20 min), followed immediately by resuspension in the chase media for measurement, allows measurement of the degradation of newly synthesized proteins, which also includes the short-lived cell proteins. Protocol Subheading 3.2 describes our standard method to label proteins using a 20-min pulse to follow the degradation of this highly heterogeneous group of proteins (Fig. 1c). Consequently, their degradation is rapid, nonlinear, and harder to follow. Initially these curves reflect the loss of most short-lived proteins and then the more stable and finally the long-lived components. Finally, Protocol Subheading 3.3 describes a method to measure the amount of proteins (1) that is degraded via the proteasome, (2) whose degradation is ubiquitin-dependent, and (3) that is degraded by the autophagy-lysosome pathway, by using highly specific inhibitors to completely block each of these pathways. Subtracting the degradation rate in the inhibitor-treated cells from that in the control (untreated) cells yields the amount of degradation catalyzed by the inhibited pathway (Fig. 1e). In our extensive experience, these rates remain constant for at least 6 h even in the presence of inhibitors of the proteasome or lysosome, which implies that blocking one pathway does not cause rapid compensatory changes in the other pathways. Figure 1 also provides examples where these protocols have been used to answer key research questions (Fig. 1b, d, f).

3.1 Measurement of the Degradation of Long-Lived Proteins

Day 1. Radiolabeling of long-lived proteins with ^{3}H-phenylalanine (*see* **Note 2**).

1. Cells are grown to 60–80% confluence in six-well plates. To obtain sufficient replicates, 4 wells of cells for each condition are seeded.
2. Prepare complete media supplemented with ^{3}H-phenylalanine at a final concentration of 5 μCi/mL. Add 1 mL to each well (*see* **Note 3**).
3. Incubate cells for 20 h to allow incorporation of ^{3}H-phenylalanine into cellular proteins.

Day 2. Sample collection for the measurement of degradation of long-lived proteins.

4. Prepare complete media containing 2 mM nonradioactive phenylalanine (chase medium). This chase medium will be used in all subsequent steps (*see* **Note 4**).
5. Remove radioactive media and thoroughly wash cells twice with 2 mL chase medium. It is essential to thoroughly wash the cells in order to remove as completely as possible the ^{3}H-phenylalanine remaining in the cells and media. The amount of unincorporated ^{3}H-phenylalanine is much larger than the amount incorporated into cell proteins. Therefore, if not

removed thoroughly, it can cause a high background radioactivity that will prevent accurate assay of proteolysis.

6. Add 2 mL of the chase medium to each well and incubate for 2 h. This step allows short-lived radiolabeled proteins to be degraded and makes possible the measurement specifically of the degradation of long-lived proteins.
7. Remove media from cells completely after **step 6**. Wash once more with 2 mL chase medium and remove the wash media completely.
8. Add 2 mL chase medium to each well, and begin sample collection for the measurements of degradation. The volumes of the samples taken must be very precise and consistent for accurate determination of degradation rates.
9. At 0, 1–4 h (or longer), collect precisely 200 μL aliquots of the media from each well, and transfer to a clean microcentrifuge tube. Briefly centrifuge the media samples (if necessary) to ensure that they are free of floating cells. These samples can be kept at 4 °C for up to 3 days before subjecting to trichloroacetic acid (TCA) precipitation (*see* **Note 5**).
10. After the last time point (4 h), completely remove the media from each well.
11. Prepare 0.2 M NaOH solution, and add 2 mL to each well to dissolve the cells. Make sure that the cell lysate is homogeneous by pipetting up and down 10–20 times until the lysate is no longer viscous.

Day 3. Preparation of samples of degradation products by precipitation with trichloroacetic acid (TCA).

12. Prepare scintillation vials containing 3 mL scintillation fluid before TCA precipitation.
13. To each microcentrifuge tube containing the 200 μL media samples (from **step 9**), add 20 μL TCA (100% W/V, 6.12 M), which precipitates proteins, but not amino acids. Therefore, the radioactivity that is TCA-soluble reflects ^{3}H-phenylalanine generated by protein degradation.
14. Mix well and incubate on ice for 30 min.
15. Centrifuge at 20,000 × *g* for 15 min at 4 °C in a table-top centrifuge.
16. Take the supernatants, being careful not to disturb the pellets. Add supernatants to a vial containing 3 mL scintillation fluid for measurement of radiolabeled amino acids released from proteins.
17. Take 100 μL dissolved cell lysate (from **step 11**), and add to a scintillation vial containing 3 mL scintillation fluid. Amounts of

total radiolabeled cell proteins are measured with these samples.

18. Prepare four blank control vials that contain only scintillation fluid, but no samples.

Measurement of protein degradation rates by liquid scintillation counting.

19. Place scintillation vials on an 18-vial cassette, and place the cassette in the chamber of a liquid scintillation analyzer.
20. With the software of the liquid scintillation analyzer (we typically use the QuantaSmart™ software), set up a protocol that specifies the optimal counting time for ^{3}H (e.g., 2 min) and output measurement (counts per minute, cpm).
21. Place the correct protocol flag (with protocol number) on the first cassette, and start running the program from the software.

    Calculation of degradation rates.

    To calculate the total degradation rate (percent per hour), we need to plot the percentage of cell proteins hydrolyzed at each time point against the time. The slope of this plot, which should be linear, gives the rate. To calculate the percentage of degradation at each time point, we need to divide the ^{3}H-phenylalanine cpm released into the medium by the cpm initially incorporated into protein (which is the sum of the cpm in the cell sample plus the total in the medium at the final time point). Figure 2 demonstrates how to calculate the rates of degradation from raw cpm values (Fig. 2a).

22. Follow equations in Fig. 2b, c to calculate the cumulative release of ^{3}H-phenylalanine cpm at 1–4 h, which reflects the amount of degraded radiolabeled proteins at each time (D0, D1, D2, D3, D4 values in Fig. 2c) (*see* **Note 6**).
23. Follow equations in Fig. 2d to calculate the remaining ^{3}H-phenylalanine cpm (R) in cells, which reflects the amount of remaining radiolabeled proteins. Then sum up R and D4 to obtain the total radioactivity (cpm) initially in cell proteins.
24. The percentage of protein degradation at each hour can now be calculated by dividing D0, D1, D2, D3, and D4 values by the total cpm as in Fig. 2e. These percentages are plotted over time to get a linear degradation curve (characteristic for long-lived proteins), whose slope is the degradation rate in these specific cell lines under the tested incubation conditions.
25. Calculate the degradation rate for the other 3 wells of cells, in order to obtain the average and standard deviation of degradation rate for each conditions. With this approach, degradation rates can now be compared following different physiological conditions or drug treatments.

A Raw Cpm (Count per minute)

Blank (average of four)	TCA supernatant from 200 µl media sample					100 µl cell lysate in 0.2 N NaOH
	0h	1h	2h	3h	4h	
Blank	T0	T1	T2	T3	T4	Lysate

B Calculate how many Cpm is released in each hour due to protein degradation

	0-1h	1-2h	2-3h	3-4h
Cpm increase	T1-T0	T2-T1	T3-T2	T4-T3
Sample volume (ml)	0.2	0.2	0.2	0.2
Media volume (ml)	1.8	1.6	1.4	1.2
Total Cpm released in this hour	9×(T1 -T0)	8×(T2 -T1)	7×(T3 -T2)	6×(T4 -T3)

C Calculate the cumulative Cpm release after 1, 2, 3, 4 hours

	0h	1h	2h	3h	4h
Cumulative Cpm released (degraded)	D0 = 0	D1 = 9×T1- 9×T0	D2 = 8×T2 + T1 - 9×T0	D3 = 7×T3 + T1 + T2 - 9×T0	D4 = 6×T4 + T1 + T2 + T3 - 9×T0

D Cpm in remaining protein (R) $= \dfrac{(\text{Lysate} - \text{Blank}) \times 2\ \text{ml}}{100\ \mu\text{l}} = 20 \times (\text{Lysate-Blank})$

Total Cpm (Total) = R + D4

E Calculate percent degradation at 0, 1, 2, 3, 4 hours

	0h	1h	2h	3h	4h
Percent Degradation	0	$\frac{100 \times D1}{\text{Total}}$	$\frac{100 \times D2}{\text{Total}}$	$\frac{100 \times D3}{\text{Total}}$	$\frac{100 \times D4}{\text{Total}}$

Fig. 2 Calculation of the degradation rate of long-lived proteins. (**a**) Raw cpm counts for blank, TCA-soluble media samples and cell lysate samples. (**b**) To calculate how much ^{3}H-phenylalanine is released by protein degradation in each hour, multiply cpm increments by the volume of media during that hour. (**c**) Combine the values from (**b**) to calculate cumulative ^{3}H-phenylalanine released in 1–4 h, which reflects the amount of degraded proteins (D0, D1, D2, D3, and D4). (**d**) Calculate the ^{3}H-phenylalanine cpm remaining in proteins (R) and in total proteins (Total). (**e**) Divide D0–D4 values by total cpm to calculate the percent degradation over 0–4 h. Plotting these values (% cell protein degraded) will generate a linear curve, whose slope is the rate of protein degradation (% cell protein degraded/h)

3.2 Measurement of the Degradation of Mostly Short-Lived Proteins Labeled in a Brief Pulse

The protocol for measuring the degradation of short-lived proteins is almost the same as that for long-lived proteins with the exception of the length of cell exposure to the radioactive precursor, the lack of a chase period, and other precautions, as stated below.

1. Cells are grown to 60–80% confluence in six-well plates. To obtain sufficient replicates, 4 wells of cells were seeded.
2. Prepare complete media supplemented with ^{3}H-phenylalanine at a final concentration of 5 µCi/mL. Add 1 mL to each well.
3. Incubate cells for 5–20 min to allow cellular proteins to be radiolabeled (shorter pulses cause fewer proteins to be labeled but increase the fraction that is rapidly degraded).
4. Prepare nonradioactive chase media as in the long-lived protein degradation in Subheading **3.1**, except that 100 µg/mL cycloheximide should be added (*see* **Note 7**) to prevent reincorporation of released amino acids into proteins. Cycloheximide does not significantly reduce degradation rates for 1–2 h, but with longer incubations, cycloheximide reduces the degradation of long-lived proteins.
5. Remove radioactive media and thoroughly wash cells twice with 2 mL chase medium.
6. Add 2 mL chase medium to each well, and immediately begin sample collection for the measurements of degradation. The volumes of media taken must be very precise.

7. At 0, 15, 30, 45, and 60 min, collect 200 μL aliquots of media from each well and transfer to a clean microcentrifuge tube. Briefly centrifuge the media, if necessary, to make sure it is free of floating cells. The media can be kept at 4 °C for up to 3 days before subjecting to TCA precipitation.
8. After the last time point (60 min), completely remove media from each well, and dissolve cells with 0.2 M NaOH solution as in **step 11** in Subheading 3.1.
9. TCA precipitation and liquid scintillation counting are performed as in protocol Subheading 3.1 for measurements of the degradation of long-lived proteins (*see* **Note 8**).

3.3 Measuring the Degradation of Long-Lived Proteins by the UPP and the Autophagy-Lysosomal System (See Note 9)

To quantitate the amount of protein degradation by a particular pathway under various conditions, we use small molecule inhibitors to block selectively and completely proteasome function (with 1 μM bortezomib to block its multiple peptidase activities), ubiquitin conjugation (with 10 μM TAK243, which inactivates the ubiquitin-activating enzyme, E1), or lysosomal degradation (with 200 nM concanamycin A to block the lysosomal V-ATPase and prevent lysosomal acidification) (*see* **Note 10**). The net decrease in protein degradation that is blocked by these inhibitors indicates the rate of degradation through the corresponding pathway (*see* **Note 11**). Prior studies [15] have indicated that these concentrations cause maximal selective inhibition of these pathways. The bortezomib-sensitive (proteasomal) and concanamycin A-sensitive (lysosomal) degradation rates should add up to account for nearly all protein degradation occurring in cells.

1. Cells are grown to 60–80% confluent in six-well plates. 4 wells of cells should be seeded for each inhibitor-treated condition as well as the untreated cells.
2. Radiolabel cellular proteins with ^{3}H-phenylalanine for 20 h, as described above.
3. Remove radioactive media, thoroughly wash cells twice with 2 mL chase medium, and chase away short-lived proteins by incubating cells for 2 h with 2 mL chase medium as in Subheading 3.1.
4. Remove chase medium, thoroughly wash cells again, and add 2 mL chase medium with or without bortezomib (1 μM), TAK243 (10 μM), or concanamycin A (200 nM).
5. Allow cells to be pretreated for 2 h to allow efficient inhibition.
6. After the pretreatment, begin media collection (0 h). At 0, 1–4 h, collect 200 μL (must be very precise) aliquots of media from each well, and transfer to a clean microcentrifuge tube. Briefly centrifuge the media if necessary to make sure it is

free of floating cells. The media can be kept at 4 °C for up to 3 days before subjecting to TCA precipitation.

7. After the last time point (4 h), completely remove the media from each well, and dissolve cells with 0.2 M NaOH solution.
8. TCA precipitation and liquid scintillation analysis are performed, as in Subheading 3.1 for the measurement of the degradation of long-lived proteins.
9. Calculate the degradation rate in cells treated with bortezomib as in Fig. 2. Subtract this rate from the degradation rate in untreated cells to obtain the rate of proteasome-dependent degradation. In the same way, calculate the rates of ubiquitin-dependent degradation and degradation by the autophagy-lysosome system.

4 Conclusions

The widely used methods to evaluate these degradative processes (i.e., measurement of a proteasomal peptidase activity, the cellular level of ubiquitinated proteins, the amount of autophagosomes visualized, or the levels of lipidated LC3), although often useful for estimating the activity of these pathways, are not direct measurements of the amount of proteins being degraded by these systems and under many conditions can yield misleading impressions. This chapter describes methods to directly measure cellular protein degradation rates by radiolabeling long-lived or short-lived proteins with ^{3}H-phenylalanine. By further selective inhibition of ubiquitination, proteasome activity, or lysosomal function, it is possible to precisely calculate the fraction degraded by each system under different experimental conditions. Therefore, this powerful and versatile approach makes it possible to accurately test whether protein degradation by the UPP or autophagy-lysosomes is altered under specific conditions or efficiently inhibited upon treatment with different drugs.

5 Notes

1. The methods described in this protocol can be used to measure degradation rate in any cell in culture, and related methods have been used in isolated tissues.
2. Several other radioactive amino acids, especially ^{3}H-leucine, ^{14}C-leucine, and ^{35}S-methionine, have long been used to follow turnover of cell proteins. In the past, we and others have also used ^{3}H-tyrosine especially for measuring the degradation of long-lived cell proteins in muscle [4]. However, since

treatment of cells with high concentrations of leucine (as in the chase media) can activate mTORC1 and suppress protein degradation by both autophagy and the UPP [15, 30], it is inappropriate to use this amino acid for measurements of protein turnover. Our extensive experience comparing different amino acids in pulse-chase protocols ([15] and Garcia, Zhao, et al., unpublished) further demonstrated that after pulse labeling, the amount of intracellular radioactive-free methionine and tyrosine declines over time much more slowly than phenylalanine. Since these intracellular radioactive amino acids can continue to be incorporated into newly synthesized proteins, the higher amounts of intracellular labeled methionine and tyrosine are more prone than phenylalanine to cause the calculation of an erroneously low degradation rate, particularly after a short pulse. Therefore, phenylalanine is a much better choice than leucine, methionine, or tyrosine as the radioactive precursor for such experiments.

3. Cell proteins can be efficiently labeled with ^{3}H-phenylalanine despite the presence of cold phenylalanine in the standard tissue culture media and FBS. Removal of this nonradioactive phenylalanine to enhance specific activity of the ^{3}H-precursor during radiolabeling and the cpm incorporated is not recommended in order to avoid the potential cell stress caused by depletion of phenylalanine and reduction of protein synthesis, especially in the 20-h labeling protocols.
4. Excessive nonradioactive phenylalanine in the media is essential to competitively prevent the reincorporation of ^{3}H-phenylalanine back into cellular proteins. The excess amount of nonradioactive phenylalanine in the media also stimulates exchange with ^{3}H-phenylalanine that is generated by cells due to degradation.
5. TCA precipitation can be performed immediately after media collection or up to 3 days after collection if the media is stored at 4 °C. However, before storing the media, remove any floating cells by centrifugation. These floating cells may lyse during storage, thus raising the radioactivity in the media.
6. The TCA-soluble cpm in the media at 0 hour (T0, Fig. 2a) is caused by ^{3}H-phenylalanine that failed to be washed out and thus should be subtracted so that D0 is set to 0 (Fig. 2c). When calculating the cumulative release of ^{3}H-phenylalanine cpm, take into account the reduction of total media volume over time due to the removal of 200 μL samples at each hour (Fig. 2b).
7. Inhibitors of protein synthesis, such as cycloheximide, should be added during the chase to eliminate the incorporation of radioactive amino acid into cell proteins during the chase after

short pulse labeling. Cycloheximide should not be used in measuring the degradation of long-lived proteins, because with time it reduces the overall degradation of these proteins.

8. Because the plot of degradation of long-lived proteins remains linear for many hours, precise degradation rates can be measured. However, with a short pulse, the proteins labeled at initial time points are primarily short-lived. Because they are rapidly degraded, the remaining proteins labeled at later time points will be increasingly dominated by proteins with longer half-lives. Consequently, the degradation rate drops significantly during the 1 h measurement period, and mean half-lives of the short-lived fraction cannot be precisely calculated. In addition, although 5–20 min pulses label differentially short-lived proteins, still a considerable amount of long-lived proteins are radiolabeled. Therefore, the degradation of pulse-labeled proteins is a combination of the rapid hydrolysis of short-lived components and the slower degradation of long-lived proteins, which predominate with time. Therefore, rigorous conclusions about the changes in degradation rates of short-lived protein can only be made under conditions, where rates of degradation of long-lived proteins are unchanged (as measured in Subheading **3.1**).
9. One difficulty in measuring the role of proteasomes in the degradation of short-lived proteins is that the proteasome inhibitors take time to enter the cells and to achieve maximal inhibition (especially at low inhibitor concentrations). Also, the degradation assay for short-lived proteins is generally complete within 1 h after radiolabeling, which is too short for maximal inhibition. Therefore, it is necessary to pretreat the cells with the inhibitor for at least 30 min before exposure to the ^{3}H-phenylalanine.
10. Other inhibitors of these pathways may also be used, but pilot experiments with different inhibitor concentrations or durations should be done to ensure that the degradation pathway of interest is completely inhibited.
11. When choosing the type of inhibitor, the concentration, and the duration to be used, there are several important considerations. (A) Cell death must be avoided. Dying cells frequently lose membrane integrity and spill radiolabeled small peptides or proteins out into the media or allow Ca^{++} to enter the cells triggering calpain activation and breakdown of cell proteins. Dead cells also elevate protein degradation by caspases. These artifacts erroneously increase the apparent rates of degradation and make data uninterpretable. Therefore, pilot experiments with LDH release assay must be performed to ensure that there is negligible cell death during inhibitor treatment. (B) It is

important to ensure that efficient inhibition of the degradation pathway is achieved before collecting media samples for TCA precipitation, because a certain duration of inhibitor treatment is required to achieve maximal inhibition of the cellular process. For example, concanamycin A inhibits lysosomal degradation by raising the pH of lysosomes, and at least 90 min of treatment is required for lysosomal proteases to be efficiently inhibited by this agent.

References

1. Voges D, Zwickl P, Baumeister W (1999) The 26S proteasome: a molecular machine designed for controlled proteolysis. Annu Rev Biochem 68:1015–1068
2. Klionsky DJ, Ohsumi Y (1999) Vacuolar import of proteins and organelles from the cytoplasm. Annu Rev Cell Dev Biol 15:1–32. https://doi.org/10.1146/annurev.cellbio.15.1.1
3. Myeku N, Clelland CL, Emrani S, Kukushkin NV, Yu WH et al (2016) Tau-driven 26S proteasome impairment and cognitive dysfunction can be prevented early in disease by activating cAMP-PKA signaling. Nat Med 22(1):46–53. https://doi.org/10.1038/nm.4011
4. Zhao J, Brault JJ, Schild A, Cao P, Sandri M et al (2007) FoxO3 coordinately activates protein degradation by the autophagic/lysosomal and proteasomal pathways in atrophying muscle cells. Cell Metab 6(6):472–483. https://doi.org/10.1016/j.cmet.2007.11.004
5. Sandri M, Sandri C, Gilbert A, Skurk C, Calabria E et al (2004) Foxo transcription factors induce the atrophy-related ubiquitin ligase atrogin-1 and cause skeletal muscle atrophy. Cell 117(3):399–412
6. Demo SD, Kirk CJ, Aujay MA, Buchholz TJ, Dajee M et al (2007) Antitumor activity of PR-171, a novel irreversible inhibitor of the proteasome. Cancer Res 67(13):6383–6391
7. Cusack JC Jr, Liu R, Xia L, Chao TH, Pien C et al (2006) NPI-0052 enhances tumoricidal response to conventional cancer therapy in a colon cancer model. Clin Cancer Res 12 (22):6758–6764
8. Petroski MD (2008) The ubiquitin system, disease, and drug discovery. BMC Biochem 9 (Suppl 1):S7
9. Kisselev AF, Goldberg AL (2001) Proteasome inhibitors: from research tools to drug candidates. Chem Biol 8(8):739–758
10. Gu C, Kolodziejek I, Misas-Villamil J, Shindo T, Colby T et al (2010) Proteasome activity profiling: a simple, robust and versatile method revealing subunit-selective inhibitors and cytoplasmic, defense-induced proteasome activities. Plant J 62(1):160–170. https://doi.org/10.1111/j.1365-313X.2009.04122.x
11. de Bruin G, Xin BT, Florea BI, Overkleeft HS (2016) Proteasome subunit selective activity-based probes report on proteasome core particle composition in a native polyacrylamide gel electrophoresis fluorescence-resonance energy transfer assay. J Am Chem Soc 138 (31):9874–9880. https://doi.org/10.1021/jacs.6b04207
12. Tanaka K (1994) Role of proteasomes modified by interferon-gamma in antigen processing. J Leukoc Biol 56(5):571–575
13. Peth A, Besche HC, Goldberg AL (2009) Ubiquitinated proteins activate the proteasome by binding to Usp14/Ubp6, which causes 20S gate opening. Mol Cell 36(5):794–804. https://doi.org/10.1016/j.molcel.2009.11.015
14. Collins GA, Goldberg AL (2017) The logic of the 26S proteasome. Cell 169(5):792–806. https://doi.org/10.1016/j.cell.2017.04.023
15. Zhao J, Goldberg AL (2016) Coordinate regulation of autophagy and the ubiquitin proteasome system by MTOR. Autophagy 12 (10):1967–1970. https://doi.org/10.1080/15548627.2016.1205770
16. Klionsky DJ, Abdelmohsen K, Abe A, Abedin MJ, Abeliovich H et al (2016) Guidelines for the use and interpretation of assays for monitoring autophagy (3rd edition). Autophagy 12 (1):1–222. https://doi.org/10.1080/15548627.2015.1100356
17. Guo X, Wang X, Wang Z, Banerjee S, Yang J et al (2016) Site-specific proteasome phosphorylation controls cell proliferation and tumorigenesis. Nat Cell Biol 18(2):202–212. https://doi.org/10.1038/ncb3289
18. Lokireddy S, Kukushkin NV, Goldberg AL (2015) cAMP-induced phosphorylation of 26S proteasomes on Rpn6/PSMD11 enhances their activity and the degradation of misfolded

proteins. Proc Natl Acad Sci U S A 112(52): E7176–E7185. https://doi.org/10.1073/pnas.1522332112

19. Kisselev AF, Callard A, Goldberg AL (2006) Importance of the different proteolytic sites of the proteasome and the efficacy of inhibitors varies with the protein substrate. J Biol Chem 281(13):8582–8590. https://doi.org/10.1074/jbc.M509043200
20. Goldberg AL (1972) Degradation of abnormal proteins in Escherichia coli (protein breakdown-protein structure-mistranslation-amino acid analogs-puromycin). Proc Natl Acad Sci U S A 69(2):422–426
21. Goldberg AL, Dice JF (1974) Intracellular protein degradation in mammalian and bacterial cells. Annu Rev Biochem 43(0):835–869. https://doi.org/10.1146/annurev.bi.43.070174.004155
22. Goldberg AL (1971) A role of aminoacyl-tRNA in the regulation of protein breakdown in Escherichia coli. Proc Natl Acad Sci U S A 68 (2):362–366
23. Etlinger JD, Goldberg AL (1977) A soluble ATP-dependent proteolytic system responsible for the degradation of abnormal proteins in reticulocytes. Proc Natl Acad Sci U S A 74 (1):54–58
24. Goldberg AL, Jablecki C, Li JB (1974) Trophic functions of the neuron. 3. Mechanisms of neurotrophic interactions. Effects of use and disuse on amino acid transport and protein turnover in muscle. Ann N Y Acad Sci 228:190–201
25. Cohen S, Nathan JA, Goldberg AL (2015) Muscle wasting in disease: molecular mechanisms and promising therapies. Nat Rev Drug Discov 14(1):58–74. https://doi.org/10.1038/nrd4467
26. Lecker SH, Goldberg AL, Mitch WE (2006) Protein degradation by the ubiquitin-proteasome pathway in normal and disease states. J Am Soc Nephrol 17(7):1807–1819. https://doi.org/10.1681/ASN.2006010083
27. Zhang J, Wang J, Ng S, Lin Q, Shen HM (2014) Development of a novel method for quantification of autophagic protein degradation by AHA labeling. Autophagy 10 (5):901–912. https://doi.org/10.4161/auto.28267
28. Zhang J, Wang J, Lee YM, Lim TK, Lin Q et al (2017) Proteomic profiling of de novo protein synthesis in starvation-induced autophagy using bioorthogonal noncanonical amino acid tagging. Methods Enzymol 588:41–59. https://doi.org/10.1016/bs.mie.2016.09.075
29. McShane E, Sin C, Zauber H, Wells JN, Donnelly N et al (2016) Kinetic analysis of protein stability reveals age-dependent degradation. Cell 167(3):803–815 e821. https://doi.org/10.1016/j.cell.2016.09.015
30. Mordier S, Deval C, Bechet D, Tassa A, Ferrara M (2000) Leucine limitation induces autophagy and activation of lysosome-dependent proteolysis in C2C12 myotubes through a mammalian target of rapamycin-independent signaling pathway. J Biol Chem 275 (38):29900–29906. https://doi.org/10.1074/jbc.M003633200
31. Zhao J, Zhai B, Gygi SP, Goldberg AL (2015) mTOR inhibition activates overall protein degradation by the ubiquitin proteasome system as well as by autophagy. Proc Natl Acad Sci U S A 112(52):15790–15797. https://doi.org/10.1073/pnas.1521919112

Chapter 18

Methods to Rapidly Prepare Mammalian 26S Proteasomes for Biochemical Analysis

Chueh-Ling Kuo, Galen Andrew Collins, and Alfred L. Goldberg

Abstract

Rapid, gentle isolation of 26S proteasomes from cells or tissues is an essential step for studies of the changes in proteasome activity and composition that can occur under different physiological or pathological conditions and in response to pharmacological agents. We present here three different approaches to affinity purify or to prepare proteasome-rich cell fractions. The first method uses affinity tags fused to proteasome subunits and has been useful in several cell lines for studies of proteasome structure by cryo-electron microscopy and composition by mass spectrometry. A second method uses the proteasome's affinity for a ubiquitin-like (UBL) domain and can be used to purify these particles from any cell or tissue. This method does not require expression of a tagged subunit and has proven to be very useful to investigate how proteasomal activity changes in different physiological states (e.g., fasting or aging), with neurodegenerative diseases, and with drugs or hormones that cause subunit phosphorylation. A third, simple method that is based on the 26S proteasome's high molecular weight (about 2.5 MDa) concentrates these particles greatly by differential centrifugation. This method maintains the association of proteasomes with ubiquitin (Ub) conjugates and many other loosely associated regulatory proteins and is useful to study changes in proteasome composition under different conditions.

Key words 26S proteasome, Affinity-based purification, Ubiquitin-like domain (UBL), Ubiquitin-interacting motif (UIM), Ubiquitin (Ub), Differential centrifugation

1 Introduction

Rigorous studies of proteasome composition, regulation, and its detailed mechanisms require proteasome purification or at least isolation away from other cytosolic enzymes. The ATP-dependent degradation of ubiquitinated proteins proceeds through multiple steps [1], which cannot be analyzed without purification of the 26S complex. Also, analysis of its regulation, the effects of pharmacological and physiological stimuli, and disease processes requires rapid and gentle purification of 26S proteasomes, without which superficial or even misleading conclusions about proteasome activity may be obtained (e.g., through reliance on peptidase activities in crude extracts). Purification is also essential to analyze proteasomal

Thibault Mayor and Gary Kleiger (eds.), *The Ubiquitin Proteasome System: Methods and Protocols*, Methods in Molecular Biology, vol. 1844, https://doi.org/10.1007/978-1-4939-8706-1_18, © Springer Science+Business Media, LLC, part of Springer Nature 2018

activities free of cytosolic, deubiquitinating enzymes (DUBs), Ubiquitin (Ub) ligases, proteases, and ATPases that all can mask or alter 26S activity. Moreover, it is increasingly clear that 26S proteasomes within cells are a heterogeneous mixture of active and inactive forms [2] and may contain one or two 19S regulatory particles per 20S core, alternative activating complexes (e.g., PA28αβ or PA200), or sub-stoichiometric amounts of different associated proteins that can alter degradation rates (e.g., shuttling factors, Ub ligases, and DUBs). Another important source of heterogeneity is the presence of ubiquitinated substrates on some cellular proteasomes [2] since Ub conjugates activate the proteasomes [3] and alter the levels of other associated proteins [4]. Finally, several post-synthetic modifications of proteasome subunits (e.g., phosphorylation [5–7] and ubiquitination [8]) can change the proteasomes' activity (*see* Chapter 20). Thus, with the proteasomes (as with other biochemical preparations), clean results and conclusions require clean preparations.

Until the past decade, biochemical and structural studies of 26S proteasomes utilized various classical multistep purification methods (e.g., ion exchange and size exclusion chromatography). Aside from being lengthy and time-intensive methods, these efforts strip away 26S-associated proteins, decrease the proteasomes' ability to degrade Ub conjugates, and result in the loss of certain regulatory properties. Therefore, we have developed several methods to rapidly and gently isolate 26S particles that maintain proteasome activity and regulatory features better than conventional purification methods. These approaches have proven very useful for various studies of proteasome mechanisms and regulation (*see* Chapters 19 and 20).

1.1 Affinity Purification of Proteasomes Bearing a Tagged Subunit

Several laboratories have fused affinity tags to various proteasome subunits, including β4 of the 20S core [9] and Rpn11 [10–12], and the sub-stoichiometric subunit Dss1 [13] of the 19S particle and used them to purify active proteasomes (Table 1). Since the intracellular pools of free 26S subunits are very small, affinity-tag methods primarily yield intact proteasome particles. However, depending on the tagged subunits and the cell source, the resulting 26S proteasomes may be co-purified with free 20S proteasomes or 19S regulatory particles. For example, if tagged Rpn11 is used, one may purify a mixture of singly and doubly capped 26S particles and some free 19S, especially in certain cell types or experimental conditions (e.g., ATP-deficient cells) where there are abundant free 19S complexes. Using tagged β4 normally purifies a mixture of 26S and 20S proteasomes, but 20S particles can be specifically purified using tagged β4 if ATP is omitted from the buffers to promote dissociation of 26S into 19S and 20S components.

In yeast, the affinity-tagged subunits are integrated into the endogenous genomic loci, and thus all cellular proteasomes are

Table 1
Cell lines expressing tagged proteasome subunits

Tag	Tagged subunit	Strain	Source	References
3× FLAG	Dss1	HEK293F	Human	[13]
3× FLAG	β4	Myoblast	Human	
HTBH[a]	Rpn11	HEK293	Human	[12]
3× FLAG	Rpn11	YYS40	*S. cerevisiae*	[10]
Protein A	Rpn11	SDL66	*S. cerevisiae*	[11]
Protein A	β4	Y-SDL135	*S. cerevisiae*	[9]

aThe HTBH tag consists of two hexahistidine tags, a TEV cleavage site, and a bacterially derived peptide that induces biotinylation in vivo [12]

tagged and can be purified by this approach. The mammalian cell lines presently in use stably overexpress the tagged subunit of a multi-copy plasmid and therefore still have an endogenous untagged copy of the subunit gene. Because only the proteasomes that incorporated the tagged subunits can be isolated by this approach, many cytosolic proteasomes are not purified. It is very likely that the increased use of CRISPR-Cas9 genome editing methods should change this situation and introduce tags on both genomic copies of a given subunit, and thereby improve the purification yields. These affinity-tag approaches co-purify more of the proteasome-associated proteins, even the weakly bound ones and Ub conjugates than the UBL method (discussed below). Consequently, these tagged particles are particularly useful for studies of the changes in 26S composition under different conditions, as many 26S-associated proteins cycle rapidly on and off the particles at 37°C [4].

1.2 The UBL Affinity Method

The proteasome's strong affinity for proteins containing UBL domains is the basis for another very useful affinity purification method. The UBL method described by Besche and Goldberg [14] uses a GST-tag fused to the UBL domain of the shuttling factor Rad23b. 26S proteasomes bind to GST-UBL columns without the need for tagging proteasome subunits, and then can be eluted from the GST-UBL resin by the addition of high concentrations of the ubiquitin-interacting motif (UIM) derived from the Ub-binding subunit Rpn10/S5a/PSMD4. The chief advantage of the UBL affinity method is that it can gently and rapidly purify proteasomes or immunoproteasomes from a wide diversity of cells and tissues without any genetic manipulations. Therefore, it has been successfully used to characterize proteasomes from various cultured cells and yeast, as well as rabbit and mouse muscles [15], spleen [16], peripheral neurons, and brains [17–19], and also to identify proteasome adaptations during different physiological or pathological states (e.g., fasting or after hormone treatments) [7, 19].

It is not clear which of the several potential UBL binding sites on the proteasome is targeted by this method. In yeast, the Rad23 UBL domain binds to the T1 site of Rpn1 [20]. In mammalian cells, both the Pru domain of Rpn13/ADRM1 and the two UIM regions of Rpn10 have been reported to be important for the binding of Rad23b to the proteasome [21]. All these potential UBL binding sites also can bind Ub conjugates and some other UBL-containing proteins. Consequently, this method dislodges several proteins from the proteasomes, including most ubiquitinated substrates and proteins that depend on the Ub conjugates to associate with proteasomes (e.g., Ube3c/Hul5). However, this method co-purifies other important 26S-associated proteins (e.g., the DUB Usp14) and has been very valuable for mechanistic studies of the multiple steps during Ub conjugate degradation (e.g., ATP hydrolysis, activation by substrates, and proteasome regulation).

1.3 Isolation of Proteasomes by Differential Centrifugation

Because of their high molecular weights, proteasomes can be quickly isolated from cells or tissue extracts by differential centrifugation. This method can capture more than 99% of the proteasomes in the enriched fraction (*see* Table 2) and is thus advantageous for

Table 2
Nearly all proteasomes can be isolated by differential centrifugation at 150,000 × *g* for 1 h

Subunits and proteasome-associated proteins	Untreated control (%)	After depletion of Ub conjugates (%)
Ub conjugates	81 ± 4	Not detectable
Rpt6 (19S)	96	98
β5 (20S)	100 ± 0	100 ± 0
Rad23b	33 ± 4	9 ± 2
Ddi2	23 ± 4	11
p97 (VCP)	95 ± 5	98

Ub conjugates and shuttling factors (e.g., Rad23b and Ddi2) are co-isolated from typical cell lines. However, in the absence of Ub conjugates (after treatment with the E1 Ub-activating enzyme inhibitor MLN7243), the levels of Rad23b and Ddi2 in the proteasome-rich fraction decrease markedly. HEK293 cells were treated with 5 μM of the E1 inhibitor (MLN7243) for 1 h. Proteasome-rich fractions were prepared by the differential centrifugation method. Levels of proteins in total cell extracts and proteasome-rich fractions were analyzed by infrared Western blot. These data represent proteins recovered in the proteasome-rich fraction as a percentage of proteins from the cell extract large-scale preparations. This rapid preparation of proteasome-rich fractions without any prior genetic manipulations also captures most Ub conjugates and many other proteins that associate with the proteasomes. However, this approach inevitably co-isolates, in the proteasome-rich fraction, other high molecular weight complexes (e.g., p97/VCP), and 26S and 20S particles cannot be reliably separated by this method. Thus, it is an enrichment method rather than a means of purification. However, differential centrifugation, if used prior to an affinity purification method, may improve the purity of the proteasomes obtained.

2 Materials

2.1 Purification of FLAG-Dss1-Tagged Proteasomes Using Anti-FLAG Affinity Gels

1. HEK293 cells stably expressing FLAG-Dss1 [13].
2. Cell growth media: Dulbecco's Modification of Eagle's Medium (contains 4.5 g/L glucose, L-glutamine, and sodium pyruvate), 10% fetal bovine serum.
3. Phosphate-buffered saline (PBS): 10 mM Na_2HPO_4, 1.8 mM KH_2PO_4, 137 mM NaCl, 2.7 mM KCl (pH 7.4).
4. Cell lysis buffer: 25 mM HEPES-Na (pH 7.4), 1× protease inhibitors cocktail (Roche), 10% (v/v) glycerol, 1 mM ATP-Na (pH 7.4), and 5 mM $MgCl_2$ (*see* **Note 1**).
5. Wash buffer: the same as the cell lysis buffer but without protease inhibitors.
6. Anti-FLAG M2 affinity gel (Sigma).
7. 0.2 μm nylon membrane centrifugal filters (VWR).
8. Sonicator instrument.
9. Ultracentrifuge and rotor capable of 100,000 × *g* force.
10. 3× FLAG peptide.
11. Suc-LLVY-amc.

2.2 Proteasome Purification by the UBL Method

1. His_{10}-UIM [15].
2. GST-UBL [15].
3. 150 mL affinity purification buffer (APB): 25 mM HEPES-K (pH 7.4), 10% (v/v) glycerol, 5 mM $MgCl_2$, 1 mM ATP-Na (pH 7.4), and 1 mM DTT. Always add ATP and DTT immediately before the purification (*see* **Note 2**).
4. Glutathione-sepharose (GE Healthcare).
5. Ni-NTA agarose (Qiagen).
6. 20 mL empty column pack (Bio-Rad).
7. Bradford reagent.
8. 100 mM ATP-Na (pH 7.4).
9. 0.45 μm nylon membrane centrifugal filters (VWR).

2.3 Isolation of Proteasomes by Differential Centrifugation

1. Cells or tissues of interest.
2. Cell lysis buffer: 25 mM HEPES-Na (pH 7.4), 1× protease inhibitors cocktail (Roche), 10% (v/v) glycerol, 1 mM ATP-Na (pH 7.4), and 5 mM $MgCl_2$.
3. Wash buffer: the same as the cell lysis buffer but without protease inhibitors.
4. Thick-wall, polycarbonate centrifuge tubes (11 × 34 mm) (Beckman Coulter).
5. Ultracentrifuge and rotor capable of 150,000 × *g* force.

3 Methods

3.1 Purification of FLAG-Dss1 Tagged Proteasomes Using Anti-FLAG Affinity Gels

3.1.1 Preparation of Crude Extract

Perform all steps on ice or at 4 °C.

1. HEK293 cells stably expressing FLAG-Dss1 are grown to 90% confluency in one 100 mm culture dish (*see* **Notes 3** and **4**), washed with 10 mL PBS, and then harvested by scraping into 4 mL ice-cold PBS.
2. Cell pellets were collected by centrifugation at 350 × *g* for 5 min and then resuspended in 0.4 mL cell lysis buffer.
3. Cells were lysed by sonication with 4 rounds of 10 s bursts (12 μm amplitude) and 50 s rests on ice.
4. Cell debris were cleared by centrifugation at 10,000 × *g* for 10 min, and the resulting supernatants were further centrifuged at 100,000 × *g* for 30 min (*see* **Note 5**). The supernatants (S100 cell extracts) were then utilized in further purification steps.

3.1.2 Affinity Purification

Perform all steps on ice or at 4 °C.

1. Prepare the FLAG affinity gel as advised by the manufacturer. An affinity gel volume that binds approximately 4 μg FLAG fusion protein was used for cell extracts from one 100 mm dish.
2. Incubate the cell extracts with the affinity gel at 4 °C for 1 h.
3. Pellet the protein-bound affinity gel by centrifugation at 300 × *g* for 1 min at 4 °C. Collect the supernatants to monitor the amount of unbound proteins.
4. The protein-bound affinity gels are washed four times with 150 μL of wash buffer. Spin at 300 × *g* for 1 min (*see* **Notes 6** and **7**).
5. The bound FLAG-tagged proteasomes are eluted with 45 μL of elution buffer (200 ng/mL 3× FLAG peptide) for 30 min. Repeat (*see* **Note 8**).

6. Combine the eluted proteasomes, and apply them to 0.2 μm nylon membrane centrifugal filters to remove the trace amount of affinity gel. Spin at 1500 × *g* for 1 min. Purified proteasomes are collected at the bottom of the 1.5 mL tube.

3.1.3 Assay and Storage

1. Measure proteasome peptidase activity usually using Suc-LLVY-amc as the substrate (*see* Chapter 19) (*see* **Note 9**).
2. Store purified proteasomes at −20 °C for short-term storage or − 80 °C for prolonged storage.

3.2 Proteasome Purification by the UBL Method

3.2.1 Preparation of Crude Extract

Perform all steps on ice or at 4°C.

1. Collect cells from about ten to thirty 150 mm dishes (*see* **Note 10**) by scraping in 1× PBS and pooling the cells from three 150 mm dishes in a 15 mL conical tube. The cells are spun at 100 × g for 5 min and the supernatants removed by aspiration.
2. Add 2.5 mL APB to each of the tubes containing cell pellets. Vortex to resuspend the cells.
3. Lyse the cells by sonication (*see* **Note 11**). We use six 10 s bursts (12 μm amplitude) and at least 10 s recovery periods on ice. Pool the lysates.
4. Clear the lysates by centrifugation at 100,000 × *g* for 30 min (*see* **Note 12**).
5. Aliquot a small sample of clarified lysate (S100) to measure the protein concentration by a Bradford assay.

3.2.2 Affinity Purification

Perform all steps on ice or at 4 °C.

1. Equilibrate 500 μL of a 50% slurry of glutathione-sepharose by washing three times in 2.5 mL APB. Collect the resin at 100 × g for 5 min, discarding the supernatant each time.
2. Equilibrate 5 × 60 μL 50% slurries of Ni-NTA agarose by washing three times in 300 μL APB. The agarose sediments quickly by gravity. Carefully aspirate off the supernatant.
3. Combine 80–100 mg of protein from the S100 lysate in 10–20 mL with 1 mg GST-UBL and a 250 μL bed volume of glutathione-sepharose.
4. Gently rotate the S100, GST-UBL, and glutathione-sepharose mixture for 2 h.
5. Pour the S100, GST-UBL, and glutathione-sepharose mixture into an empty 20 mL column. Set aside the flow-through for later analysis of unbound proteins.
6. Wash the resin twice with 10 mL (40× bed volume) APB. Next, close the column with a plastic cap.

7. Cut the end of a P1000 pipet tip, and add 250 μL of HIS_{10}-UIM (0.5 mg/mL) (*see* **Note 13**) to the 250 μL GSH-resin in the column body. Gently pipet the resin up and down to mix. Incubate for 15 min.
8. Open the column, and collect the HIS_{10}-UIM eluate into a 1.5 ml Eppendorf tube containing 30 μL bed volume of washed Ni-NTA agarose (*see* **Note 14**).
9. Repeat the elution with another 250 μL of HIS_{10}-UIM, and collect into another tube with 30 μL bed volume of washed Ni-NTA agarose (*see* **Note 15**). Perform a total of five elutions.
10. To remove the UIM, incubate the 500 μL HIS_{10}-UIM eluate with the Ni-NTA for 20 min on a rotating wheel. Spin for 1 min at 500 × *g* to sediment Ni-NTA.
11. Transfer the supernatant into a 500 μL filter unit (0.45 μm), and spin for another 5 min at 10,000 × *g* to remove the Ni-NTA. The flow-through contains the purified 26S proteasomes (but not the eluting peptide His_{10}-UIM).
12. Determine the protein concentration of the 26S preparation by Bradford assay. Typically, the first elution has very little 26S. The second and third elution fractions contain the bulk of the proteasomes and can be combined. Aliquot and store at −80 °C. To maintain the structural integrity of the 26S as well as possible, avoid freeze-thaw cycles.

3.3 Isolation of Proteasomes by Differential Centrifugation

Perform all steps on ice or at 4 °C.

1. Cells grown to 90% confluency in one 100 mm cell culture dish are washed with 10 mL PBS and then collected by scraping into 4 mL ice-cold PBS.
2. Cell pellets are collected by centrifugation at 350 × *g* for 5 min and then resuspended in 0.4 mL cell lysis buffer.
3. Cells are lysed by sonication with four rounds of 10 s (12 μm amplitude) bursts and 50 s rests on ice. Cell debris are cleared by centrifugation at 10,000 × *g* for 10 min, and the cell extracts (S100) are prepared by further centrifugation of the resulting supernatants at 100,000 × *g* for 30 min.
4. Pellet proteasomes by centrifuging 200 μL S100 at 150,000 × *g* for 1 h (*see* **Note 16**). Collect supernatant (S150) (*see* **Note 17**).
5. Rinse proteasome pellets with 200 μL wash buffer once, and resuspend the proteasome pellet in 200 μL wash buffer by sonication with four rounds of 10 s (12 μm amplitude) bursts and 50 s rests on ice (*see* **Note 18**).

4 Notes

1. DTT (1 mM) was omitted in the cell lysis and wash buffers because reducing agents disrupt the disulfide bridges holding the antibody chains of the FLAG M2 affinity gel (Sigma) together.
2. The purification buffer can also be prepared with 150 mM NaCl. The GST-UBL also isolates a complex containing the hexameric ATPase p97/VCP and associated factors, but the use of NaCl (150 mM) prevents the p97 complex from binding to the GST-UBL [14]. However, this salt concentration also reduces the amount of proteasomes and certain proteasome-associated proteins [14]. We do not recommend adding NaCl when purifying from the liver or spleen. For reasons that are unclear, 150 mM NaCl increases the amount of free 19S in the purified proteasomes from liver extract, although adding NaCl to purified 26S particles from the liver does not cause their disassembly. Yeast proteasomes seem to have a lower affinity for the UBL domain of mammalian Rad23b and do not bind in the presence of salt [15].
3. These purification procedures are used for small-scale analytical preparations, and thus protein concentrations are too low to be reliably measured by standard assays, such as the Bradford assay. To quantify the concentration of purified proteasomes, we used the FLAG peptides or a 19S or 20S proteasome subunit expressed in *E. coli* and purified as a reference. The concentrations of these proteins can be determined and then are used to create a standard curve to determine the concentrations of the purified proteasomes by Western blot [4].
4. To scale up the preparations, one should determine the optimal affinity gel and cell extract ratio.
5. Longer spins of the cell extracts may precipitate proteasome particles, leading to a loss of proteasomes from the cell extracts.
6. To study how proteasome composition changes during in vitro incubations and the effects on Ub conjugate binding, protein-bound affinity gels were washed three times with 150 μL wash buffer. Spun at 300 × *g* for 1 min. The bound FLAG-tagged proteasomes were incubated at either 4° or 37 °C for 20 min (*see* **Note 7**) before the final wash and elution.
7. After incubation of the gel-bound FLAG-tagged proteasomes at either 4° or 37 °C for 20 min, spin at 300 × *g*, for 1 min, and collect the supernatant. These supernatants can be used to monitor the dissociation of proteins during incubation [4].

8. Compared to a single elution, multiple elutions using smaller buffer volumes each time can result in a more complete elution.
9. Perform the proteasome activity assays on the same day because repeated freeze-thaw cycles decrease proteasome activity.
10. The necessary amount of cells varies with the cell type and should be optimized for each cell line. The goal should be to harvest 10–20 mL of lysate containing 80–100 mg total protein, assuming that proteasomes represent 1–2% of the cell protein content, which is an approximation not necessarily true for all cell types or tissues. This procedure can be scaled down to a smaller size, if necessary for particularly valuable tissue sources, although such a modification tends to decrease the concentration of proteasomes obtained and consequently results in a less stable preparation.
11. The lysis of tissues is discussed in a previous description of this method [15].
12. We have found that 30 min is sufficient to separate the proteasomes from larger microsomal components. Longer spin times begin to pellet most proteasomes.
13. Add ATP to 1 mM in the HIS_{10}-UIM solution immediately prior to elution.
14. We have found that Qiagen Ni-NTA provides consistently more active proteasome preparations than several other sources of this resin.
15. Immediately after eluted proteins are added to the Ni-NTA agarose, begin to rotate for 20 min (**step 8**). Prolonged incubation with Ni-NTA agarose seems to decrease proteasome activity. Conversely, a shorter incubation causes HIS_{10}-UIM to be insufficiently cleared, and the presence of HIS_{10}-UIM in the 26S preparation can also decrease proteasome activity. Therefore, it is important to incubate specifically for 20 min. This elution strategy is modified by Dr. Donghoon Lee from Besche and Goldberg's original method [14] to improve proteasome yield.
16. The centrifugal force was determined to pellet more than 99% of proteasomes while minimizing the co-isolation of other proteins. The optimal centrifugal force may vary with preparations of different sources.
17. To determine if the centrifugal force used is sufficient to enrich proteasomes, check the presence of proteasome subunits (e.g., β5) in the collected supernatant (S150) by Western blot, or measure the proteasomes' peptidase activity in this S150 fraction using Suc-LLVY-amc as the substrate (*see* Chapter 19).

18. To reduce contaminating proteins in the proteasome-rich fraction, resuspend and pellet proteasomes again by further centrifugation at 150,000 × *g* for 1 h, and repeat **step 5**.

Acknowledgments

The methods described here were developed through research support to our laboratory from the NIH-NIGMS (R01 GM51923), Cure Alzheimer's Fund, Muscular Dystrophy Association (MDA-419143), and Project A.L.S (2015-06).

References

1. Collins GA, Goldberg AL (2017) The Logic of the 26S Proteasome. Cell 169:792–806
2. Asano S, Fukuda Y, Beck F, Aufderheide A, Forster F, Danev R, Baumeister W (2015) Proteasomes. A molecular census of 26S proteasomes in intact neurons. Science 347: 439–442
3. Peth A, Besche HC, Goldberg AL (2009) Ubiquitinated proteins activate the proteasome by binding to Usp14/Ubp6, which causes 20S gate opening. Mol Cell 36:794–804
4. Kuo CL, Goldberg AL (2017) Ubiquitinated proteins promote the association of proteasomes with the deubiquitinating enzyme Usp14 and the ubiquitin ligase Ube3c. Proc Natl Acad Sci U S A 114:E3404–E3413
5. Lokireddy S, Kukushkin NV, Goldberg AL (2015) cAMP-induced phosphorylation of 26S proteasomes on Rpn6/PSMD11 enhances their activity and the degradation of misfolded proteins. Proc Natl Acad Sci U S A 112: E7176–E7185
6. Guo X, Wang X, Wang Z, Banerjee S, Yang J, Huang L, Dixon JE (2016) Site-specific proteasome phosphorylation controls cell proliferation and tumorigenesis. Nat Cell Biol 18:202–212
7. VerPlank JJS, Goldberg AL (2017) Regulating protein breakdown through proteasome phosphorylation. Biochem J 474:3355–3371
8. Jacobson AD, MacFadden A, Wu Z, Peng J, Liu CW (2014) Autoregulation of the 26S proteasome by in situ ubiquitination. Mol Biol Cell 25:1824–1835
9. Leggett DS, Glickman MH, Finley D (2005) Purification of proteasomes, proteasome subcomplexes, and proteasome-associated proteins from budding yeast. Methods Mol Biol 301:57–70
10. Saeki Y, Isono E, Toh EA (2005) Preparation of ubiquitinated substrates by the PY motif-insertion method for monitoring 26S proteasome activity. Methods Enzymol 399:215–227
11. Leggett DS, Hanna J, Borodovsky A, Crosas B, Schmidt M, Baker RT, Walz T, Ploegh H, Finley D (2002) Multiple associated proteins regulate proteasome structure and function. Mol Cell 10:495–507
12. Wang X, Chen CF, Baker PR, Chen PL, Kaiser P, Huang L (2007) Mass spectrometric characterization of the affinity-purified human 26S proteasome complex. Biochemistry 46:3553–3565
13. Besche HC, Sha Z, Kukushkin NV, Peth A, Hock EM, Kim W, Gygi S, Gutierrez JA, Liao H, Dick L, Goldberg AL (2014) Autoubiquitination of the 26S proteasome on Rpn13 regulates breakdown of ubiquitin conjugates. EMBO J 33:1159–1176
14. Besche HC, Haas W, Gygi SP, Goldberg AL (2009) Isolation of mammalian 26S proteasomes and p97/VCP complexes using the ubiquitin-like domain from HHR23B reveals novel proteasome-associated proteins. Biochemistry 48:2538–2549
15. Besche HC, Goldberg AL (2012) Affinity purification of mammalian 26S proteasomes using an ubiquitin-like domain. Methods Mol Biol 832:423–432
16. Nathan JA, Spinnenhirn V, Schmidtke G, Basler M, Groettrup M, Goldberg AL (2013) Immuno- and constitutive proteasomes do not differ in their abilities to degrade ubiquitinated proteins. Cell 152:1184–1194
17. VerPlank JJS, Lokireddy S, Feltri ML, Goldberg AL, Wrabetz L (2018) Impairment of protein degradation and proteasome function in hereditary neuropathies. Glia 66:379–395

18. Tai HC, Besche H, Goldberg AL, Schuman EM (2010) Characterization of the brain 26S proteasome and its interacting proteins. Front Mol Neurosci 3:12
19. Myeku N, Clelland CL, Emrani S, Kukushkin NV, Yu WH, Goldberg AL, Duff KE (2016) Tau-driven 26S proteasome impairment and cognitive dysfunction can be prevented early in disease by activating cAMP-PKA signaling. Nat Med 22:46–53
20. Shi Y, Chen X, Elsasser S, Stocks BB, Tian G, Lee BH, Shi Y, Zhang N, de Poot SA, Tuebing F, Sun S, Vannoy J, Tarasov SG, Engen JR, Finley D, Walters KJ (2016) Rpn1 provides adjacent receptor sites for substrate binding and deubiquitination by the proteasome. Science 351(6275):aad9421
21. Hamazaki J, Hirayama S, Murata S (2015) Redundant roles of Rpn10 and Rpn13 in recognition of ubiquitinated proteins and cellular homeostasis. PLoS Genet 11:e1005401

Chapter 19

Measurement of the Multiple Activities of 26S Proteasomes

Hyoung Tae Kim, Galen Andrew Collins, and Alfred L. Goldberg

Abstract

Because proteasomes catalyze most of the protein degradation in mammalian cells, and their functioning is essential for cellular homeostasis, proteasome structure, biochemical mechanisms, and regulation in normal and disease states are now widely studied and are of major importance. In addition, inhibitors of the proteasome's peptidase activity have proven to be very valuable as research tools and in the treatment of hematologic malignancies, and a number of newer pharmacological agents that alter proteasome function are being developed. The rapid degradation of ubiquitinated proteins by the 26S proteasome involves multiple enzymatic and non-enzymatic steps, including the binding of ubiquitinated substrates to the 19S particle (Subheading 3.2), opening the gated substrate entry channel into the 20S particle (Subheading 3.3), disassembly of the Ub chain (Subheading 3.4), ATP hydrolysis (Subheading 3.5), substrate unfolding and translocation, and proteolysis within the 20S particle (Subheadings 3.3 and 3.7). Assaying each of these processes is important if we are to fully understand the physiological regulation of proteasome function and the effects of disease or drugs. Here, we describe several methods that we have found useful to measure many of these individual activities using purified proteasomes. Studies using these approaches have already provided valuable new insights into the effects of post-synthetic modifications to 26S subunits, the physiological regulation of the ubiquitin-proteasome system, and the impairment of proteasome activity in neurodegenerative disease. These advances would not have been possible if only the standard assays of peptidase activity were used.

Key words 26S proteasomes, Ubiquitin conjugate degradation, Peptide hydrolysis, ATP hydrolysis, Deubiquitination, Ubiquitin-proteasome pathway, Gate opening, Ubiquitin binding

1 Introduction

The 26S proteasome consists of the 20S core proteolytic particle and one or two 19S regulatory particles [1, 2]. The 20S proteasome is a hollow four-ring cylinder containing in each of its central β-rings three types of peptidase sites, which have chymotrypsin-like, trypsin-like, and caspase-like specificities [3]. The outer α-rings contain a gated central channel for substrate entry [4, 5]. The core particle may be capped on one or both ends with a 19S

Hyoung Tae Kim and Galen Andrew Collins contributed equally to this work.

Thibault Mayor and Gary Kleiger (eds.), *The Ubiquitin Proteasome System: Methods and Protocols*, Methods in Molecular Biology, vol. 1844, https://doi.org/10.1007/978-1-4939-8706-1_19,

regulatory particle, which binds ubiquitin (Ub) conjugates, disassembles the Ub chain, hydrolyzes ATP, and unfolds and translocates the substrate protein into the 20S particle [1, 2]. Most steps in the degradation process are linked to ordered cycles of ATP binding and hydrolysis [6], catalyzed by the six 19S ATPase subunits which unfold and translocate the polypeptide through their central channel and through the gated entry pore into the 20S particle, within which proteolysis occurs [1, 7]. Assaying these multiple steps is therefore essential for a full understanding of the degradative process, the actions of drugs, and the alterations in rates of protein breakdown occurring in different physiological and pathological states.

Recent studies have revealed that proteasome function is regulated in multiple ways including its activation by ubiquitinated substrates [8], by covalent modifications of subunits (e.g., by phosphorylation) [9], and by various proteasome-associated regulatory proteins [10]. However, generally proteasome activity in cell or tissue extracts has been determined by measurements of the hydrolysis of a fluorogenic peptide, most often Suc-LLVY-amc, a substrate of the 20S's chymotrypsin-like site. Although this approach is simple and convenient, it does not actually measure the proteasomes' ability to degrade its physiological substrates, Ub conjugates. Also, these peptidase activities in cell lysates can be altered by various factors, especially PA28αβ/11S, PA28γ, PA200/BLM10, the inhibitory factor PI31/PSMF1, or even by exposure to salts or detergents [11–13]. The capacity of 26S particles to degrade ubiquitinated proteins depends on multiple actions of the 19S regulatory particle, including (1) the extent of binding of a Ub conjugate, (2) the rate of substrate deubiquitination, and (3) the rate of ATP hydrolysis [10]. Moreover, these steps can be activated by physiological stimuli (e.g., hormones or nutrient deprivation) or pharmacological agents. For example, phosphorylation of the 19S subunit, Rpn6, is stimulated by cAMP via protein kinase A (PKA), and during the S-M phases of the cell cycle, the Rpt3 subunit is phosphorylated by Dyrk2. These modifications activate not only 20S gate opening and peptide hydrolysis but also the 19S ATPases [14, 15] and the breakdown of Ub conjugates (*see* Chapter 20). Also, the ubiquitination of the 19S subunit Rpn13 blocks the binding of Ub conjugates to the particle and prevents Ub conjugate degradation without affecting the proteasome's peptidase activities [16]. The recent elucidation of these regulatory mechanisms has been made possible by the development of methods to assay these specific steps in conjugate degradation, which can only be analyzed once the 26S particles are purified free of exogenous regulatory factors and soluble enzymes, especially other ATPases and DUBs present in cell lysates. We describe below the protocols that we have found useful to measure several key steps in the degradation of ubiquitinated proteins.

1.1 Binding of Ub Proteins to the Proteasome

In selectively degrading ubiquitinated proteins, the proteasome must recognize both the Ub chain on the substrate and the presence of a loosely folded region within the protein [17]. This recognition of Ub conjugates proceeds by a two-step mechanism. Initially, the proteasome reversibly binds a Ub chain, and then the protein's loosely folded region becomes tightly bound to the ATPases, committing the substrate to degradation. Our lab has developed an assay that distinguishes these two binding steps and that has proven useful in clarifying the roles in binding of ATP and the receptor subunits Rpn10 and Rpn13 [18]; the effects of the shuttling factors, Rad23a and Rad23b [19]; and the impairment of binding upon Rpn13 ubiquitination [16].

The initial step in proteasome degradation of ubiquitinated substrates is the reversible association of the Ub chain with the Pru domain of Rpn13/ADRM1 and the Ub-interacting motifs (UIM) of Rpn10/S5a [20, 21] (although other sites on the proteasome can also contribute to Ub binding, notably the T1 region of Rpn1 [22]). This initial step does not require ATP hydrolysis [18] and therefore can occur at 4 °C. This temperature also greatly reduces the activities of contaminating cytosolic enzymes (e.g., DUBs or ATPases). This binding step is easily reversed by competition with an excess of free UIM or by washing with a buffer containing relatively high concentrations of salt (300 mM). The second, tighter binding step cannot be reversed by a UIM or a salt wash and depends on the presence of a loosely folded region in the substrate [18]. This transition to tight binding requires ATP hydrolysis and thus can only occur at higher temperatures (e.g., 37 °C). This step, most likely, involves the capture of the protein by the six ATPase ring, whose pore loops appear to grab the substrate and drive its translocation [23].

Our method to measure the binding of Ub conjugates enables the separation of the initial and tighter binding steps. In these assays, we have used several different immobilized ubiquitinated protein substrates DHFR, Nedd4, and E6AP/Ube3a [16, 18, 19], but because of its ease of preparation, ubiquitinated E6AP is preferred and will be discussed here. After they bind to ubiquitinated E6AP immobilized on resins, the proteasomes can be detected either by immunoblot or more rapidly, simply, and quantitatively with a peptidase assay using a small fluorogenic substrate (*see* Subheading 2). The values obtained by the two methods agree closely [18]. The initial reversible and tighter binding steps can be distinguished by incubation at different temperatures. At 4 °C only the initial binding step can occur, while incubation at 37 °C allows the tighter binding. Also, these two steps can be distinguished by the sensitivity of the reversible proteasome binding to washing with high salt (300 mM) or UIM solutions.

1.2 Peptide Hydrolysis and Opening of the Substrate Entry Channel into the 20S Particle

The eukaryotic 20S proteasome contains three proteolytic sites on each of its central β-rings. These active sites differ in their substrate specificities. Two sites (located on β5/PSMB5) are "chymotrypsin-like" and cleave after hydrophobic residues; two sites (β2/PSMB7) are "trypsin-like" and cleave after basic residues; and two (located on β1/PSMB6) are "caspase-like" and cleave after acidic residues [24]. In the immunoproteasomes, these constitutive β-subunits are replaced by homologous β-subunits with different specificities that favor antigen presentation [1]. The activities of these individual sites can be monitored by measuring the hydrolysis of fluorogenic peptide substrates specific for each active site [25]. We typically use Suc-LLVY-amc to monitor the chymotrypsin-like activity, Boc-LRR-amc for the trypsin-like activity, and Ac-nLPnLD-amc for the caspase-like activity of the proteasome.

To reach the proteolytic sites within the 20S particle, protein substrates are actively translocated through the ATPases' central channel [7, 26] and the gated central pore in the 20S α-ring [5]. Upon nucleotide binding, the ATPases' C-terminal HbYX residues bind to intersubunit pockets in the 20S's α-ring [27]. Under many experimental conditions (e.g., with ATP or especially the non-hydrolyzable nucleotide, ATPγS), the hydrolysis of substrates by all three active sites is stimulated simultaneously, due to the opening of the gated entry channel. Therefore, it is important in studies of proteasome function to measure the rates of substrate cleavage by all three active sites. Often in the literature, the rate of hydrolysis of substrates of the chymotrypsin-like site is taken to indicate proteasome activity. However, this assumption is not valid. In fact, the proteasome inhibitors used most widely as research tools, MG132, or in the clinic for treatment of multiple myeloma, bortezomib, preferentially block the chymotrypsin-like activity and only at high doses does it also inhibit the caspase-like site, but it does not block the trypsin-like activity [28]. Therefore, bortezomib at nanomolar concentrations completely inactivates the chymotrypsin-like site but slows protein and Ub conjugate degradation by only about 50% due to the continued functioning of the trypsin-like and caspase-like sites. Consequently, to effectively block proteolysis, it is essential to use much higher concentrations of these agents to inhibit multiple sites, and to evaluate proteasome function, it is important to assay Ub conjugate degradation rather than just peptidase activities.

1.3 Disassembly of Ub Chains by Proteasomal DUBs

During the degradation of a Ub conjugate, the Ub chain is disassembled in order that the Ub molecules can be reutilized and that substrates can be readily translocated into the 20S particle. Three different deubiquitinating enzymes (DUBs) are associated with the 19S regulatory particle in higher eukaryotes. Two are cysteine proteases, Usp14/Ubp6 [29] and Uch37/UchL5 [30, 31], and the third, Rpn11/PSMD14, is a metalloprotease [32, 33]. Rpn11 is an integral 19S subunit, but Usp14 and Uch37 associate reversibly with the proteasome and usually are present on the 26S in substoichiometric amounts. Deubiquitination of substrates by Rpn11 is an ATP-stimulated process [32, 33] that removes the Ub chain by cleaving the isopeptide bond near to the substrate and thus facilitates the translocation of the polypeptide through the ATPase ring into the 20S particle [32].

Several approaches have been utilized to evaluate the activities of these three DUBs, but they are at best semiquantitative. Selective assays of the contributions of these enzymes are desirable because the proteasomal DUBs have emerged as attractive drug targets for multiple diseases. However, their actual importance in conjugate degradation is quite difficult to measure. For example, Usp14 levels on the proteasome depend on the supply of Ub conjugates [34]; its DUB activity increases dramatically upon proteasome association [35]. Usp14 prefers substrate proteins bearing multiple Ub chains [36], and in addition to their DUB activities, Usp14 and Uch37 have major roles in regulating the other proteasomal activities [8, 37, 38]. The most widely used substrate for monitoring DUB activities is Ub-amc, and both Usp14 and Uch37 rapidly cleave it. Alternatively, the formation of a covalent linkage between the suicide substrate HA-Ub-vinyl sulfone (VS) and the DUB active site cysteines can be used to evaluate their activities [39]. However, these Ub-based substrates are not physiological, are not Ub chains, and do not accurately reflect the deubiquitination capacity of the 26S particle because the affinities of these enzymes for free Ub and Ub chains differ widely. Also Rpn11 hydrolyzes Ub-amc poorly and does not react with HA-Ub-VS. We have found that measuring the disassembly of short poly-Ub chains, which are homogeneous in length and contain only K48 or K63 Ub chains, is quite useful to monitor Rpn11's activity [38]. This activity requires an active site Zn^{2+} ion and thus can be selectively inhibited by o-phenanthroline, 8-thioquinoline, or its derivative capzimin [40, 41].

1.4 ATP Hydrolysis

The 26S proteasome is an ATP-dependent proteolytic machine, whose function depends on the ring of six ATPase subunits (Rpt1-6) at the base of the 19S complex. These ATPases are homologous to the bacterial ATP-dependent proteases and archaeal PAN ATPase, which are also hexameric complexes. These AAA ATPases all bind and translocate polypeptide substrates into an associated proteolytic compartment, and their ATPase activities are stimulated upon binding a loosely folded polypeptide substrate [42–44]. Unlike these prokaryotic enzymes, the 26S ATPases are stimulated by ubiquitinated proteins or by the simultaneous binding of a Ub chain to Usp14 or Uch37, in addition to the unstructured region of the polypeptide (which need not be covalently linked) [37]. The binding of a substrate or ATPγS to the proteasome induces a massive change in the ATPases' conformations. In the resulting degradation-competent state [7, 26], the ATPases' substrate translocation channel is wider and aligned with the gated 20S pore. Also, Rpn11 becomes positioned directly above the central pore so as to facilitate the degradation of the ubiquitinated substrate. ATP binding and hydrolysis thus drive almost all proteasomal steps, including substrate binding, gate opening, deubiquitination, unfolding, and translocation. Consequently, ATP consumption is directly proportional to the rate of conjugate degradation and depends on both the folding and size of the substrate [45]. Here we will describe a protocol to assay the activity of the 26S ATPases using a modification of the malachite green assay [46] which measures the release of inorganic phosphate from ATP.

1.5 Degradation of Ubiquitinated Proteins

In studying the effects of inhibitors, post-synthetic modifications, physiological stimuli, or disease states on proteasome function, the capacity of the 26S particles to degrade ubiquitinated proteins should be the major concern. The main challenge in such studies has been to generate defined ubiquitinated substrates that are easy to prepare, have defined ubiquitin chains, are easily quantitated, and are degraded rapidly by isolated proteasomes. Unfortunately, most labs have followed the breakdown of only one or two ubiquitinated proteins (DHFR, cyclins, troponin I, Sic1) and non-ubiquitinated ones (e.g., casein, Sic1) as substrates.

Here, we describe a protocol for one model substrate, PY-Sic1, based on a method originally described by Toh-E and colleagues [47] that we have modified by labeling Sic1 with ^{32}P through phosphorylation with casein kinase II and ^{32}P-ATP. By following the hydrolysis of this labeled substrate to radioactive peptides, it is possible to precisely quantitate the degradative process. Multiple other methods are available for radiolabeling or adding fluorescent tags to the substrate that presumably could also be used in similar assays. However, this approach allows precise measurement of degradation rates, is very sensitive, and offers advantages for kinetic studies. Most often, the degradation of ubiquitinated proteins has been assayed by monitoring the loss of the ubiquitinated protein by anti-ubiquitin or anti-substrate immunoblot. Although simple, this approach lacks sensitivity, and the ubiquitinated substrates (usually a heterogenous smear) may disappear through either proteasomal degradation or deubiquitination without degradation. Although it can be valuable in some studies to monitor simultaneously these two processes, quantitating each and identifying which process is changing can be problematic. In addition, the size heterogeneity of Ub conjugates makes quantitative immunoblots especially challenging. Moreover to measure the loss of the ubiquitinated substrate, it is often necessary to use low concentrations of the substrate (that are well below the Vmax of the enzyme) to increase the fraction of the substrate that is degraded. Such conditions generally do not yield linear degradative rates. Finally, having to run samples on SDS-PAGE limits the number of experimental variables that can be studied in an experiment.

The development of fluorescently tagged substrates [48] is a useful alternative to relying on antibodies against the substrate. Even though one can still run the reactions on gels, the great advantage of fluorescence is the ability to assay degradation in real time on a multi-well plate as described by Matouschek and colleagues (*see* Chapter 21) [48]. That method has the great potential for larger-scale screening of proteasome activity. In using anisotropy to assay proteolysis, the substrate ideally should be labeled at a single residue, and one must ensure that the change in rotational size is the consequence of degradation rather than deubiquitination of the substrate.

The "classical" method for following degradation of ubiquitinated protein, which we describe here, is to measure the conversion of the TCA-precipitable radioactive substrate to TCA-soluble (peptide) products. Similar approaches were used originally in the studies of ATP-dependent proteases in bacteria [49] and of the Ub proteasome pathway [16]. Although working with radioactive tracers requires certain special precautions, it is a very sensitive and precise assay. Because this method directly measures the peptide products, radioactively labeled substrates can be used at high concentrations that are suitable for kinetic analysis and make possible the establishment of linear assay conditions. Because of the prior difficulties in preparing ubiquitinated substrates and quantitating their degradation, there have been surprisingly few kinetic studies of Ub conjugate degradation despite their potential in clarifying proteasome mechanisms and regulation.

2 Materials

2.1 Purified 26S Proteasomes (100 nM) (See Note 1) as Described by Kuo et al. (See Chapter 18)

2.2 Purification and Ubiquitination of E6AP

1. Ubiquitin (Boston Biochem U-100H).
2. Ube1 (Boston Biochem E-304).
3. UbcH5b (Boston Biochem E2-622).
4. Emulsiflex (Avestin) (*see* **Note 2**).
5. Glutathione-Sepharose (GE Health Sciences).
6. IPTG.
7. Ubiquitination buffer: 20 mM Tris-HCl (pH 7.5), 20 mM KCl, and 5 mM $MgCl_2$.

2.3 Ub Conjugate Binding Assay

1. Binding buffer: 25 mM HEPES-KOH (pH 8), 125 mM KAc, 1 mM DTT, 2.5 mM $MgCl_2$, 0.1 g/L BSA, 0.05% Triton X-100, and 1 mM ATP.
2. Wash buffer I: 25 mM HEPES-KOH (pH 8), 125 mM KAc, 1 mM DTT, 2.5 mM $MgCl_2$, 0.05% Triton X-100, and 1 mM ATP.
3. Wash buffer IIa: 50 mM Tris-HCl (pH 7.5), 10 mM $MgCl_2$, 1 mM DTT, and 1 mM ATP.
4. Wash buffer IIb: 50 mM Tris-HCl (pH 7.5), 300 mM NaCl, 10 mM $MgCl_2$, 1 mM DTT, and 1 mM ATP.
5. Wash buffer IIc: 50 mM Tris-HCl (pH 7.5), 5 μM UIM, 10 mM $MgCl_2$, 1 mM DTT, and 1 mM ATP.

2.4 Peptide Hydrolysis Assay

1. Fluorogenic substrate stock solutions: dissolve Suc-Leu-Leu-Val-Tyr-amc (Suc-LLVY-amc for the chymotrypsin-like activity), Boc-Leu-Arg-Arg-amc (Boc-LRR-amc for the trypsin-like activity), or Ac-Nle-Pro-Nle-Asp-amc (Ac-nLPnLD-amc for the caspase-like activity) in DMSO (final concentration of 20 mM).
2. Peptidase assay buffer: 50 mM Tris-HCl (pH 7.6), 100 mM KCl, 0.1 mM ATP (or ATPγS), 0.5 mM $MgCl_2$, and 25 ng/μL bovine serum albumin with 10 μM selected fluorogenic peptide substrate.
3. Bovine serum albumin (Sigma).
4. Flat-bottomed black 96-well plate (Greiner Bio-One).
5. Fluorometer.

2.5 Deubiquitination Assay

1. Deubiquitination assay buffer (5×): 125 mM HEPES-KOH, pH 7.6, 500 mM KCl, 25 mM $MgCl_2$, 5 mM DTT, and 5 mM ATP.
2. 0.25 μg/μL tetramer Ub chains (Boston Biochem) in 50 mM Tris-HCl (pH 7.6).
3. Usp14 and Uch37 inhibitor: 6 μM Ub-vinylsulfone (Boston Biochem).
4. Rpn11 inhibitor: 10 mM o-phenanthroline in methanol.
5. Mini SDS-PAGE gel (4–12% gradient) (*see* **Note 3**).
6. Kit for SDS-PAGE and protein transfer to a membrane (*see* **Note 3**).
7. PVDF membrane (GE Healthcare) (*see* **Note 3**).
8. PBST: dilute 100 mL of 10× PBS with 890 mL of H_2O, and add 10 mL of 20% Triton X-100.
9. PBSTB: dissolve 3 g of BSA in 100 mL of PBST.

10. 4× SDS-PAGE sample buffer: 160 mM Tris-HCL (pH 6.8), 4% SDS, 20% glycerol, and 0.0012% bromophenol blue.
11. Anti-ubiquitin antibody (Santa Cruz Biotechnology P4D1 or Life Sensors VU-1) (*see* **Note 4**).
12. SuperSignal™ West Pico PLUS Chemiluminescent substrate (Thermo Fisher Scientific).

2.6 ATPase Assay (see Note 5)

1. Solution A (5× ATP hydrolysis reaction buffer): 125 mM HEPES-KOH (pH 7.6), 625 mM KAc, 0.13% Triton X-100, 12.5 mM $MgCl_2$, 2.5 mM ATP, 0.5 mg/mL BSA, and 2.5 mM DTT.
2. Solution B (malachite green solution): dissolve 45 mg of malachite green hydrochloride in 100 mL of H_2O with vigorous stirring for 30 min at room temperature. Pass the solution through 0.45 μm filter. This solution is stable at 4 °C for at least a few months (*see* **Note 6**).
3. Solution C (4.2% ammonium molybdate in 4 N HCl): 4.3 g of ammonium molybdate tetrahydrate is dissolved in 50 mL of H_2O. Slowly add 34.5 mL of concentrated HCl. Add H_2O to 100 mL. This solution is stable at 4 °C for a few months.
4. 34% citric acid in H_2O.
5. Flat-bottomed clear 96-well plate (Greiner Bio-One).
6. Spectrophotometer.
7. Standard inorganic phosphate solution (*see* Table 1).

2.7 Expression, Purification, and Labeling of Sic1 (see Note 7)

1. BL21 Star (DE3) E. coli (Thermo Fisher Scientific).
2. Emulsiflex (Avestin) (*see* **Note 2**).
3. Ni-NTA resin (Qiagen).
4. Disposable 20 mL column body.
5. Protein kinase CK2 (NEB).
6. ATP, [γ-^{32}P]—6000 Ci/mmol 10 mCi/mL, 250 μCi (PerkinElmer).
7. Bio-Spin 30 column (Bio-Rad).
8. Ubiquitin (Boston Biochem).
9. Ube1 (Boston Biochem).
10. Ubc4 [47, 50].
11. Rsp5 [47].
12. Sic1 lysis buffer: 50 mM Na_3PO_4 (pH 8.0), 300 mM NaCl, 10 mM imidazole, 0.5 mM DTT, and 1 mM PMSF.
13. Sic1 wash buffer: 50 mM Na_3PO_4 (pH 8.0), 300 mM NaCl, and 20 mM imidazole.

Table 1
Preparation of standard inorganic phosphate solutions

Phosphate (μM)	100 μM KH_2PO_4 (μL)	H_2O (μL)
0	0	1000
5	50	950
10	100	900
20	200	800
30	300	700
40	400	600
50	500	500
60	600	400
70	700	300
80	800	200
90	900	100
100	1000	0

14. Sic1 second wash buffer: 7.5 mL wash buffer and 2.5 mL elution buffer.
15. Sic1 elution buffer: 50 mM Na_3PO_4 (pH 8.0), 300 mM NaCl, and 250 mM imidazole.
16. Sic1 storage buffer: 50 mM Tris-HCl (pH 7.5), 100 mM NaCl, 10% glycerol, and 1 mM DTT.

2.8 Ubiquitinated (Ub (n)) Sic1 Degradation

1. 2× degradation buffer: 100 mM Tris-HCl (pH 7.5), 10 mM $MgCl_2$, 4 mM ATP, 2 mM DTT, and 0.02 g/L BSA.
2. 10 g/L BSA.
3. 40% TCA solution chilled on ice.
4. BioSafe II liquid scintillation fluid (RPI).
5. Liquid scintillation counter.

3 Methods

3.1 Ub(n) E6AP

1. Transform pGEX4T-E6AP into BL21(DE3). Select one Amp^R colony.
2. Grow at 37 °C in 2 L LB + ampicillin to OD_{600} 0.8.
3. Induce with 0.5 mM IPTG at 16 °C overnight.
4. Collect the bacterial pellet by centrifugation (3000 × *g* for 15 min), wash the pellet with 1× PBS, centrifuge again

(3000 × *g* for 20 min), and store the pellet at −80 °C until ready to use.

5. Perform lysis and purification **steps 5–8** at 4 °C. Lyse pellet in 20 mL 1× PBS. Clear lysate by centrifugation at 10,000 × *g* for 20 min.
6. Bind lysate to 1.25 mL (bed volume) equilibrated glutathione-Sepharose beads by rotation for 90 min.
7. Wash twice with 12.5 mL 1× PBS (10 bed volumes).
8. Wash twice with ubiquitination buffer.
9. Set up the ubiquitination reaction in 15 mL with 20 mM Tris-HCl (pH 7.5), 20 mM KCl, 5 mM $MgCl_2$, 2 mM ATP, 1 mM DTT, 30 μM ubiquitin, 5 nM Ube1, 140 nM UbcH5b, and 300 nM GST-E6AP on glutathione-Sepharose (*see* **Note 8**).
10. Incubate overnight at 37 °C with rotation. Collect the resin by centrifugation at 100 × *g* for 3 min.
11. Wash five times with 20 bed volumes of TNG buffer (50 mM Tris-HCl (pH 7.5), 50 mM NaCl, 10% (v/v) glycerol). Collect the resin by centrifugation at 100 × *g* for 3 min.

3.2 Ub Conjugate Binding Assay

1. Set up a 60 μL reaction with 30 nM Ub(n)E6AP (*see* **Note 9**) and 10 nM 26S in 1× binding buffer.
2. Rotate at 4 °C for 30 min (*see* **Note 10**). Spin at 500 × *g* at 4 °C for 3 min. Set aside supernatant (*see* **Note 11**).
3. Wash the resin once with 400 μL wash buffer I, rotating at 4 °C for 5 min. Spin at 500 × *g* at 4 °C for 3 min. Remove the supernatant by pipette.
4. Wash the resin twice with 400 μL wash buffer II (*see* **Note 12**), rotating at 4 °C for 5 min. Spin at 500 × *g* at 4 °C for 3 min. Remove the supernatant by pipette.
5. Resuspend the resin in 310 μL peptidase assay buffer, aliquot 3 × 100 μL into a 96-well plate, and assay as described in Subheading 3.3 starting at **step 2**.
6. Analysis involves determining the slopes for each of the binding conditions. It often takes 10–20 min for the slopes to stabilize to a linear rate. Once the slopes are determined, the background binding to non-ubiquitinated E6AP is subtracted from the initial binding, salt-/UIM-disrupted initial binding, and the tighter binding.

3.3 Peptide Hydrolysis Assay

1. In a black 96-well plate, mix 1 μL of the purified 26S proteasome (1 nM final) with 99 μL of the peptidase assay buffer.
2. Measure generation of cleaved amc fluorescent signal at 37 °C for up to 1 h using a fluorometer (λ_{ex} = 380 nM; λ_{em} = 460 nm).

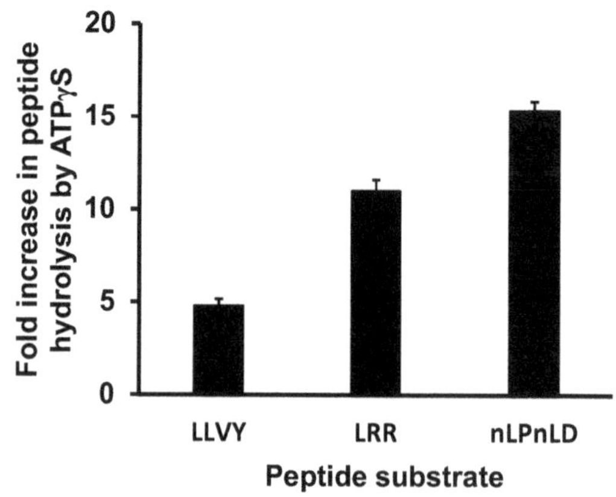

Fig. 1 Stimulation of peptidase activities of purified mouse embryonic fibroblast 26S proteasomes by ATPγS. Activities were assayed by measuring hydrolysis of fluorogenic peptide-amc substrates (10 μM) in the presence of 100 μM ATP or ATPγS. The markedly stimulated hydrolysis of all three substrate in the presence of ATPγS over the basal activities indicates gate opening induced by the non-hydrolyzable nucleotide. The basal peptidase activity of the proteasomes was measured in the presence of ATP. Peptidase activities are presented as the mean ± SD of three independent measurements (Modified from Kim and Goldberg, 2017 [38])

3. Calculate the peptidase activity of the proteasome by plotting the gain of fluorescence as a function of time. Assuming that the data are linear with respect to time, measure the rate of peptidase activity by linear fitting of the data and estimation of the slope (Fig. 1).

3.4 Deubiquitination Assay

1. Assemble deubiquitination reactions on ice: 1 μL of 26S proteasomes (final 5 nM), 4 μL of deubiquitination reaction buffer, and 14.5 μL H_2O.
2. If desired, add o-phenanthroline or Ub-vinylsulfone to inhibit different proteasomal deubiquitinases, subtract the H_2O volume by the volume of added factors, and preincubate for 15 min at room temperature before the reaction.
3. Start the reactions by adding 0.5 μL of tetra-Ub solution, and incubate the reaction at 37 °C for 20 min.
4. Stop reactions by adding 1 μL of DTT, 7.5 μL of 4× SDS-PAGE sample buffer, and 1.5 μL of H_2O. Incubate the samples at 70 °C for 10 min.
5. Run SDS-PAGE (4–12% gradient) to resolve the proteins.
6. Transfer proteins onto a PVDF membrane (*see* **Note 13**).
7. Block membrane with 10 mL of PBSTB at room temperature for 1 h.

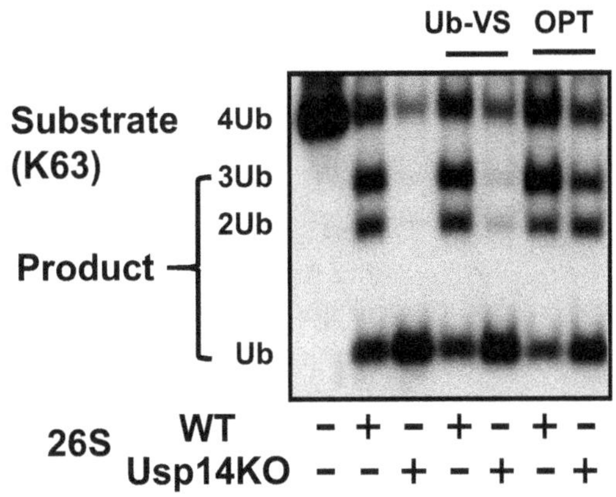

Fig. 2 Disassembly of K63 tetra-Ub chain (368 nM) by WT 26S proteasomes and ones lacking Usp14 (Usp14KO) was assayed after incubation at 37 °C for 20 min. Proteasomes were purified from mouse embryonic fibroblast. The generation of tri-, di-, and mono-Ubs was analyzed by Western blotting after electrophoresis. Usp14KO 26S proteasomes disassembled tetra-Ub chains much faster than WT 26S particles. Although Ub-vinylsulfone (Ub-VS, 1.5 μM) did not inhibit Ub chain hydrolysis, o-phenanthroline (OPT, 1 mM) inhibits markedly and reduces deubiquitination by Usp14KO 26S to the level of WT 26S. Thus, the faster disassembly of Ub chains by the Usp14KO 26S involves the Zn^{2+}-metalloprotease Rpn11 and not Uch37 (Adapted data from Kim and Goldberg, 2017 [38])

8. Remove the PBSTB, and incubate the membrane with 10 mL of PBST with 10 μL of anti-ubiquitin (1/1000 dilution) for 2 h at room temperature or overnight at 4 °C (*see* **Note 14**).
9. Wash the membrane with 20 mL of PBST for 10 min three times.
10. Incubate the membrane with 10 mL of PBST with 1 μL of anti-mouse IgG HRP conjugated antibody at room temperature for 1 h (*see* **Note 15**).
11. Wash the membrane as in **step 8**.
12. Incubate the membrane with 4 mL of SuperSignal West Pico Plus chemiluminescent substrate for 5 min.
13. Expose the membrane to X-ray film until bands of deubiquitinated products (tri-, di-, and mono-ubiquitin) become evident (Fig. 2).

3.5 ATPase Assay (See Notes 16 and 17)

1. Prepare malachite green assay reagent by adding 25 mL of solution C into 75 mL of solution B with vigorous stirring for 20 min or until precipitate disappears. Pass the solution through a 0.45 μm filter. Store light-sealed solution at room temperature.

2. On ice, prepare 1.5 mL microcentrifuge tubes containing master reaction mixture (for three replicates) containing 8 μL of solution A, X μL of required drug or additional factors, and H_2O to 28-X μL for each reaction time point (e.g., 0, 15, 30, 45, and 60 min) (*see* **Note 18**).
3. Start the reaction by adding 4 μL of proteasomes (10 nM final), mix briefly, and place the reaction at 37 °C. Reactions run for different times are started at different times. Thus, the 60-min reaction starts at 0 min, the 45-min reaction at 15 min, and so on (*see* **Note 19**). Repeat this step until the last reaction.
4. At the end of the reaction (60 min), transfer all the tubes from the water bath to ice. Let them cool down and centrifuge them briefly.
5. Put three 10 μL aliquots of each reaction and the phosphate standard into wells of the 96-well plate.
6. Add 170 μL of malachite green solution to each well, and wait until the color of the sample becomes dark green.
7. Add 20 μL of 34% citric acid to each well to stop the reactions.
8. Incubate the samples and the standard for 15–30 min at room temperature.
9. Measure the absorbance at 650 nm.
10. Calibrate readings to phosphate amounts, using a standard curve derived from Table 1. Plot the increase in free phosphate over time, and calculate the slopes derived from fitting the data to the equation for a line, which represent ATPase activity (Fig. 3).

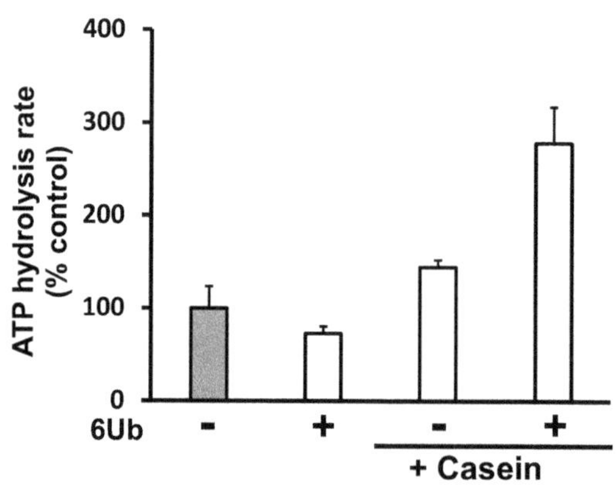

Fig. 3 ATP hydrolysis by mouse embryonic fibroblast 26S proteasomes was measured using a modified malachite green assay [46]. Upon addition of casein (1 μM) and a hexa-Ub chain (1 μM), which together mimic the binding of a Ub conjugate [37], ATP hydrolysis by WT 26S increased two- to threefold. However, hexa-Ub alone or casein alone did not increase ATP hydrolysis. 6Ub: a linear hexa-Ub chain. Data are represented as mean ± SD (Modified from Kim and Goldberg, 2017 [38])

3.6 Expression, Purification, and Labeling of Sic1

1. Transform pET21-Sic1 into BL21(DE3).
2. Select one AmpR colony. Grow at 37 °C in 2 L LB + ampicillin to OD_{600nm} 0.7.
3. Induce with 0.5 mM IPTG at 30 °C 3 h.
4. Collect the bacterial pellet by centrifugation (3000 × *g* for 15 min), wash the pellet with 1× PBS, centrifuge again (3000 × *g* 20 min), and store the pellet at −80 °C until ready to use.
5. Perform lysis and purification steps at 4 °C. Lyse the bacteria pellet in 30 mL Sic1 lysis buffer by passing through the emulsiflex at 10,000 psi (*see* **Note 2**). Add Triton X-100 to 0.2%, RNase A to 10 μg/mL, and DNase I to 5 μg/mL. Rotate for 10 min.
6. Clarify the lysate by centrifugation at 10,000 × *g* for 20 min.
7. Add clarified lysate to 1 mL bed volume equilibrated Ni-NTA resin. Rotate for 1 h.
8. Pour the slurry into an empty 20 mL disposable column body. Wash the resin twice with 10 mL Sic1 wash buffer.
9. Wash the resin once with 10 mL Sic1 second wash buffer.
10. Elute four times with 1 mL Sic1 elution buffer for 20 min at 4 °C.
11. Determine Sic1 fractions by SDS-PAGE/Coomassie blue staining. Pool Sic1 fractions and dialyze against Sic1 storage.
12. Set up phosphorylation reaction with 9 μL of 10× protein kinase buffer (NEB), 7.5 μL 500,000 U/mL CK2 (NEB), 15 μL 6000 Ci/mmol 10 mCi/mL, 250 μCi [γ-^{32}P] ATP, and 58.5 μL Sic1 (*see* **Note 20**).
13. Incubate at 30 °C for 1 h.
14. Remove unincorporated nucleotides by adding the phosphorylation reaction to a Bio-Spin 30 (Bio-Rad) column. Spin at 1000 × *g*, 4 °C for 3 min. Repeat the clean up by applying the flow-through of the spin column to a new Bio-Spin 30 column. Collect the flow-through from the second spin.
15. Set up ubiquitination reaction mixture 1 with 1× ATP solution, 200 nM Ube1, 6 μM Ubc4, and 120 μM ubiquitin in ubiquitination buffer.
16. Set up ubiquitination reaction mixture 2 with 1× ATP solution, 1 μM Rsp5, and 1 μM phosphorylated PY-Sic1.
17. Incubate ubiquitination mixture 1 at 25 °C for 5 min. Add ubiquitination mixture 2 to Ub mixture 1. Incubate at 25 °C for 3 h.

3.7 Degradation of Ub(n) Sic1

1. Prepare the reaction stop tubes (one for each time point) ahead of time by aliquoting 13.3 μL 10 g/L BSA per tube. Keep on ice.
2. Set up the degradation reaction with 0.5 nM 26S and 100 nM ubiquitinated Sic1 in degradation buffer (*see* **Note 21**).
3. Incubate the reactions at 37 °C for 0, 10, 20, 30, 40, 50, and 60 min. At each time point, collect 6.7 μL of the sample and add to the stop tubes containing 13.3 μL 10 g/L BSA on ice. Add 6.7 μL 40% TCA. Mix and incubate on ice for at least 30 min.
4. Separate TCA soluble, and pellet fractions by centrifugation at 10,000 × *g* at 4 °C for 10 min. Transfer the supernatant to 4 mL liquid scintillation fluid.
5. Re-solubilize the TCA precipitate for the 0-time samples in 50 μL 0.2 M NaOH. Add this to 4 mL of liquid scintillation fluid.
6. Measure counts per minute on a liquid scintillation counter.
7. Analysis and quantification of degradation rates. Subtract from all samples the TCA soluble 0-time counts as background signal. If measuring percent substrate degraded, calculate the sample degraded per TCA-precipitated counts at time 0. Otherwise relative rates within an experiment can be compared by just determining the slope of counts per minute in the TCA-soluble fractions.

4 Notes

1. To maintain the stability of the purified proteasomes during storage at −80 °C, add 10% (v/v) glycerol to the proteasome solution.
2. Alternative methods to lyse bacteria, such as adding lysozyme and sonicating, should also work.
3. Various kits for SDS-PAGE and for protein transfer to a membrane can be used, according to the instructions of the manufacturer. This protocol was established with Thermo Fisher Scientific's Bolt Bis-Tris Plus Mini Gel (4–12% gradient), XCell SureLock Mini-Cell, and XCell II Blot Module.
4. This protocol was originally established using Boston Biochem's polyclonal anti-ubiquitin antibody (A100), which has been discontinued. Anti-ubiquitin antibodies which detect mono-ubiquitin efficiently, such as P4D1 (from our experience) and VU-1 (39), can be used is this assay.
5. Do not use glassware during the preparation of these reagents or for the measurements when performing the ATPase assay.

6. This solution should be protected from light.
7. This protocol is adapted with minor modifications from Saeki et al. [47].
8. It is important to also prepare a non-ubiquitinated substrate as a control. This sample does not require as many reactions, and therefore a reaction volume that is half the size of the ubiquitination reaction can be prepared.
9. To analyze both the initial and tighter binding of proteasomes to a Ub conjugate, five reactions are measured: (1) the background initial binding to a non-ubiquitinated version of the substrate (e.g., E6AP), (2) the initial binding to the ubiquitinated substrate (e.g., Ub(n)E6AP), (3) the degree of salt-sensitive binding (or UIM-disrupted binding) determined by binding at 4 °C and washing with high salt, (4) the tighter binding level to the ubiquitinated substrate, and (5) any non-specific tighter binding to the non-ubiquitinated substrate.
10. A 4 °C incubation will allow only initial conjugate binding, nearly all of which is sensitive to a high salt wash or competition with UIM. To achieve tighter binding, incubation at 37 °C instead of 4 °C is necessary.
11. It is best to remove the supernatant gently by pipette, often leaving a few microliters of supernatant to avoid losing any of the resin.
12. The initial binding assay uses a wash buffer without salt (wash buffer IIa). To disrupt initial binding or confirm the transition to tighter binding, one can use either salt disruption (wash buffer IIb) or UIM competition (wash buffer IIc).
13. PVDF binds to free and di-Ub chains much better than nitrocellulose.
14. An appropriate dilution factor for the anti-ubiquitin antibody should be determined before the assay.
15. Selection of secondary antibody and detection method can be modified.
16. For proteins purified from *E. coli*, Hsp70 contamination is a major source of background activity. Incubating the affinity resin with a wash buffer containing 5 mM $MgCl_2$ and 1 mM ATP and then washing again with the normal wash buffer usually helps to reduce contaminating Hsp70 from the elution.
17. High concentrations of glycerol can interfere with the reaction. Use less than 10% glycerol in the final reaction conditions.
18. To test for ATPase contamination, prepare the sample containing 8 μL of solution A, X μL of drug or additional factors, and 32-X μL of H_2O without addition of proteasomes.

19. Start the reaction with the longest reaction time first. The last reaction (0 min reaction) should be mixed and used at the next step in the protocol without incubation.
20. It is recommended that a mock phosphorylation reaction be set up to (1) estimate the concentration of Sic1 eluted through the Bio-Spin columns and (2) to set up a "cold" ubiquitination reaction to check the effectiveness of the ubiquitination reaction.
21. Typically, the degradation of Ub(n) Sic1 is measured every 10 min over the course of an hour. It is very important to collect the initial (at 0 time) point as quickly as possible, and therefore it is often advisable to stagger the start of different reactions by about 30–45 s to ensure an accurate time course.

Acknowledgments

The methods described here were developed through research support to our laboratory from the NIH-NIGMS (R01 GM51923), Cure Alzheimer's Fund, Muscular Dystrophy Association (MDA-419143), and Project A.L.S (2015-06).

References

1. Finley D (2009) Recognition and processing of ubiquitin-protein conjugates by the proteasome. Annu Rev Biochem 78:477–513
2. Smith DM, Benaroudj N, Goldberg A (2006) Proteasomes and their associated ATPases: a destructive combination. J Struct Biol 156:72–83
3. Kisselev AF, Garcia-Calvo M, Overkleeft HS, Peterson E, Pennington MW, Ploegh HL, Thornberry NA, Goldberg AL (2003) The caspase-like sites of proteasomes, their substrate specificity, new inhibitors and substrates, and allosteric interactions with the trypsin-like sites. J Biol Chem 278:35869–35877
4. Groll M, Huber R (2003) Substrate access and processing by the 20S proteasome core particle. Int J Biochem Cell Biol 35:606–616
5. Groll M, Bajorek M, Köhler A, Moroder L, Rubin DM, Huber R, Glickman MH, Finley D (2000) A gated channel into the proteasome core particle. Nat Struct Biol 7:1062–1067
6. Smith DM, Fraga H, Reis C, Kafri G, Goldberg AL (2011) ATP binds to proteasomal ATPases in pairs with distinct functional effects, implying an ordered reaction cycle. Cell 144:526–538
7. Matyskiela ME, Lander GC, Martin A (2013) Conformational switching of the 26S proteasome enables substrate degradation. Nat Struct Mol Biol 20:781–788
8. Peth A, Besche HC, Goldberg AL (2009) Ubiquitinated proteins activate the proteasome by binding to Usp14/Ubp6, which causes 20S gate opening. Mol Cell 36:794–804
9. Cui Z, Scruggs SB, Gilda JE, Ping P, Gomes AV (2014) Regulation of cardiac proteasomes by ubiquitination, SUMOylation, and beyond. J Mol Cell Cardiol 71:32–42
10. Collins GA, Goldberg AL (2017) The logic of the 26S proteasome. Cell 169:792–806
11. Li X, Thompson D, Kumar B, DeMartino GN (2014) Molecular and cellular roles of PI31 (PSMF1) protein in regulation of proteasome function. J Biol Chem 289:17392–17405
12. Ortega J, Heymann JB, Kajava AV, Ustrell V, Rechsteiner M, Steven AC (2005) The axial channel of the 20S proteasome opens upon binding of the PA200 activator. J Mol Biol 346:1221–1227
13. Ma CP, Slaughter CA, DeMartino GN (1992) Identification, purification, and characterization of a protein activator (PA28) of the 20 S

proteasome (macropain). J Biol Chem 267:10515–10523
14. Lokireddy S, Kukushkin NV, Goldberg AL (2015) cAMP-induced phosphorylation of 26S proteasomes on Rpn6/PSMD11 enhances their activity and the degradation of misfolded proteins. Proc Natl Acad Sci U S A 112: E7176–E7185
15. Guo X, Wang X, Wang Z, Banerjee S, Yang J, Huang L, Dixon JE (2016) Site-specific proteasome phosphorylation controls cell proliferation and tumorigenesis. Nat Cell Biol 18:202–212
16. Besche HC, Sha Z, Kukushkin NV, Peth A, Hock EM, Kim W, Gygi S, Gutierrez JA, Liao H, Dick L, Goldberg AL (2014) Autoubiquitination of the 26S proteasome on Rpn13 regulates breakdown of ubiquitin conjugates. EMBO J 33:1159–1176
17. Lee C, Schwartz MP, Prakash S, Iwakura M, Matouschek A (2001) ATP-dependent proteases degrade their substrates by processively unraveling them from the degradation signal. Mol Cell 7:627–637
18. Peth A, Uchiki T, Goldberg AL (2010) ATP-dependent steps in the binding of ubiquitin conjugates to the 26S proteasome that commit to degradation. Mol Cell 40:671–681
19. Nathan JA, Kim HT, Ting L, Gygi SP, Goldberg AL (2013) Why do cellular proteins linked to K63-polyubiquitin chains not associate with proteasomes? EMBO J 32:552–565
20. Deveraux Q, Ustrell V, Pickart C, Rechsteiner M (1994) A 26 S protease subunit that binds ubiquitin conjugates. J Biol Chem 269:7059–7061
21. Qiu XB, Ouyang SY, Li CJ, Miao S, Wang L, Goldberg AL (2006) hRpn13/ADRM1/GP110 is a novel proteasome subunit that binds the deubiquitinating enzyme, UCH37. EMBO J 25:5742–5753
22. Shi Y, Chen X, Elsasser S, Stocks BB, Tian G, Lee BH, Shi Y, Zhang N, de Poot SA, Tuebing F, Sun S, Vannoy J, Tarasov SG, Engen JR, Finley D, Walters KJ (2016) Rpn1 provides adjacent receptor sites for substrate binding and deubiquitination by the proteasome. Science 351(6275):aad9421
23. Erales J, Hoyt MA, Troll F, Coffino P (2012) Functional asymmetries of proteasome translocase pore. J Biol Chem 287:18535–18543
24. Kisselev AF, Goldberg AL (2001) Proteasome inhibitors: from research tools to drug candidates. Chem Biol 8:739–758
25. Kisselev AF, Goldberg AL (2005) Monitoring activity and inhibition of 26S proteasomes with fluorogenic peptide substrates. Methods Enzymol 398:364–378
26. Sledz P, Unverdorben P, Beck F, Pfeifer G, Schweitzer A, Forster F, Baumeister W (2013) Structure of the 26S proteasome with ATP-gammaS bound provides insights into the mechanism of nucleotide-dependent substrate translocation. Proc Natl Acad Sci U S A 110:7264–7269
27. Smith DM, Chang SC, Park S, Finley D, Cheng Y, Goldberg AL (2007) Docking of the proteasomal ATPases' carboxyl termini in the 20S proteasome's alpha ring opens the gate for substrate entry. Mol Cell 27:731–744
28. Kisselev AF, Callard A, Goldberg AL (2006) Importance of the different proteolytic sites of the proteasome and the efficacy of inhibitors varies with the protein substrate. J Biol Chem 281:8582–8590
29. Borodovsky A, Kessler BM, Casagrande R, Overkleeft HS, Wilkinson KD, Ploegh HL (2001) A novel active site-directed probe specific for deubiquitylating enzymes reveals proteasome association of USP14. EMBO J 20:5187–5196
30. Li T, Naqvi NI, Yang H, Teo TS (2000) Identification of a 26S proteasome-associated UCH in fission yeast. Biochem Biophys Res Commun 272:270–275
31. Stone M, Hartmann-Petersen R, Seeger M, Bech-Otschir D, Wallace M, Gordon C (2004) Uch2/Uch37 is the major deubiquitinating enzyme associated with the 26S proteasome in fission yeast. J Mol Biol 344:697–706
32. Yao T, Cohen RE (2002) A cryptic protease couples deubiquitination and degradation by the proteasome. Nature 419:403–407
33. Verma R, Aravind L, Oania R, McDonald WH, Yates JR 3rd, Koonin EV, Deshaies RJ (2002) Role of Rpn11 metalloprotease in deubiquitination and degradation by the 26S proteasome. Science 298:611–615
34. Kuo CL, Goldberg AL (2017) Ubiquitinated proteins promote the association of proteasomes with the deubiquitinating enzyme Usp14 and the ubiquitin ligase Ube3c. Proc Natl Acad Sci U S A 114:E3404–E3413
35. Lee BH, Lee MJ, Park S, Oh DC, Elsasser S, Chen PC, Gartner C, Dimova N, Hanna J, Gygi SP, Wilson SM, King RW, Finley D (2010) Enhancement of proteasome activity by a small-molecule inhibitor of USP14. Nature 467:179–184
36. Lee BH, Lu Y, Prado MA, Shi Y, Tian G, Sun S, Elsasser S, Gygi SP, King RW, Finley D (2016) USP14 deubiquitinates proteasome-bound

substrates that are ubiquitinated at multiple sites. Nature 532:398–401
37. Peth A, Kukushkin N, Bossé M, Goldberg AL (2013) Ubiquitinated proteins activate the proteasomal ATPases by binding to Usp14 or Uch37 homologs. J Biol Chem 288:7781–7790
38. Kim HT, Goldberg AL (2017) The deubiquitinating enzyme Usp14 allosterically inhibits multiple proteasomal activities and ubiquitin-independent proteolysis. J Biol Chem 292:9830–9839
39. VerPlank JJS, Lokireddy S, Feltri ML, Goldberg AL, Wrabetz L (2018) Impairment of protein degradation and proteasome function in hereditary neuropathies. Glia 66:379–395
40. Li J, Yakushi T, Parlati F, Mackinnon AL, Perez C, Ma Y, Carter KP, Colayco S, Magnuson G, Brown B, Nguyen K, Vasile S, Suyama E, Smith LH, Sergienko E, Pinkerton AB, Chung TDY, Palmer AE, Pass I, Hess S, Cohen SM, Deshaies RJ (2017) Capzimin is a potent and specific inhibitor of proteasome isopeptidase Rpn11. Nat Chem Biol 13:486–493
41. Perez C, Li J, Parlati F, Rouffet M, Ma Y, Mackinnon AL, Chou TF, Deshaies RJ, Cohen SM (2017) Discovery of an inhibitor of the proteasome subunit Rpn11. J Med Chem 60:1343–1361
42. Benaroudj N, Zwickl P, Seemuller E, Baumeister W, Goldberg AL (2003) ATP hydrolysis by the proteasome regulatory complex PAN serves multiple functions in protein degradation. Mol Cell 11:69–78
43. Goldberg AL (1992) The mechanism and functions of ATP-dependent proteases in bacterial and animal cells. Eur J Biochem 203:9–23
44. Menon AS, Goldberg AL (1987) Protein substrates activate the ATP-dependent protease La by promoting nucleotide binding and release of bound ADP. J Biol Chem 262:14929–14934
45. Peth A, Nathan JA, Goldberg AL (2013) The ATP costs and time required to degrade ubiquitinated proteins by the 26 S proteasome. J Biol Chem 288:29215–29222
46. Lanzetta PA, Alvarez LJ, Reinach PS, Candia OA (1979) An improved assay for nanomole amounts of inorganic phosphate. Anal Biochem 100:95–97
47. Saeki Y, Isono E, Toh EA (2005) Preparation of ubiquitinated substrates by the PY motif-insertion method for monitoring 26S proteasome activity. Methods Enzymol 399:215–227
48. Bhattacharyya S, Renn JP, Yu H, Marko JF, Matouschek A (2016) An assay for 26S proteasome activity based on fluorescence anisotropy measurements of dye-labeled protein substrates. Anal Biochem 509:50–59
49. Chung CH, Goldberg AL (1981) The product of the lon (capR) gene in Escherichia coli is the ATP-dependent protease, protease La. Proc Natl Acad Sci U S A 78:4931–4935
50. Saeki Y, Tayama Y, Toh-e A, Yokosawa H (2004) Definitive evidence for Ufd2-catalyzed elongation of the ubiquitin chain through Lys48 linkage. Biochem Biophys Res Commun 320:840–845

Chapter 20

Exploring the Regulation of Proteasome Function by Subunit Phosphorylation

Jordan J. S. VerPlank and Alfred L. Goldberg

Abstract

Rates of degradation by the ubiquitin proteasome system depend not only on rates of ubiquitination, but also on the level of proteasome activity which can be regulated through phosphorylation of proteasome subunits. Many protein kinases have been proposed to influence proteasomal activity. However, for only two is there strong evidence that phosphorylation of a specific 26S subunit enhances the proteasome's capacity to degrade ubiquitinated proteins and promotes protein breakdown in cells: (1) protein kinase A (PKA), which after a rise in cAMP phosphorylates the 19S subunit Rpn6, and (2) dual tyrosine receptor kinase 2 (DYRK2), which during S through M phases of the cell cycle phosphorylates the 19S ATPase subunit Rpt3. In this chapter, we review and discuss the different methods used to assess the impact of phosphorylation by these two kinases on proteasomal activity and intracellular protein degradation. In addition, we present one method to determine if phosphorylation is responsible for an observed increase in proteasomal activity and another to evaluate by Phos-tag gel electrophoresis whether a specific proteasome subunit is modified by phosphorylation. The methods reviewed and presented here should be useful in clarifying the roles of other kinases and other posttranslational modifications of proteasome subunits.

Key words Proteasome phosphorylation, Protein degradation, Protein kinase, Proteasome activation, Protein homeostasis, Protein kinase A, DYRK2, Ubiquitin

1 Introduction

Phosphorylation of 26S proteasome subunits is a newly appreciated mechanism for regulating protein degradation by the proteasome [1]. Although more than 300 phosphorylation sites in proteasome subunits have been detected [2], for nearly all the functional consequences of these modifications (if there are any) and the responsible kinases are unknown. While many kinases have been proposed to regulate proteasome activity, for only two are there well established sites of phosphorylation with clear effects on proteasome function and protein degradation: protein kinase A (PKA) [3] and dual receptor tyrosine kinase 2 (DYRK2) [4]. Pharmacological agents and physiological stimuli that raise cAMP and activate PKA enhance proteasome activity by phosphorylating the 19S subunit

Thibault Mayor and Gary Kleiger (eds.), *The Ubiquitin Proteasome System: Methods and Protocols*, Methods in Molecular Biology, vol. 1844, https://doi.org/10.1007/978-1-4939-8706-1_20,

Rpn6 at serine 14 [3]. During S through M phases of the cell cycle, DYRK2 activates the proteasome by phosphorylating the 19S ATPase subunit Rpt3 at threonine 25, which promotes cell proliferation [4]. However, there is also evidence suggesting that phosphorylation of proteasome subunits occurs after a rise in cGMP [5, 6], following calcium influx following synaptic depolarization [7], and during apoptosis [8].

Our understanding of proteasome regulation by PKA and by DYRK2 originated through quite different studies. Several groups had suggested effects of cAMP on proteasome activity [9, 10], but the systematic studies of Lokireddy et al. [3], using pharmacological agents affecting adenylate cyclase and phosphodiesterase 4 (PDE4), demonstrated that cAMP via PKA enhances multiple proteasome activities. This work and related studies in mice by Myeku et al. [11] demonstrated that these agents promote the degradation of mutant proteins that cause neurodegenerative diseases. In this particular case, the initial motivation for the work was to understand proteasome regulation and also mechanisms of proteotoxic diseases. The discovery of the role of DYRK2 came from an analysis by Guo et al. of proteins phosphorylated during different stages of the cell cycle [4]. The authors found in several published proteomic studies that phosphorylation of Rpt3 on T25 increases during S through M phases. Therefore, they screened a library of all human kinases and found DYRK2 could catalyze this modification. Further experiments showed that phosphorylation by DYRK2 enhanced proteasome function and protein breakdown during these phases of the cell cycle [4].

The investigators of both studies used similar approaches to analyze how subunit phosphorylation influenced proteasomal activities and intracellular protein degradation. First, 26S proteasomes were purified, and the particles' multiple enzymatic activities were assayed. Mass spectrometry was performed on the proteasomes to identify the sites of phosphorylation, and phosphosite-specific antibodies were generated to confirm modification of the sites when the particles were activated. Finally, site-directed mutagenesis determined whether the identified phosphorylation sites were in fact necessary and/or sufficient for activation of the proteasome. Measurement of overall rates of protein breakdown by pulse-chase labeling of cell proteins demonstrated that these 26S modifications have clear effects on protein half-lives.

Initial studies of proteasome activation by PKA and DYRK2 indicated that both stimulated the particles' chymotrypsin-like activity in crude cell lysates [3, 4]. For cAMP and PKA, this stimulation was studied in different cell types using pharmacological agents (e.g., forskolin and the PDE4 inhibitor rolipram) and subsequently with neurotransmitters and hormones (e.g., epinephrine and glucagon) that raise cAMP levels [3, 12]. Affinity purification of 26S proteasomes by the UBL method of Besche and Goldberg

[13] (*see* also Chapter 18) and subsequent measurement of its multiple enzymatic activities (*see* also Chapter 19) revealed an increase in the proteasome's capacity to hydrolyze polyubiquitinated proteins, ATP, and short peptides [3]. The UBL purification method [13] has proven particularly useful for such studies, because it can be used in all cells and tissues without prior genetic manipulations. For example, this approach was used to show proteasome impairment in the nervous system of mouse models of frontotemporal dementia (FTD) [11] and Charcot-Marie-Tooth 1B peripheral neuropathy [14] and that treatment of the FTD mice with a PDE4 inhibitor could reduce the impairment of proteasome function and the accumulation of the mutant tau in the brain [11].

To provide stronger evidence that phosphorylation was responsible for the PKA- or DYRK2-dependent increase in proteasomal activities, phosphorylation was prevented by inhibiting PKA in cells with the kinase inhibitor H89 [3] or by knocking down DYRK2 with shRNA [4]. These treatments reduced proteasomal peptidase activity, presumably by preventing its modification by PKA or DYRK2. To further test if phosphorylation was responsible for the cAMP-mediated increase in proteasome activity, the purified 26S proteasomes were incubated with protein phosphatase 1 (PP1), which reduced the peptidase and ATPase activities toward control levels [3].

To identify the critical site of phosphorylation by PKA, 26S proteasomes were purified and analyzed by mass spectrometry. These studies suggested that Rpn6-Serine14 was phosphorylated when cAMP rose in cells [3]. The PKA-mediated phosphorylation of Rpn6 was verified with Phos-tag SDS-PAGE using the method described below [15]. Of particular value was the subsequent production of an anti-phosphopeptide rabbit polyclonal antibody that recognizes specifically the phosphorylated Rpn6-S14 (Fig. 1) [12]. This antibody has proven very useful for the rapid and reliable demonstration that Rpn6-S14 is phosphorylated in different cells and tissues in various physiological states [12]. An antibody recognizing the phosphorylated Rpt3-Thr25 was generated by Guo et al. [4] and was used in a screen of cells overexpressing human kinases. This screen identified DYRK2 as the kinase that is likely to phosphorylate Rpt3 during S through M phases of the cell cycle [4].

Identifying by mass spectrometry phosphorylation sites that regulate proteasome function has proven quite difficult for reasons that are not entirely clear. In fact, several proteasomal sites reported to be phosphorylated by kinases could not be confirmed in later studies [3, 8, 10, 16, 17], even though changes in proteasome activity were reproduced. Detecting a critical phosphorylation site on the proteasome may be complicated because a kinase may modify hundreds of cellular targets that could also co-purify with proteasomes as contaminating phosphorylated proteins. Also, in addition to the more than 30 unique subunits that comprise the

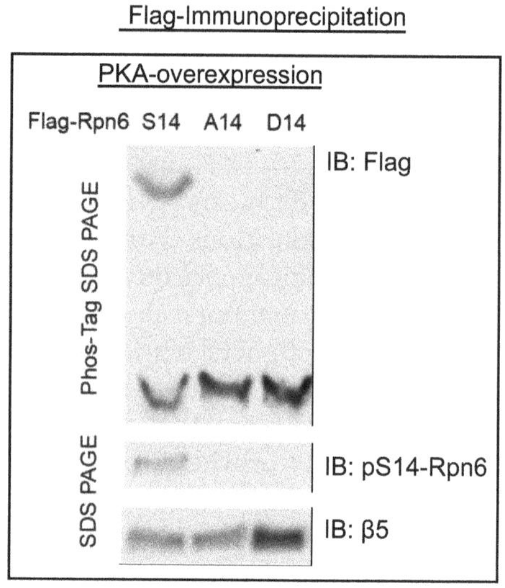

Fig. 1 26S proteasomes are phosphorylated on Rpn6 by PKA, but not when "phosphomimetic" or "phosphodead" mutations at serine 14 of Rpn6 are overexpressed in HEK293 cells. Cells were co-transfected with 0.5 μg/mL PKA and FLAG-Rpn6-S14, FLAG-Rpn6-A14, or FLAG-Rpn6-D14 using the FuGENE HD transfection reagent. After 24 h, 26S proteasomes were purified by the FLAG-tag, and Phos-tag SDS-PAGE and Western blot analyses were performed for FLAG. Additionally, SDS-PAGE and Western blot analysis were performed with our antibody that specifically detects Rpn6 phosphorylated at serine 14, along with the 20S subunit β5 to detect the number of proteasomes isolated in each preparation

proteasome, several deubiquitinases, ubiquitin ligases, shuttling factors, and substrates associate in sub-stochiometric amounts with the proteasome and co-purify with it [18]. Consequently, proteasome preparations purified from HEK293 cells after kinase overexpression contain many p-Ser bands detected by Western blot (Fig. 2). Furthermore, phosphatases and kinases, including PKA, co-purify with the 26S proteasome [3] and may remove or add phosphate groups during particle isolation and analysis. Therefore, it is important in such studies to verify a site of phosphorylation by multiple approaches.

To conclusively demonstrate the biochemical importance of Rpn6-S14 phosphorylation, site-directed mutagenesis was used to generate "phosphodead" (serine to alanine) and "phosphomimetic" (serine to aspartic acid) mutations at serine 14 of human Rpn6 [3]. The DNA constructs encoding the FLAG-tagged Rpn6 with the "phosphodead," "phosphomimetic," or no mutation were transiently over-expressed in HEK293 cells. The proteasomes were then purified using the FLAG-tag (*see* Chapter 18). The phosphomimetic mutation stimulated the capacity of the 26S proteasomes to hydrolyze ATP and small peptides, and the "phosphodead"

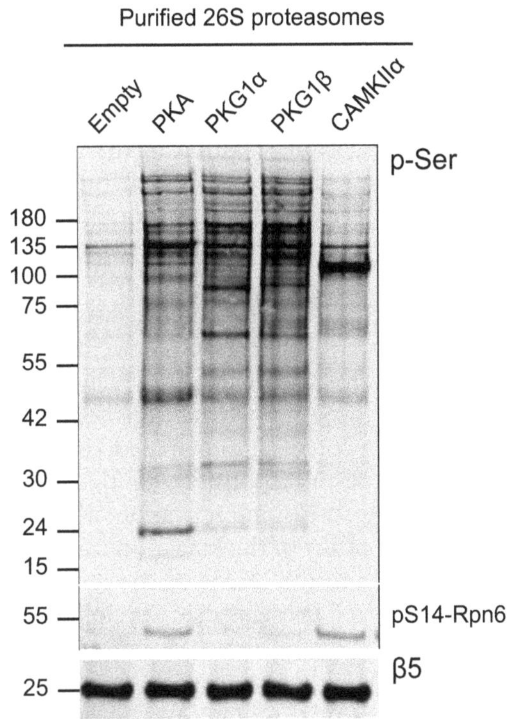

Fig. 2 26S proteasome preparations purified from HEK293 cells after overexpression of different kinases contain many phospho-serine positive bands by Western blot analysis. The pattern of phosphoprotein bands differs with each kinase. HEK293A cells stably overexpressing Rpn11 with a high-affinity biotin tag (*see* Chapter 18) were transfected with 0.5 μg/mL PKA catalytic subunit, PKG1α, PKG1β, or CAMKIIα, each of which has been reported to phosphorylate proteasome subunits [3, 5, 7]. After 24 h, 26S proteasomes were purified by the biotin tag, and SDS-PAGE and Western blot analysis were performed for phosphorylated serine residues, phosphorylated-Rpn6-S14, and 20S proteasome subunit β5

mutation repressed these activities [3]. Moreover, overexpression of the "phosphomimetic" mutation in cells, like treatments that raise cAMP, stimulated the degradation of aggregation-prone mutant proteins, and overexpressing the phosphodead mutation decreased it [3]. Site-directed mutagenesis was also used to test the importance of the modification of Rpt3-T25 by DYRK2 except that CRISPR-Cas9 gene editing was used to generate the site-specific "phosphodead" and "phosphomimetic" mutants. Most importantly, the phosphodead modification was found to reduce proteasome activity and to slow cell proliferation [4].

Another major challenge in such studies is to determine whether a posttranslational modification of a proteasome subunit that changes its activity in vitro actually leads to detectable effects on intracellular protein breakdown. For PKA and DYRK2, a pulse-chase isotopic labeling of cell proteins was used to measure the overall rates of breakdown of long-lived and short-lived fractions of cell proteins [19, 20]. These experiments were performed similarly to the method described in Chapter 17. In both studies, phosphorylation of the 26S was shown to accelerate degradation of proteins in the cell. Surprisingly proteasome phosphorylation by DYRK2 or PKA, though seeming to activate proteasomes similarly, potentially stimulated the degradation of a different class of proteins by the UPS. cAMP and PKA increased the degradation only of short-lived proteins, which are misfolded and regulatory proteins [3], while DYRK2 seemingly enhanced the breakdown of long-lived proteins [4], which are the bulk of cellular proteins.

An important goal of future studies will be to determine which proteins are degraded faster in each case and whether the mechanism responsible for this accelerated proteolysis is solely through proteasome activation or also involves some unknown concomitant effect of the kinase on ubiquitination of certain proteins [21]. PKA and DYRK2 were shown to promote the degradation of individual cell proteins using cycloheximide to block translation and Western blot to follow disappearance of the protein. Using this approach, cAMP was shown to stimulate the intracellular degradation of several mutant and aggregation-prone proteins that cause neurodegenerative diseases [3, 11]. Similarly, DYRK2 was shown to promote the degradation of certain cell cycle regulatory proteins [4]. The use of cycloheximide in such studies serves another important function: to ensure that the observed changes in the level of the plasmid-encoded proteins were due to degradation and not to a different kinase-dependent mechanism. For example, PKA (and other kinases) can alter plasmid number and protein levels through its ability to stimulate gene transcription and translation.

This chapter describes two methods to confirm phosphorylation by PKA on serine 14 of Rpn6, but similar approaches can be applied to studies of other kinases and proteasome subunits. These approaches could also be useful to study the many other reported posttranslational modifications of the proteasome (e.g., O-GlcNAC [22], ADP-ribosylation [23], ubiquitination [24, 25]), whose biological significance remains unclear.

2 Materials

1. Affinity purification buffer (APB): 25 mM Hepes-KOH (pH 7.5), 150 mM NaCl, 5 mM MgCl 1 mM ATP, and 1 mM DTT.
2. 1 M sodium fluoride (NaF).
3. 1 M β-glycerophosphate.
4. Lambda phosphatase (NEB).

5. Proteasome activity buffer: 50 mM Tris-HCl (pH 7.5), 5 mM $MgCl_2$, 1 mM ATP, 1 mM DTT, and 0.05 mg/mL BSA.
6. 1 M $MnCl_2$.
7. Mouse anti-phosphorylated serine antibody (Santa Cruz, sc-81514).
8. 4× SDS-PAGE sample buffer with 10% beta-mercaptoethanol.
9. Precast Zn^{2+} Phos-tag polyacrylamide gels (WAKO).
10. Precast Bis-Tris polyacrylamide gels.
11. Transfer buffer: 25 mM Tris, 192 mM glycine, pH 8.0, 10% methanol (v/v).
12. 250 mM EDTA.
13. PVDF membrane.
14. Odyssey blocking buffer (LiCor).
15. Rabbit anti-Rpn6 antibody (Cell Signaling, #14303).
16. 1× TBS: 50 mM Tris (pH 7.5) and 150 mM NaCl.
17. 1× TBS/Tween: 50 mM Tris (pH 7.5), 150 mM NaCl, and 0.1% Tween.
18. 5% BSA in 1× TBS/Tween.
19. 50 mM rolipram.
20. 50 mM forskolin.
21. 20× MOPS running buffer.
22. PBS (pH 7.4): 137 mM NaCl, 2.7 mM KCl, 10 mM Na_2HPO_4, and 1.8 mM KH_2PO_4.
23. Dulbecco's Modified Eagle Medium (DMEM) supplemented with 10% FBS.

3 Methods

The procedures described below were used to demonstrate that PKA can induce an increase of proteasome activity by phosphorylation. This protocol can be adapted for other kinases.

3.1 Measuring PKA-Mediated Stimulation of 26S Proteasome Activities

1. To raise cAMP levels, treat HEK293 cells for 6 h with either 50 μM rolipram to raise cAMP by blocking its breakdown by phosphodiesterase 4 (PDE4) or 50 μM forskolin to raise cAMP by stimulating its synthesis by adenylyl cyclases (*see* **Notes 1** and **2**) [3].
2. To isolate 26S proteasomes from treated and control cells, wash the cells in cold PBS and then collect by scraping into cold PBS and centrifuging at 100 × *g* for 5 min at 4 °C. Decant the PBS, and resuspend in 250 μL/100 mm plate of APB

including phosphatase inhibitors: 10 mM NaF and 25 mM β-glycerophosphate (*see* **Note 3**). Lyse the cells by sonication (three 10 s bursts at 10 μm), and purify 26S proteasomes using the UBL method as described in Chapter 18.

3. To characterize the enzymatic properties of the purified proteasomes, measure the hydrolysis of fluorogenic peptides, ATP, and polyubiquitinated proteins as described in Chapter 19.
4. To determine whether the enhanced proteasome ATPase and peptidase activities are due to phosphorylation, incubate 10 nM 26S proteasomes with 400 units lambda phosphatase for 30 min at 30 °C in a 10 μL volume in proteasome assay buffer supplemented with 1 mM $MnCl_2$ (*see* **Note 4**). The phosphatase will hydrolyze phosphorylated serine and threonine residues. The proteasomes can then be diluted in proteasome assay buffer to the desired concentration for measurement of peptidase activity (we commonly use 1 nM). Comparison of the activity before and after dephosphorylation will indicate whether the stimulation is caused by kinase action. It is also advisable to verify the extent of dephosphorylation by performing Western blot analysis with a phospho-serine-specific antibody.

3.2 Phos-tag SDS-PAGE to Confirm Phosphorylation of a Specific Subunit

1. Treat cells with agents to raise cAMP levels and purify proteasomes as described above.
2. Add 4× SDS-PAGE sample buffer to the purified proteasomes (*see* **Note 5**) to a final concentration of 1×, and incubate at 95°C for 5 min.
3. Load 1–2 μg 26S proteasome preparation into precast Zn^{2+} Phos-tag polyacrylamide gels (7%). Run SDS-PAGE in MOPS running buffer at 155 V until the dye front runs off the bottom of the gel (approximately 2 h). The same samples should also be run in a precast Bis-Tris polyacrylamide gel, to compare the amount of phosphorylated Rpn6 to the total amount of Rpn6.
4. Incubate the Phos-tag gel with gentle shaking for 30 min in transfer buffer plus 1 mM EDTA to chelate the zinc ions from the gel. Replace the buffer every 5 min.
5. Incubate the Phos-tag gel with gentle shaking for 5 min in transfer buffer without EDTA.
6. Transfer proteins in transfer buffer without EDTA to PVDF at 75 V for 1 h, or 15 V overnight, at 4°C.
7. Incubate PVDF membranes in 5 % BSA/TBS or Odyssey blocking solution for 1 h at room temperature, followed by an overnight incubation with a Rpn6 antibody diluted in 3% BSA/TBST (rabbit anti-Rpn6, Cell Signaling Technology).
8. Wash PVDF membranes in TBST three times for 5 min each.

9. Incubate for 45 min at room temperature in secondary antibody (goat anti-rabbit, 800) diluted 1:10,000 in 5% BSA/TBST or Odyssey blocking solution.
10. Wash PVDF membranes in TBST three times for 5 min each.
11. Wash in TBS once for 5 min.
12. Image membranes on Odyssey (*see* **Note 6**). To quantify the percent of Rpn6 phosphorylated, normalize the amount of the up-shifted Rpn6 on the Phos-tag gel to the total amount of Rpn6 detected on the Bis-Tris gel.

4 Notes

1. Many pharmacological agents are known to raise cAMP and activate PKA. These small molecules are useful tools to investigate how raising cAMP stimulates proteasomal activities and intracellular degradation of some proteins. Other signaling systems may not offer this advantage.
2. 6 h treatments were used by Lokireddy et al. [3], but enhanced proteasomal activities can also be seen with shorter treatment times. Additionally, the cAMP-mediated increase in proteasome activity is not limited to HEK293 cells and has now been observed in many cell lines and mouse primary cell cultures.
3. NaF and β-glycerophosphate, which inhibit Ser/Thr phosphatases, can also be added to the reaction buffers used for measuring diverse proteasome activities (as described in Chapter 19). The inclusion of these inhibitors during the assay does not reduce proteasome peptidase activities and can even result in a greater measured stimulation of peptidase activity. 50 nM calyculin A, an inhibitor of the PP1 and PP2A phosphatases, can also be added to lysis and reaction buffers [4].
4. The PP1 phosphatase, another manganese-dependent protein phosphatase, can also be used [3].
5. Phos-tag SDS-PAGE analysis can also be performed on the cell lysates. In our experience, the phosphorylation-dependent shift is less detectable in lysates than with purified 26S proteasomes.
6. We use an Odyssey CLX imager to visualize Western blots, but chemiluminescent and film methods can also be used.

Acknowledgments

The methods described here were developed through research support to our laboratory from the NIH-NIGMS (R01 GM51923), Cure Alzheimer's Fund, Muscular Dystrophy Association (MDA-419143), and Project A.L.S (2015-06).

References

1. VerPlank JJS, Goldberg AL (2017) Regulating protein breakdown through proteasome phosphorylation. Biochem J 474(19):3355–3371. https://doi.org/10.1042/bcj20160809
2. Guo X, Huang X, Chen MJ (2017) Reversible phosphorylation of the 26S proteasome. Protein Cell 8(4):255–272. https://doi.org/10.1007/s13238-017-0382-x
3. Lokireddy S, Kukushkin NV, Goldberg AL (2015) cAMP-induced phosphorylation of 26S proteasomes on Rpn6/PSMD11 enhances their activity and the degradation of misfolded proteins. Proc Natl Acad Sci U S A 112(52): E7176–E7185. https://doi.org/10.1073/pnas.1522332112
4. Guo X, Wang X, Wang Z, Banerjee S, Yang J, Huang L, Dixon JE (2016) Site-specific proteasome phosphorylation controls cell proliferation and tumorigenesis. Nat Cell Biol 18 (2):202–212. https://doi.org/10.1038/ncb3289
5. Ranek MJ, Terpstra EJ, Li J, Kass DA, Wang X (2013) Protein kinase g positively regulates proteasome-mediated degradation of misfolded proteins. Circulation 128(4):365–376. https://doi.org/10.1161/circulationaha.113.001971
6. Ranek MJ, Kost CK Jr, Hu C, Martin DS, Wang X (2014) Muscarinic 2 receptors modulate cardiac proteasome function in a protein kinase G-dependent manner. J Mol Cell Cardiol 69:43–51. https://doi.org/10.1016/j.yjmcc.2014.01.017
7. Djakovic SN, Schwarz LA, Barylko B, DeMartino GN, Patrick GN (2009) Regulation of the proteasome by neuronal activity and calcium/calmodulin-dependent protein kinase II. J Biol Chem 284(39):26655–26665
8. Um JW, Im E, Park J, Oh Y, Min B, Lee HJ, Yoon JB, Chung KC (2010) ASK1 negatively regulates the 26 S proteasome. J Biol Chem 285(47):36434–36446. https://doi.org/10.1074/jbc.M110.133777
9. Asai M, Tsukamoto O, Minamino T, Asanuma H, Fujita M, Asano Y, Takahama H, Sasaki H, Higo S, Asakura M, Takashima S, Hori M, Kitakaze M (2009) PKA rapidly enhances proteasome assembly and activity in in vivo canine hearts. J Mol Cell Cardiol 46 (4):452–462. https://doi.org/10.1016/j.yjmcc.2008.11.001
10. Zhang F, Hu Y, Huang P, Toleman CA, Paterson AJ, Kudlow JE (2007) Proteasome function is regulated by cyclic AMP-dependent protein kinase through phosphorylation of Rpt6. J Biol Chem 282(31):22460–22471. https://doi.org/10.1074/jbc.M702439200
11. Myeku N, Clelland CL, Emrani S, Kukushkin NV, Yu WH, Goldberg AL, Duff KE (2016) Tau-driven 26S proteasome impairment and cognitive dysfunction can be prevented early in disease by activating cAMP-PKA signaling. Nat Med 22(1):46–53. https://doi.org/10.1038/nm.4011
12. Lokireddy S, VerPlank JJS, Zhao J, Davogusotto G, Parker B, James D, Richter E, Taegetmeyer H, Goldberg AL. Hormones, exercise, and fasting activate 26S proteasome function via cAMP-PKA pathway. Manuscript in revision
13. Besche HC, Goldberg AL (2012) Affinity purification of mammalian 26S proteasomes using an ubiquitin-like domain. Methods Mol Biol 832:423–432. https://doi.org/10.1007/978-1-61779-474-2_29
14. VerPlank JJS, Lokireddy S, Feltri ML, Goldberg AL, Wrabetz L (2018) Impairment of protein degradation and proteasome function in hereditary neuropathies. Glia 66 (2):379–395. https://doi.org/10.1002/glia.23251
15. Kinoshita E, Kinoshita-Kikuta E, Koike T (2009) Separation and detection of large phosphoproteins using Phos-tag SDS-PAGE. Nat Protoc 4(10):1513–1521. https://doi.org/10.1038/nprot.2009.154
16. Lee SH, Park Y, Yoon SK, Yoon JB (2010) Osmotic stress inhibits proteasome by p38 MAPK-dependent phosphorylation. J Biol Chem 285(53):41280–41289. https://doi.org/10.1074/jbc.M110.182188
17. Leestemaker Y, de Jong A, Witting KF, Penning R, Schuurman K, Rodenko B, Zaal EA, van de Kooij B, Laufer S, Heck AJR, Borst J, Scheper W, Berkers CR, Ovaa H (2017) Proteasome activation by small molecules. Cell Chem Biol 24(6):725–736.e7. https://doi.org/10.1016/j.chembiol.2017.05.010
18. Collins GA, Goldberg AL (2017) The logic of the 26S proteasome. Cell 169(5):792–806. https://doi.org/10.1016/j.cell.2017.04.023
19. Zhao J, Zhai B, Gygi SP, Goldberg AL (2015) mTOR inhibition activates overall protein degradation by the ubiquitin proteasome system as well as by autophagy. Proc Natl Acad Sci U S A 112(52):15790–15797. https://doi.org/10.1073/pnas.1521919112
20. Zhao J, Brault JJ, Schild A, Cao P, Sandri M, Schiaffino S, Lecker SH, Goldberg AL (2007)

FoxO3 coordinately activates protein degradation by the autophagic/lysosomal and proteasomal pathways in atrophying muscle cells. Cell Metab 6(6):472–483. https://doi.org/10.1016/j.cmet.2007.11.004

21. Filipcik P, Curry JR, Mace PD (2017) When worlds collide-mechanisms at the Interface between phosphorylation and ubiquitination. J Mol Biol 429(8):1097–1113. https://doi.org/10.1016/j.jmb.2017.02.011
22. Zhang F, Su K, Yang X, Bowe DB, Paterson AJ, Kudlow JE (2003) O-GlcNAc modification is an endogenous inhibitor of the proteasome. Cell 115(6):715–725
23. Cho-Park PF, Steller H (2013) Proteasome regulation by ADP-ribosylation. Cell 153(3):614–627. https://doi.org/10.1016/j.cell.2013.03.040
24. Besche HC, Sha Z, Kukushkin NV, Peth A, Hock EM, Kim W, Gygi S, Gutierrez JA, Liao H, Dick L, Goldberg AL (2014) Autoubiquitination of the 26S proteasome on Rpn13 regulates breakdown of ubiquitin conjugates. EMBO J 33(10):1159–1176. https://doi.org/10.1002/embj.201386906
25. Marshall RS, McLoughlin F, Vierstra RD (2016) Autophagic turnover of inactive 26S proteasomes in yeast is directed by the ubiquitin receptor Cue5 and the Hsp42 chaperone. Cell Rep 16(6):1717–1732. https://doi.org/10.1016/j.celrep.2016.07.015

Chapter 21

Scalable In Vitro Proteasome Activity Assay

Amit Kumar Singh Gautam, Kirby Martinez-Fonts, and Andreas Matouschek

Abstract

We developed a degradation assay based on fluorescent protein substrates that are efficiently recognized, unfolded, translocated, and hydrolyzed by the proteasome. The substrates consist of three components: a proteasome-binding tag, a folded domain, and an initiation region. All the components of the model substrate can be changed to modulate degradation, and the assay can be performed in parallel in 384-well plates. These properties allow the assay to be used to explore a wide range of experimental conditions and to screen proteasome modulators.

Key words Proteasome, Ubiquitin-proteasome system, High-throughput degradation assay

1 Introduction

The proteasome adjusts the concentrations of thousands of proteins in the cytoplasm and nucleus of eukaryotic cells by hydrolyzing them into small peptides. The 26S proteasome is composed of a 20S proteolytic core particle sandwiched between one or two 19S regulatory particles [1, 2]. The 20S particle consists of four heptameric rings that are stacked on top of each other and form a central cavity. Three pairs of proteolytic sites are located on a surface facing the cavity near the center of the particle [3, 4]. Only unfolded proteins can reach the proteolytic sites because access is through 13 Å diameter pores at the ends of the particle, which can be opened by the interaction of the 20S core with the 19S regulatory particle [5, 6]. The 19S regulatory particle provides specificity to degradation by binding substrates and then unfolding and translocating them into the 20S core particle [7].

The proteasome regulates many cellular processes such as the cell cycle, transcription of some genes, and membrane trafficking [8–11]. It also generates peptides for antigen presentation and recycles amino acids [12, 13]. Hence, it is not surprising that defects in the proteasome machinery can result in cardiovascular

Thibault Mayor and Gary Kleiger (eds.), *The Ubiquitin Proteasome System: Methods and Protocols*, Methods in Molecular Biology, vol. 1844, https://doi.org/10.1007/978-1-4939-8706-1_21, © Springer Science+Business Media, LLC, part of Springer Nature 2018

and neurodegenerative diseases, cancer, and other pathologies [14–16]. Several therapeutic approaches target the proteasome [17, 18], and understanding the mechanism of substrate degradation by the proteasome at the molecular level would be helpful in expanding these. Thus a robust, high-throughput assay would be valuable. The following properties in this assay are desirable:

1. Assay reports on the activity of the 19S regulatory particle, as well as proteolysis, because it is based on substrates that are recognized by the 19S regulatory particle.
2. Assay allows degradation to be followed directly and continuously.
3. Assay is sensitive and reports on proteasome function quantitatively.
4. Assay is simple, robust, and reproducible.

A common proteasome assay is based on fluorogenic peptide substrates. Peptide cleavage by the proteasome releases a fluorophore, which can be detected by fluorescence spectroscopy [19, 20]. This assay is robust and convenient. However, the peptide substrates can side step the recognition and unfolding steps required for the degradation of proteins and reports primarily on the activity of the proteolytic sites and thus of the 20S core particle.

Other in vitro degradation assays investigate the 26S proteasome by employing substrate proteins synthesized by in vitro transcription and translation or purified from cells [21–24]. Degradation is monitored discontinuously by taking samples at different time points and measuring substrate abundance by SDS-PAGE and protein detection by autoradiography, in gel fluorescence imaging, or Western blotting [21, 23, 25–28]. This approach is quantitative, is robust, and can detect degradation intermediates and thus provide information about the processivity of the proteasome [29, 30]. It has provided fundamental understanding of proteasomal degradation. However, these assays are laborious and cannot easily be scaled up for multiplexing.

We and others have developed proteasome substrates based on fluorescent proteins that make it possible to monitor the activity of the 26S proteasome continuously [22, 31–35]. Other assays follow the anisotropy of a fluorescence signal of a dye covalently attached to a substrate protein [36]. The substrates can be targeted to the proteasome by artificial ubiquitin tags, ubiquitin-like domains, or polyubiquitin tags [22, 23, 31, 33, 36–38]. Their degradation reports on their recognition, unfolding, and translocation by the 19S regulatory particle into the proteolytic core particle. In this chapter, we will provide a detailed protocol for the determination of the degradation kinetics of purified proteasome substrates in three examples (Fig. 1a, b).

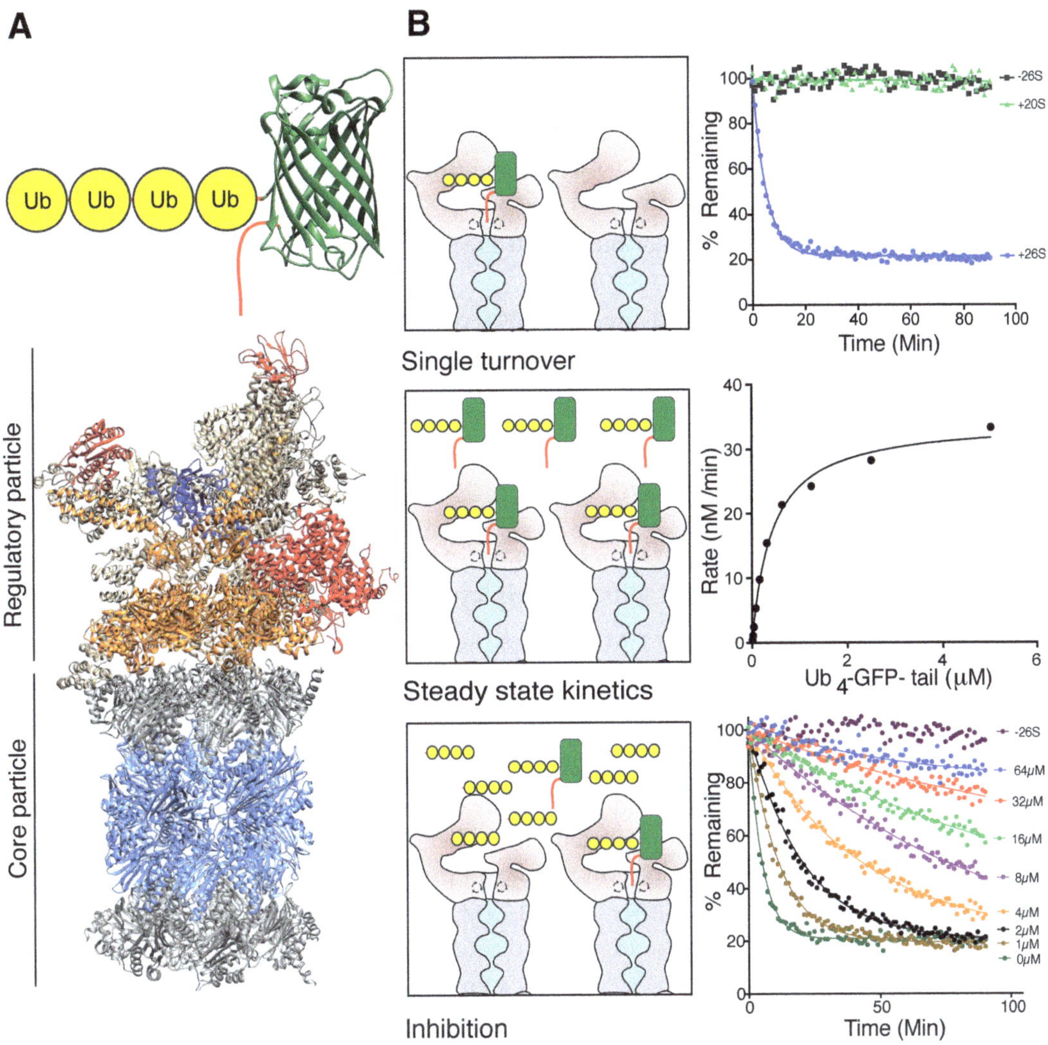

Fig. 1 An overview of a high-throughput proteasome degradation assay. (**a**) Top: Schematic representation of model substrate with three components, linear tetra ubiquitin (yellow) connected through a short linker, a folded domain (GFP), and a disordered tail derived from cytochrome b_2 (red). Bottom: 26S proteasome (3JOC) is composed of a 20S core particle and a 19S regulatory particle. The proteasome binds ubiquitinated proteins via a least three subunits (red); substrates are engaged and unfolded by ATPases (brown); ubiquitin chains are recycled via deubiquitinating enzyme (blue). (**b**) Top: For single-turnover kinetics, high ratio of proteasome to substrate concentration was used, and the intensity of GFP fluorescence was monitored over time. Middle: For steady-state kinetics, high substrate to proteasome concentration ratios were used, and the initial degradation rate was plotted against substrate concentration fit to the Michaelis-Menten equation. Bottom: Substrate was competed with different concentrations of Ub_4 for proteasome binding

2 Materials

2.1 Purification of Proteasome Substrate

1. Ni-NTA agarose (Qiagen).
2. 1M IPTG (*see* **Note 1**).
3. Protease Inhibitor Cocktail Set V (EDTA-free) (*see* **Note 2**).

4. NPI-10: 50 mM sodium phosphate (pH 8.0), 300 mM NaCl, and 10 mM imidazole (*see* **Note 3**).
5. NPI-25: 50 mM sodium phosphate (pH 8.0), 300 mM NaCl, and 25 mM imidazole.
6. NPI-250: 50 mM sodium phosphate (pH 8.0), 300 mM NaCl, and 250 mM imidazole.
7. Storage buffer: 50 mM Tris–HCl (pH 7.6), 1 mM DTT, and 5% glycerol.
8. 2× YT media: 1.6% tryptone, 1% yeast extract, and 0.5% NaCl.
9. Instruments: Centrifuge, homogenizer (EmulsiFlex-C3 from Avestin), rocker or rotor at 4 °C (cold room).

2.2 Affinity Purification of the Proteasome

Yeast strains to purify the 19S regulatory particle or 26S proteasome (YYS40: *MATa, leu2, trp1, ura3, his3, ade2, can1, rpn11::RPN11-3× FLAG::HIS3*) and the 20S proteasome (YYS37: *MATa, leu2, trp1, ura3, his3, ade2, can1, pre1::PRE1-3× FLAG::HIS3*) were kind gifts from Hideyoshi Yokosawa [39].

1. Anti-FLAG M2-agarose (Sigma).
2. 100 mM ATP (*see* **Note 4**).
3. 3× FLAG peptide (Sigma).
4. 50× ATP regeneration system (ARS): 500 mM creatine phosphate and 5 mg/mL creatine phosphokinase. 50× ATP regeneration system and other buffers containing ATP should be prepared on the day of proteasome purification.
5. 2× buffer A: 100 mM Tris–HCl (pH 7.5) and 20% glycerol. Can be made in advance and stored at 4 °C.
6. Buffer B: 2 mM ATP, 5 mM $MgCl_2$, and 1 mM DTT are prepared in 1× buffer A.
7. Buffer C: 4 mM ATP, 10 mM $MgCl_2$, 1 mM DTT, and 2× ARS are prepared in 1× buffer A.
8. Buffer D: 2 mM ATP, 5 mM $MgCl_2$, 1 mM DTT, and 500 mM NaCl are prepared in 1× buffer A.
9. Elution buffer: 0.15 mg/mL 3× FLAG peptide in buffer B.
10. Yeast nutrient-rich medium (YPD): 2% polypeptone, 1% yeast extract, and 2% glucose.
11. Instruments: Centrifuge, homogenizer (EmulsiFlex-C3 from Avestin), rocker or rotor at 4 °C (cold room).

2.3 High-Throughput In Vitro Degradation Assay

1. 25× ARS: 25 mM ATP, 250 mM creatine phosphate, and 2.5 mg/mL creatine phosphokinase.
2. 10× degradation buffer: 500 mM Tris–HCl (pH 7.5), 50 mM $MgCl_2$, and 10% glycerol.

3. Substrate buffer: 1× degradation buffer, 1 mg/mL BSA, and 1 mM DTT.
4. 2× Proteasome buffer: 1× degradation buffer, 1 mM DTT, 1 mg/mL BSA, and 2× ARS.

3 Methods

3.1 One-Step Purification of Proteasome Substrate

An efficient proteasome substrate has a two-part degron consisting of a proteasome-binding tag and an initiation region [23]. We have used different proteasome-binding tags and initiation regions of varying strength to explore proteasomal degradation kinetics [22, 31, 36, 38]. Here, we use a simple proteasome substrate consisting of four ubiquitin molecules connected by short linker (*see* below), a fluorescent protein domain, a disordered region, and finally a C-terminal 6XHis for purification (Ub_4(lin)-GFP-tail; Fig. 1a). The first three components of the substrate, which affect recognition and degradation by the proteasome, are modified individually to tune the system according to the scientific question asked:

1. Proteasome binding tag: We have tested various proteasome-binding tags including ubiquitin-like domains and polyubiquitin chains of different lengths and linkages (K48, K63, K11, and Ub_4(lin)) [22, 31, 33]. For simplicity, here we use a tag that mimics polyubiquitin chains but is encoded directly within the primary sequence of the substrate, which we call Ub_4(lin). It consists of four ubiquitin molecules in sequence connected to each other via short linkers (GSGGGG) and then to the N-terminus of the fluorescent protein. G76 of each ubiquitin moiety is mutated to V to prevent cleavage by C-terminal ubiquitin hydrolases. This binding tag has been widely adopted by us as well as by others groups in the field [23, 37, 40, 41].
2. Fluorescent protein: As a folded domain, we use a circular permutant of GFP, which is monomeric, folded into a compact domain, and readily unfolded by ATP-dependent proteases [22, 42]. We have also used monomeric fluorescent proteins of different stabilities such as superfolder GFP and YFP.
3. Initiation region: Effective degradation by the proteasome requires the presence of a disordered region in the substrate from which the proteasome can engage the substrate and initiate degradation. The properties of the initiation region can tune degradation rates over at least two orders of magnitude [31, 43]. Here, we use an initiation region derived from cytochrome b_2, which is easily recognized by the proteasome.

Cells expressing Ub_4(lin)-GFP-tail and the cell lysate will appear green due to the high level of GFP. It is a good indication

Table 1
Troubleshooting substrate purification

Problem	Probable reason	Tentative solution
Low or no growth in starter culture	Wrong selection marker	Cross-check the antibiotic marker and use an appropriate concentration
	Long storage of cells	Retransform the construct and use individual colony for starter culture
Low yield	IPTG induction related	Run a pilot experiment to test appropriate cell density, IPTG concentration, and induction time. Quality of IPTG also can affect yield and viability of cells
	Starter culture over-grown	Try growing starter culture for 8–10 h in 2XYT; using appropriate dilution is also crucial for the induction
	Low Ni-NTA binding	Check the pH of the lysate; it should be >7; avoid using high concentration of DTT and EDTA in purification buffer. If the beads are not blue (less bead associated Nickel), it can be easily charged (check provider's manual)
	Loss during washing	pH of the buffer should be >7 and limit the amount of imidazole in washing buffer
Non-fluorescence substrate	Protein is unfolded	Avoid denaturation condition like exposure to chemical denaturants and heat
	Aggregation	Modify expression and purification strategy; use GST or MBP tag for purification and cleave them afterward
Protein concentration not reliable	Nucleotide contamination	Add DNase/Benzonase and $MgCl_2$ to lysis buffer; incubate the lysate for 30 min on ice. Use DNA precipitant like polyethyleneimine or an ion exchange/heparin column to get rid of DNA
	High salts and contaminants	Desalt the purified protein prior to measuring protein concentration
Multiple fragment on SDS-PAGE	Protease contamination	Use the appropriate class and concentration of protease inhibitor. Perform all purification step at 4 °C
	Prolonged exposure to heat in SDS buffer	Some GFPs are sensitive to high heat in SDS buffer. Try running SDS-PAGE without boiling the purified protein in SDS buffer
	Alternate start site	Avoid multiple start codon after T7 promoter

that IPTG induction for protein expression has worked. Please *see* Table 1 for troubleshooting related to this section.

From a starter culture to pure proteasome substrate, the protocol requires 3 days. Days 1 and 2 are dedicated to expression (only requires ~1 h of active time). Day 3 is dedicated to purification; the entire purification can be achieved in 5–6 h.

1. Day 1: A single colony of *Escherichia coli* Rosetta (DE3) pLysS (Novagen) cells containing a plasmid encoding Ub_4(lin)-GFP-tail is inoculated in 20 mL of 2× YT medium with antibiotics and incubated at 37 °C (*see* **Note 5**).
2. Day 2: The starter culture is diluted 1:100 and grown in 2× YT medium containing antibiotics in a shaker incubator at 37 °C and 225 rpm.
3. Induce protein expression when cells reach an OD_{600} of 0.6 with 0.4 mM IPTG for 4 h.
4. Cells are collected by centrifugation at 6000 × *g* for 10 min, resuspended in NPI-10 and stored at −80 °C in an ultralow freezer.
5. Day 3: Before thawing the cells, add 1× Protease Inhibitor Cocktail Set V (EDTA-free) (*see* **Note 2**).
6. The cells are then thawed in a water bath at room temperature, and a homogeneous clump-free suspension is achieved by pipetting up and down several times.
7. Resuspended cells are lysed by passaging them twice through an EmulsiFlex-C3 (Avestin) homogenizer at 15,000 pound per square inch (psi).
8. Clear the lysate by two centrifugation steps at 30,000 × *g* for 20 min at 4 °C and by passing the lysate through a 0.4 μm filter.
9. During the centrifugation step, the Ni-NTA column is prepared by adding the 50% slurry in a PD-10 column (GE healthcare) (*see* **Note 6**).
10. Ni-NTA beads are washed with 10 column volumes (CV) of NPI-10 and nutated with clarified cell lysate at 4 °C for 1 h.
11. After collecting the unbound cell lysate, wash the beads with 20 CV of NPI-10 followed by 15 CV of NPI-25.
12. Elute the protein with 10 CV of NPI-250; collect a 5 μL sample to check substrate purity by SDS-PAGE.
13. Concentrate and buffer exchange the purified protein using a 10 kDa molecular weight cutoff Amicon Ultra centrifugation filter in storage buffer.
14. Protein concentration is estimated by measuring sample OD at 280 nm using an extinction coefficient of 26,485 M^{-1} cm^{-1}. Snap freeze the purified substrate in small aliquots and store them at −80 °C.
15. The purity of each batch of substrate is analyzed by SDS-PAGE.

3.2 Proteasome Purification

The eukaryotic 26S proteasome has been purified from various sources such as yeast, plants and rabbit reticulocytes, animal tissue, and mammalian cell cultures [44–49]. Classical tag-free proteasome purification methods rely on the acidic nature and large size of the proteasome but are laborious [44, 46]. The affinity of ubiquitin-like (UbL) domain to the proteasome simplifies purification and has been used to pull down the proteasome and associated proteins [50]. However, most currently used protocols rely on affinity tags, which have made proteasome purification relatively easy. The protein A tag [51, 52] and FLAG tags [39, 53] have been fused to the proteasomal subunits *RPN11* in the lid, *RPN1* in the base, and *PRE1* in the 20S particle. These tags are generally incorporated by replacing the chromosomal copy and have been shown to purify the entire 26S complex [39]. The standard purification protocols involve washes at low salt, which allow some proteasome-interacting proteins (PIPs) to remain associated and to co-purify with the 26S particle [51]. PIPs can be removed by high salt washes, which also lead to the dissociation of the 19S regulatory particle from the 20S core particle. These particles have then to be purified separately using tags attached to specific subunits in them. A FLAG tag on a proteasome subunit in the lid (Rpn11-FLAG) or in the base (Rpn1-FLAG) allows isolation of the 19S regulatory particle, and attachment of a tag to the α-subunit Pre1 allows isolation of the 20S core particle. We have purified and tested 26S proteasomes in up to 150 mM NaCl.

It takes around 24 h to grow yeast culture to the desired density. Cell lysis and purification of proteasomes can be completed in 6–8 h. This method typically yields 0.5–0.8 mg proteasome/liter of culture.

3.2.1 Affinity Purification of the 26S Proteasome

1. Day 1: Inoculate an isolated colony of YYS40 in 10 mL YPD media in 125 mL conical flask, and incubate overnight at 30 °C with 220 rpm shaking (*see* **Note 7**).
2. Day 2: Dilute the starter culture to 1: 1000 with YPD media and incubate at 30 °C and 220 rpm.
3. At OD_{600} ~2 (typically after 24 to 30 h of growth), cells are harvested by centrifugation at 3000 × *g* for 5 min.
4. Cell pellet is washed with cold water followed by cold buffer B.
5. After centrifugation, the cell pellet is stored at −80 °C until further use.
6. Resuspend the yeast cell pellet in cold buffer C (typically ~1–2 mL buffer C/g of cell pellet).
7. Cells resuspended in buffer C are passed twice through an EmulsiFlex-C3 (Avestin) homogenizer at 25,000–30,000 psi (*see* **Note 8**).
8. Collect cell debris by centrifuging for 30 min at 30,000 × *g* at 4 °C.

9. During centrifugation, transfer the anti-FLAG M2-agarose beads in a PD-10 column and wash with 5 CV of buffer C (for 2 L yeast culture, 2 mL of beads were used; *see* **Note 9**).
10. The cell lysate is supplemented with 5 mM ATP and 1× ARS and filtered through a 0.45 μm filter (*see* **Note 10**).
11. Incubate the cell lysate with prewashed anti-FLAG M2-agarose beads and agitate at 4 °C for 2 h.
12. Pass cell lysate and beads through the PD-10 column.
13. In order to capture maximum amount of proteasomes, the flow through is passed thought the anti-FLAG M2-agarose beads a second time.
14. Wash the proteasome-bound anti-FLAG M2-agarose twice with 30 CV of buffer B.
15. Quickly spin the PD-10 containing proteasomes at 500 × *g* for 5 s using a swinging bucket rotor to get rid of excess buffer B.
16. Incubate the beads with 500 μL of elution buffer for 15 min at room temperature (*see* **Note 11**).
17. Elute proteasomes in T15 tube by centrifugation in a swinging bucket rotor for 1 min.
18. Incubate beads with another 250 μL elution buffer, and repeat the elution process as described above.
19. Snap freeze small aliquots of purified proteasomes and store at −80 °C (*see* **Notes 12** and **13**).
20. The concentration of the proteasome is estimated using a BCA assay and converted into molarity using the molecular weights of 900 kDa for the 19S particle, 700 kDa for the 20S particle, and 2500 kDa for the 26S proteasome.

3.2.2 Affinity Purification of 19S or 20S Proteasomes

To purify 20S proteasomes, use the strain YYS37, and to purify 19S regulatory particles, use the strain YYS40.

1. Follow **steps 1–14** from Subheading 3.2.1.
2. To disrupt the 19S and 20S interaction, beads are washed with 10 CV of buffer D.
3. Agitate the beads with 10 CV of buffer D for 1 h at 4 °C.
4. After collecting the unbound fraction, the beads containing 20S proteasomes or 19S regulatory particles are further washed twice with 15 CV buffer D followed by 30 CV of buffer C.
5. Pure 20S proteasomes or 19 S regulatory complexes are eluted as described above (*see* Subheading 3.2.1, **steps 15–19**).
6. 26S proteasome can be assembled by combining the 19S and 20S particles at a molar ratio of 2:1 and incubating at 30 °C for 30 min. The assembled proteasome can be used directly for the degradation assay.

3.2.3 Assessment of Purified Proteasomes

Since the proteasome is a multienzyme complex, it is necessary to assess every batch of purified proteasomes. We list here several common methods to characterize purified proteasomes from any source. Every batch of purified proteasomes should be resolved on SDS-PAGE, and the presence of all proteasomal subunits is verified based on their molecular weights [39, 52]. Fluorogenic substrates like LLVY-AMC can be used not only to monitor the purification steps but also to characterize purified proteasomes. In-gel activity assays using native PAGE allow to resolve intact complexes to assess their proteolytic activity [54]. The ability of the 26S and 20S proteasome to degrade unstructured proteins, such as casein, and to hydrolyze ATP can also be used for simple evaluations of purified proteasomes [55, 56].

3.3 High-Throughput In Vitro Degradation Assay Platform

Substrate degradation kinetics by the proteasome can be analyzed by two standard kinetic approaches: steady-state or multiple-turnover conditions, where substrate is present in excess over enzyme, and in single-turnover conditions, where enzyme is present in excess over substrate. Both these approaches are informative in the characterization of proteasomal degradation.

We provide below streamlined protocols for the analysis of single and multiple-turnover reactions and for competition assay to characterize substrate binding using purified yeast proteasome and the substrate Ub_4(lin)-GFP-tail. These methods can be used to characterize the various steps of protein degradation, to screen the effects of proteasome mutants, and to screen modulators of proteasome activity such as small molecules that inhibit or activate proteasomes (e.g., [57]). For global fitting of larger datasets, we recommend using KinTek Explorer [58] or other fitting software to analyze proteasome kinetics. *See* Table 2 for troubleshooting related to this section.

3.3.1 Plate Reader and Selection of Plate

We use the plate reader Infinite M1000 Pro (Tecan) for our assays: 488 nm excitation wavelength and 520 nm emission wavelength (bandwidths: excitation 5 nm and emission 10 nm). We recommend optimizing the following parameters to find the best compromise between signal to noise required for the desired precision of measurements, the cost of reagents, and the breadth of conditions to be explored:

1. Excitation and emission wavelengths and bandwidths.
2. When following fluorescence changes from the top of the plate, the Z position of the optical head needs to be adjusted. We do pilot experiments with different volumes of degradation buffer to determine the optimal Z position.
3. Gain: In an independent experiment, we serially dilute the florescent substrate to determine the optimal gain

Table 2
Troubleshooting high-throughput degradation assay

Problem	Probable reason	Tentative solution
Low signal to noise ratio	Noise from plate	Use the plate according to the plate reader recommendation
	Wrong excitation and emission wavelength selection	Run excitation and emission scan for the substrate and choose appropriate slit width
	Gain and Z position are not right	In an independent experiment, optimize the gain and Z position with different substrate concentrations
	Substrate unfolded or aggregated	Remake the substrate, avoid freeze thaw cycle, use the buffer pH at least 1 pH higher or lower than pI, and include 150 mM NaCl, 1 mM DTT, and 5% glycerol when you snap freeze the substrate
	Problem related to plate reader	Test the plate reader with appropriate fluorescence dye
No degradation	Inactive proteasome	Activity of the proteasome can be tested by any of the methods mentioned in Subheading 3.2.3
	One of the two-part degrons is compromised	Discard this preparation and make a new one; avoid protease contamination
	ATP and buffer conditions not optimal	Amount and quality of ATP will affect substrate degradation
Non-reproducible results	Reaction conditions	Cross-check the buffers, ATP, and pipetting-related issues
	Well-to-well temperature variation in plate	Equilibrate the pate and sample at 30 °C for at least 30 min
	Scattering-related issue	Avoid air bubbles; filter all the buffers to avoid particulate matter
	Binding of substrate to plate	Test different BSA concentrations and use up to 150 mM sodium chloride in the degradation buffer

experimentally by gradually increasing the gain at proper Z position. We also include a "buffer-only control" to monitor the background fluorescence at every gain.

Select multi-well plate appropriate for fluorescent measurements based on the plate reader recommendations. Ideally, plates should not bind the substrate and proteasome and should have a minimal light scattering; depth of the wells should be compatible with the plate reader. We advise testing the fluorescent intensity across the plate at different substrate concentrations to determine well-to-well signal bleeding, evaporation during run, and stability of signal in the degradation buffer. Note that at temperatures far from room temperature, gradients can form, especially at the edges of plates. We use black, nonbinding 384-well plates (Corning).

3.3.2 Single-Turnover Reaction

To study single-turnover reactions, we prepare substrate and proteasome solutions at twice the desired final concentrations and start the reaction by mixing equal volumes of substrate and proteasome solutions. 5 nM of Ub_4(lin)-GFP-tail substrate and 25 nM of 26S proteasome are suitable concentrations for single-turnover reactions.

1. Turn on the plate reader and load the experiment settings (excitation 488 nm, emission 520 nm, and the experimentally determined Z position, gain, and bandwidths).
2. Set the temperature at 30 °C and leave the 384-well plate inside the plate reader to equilibrate at 30 °C (*see* **Notes 14** and **15**).
3. Make 22 μL of 2× substrate in substrate buffer and incubate at 30 °C for 1–10 min.
4. Make 22 μL 2× 26S or 20S proteasome mix and incubate at 30 °C for 1–10 min.
5. First, pipette controls A and B in the plate reader (Table 3); control A is "buffer-only control" to monitor intrinsic noise; control B monitors the stability of GFP fluorescence over time.
6. Pipette out the 20 μL of 2× substrate in 384-well plate (make sure to avoid bubbles) and 20 μL of 2× 26S or 20S proteasome and mix, for example, by pipetting up and down (*see* **Note 16**).
7. Transfer the 384-well plate in the plate reader and start data collection as soon as possible.
8. Collect GFP fluorescence every minute for a total of 90 min.
9. Export data as a CSV file for further analysis.

Table 3
Example reaction condition for single-turnover kinetics

	Test 1 (μL)	Test 2 (μL)	Control A (μL)	Control B (μL)
2× substrate	20	20		20
2× 26S proteasome mix	20			
2× 20S proteasome mix		20		
2× proteasome buffer			20	20
Substrate buffer			20	
Total	40	40	40	40

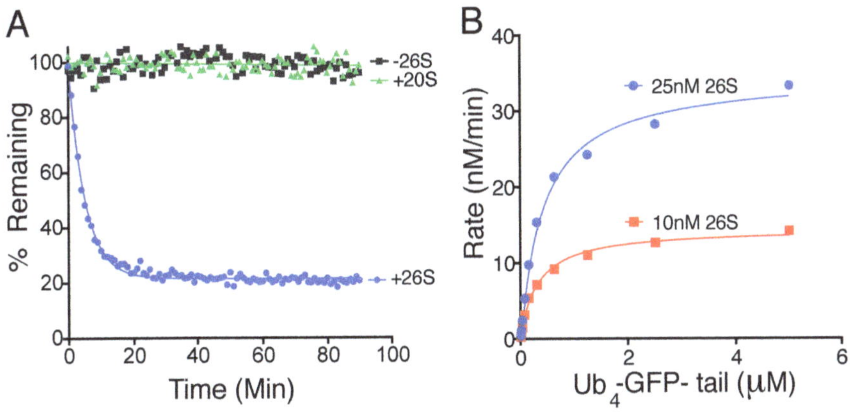

Fig. 2 Single-turnover and steady-state kinetics of proteasome degradation. (**a**) Ub_4(lin)-GFP-tail was incubated with either 26S (blue) or 20S (green) proteasome, and GFP fluorescence was monitored every minute for 90 min, and data was fitted to single-phase exponential decay to calculate the initial rate. (**b**) Different concentrations of Ub_4(lin)-GFP-tail were incubated with either 10 nM (red) or 25 nM (blue) 26S proteasome. The initial rate plotted as a function of substrate concentration was used to fit Michaelis-Menten curve

3.3.3 Data Analysis of the Single-Turnover Reaction

We monitor the conversion of substrate to product by measuring the decrease in fluorescence intensity over time. After export, the data can be analyzed using various analysis software such as GraphPad Prism, KaleidaGraph, MATLAB, R, or Kintek Explorer.

1. Data are exported and background is subtracted using values from control A.
2. Data are then normalized to the "substrate-only control" at time zero.
3. If GraphPad Prism 6 is used to fit the data to a single-phase exponential decay (Fig. 2a), the fitting equation takes the following form:

$$\Upsilon = (\Upsilon_0 - \text{Plateau}) \times \exp(-k \times X) + \text{Plateau}$$

   where Υ_0 is the GFP fluorescence intensity at time zero, k is the rate constant of the fluorescent decay, and Plateau is the GFP fluorescence at infinite time.
4. The half-time of the reaction can be calculated as *ln2/k*, and the initial rate of decay as k*(Υ_0-Plateau).
5. By using Ub_4(lin)-GFP-tail and yeast proteasome FLAG-tagged at Rpn11 as described here, we obtain a half-life of 3.5 min and an initial rate of 15 % min^{-1}. Ub_4(lin)-GFP-tail is not degraded by 20S proteasome (Fig. 2a) suggesting that the recognition, unfolding, and translocation steps are essential for Ub_4(lin)-GFP-tail degradation.

3.3.4 Steady-State Kinetics of Proteasomal Degradation

For Michaelis-Menten analysis, degradation reactions have to be followed at different substrates and proteasome concentrations, which can be done in parallel using a plate reader. In the example shown here, initial rates of degradation assay were measured at ten different substrate concentrations ranging from 10 nM to 10 μM of Ub_4(lin)-GFP-tail at 10 nM or 25 nM yeast proteasome FLAG-tagged at Rpn11 (Fig. 2b).

1. Turn on the plate reader, load the experiment settings, set the temperature at 30 °C, and leave the 384-well plate inside the plate reader to equilibrate at 30 °C.
2. 2× substrate: Ub_4(lin)-GFP-tail was serially diluted from 20 μM to 20 nM (10 steps of twofold serial dilution) in substrate buffer to yield a total volume of 65 μL for each concentration (20 μL each will be required for two different proteasome concentrations, and 20 μL will be used for the "substrate-only control").
3. 2× proteasome mix: Prepare 50 nM and 20 nM proteasome in 210 μL 2× proteasome buffer (20 μL each for ten different substrate concentrations).
4. Incubate 2× substrate, 2× proteasome, and the control solutions A and B separately at 30 °C for 1 to 10 min.
5. Transfer 20 μL of 2× substrate and 20 μL of 2× proteasome in the 384-well plate along with the controls and mix, for example, by pipetting up and down, and start the reaction as soon as possible.
6. Collect GFP fluorescence every minute for total of 90 min.
7. Export the data as a CSV file for further analysis.

3.3.5 Data Analysis of Steady-State Kinetics

1. Subtract the background from the data as described in Subheading 3.3.3.
2. Initiation rate of degradation for each substrate and proteasome pair is calculated as described above or by fitting the initial linear portion of the decay curve to a straight line.
3. To convert the degradation rate from units of fluorescence intensity per min to nM min^{-1}, the initial rate can be corrected by a conversion factor derived from a Ub_4(lin)-GFP-tail standard curve (*see* **Note 17**).
4. The initial rate plotted as a function of substrate concentration can be fitted to the Michaelis-Menten equation to obtain K_M and V_{max} values (Fig. 2b).
5. If GraphPad Prism 6 is used for curve fitting, the equation takes the form:

$$Y = V_{max} \times S/(K_M + S)$$

where V_{max} is the maximum reaction rate at infinite substrate concentration, S is the substrate concentration in nM, and K_M is the substrate concentration at which the reaction rate is the maximum reaction rate (1/2 V_{max}).

6. The following parameters are obtained under the conditions described here for Ub_4(lin)-GFP-tail and yeast proteasome FLAG-tagged at Rpn11: $V_{max} = 34.0 \pm 0.9$ nM/min and $K_M = 420 \pm 40$ nM for 25 nM proteasome and $V_{max} = 14.5 \pm 0.3$ nM/min and $K_M = 317 \pm 27$ nM for 10 nM proteasome. The errors are standard errors of the mean, and the difference in the K_M values obtained at the two proteasome concentrations suggests that the reported errors may underestimate the actual errors in the K_M determination.

3.3.6 Multiplex Competition Assay

The affinity of ligands, such as ubiquitin chains, whose binding sites overlap with the substrate binding sites, can be measured in competition assays. In this section, we used 5 nM Ub_4(lin)-GFP-tail substrate and competed it with varying concentrations of purified Ub_4(lin)-6XHis (1–64 μM) and yeast proteasome FLAG-tagged at Rpn11 (25 nM). *See* Table 4 for an example reaction conditions.

1. Ub_4(lin)-6XHis is expressed and purified as described in Subheading 3.1.
2. Turn on the plate reader, load the program, and equilibrate the 384-well plate inside the plate reader at 30 °C.
3. 4× substrate was diluted in 300 μL substrate buffer (20 nM), 147 μL (21 μL each) for seven inhibitor concentrations, and 147 μL (21 μL each) for "substrate-only control."
4. Ub_4-6XHis inhibitor was serially diluted in 150 μL substrate buffer from 256 μM to 4 μM (4× concentration, final 64 μM to 1 μM).

Table 4
Example reaction condition for inhibition assay

	Test 1–7 (μL)	Control A (μL)	Control B (μL)	Control C1–7 (μL)
4× substrate	10		10	10
4× inhibitor	10			10
2× proteasome mix	20		20	
2× proteasome buffer		20		20
Substrate buffer		20	10	
Total	40	40	40	40

5. 2× substrate and inhibitor mix were prepared by combining 21 μL of 4× substrate and 21 μL of 4× inhibitor.
6. 2× substrate inhibitor mix was incubated at 30 °C for 1–10 min.
7. 185 μL of 50 nM proteasome (2× concentration) was diluted in 2× degradation buffer and incubated for 1–10 min at 30 °C.
8. Control A is as "buffer-only control" to measure background fluorescence; control B provides substrate stability in the absence of inhibitor. Controls C1 to C7 are substrate and inhibitor mixes to monitor substrate fluorescence over time in the presence of different inhibitor concentrations.
9. 20 μL of 2× substrate or 2× substrate inhibitor mix was mixed with 20 μL of 2× proteasome, and program was started as soon as possible.
10. Measure GFP fluorescence every minute for 90 min.
11. Export data as CSV files for further analysis.

3.3.7 Data Analysis of Competition Assay

Binding competition and enzyme inhibition data can be analyzed based on the nature of inhibition. In this scenario, we established that the substrate and ligand compete for binding to the same site (not shown).

1. Subtract fluorescence intensity from background (data from control A) and plot over time.
2. Fit data to single exponential decay as described in Subheading 3.3.3 (Fig. 3a).

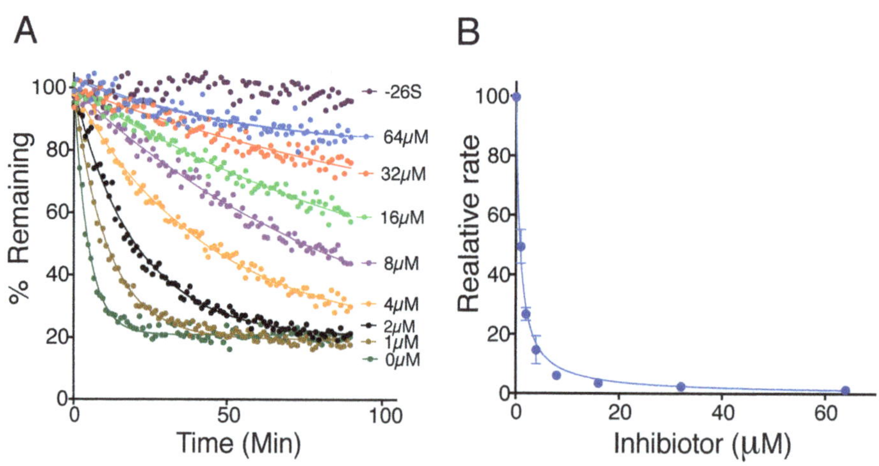

Fig. 3 High-throughput inhibition assay. (**a**) Varying concentrations of Ub_4-6XHis were used to compete Ub_4(lin)-GFP-tail for proteasomal degradation. GFP fluorescence was monitored every minute for 90 min and the data was fitted to a single-phase exponential decay. (**b**) The relative initial rate was plotted for different Ub_4-6XHis (inhibitor) concentrations to calculate inhibition constant

3. The initial rate is plotted as a function of concentration and fitted by using

$$y/(1+[I]/K_{\mathrm{i}})$$

where $y = v_0$, which is the reaction rate in the absence of inhibitor and $[I]$ is the inhibitor concentration (here the Ub_4(lin)-6XHis concentration). K_{i} is the apparent inhibition constant, which is related to the affinity of the inhibitor to the proteasome (Fig. 3b).

4. The inhibition constant (K_{i}) for Ub_4(lin)-GFP-tail and Ub_4 for proteasome from this experiment is 0.8 ± 0.04 μM.

4 Notes

1. Use good-quality IPTG; it is best to use freshly prepared and filtered sterilized IPTG. Alternatively make IPTG at 1 M and store in aliquots at −20 °C; discard remaining IPTG after single use.
2. Use EDTA-free protease inhibitor, as excess EDTA will strip Ni from Ni-NTA resin (which will turn white). High concentration of DTT or more than 20 mM β-mercaptoethanol should also be avoided as they reduce Ni ions (which turns brown).
3. Phosphate-based buffer system is preferred for Ni-NTA purification, as buffers with secondary or tertiary amines (Tris, HEPES, MOPS at >100 mM) interfere with nickel (Ni) binding.
4. ATP in water is unstable; we recommend making 100 mM ATP stock adjusted using 1 M Tris base to pH 7–7.5. ATP should be stored at −20 °C freezer or lower temperature, and freeze thawing should be minimized.
5. We use conical flasks for better aeration and maintain at least 1:5 ratio of media to total volume of the flask. Purification of Ub_4(lin)-GFP-6XHis was also optimized for FPLC system. Briefly, the clarified lysate was applied to a 5 mL HisTrap FF Crude column on an FPLC system (AKTA purifier 10) at a rate of 2 mL/min, washed with 10 CV of NPI-10 followed by washed with 10 CV of NPI-25, and eluted with 10 CV of NPI-250 [22].
6. Check the manufacturer manual for the binding capacity. This will give an idea how much beads to use.
7. When proteasome was purified from cells harvested in late stationary phase, the yield of 26S proteasomes was relatively low. Therefore, we always start growing the cells from a fresh isolated single colony (appearing white and not pink). Starting

culture from a single colony and growing cells till $OD_{600} \leq 2$ ensures consistently active and intact proteasomes.

8. Alternate yeast cell lysis method could be French press, Freezer/Mills, and liquid nitrogen-based mechanical lysis. We do not recommend using an ultrasonic cell disruptor.
9. Typically for 2 liters of yeast cells, we use 2 mL of 50% anti-FLAG M2 agarose beads, which gives 1 mL CV.
10. Before incubating the lysate with anti-FLAG M2 agarose, check the pH of lysate, and if needed adjust it to 7–7.5 with Tris base.
11. Elution step can be done at 4 °C by incubating the beads with elution buffer for longer time.
12. Purified proteasome may be dialyzed or buffer exchanged to remove FLAG peptide (buffer must contain at least 1 mM ATP, 5 mM $MgCl_2$, and 10% glycerol).
13. Don't vigorously vortex purified proteasomes.
14. For reliable florescence readout, equilibration of plates and samples at 30 °C is important.
15. We recommend doing a pilot experiment in the absence of proteasome to monitor performance of the plate, sample, and reaction conditions for the period of at least 90 min.
16. For consistency, always transfer proteasomes at the end and start the measurement as soon as possible. If you are working with multiple samples, we recommend making plate layout for easy annotation of samples and using a multichannel pipette.
17. For steady-state kinetics, we assess a range of substrate concentrations, which will have range of GFP fluorescence. We recommend to perform a pilot experiment with different gain setups, as at higher gain fluorescence, the signal might overflow, and at low gain, you may not be able to detect lower concentrations of the substrate. This experiment also provides a standard curve to convert fluorescent intensity (au) to nM substrate concentration.

Acknowledgment

This work was supported by U54 GM105816, R21 CA191664, R21 CA196456, and R01 GM124501 from the National Institutes of Health; RP140328 from the Cancer Prevention and Research Institute of Texas (CPRIT); and F-1817 from the Welch.

References

1. Baumeister W, Walz J, Zuhl F et al (1998) The proteasome: paradigm of a self-compartmentalizing protease. Cell 92:367–380. https://doi.org/10.1016/S0092-8674(00)80929-0
2. Finley D, Chen X, Walters KJ (2016) Gates, channels, and switches: elements of the proteasome machine. Trends Biochem Sci 41:77–93. https://doi.org/10.1016/j.tibs.2015.10.009
3. Beck F, Unverdorben P, Bohn S et al (2012) Near-atomic resolution structural model of the yeast 26S proteasome. Proc Natl Acad Sci U S A 109:14870–14875. https://doi.org/10.1073/pnas.1213333109
4. Lander GC, Estrin E, Matyskiela ME et al (2012) Complete subunit architecture of the proteasome regulatory particle. Nature 482:186–191. https://doi.org/10.1038/nature10774
5. Groll M, Bajorek M, Köhler A et al (2000) A gated channel into the proteasome core particle. Nat Struct Mol Biol 7:1062–1067. https://doi.org/10.1038/80992
6. Groll M, Ditzel L, Löwe J et al (1997) Structure of 20S proteasome from yeast at 2.4 A resolution. Nature 386:463–471. https://doi.org/10.1038/386463a0
7. Bhattacharyya S, Yu H, Mim C, Matouschek A (2014) Regulated protein turnover: snapshots of the proteasome in action. Nat Rev Mol Cell Biol 15:122–133. https://doi.org/10.1038/nrm3741
8. Hershko A, Ciechanover A (2003) The ubiquitin system. Annu Rev Biochem 67:425–479. https://doi.org/10.1146/annurev.biochem.67.1.425
9. DeMartino GN, Gillette TG (2007) Proteasomes: machines for all reasons. Cell 129:659–662. https://doi.org/10.1016/j.cell.2007.05.007
10. Glickman MH, Ciechanover A (2002) The ubiquitin-proteasome proteolytic pathway: destruction for the sake of construction. Physiol Rev 82:373–428. https://doi.org/10.1152/physrev.00027.2001
11. Wolf DH, Hilt W (2004) The proteasome: a proteolytic nanomachine of cell regulation and waste disposal. Biochim Biophys Acta 1695:19–31. https://doi.org/10.1016/j.bbamcr.2004.10.007
12. Suraweera A, Münch C, Hanssum A, Bertolotti A (2012) Failure of amino acid homeostasis causes cell death following proteasome inhibition. Mol Cell 48(2):242–253. https://doi.org/10.1016/j.molcel.2012.08.003
13. Goldberg AL, Rock KL (1992) Proteolysis, proteasomes and antigen presentation. Nature 357:375–379. https://doi.org/10.1038/357375a0
14. Schmidt M, Finley D (2014) Regulation of proteasome activity in health and disease. Biochim Biophys Acta 1843:13–25. https://doi.org/10.1016/j.bbamcr.2013.08.012
15. Dantuma NP, Bott LC (2014) The ubiquitin-proteasome system in neurodegenerative diseases: precipitating factor, yet part of the solution. Front Mol Neurosci 7:1345. https://doi.org/10.3389/fnmol.2014.00070
16. Powell SR, Herrmann J, Lerman A et al (2012) The ubiquitin–proteasome system and cardiovascular disease. In: The proteasomal system in aging and disease. Elsevier, Amsterdam, pp 295–346
17. Cromm PM, Crews CM (2017) The proteasome in modern drug discovery: second life of a highly valuable drug target. ACS Cent Sci 3:830–838. https://doi.org/10.1021/acscentsci.7b00252
18. Shah JJ, Orlowski RZ (2009) Proteasome inhibitors in the treatment of multiple myeloma. Leukemia 23:1964–1979. https://doi.org/10.1038/leu.2009.173
19. Liggett A, Crawford LJ, walker B et al (2010) Methods for measuring proteasome activity: current limitations and future developments. Leuk Res 34:1403–1409. https://doi.org/10.1016/j.leukres.2010.07.003
20. Kisselev AF, Goldberg AL (2005) Monitoring activity and inhibition of 26S proteasomes with fluorogenic peptide substrates. Meth Enzymol 398:364–378. https://doi.org/10.1016/S0076-6879(05)98030-0
21. Saeki Y, Isono E, Toh-E A (2005) Preparation of ubiquitinated substrates by the PY motif-insertion method for monitoring 26S proteasome activity. Meth Enzymol 399:215–227. https://doi.org/10.1016/S0076-6879(05)99014-9
22. Martinez-Fonts K, Matouschek A (2016) A rapid and versatile method for generating proteins with defined ubiquitin chains. Biochemistry 55:1898–1908. https://doi.org/10.1021/acs.biochem.5b01310
23. Prakash S, Tian L, Ratliff KS et al (2004) An unstructured initiation site is required for efficient proteasome-mediated degradation. Nat Struct Mol Biol 11:830–837. https://doi.org/10.1038/nsmb814
24. Singh Gautam AK, Balakrishnan S, Venkatraman P (2012) Direct ubiquitin independent

recognition and degradation of a folded protein by the eukaryotic proteasomes-origin of intrinsic degradation signals. PLoS One 7: e34864–e34814. https://doi.org/10.1371/journal.pone.0034864

25. Li Y, Tomko RJ, Hochstrasser M (2015) Proteasomes: isolation and activity assays. Curr Protoc Cell Biol 4:3.43.1–3.43.20. https://doi.org/10.1002/0471143030.cb0343s67
26. Verma R, McDonald H, Yates JR, Deshaies RJ (2001) Selective degradation of ubiquitinated Sic1 by purified 26S proteasome yields active S phase cyclin-Cdk. Mol Cell 8:439–448
27. Hanna J, Hathaway NA, Tone Y et al (2006) Deubiquitinating enzyme Ubp6 functions noncatalytically to delay proteasomal degradation. Cell 127:99–111. https://doi.org/10.1016/j.cell.2006.07.038
28. Kraut DA, Israeli E, Schrader EK et al (2012) Sequence- and species-dependence of proteasomal processivity. ACS Chem Biol 7:1444–1453. https://doi.org/10.1021/cb3001155
29. Tian L, Holmgren RA, Matouschek A (2005) A conserved processing mechanism regulates the activity of transcription factors Cubitus interruptus and NF-κB. Nat Struct Mol Biol 12:1045–1053. https://doi.org/10.1038/nsmb1018
30. Schrader EK, Harstad KG, Holmgren RA, Matouschek A (2011) A three-part signal governs differential processing of Gli1 and Gli3 proteins by the proteasome. J Biol Chem 286:39051–39058. https://doi.org/10.1074/jbc.M111.274993
31. Yu H, Singh Gautam AK, Wilmington SR et al (2016) Conserved sequence preferences contribute to substrate recognition by the proteasome. J Biol Chem 291:14526–14539. https://doi.org/10.1074/jbc.M116.727578
32. Wilmington SR, Matouschek A (2016) An inducible system for rapid degradation of specific cellular proteins using proteasome adaptors. PLoS One 11:e0152679–e0152616. https://doi.org/10.1371/journal.pone.0152679
33. Yu H, Kago G, Yellman CM, Matouschek A (2016) Ubiquitin-like domains can target to the proteasome but proteolysis requires a disordered region. EMBO J 35:1522–1536. https://doi.org/10.15252/embj.201593147
34. Beckwith R, Estrin E, Worden EJ, Martin A (2013) Reconstitution of the 26S proteasome reveals functional asymmetries in its AAA+ unfoldase. Nat Publ Group 20:1164–1172. https://doi.org/10.1038/nsmb.2659
35. Neefjes J, Dantuma NP (2004) Fluorescent probes for proteolysis: tools for drug discovery. Nat Rev Drug Discov 3:58–69. https://doi.org/10.1038/nrd1282
36. Bhattacharyya S, Renn JP, Yu H et al (2016) An assay for 26S proteasome activity based on fluorescence anisotropy measurements of dye-labeled protein substrates. Anal Biochem 509:50–59. https://doi.org/10.1016/j.ab.2016.05.026
37. Stack JH, Whitney M, Rodems SM, Pollok BA (2000) A ubiquitin-based tagging system for controlled modulation of protein stability. Nat Biotechnol 18:1298–1302. https://doi.org/10.1038/82422
38. Fishbain S, Prakash S, Herrig A et al (2011) Rad23 escapes degradation because it lacks a proteasome initiation region. Nat Commun 2:192–199. https://doi.org/10.1038/ncomms1194
39. Sone T, Saeki Y, Toh-e A, Yokosawa H (2004) Sem1p is a novel subunit of the 26 S proteasome from Saccharomyces cerevisiae. J Biol Chem 279:28807–28816. https://doi.org/10.1074/jbc.M403165200
40. Inobe T, Fishbain S, Prakash S, Matouschek A (2011) Defining the geometry of the two-component proteasome degron. Nat Chem Biol 7:161–167. https://doi.org/10.1038/nchembio.521
41. Thrower JS, Hoffman L, Rechsteiner M, Pickart CM (2000) Recognition of the polyubiquitin proteolytic signal. EMBO J 19:94–102. https://doi.org/10.1093/emboj/19.1.94
42. Nager AR, Baker TA, Sauer RT (2011) Stepwise unfolding of a β barrel protein by the AAA + ClpXP protease. J Mol Biol 413:4–16. https://doi.org/10.1016/j.jmb.2011.07.041
43. Fishbain S, Inobe T, Israeli E et al (2015) Sequence composition of disordered regions fine-tunes protein half-life. Nat Struct Mol Biol 22:214–221. https://doi.org/10.1038/nsmb.2958
44. Fischer M, Hilt W, Richter-Ruoff B et al (1994) The 26S proteasome of the yeast Saccharomyces cerevisiae. FEBS Lett 355:69–75. https://doi.org/10.1016/0014-5793(94)01177-X
45. Ostuka Y, Homma N, Shiga K et al (1998) Purification and properties of rabbit muscle proteasome, and its effect on myofibrillar structure. Meat Sci 49:365–378. https://doi.org/10.1016/S0309-1740(97)00141-1
46. Yang P (2003) Purification of the Arabidopsis 26 S proteasome: biochemical and molecular analyses revealed the presence of multiple

isoforms. J Biol Chem 279:6401–6413. https://doi.org/10.1074/jbc.M311977200

47. DeMartino GN, Proske RJ, Moomaw CR et al (1996) Identification, purification, and characterization of a PA700-dependent activator of the proteasome. J Biol Chem 271:3112–3118. https://doi.org/10.1074/jbc.271.6.3112
48. Hirano Y, Murata S, Tanaka K (2005) Large- and small-scale purification of mammalian 26S proteasomes. Meth Enzymol 399:227–240. https://doi.org/10.1016/S0076-6879(05)99015-0
49. Besche HC, Goldberg AL (2012) Affinity purification of mammalian 26S proteasomes using an ubiquitin-like domain. In: cDNA Libraries. Humana Press, Totowa, NJ, pp 423–432
50. Besche HC, Haas W, Gygi SP, Goldberg AL (2009) Isolation of mammalian 26S proteasomes and p97/VCP complexes using the ubiquitin-like domain from HHR23B reveals novel proteasome-associated proteins. Biochemistry 48:2538–2549. https://doi.org/10.1021/bi802198q
51. Leggett DS, Hanna J, Borodovsky A et al (2002) Multiple associated proteins regulate proteasome structure and function. Mol Cell 10:495–507
52. Leggett DS, Glickman MH, Finley D (2005) Purification of proteasomes, proteasome subcomplexes, and proteasome-associated proteins from budding yeast. Methods Mol Biol 301:57–70. https://doi.org/10.1385/1-59259-895-1:057
53. Verma R, Chen S, Feldman R et al (2000) Proteasomal proteomics: identification of nucleotide-sensitive proteasome-interacting proteins by mass spectrometric analysis of affinity-purified proteasomes. Mol Biol Cell 11:3425–3439
54. Elsasser S, Schmidt M, Finley D (2005) Characterization of the proteasome using native gel electrophoresis. Meth Enzymol 398:353–363. https://doi.org/10.1016/S0076-6879(05)98029-4
55. Peth A, Kukushkin N, Bossé M, Goldberg AL (2013) Ubiquitinated proteins activate the proteasomal ATPases by binding to Usp14 or Uch37 homologs. J Biol Chem 288:7781–7790. https://doi.org/10.1074/jbc.M112.441907
56. Benaroudj N, Zwickl P, Seemuller E et al (2003) ATP hydrolysis by the proteasome regulatory complex PAN serves multiple functions in protein degradation. Mol Cell 11:69–78. https://doi.org/10.1016/S1097-2765(02)00775-X
57. Takahashi K, Matouschek A, Inobe T (2015) Regulation of proteasomal degradation by modulating proteasomal initiation regions. ACS Chem Biol 10:2537–2543. https://doi.org/10.1021/acschembio.5b00554
58. Johnson KA (2009) Fitting enzyme kinetic data with KinTek Global Kinetic Explorer. Meth Enzymol 467:601–626. https://doi.org/10.1016/S0076-6879(09)67023-3

Part V

Proteomic Methods to Study the Ubiquitin Proteasome System

Chapter 22

Exploring the Rampant Expansion of Ubiquitin Proteomics

Amalia Rose and Thibault Mayor

Abstract

The ubiquitin proteasome system can arguably affect all cellular proteins with few exceptions. In addition to regulating many pathways such as cell cycle progression, inflammation, gene expression, DNA repair, and vesicle trafficking—to just name a few—ubiquitination can occur to any nascent or newly translated protein that misfolds. In the past years, substantial progress has been achieved in advancing our global understanding of the ubiquitinome—the ensemble of ubiquitinated proteins within a cell—using mass spectrometry-based proteomics. Notably, over 50,000 conjugation sites have now been reported. In this review, we discuss recent proteomics methods used to expand our knowledge of the ubiquitin proteasome system through the identification of ubiquitination sites, poly-ubiquitin chain types, and E3 ubiquitin ligase substrates.

Key words Ubiquitin, Proteasome, E3, Proteomics, Mass spectrometry, diGly

1 Introduction

As with phosphorylation, ubiquitination of proteins regulates or affects most cellular processes, and defects in this system are associated to many diseases [1]. There is now a great interest in developing novel therapeutics by targeting the ubiquitin proteasome system (UPS), bringing the field to a new tipping point. Considering that only one drug targeting the UPS was clinically approved less than 10 years ago—as opposed to 16 targeting kinases [1]—we have made remarkable progress in a short period of time. There are now multiple small molecules targeting the UPS used in the clinic to treat multiple myeloma and other cancers, such as three different proteasome inhibitors, thalidomide, and its derivatives [2]. In order to continue to capitalize on these early successes, it is critical to achieve a greater understanding of the UPS to nurture potential novel therapeutics to combat both cancers and neurodegenerative diseases [2, 3]. In this review, we will discuss recent advances of mass spectrometry (MS)-based proteomics methods used to unveil the "ubiquitinome" and to identify substrates of E3 ubiquitin ligases.

Thibault Mayor and Gary Kleiger (eds.), *The Ubiquitin Proteasome System: Methods and Protocols*, Methods in Molecular Biology, vol. 1844, https://doi.org/10.1007/978-1-4939-8706-1_22,

Ubiquitination, which consists of the covalent attachment of ubiquitin to a protein, targets the substrate to a specific process or component of the cell such as the proteasome where it gets degraded [4]. Ubiquitin is a highly conserved 76 amino acid long protein that can be conjugated via its C-terminus to mostly lysine residues on other polypeptides [5]. Ubiquitin is added to target substrates sequentially through a multienzyme cascade. Initially, an El enzyme activates the C-terminus of ubiquitin using ATP to form a high energy El~Ub thioester adduct. Once activated, ubiquitin is transferred from the El to an E2 enzyme, otherwise known as a ubiquitin-conjugating enzyme. The E2 then transfers ubiquitin to the catalytic cysteine of an E3 ubiquitin ligase or directly to the substrate associated with an E3 [5]. Ubiquitination is a highly dynamic process that can be reverted by deubiquitinases (DUBs), which remove ubiquitin chains from their substrates [6]. In humans, there are two Els, over 30 E2s, more than 600 - different E3s or E3 complexes, and approximately 90 DUBs, creating a complex and intertwined system that is challenging to decipher [7].

2 Global Identification of Ubiquitin Sites

Over the last 15 years, significant strides have been made toward obtaining a system-wide view of the "ubiquitinome"—the ensemble of ubiquitinated proteins in a given cell or tissue. Most methods have relied on proteomic approaches, which will be the focus of this review. As for most protein MS-based methods, a bottom-up approach is typically used to analyze ubiquitinated proteins, where proteins are indirectly analyzed through their corresponding peptides that are generated most commonly by trypsin digestion [8]. Upon trypsin digestion of a ubiquitin-conjugated protein, a signature peptide is produced consisting of the last two C-terminal glycines of ubiquitin (Gly-Gly, also called a diGly or the ubiquitin remnant) that remain covalently attached to the lysine residue of the target protein. As trypsin is unable to cleave after the conjugated lysine, the signature peptide contains both a missed cleavage site and the 114.04 kDa diGly mass shift at the modified lysine, which can then be readily identified during the database search [9].

Although trypsinized ubiquitinated proteins produce unique peptides, a major challenge for the identification of these conjugation sites is the substoichiometric nature of these modifications [10]. Indeed, a large proportion of the modified proteins are either promptly targeted for proteolysis, or, when not degraded, DUBs often reverse the modification. Therefore, the transient nature of ubiquitination makes it particularly difficult for identification by mass spectrometry. The enrichment of ubiquitinated proteins or peptides prior to MS analysis is therefore critical for a successful large-scale analysis of the ubiquitin proteome. We will first review different methods of isolating ubiquitin conjugates for MS analysis below.

2.1 Enrichment with Tagged Ubiquitin

One of the first attempts at a global identification of ubiquitin conjugates in *Saccharomyces cerevisiae* was carried out by Stephen Gygi's group. In their milestone study, they enriched for proteins conjugated to 6xHis-tagged ubiquitin by affinity chromatography purification and identified 110 ubiquitin sites in 72 different proteins, of which only 3 were previously known to be ubiquitinated [9]. Although only a small subset of the ubiquitin proteome was identified, this study demonstrated the ability to enrich for tagged ubiquitin for the general identification of ubiquitination sites using MS. With the advancement of the sensitivity of the MS instruments and the refinement of enrichment methods, the number of reported ubiquitinated proteins and conjugation sites increased to several hundreds in the following years [11–14], though this number remained a modest representation of the ubiquitinome before the advent of a better approach.

2.2 Enrichment Using Gly-Gly Remnant Antibody

In past years, major breakthroughs have been achieved in the identification of ubiquitination sites by enriching for conjugates at the peptide level as opposed to at the protein level. The purification of conjugated peptides has been enabled by the introduction of antibodies that recognize the signature Gly-Gly motif that remains bonded to lysine of ubiquitinated proteins after tryptic digestion. Xu et al. generated an anti-diglycyl-lysine (K-ε-GG) antibody to immunoprecipitate the Gly-Gly-modified lysine at a high yield, leading to the discovery of more than 150 previously unknown ubiquitinated proteins [15]. Importantly, this purification method hinges on endogenous ubiquitin and thus avoids potential off-target effects from the ectopic expression of tagged ubiquitin. However, the anti-K-ε-GG antibodies do not recognize non-lysine ubiquitination sites like on serine, threonine, and the N-terminus of modified proteins. Furthermore, two other ubiquitin-like proteins—Nedd8 and ISG15—generate identical Gly-Gly signatures after trypsin digest, making it impossible to distinguish between the three modifications without additional experiments. Fortunately, Nedd8 conjugation is mostly restricted to the cullin proteins in normal conditions, and ISG15 is only induced in response to interferons [16, 17].

The introduction of the antibody-based enrichment approach led to the identification of thousands of new ubiquitination sites in mammalian tissue culture cells and tissues. For instance, Dr. Steve Gygi and colleagues used the antibody coupled with an improved enrichment method to identify approximately 19,000 ubiquitin sites from more than 5000 unique proteins from HCT116 cells, which was a marked improvement on the approximately 1000 previously identified sites [16]. It should be noted that a relatively high amount of starting material (10–20 mg of proteins) is required for this approach. Following the chemical inhibition of the Nedd8 conjugation machinery, the same authors identified many potential substrates of the cullin-RING ligases (CRLs), the largest family of E3s in the human proteome. Changes were also quantified in the ubiquitin proteome in response to proteasome inhibition, and it was found that many proteasome targets were newly synthesized proteins. Importantly, a series of experiments were performed to show that the vast majority of the identified sites (>95%) were derived from ubiquitin and not Nedd8 conjugation.

Using an independently generated anti-K-ε-GG antibody, the Choudhary lab was able to identify approximately 20,000 ubiquitination sites in mice, as well as some 11,000 sites in HEK293 cells [18, 19]. By comparing the amino acid sequences surrounding the modified lysine for this considerable number of sites, the authors of these studies showed that these antibodies do not have any strong bias toward any particular sequence that surrounds the diGly motif, illustrating the efficacy of this approach [16, 18]. As of the beginning of 2018, the Phosphosite website reports the identification of almost 58,000 ubiquitination sites [20]. The method to enrich for ubiquitination sites using the anti-K-ε-GG antibody is described in Chapter 23 of this book, which also provides a summary table of most proteomics studies that have used this approach.

2.2.1 Gly-Gly Enrichment Coupled with Quantitative Approaches

The diGly enrichment method is often coupled with quantitative approaches in order to measure changes in the ubiquitination levels. Stable-isotope labeling by amino acids in cell culture (SILAC) is perhaps the most common methodology used for the quantitation of ubiquitination [15, 16, 21]. Two or three distinct populations of cells are grown in media containing either "light"/normal or "heavy" amino acids [22]. The peptides generated following SILAC have a distinct mass shift that can be discerned upon MS analysis. This approach enabled the characterization of global changes of the ubiquitinome, such as after heat or oxidative stresses [23, 24], or of more targeted perturbations, as reviewed in a later section (*see* Subheading 4.1).

One issue with the bottom-up approach is that information is gathered at the peptide level. If a protein has several ubiquitination sites that differentially fluctuate in a given condition (e.g., one site is reduced whereas a second site displays an increased level), it is difficult to infer how the overall ubiquitination level of that protein is altered.

Recently, the laboratories of Drs. Gygi and Harper have developed a method that couples diGly enrichment with isobaric labeling that increases the number of samples that can be compared in one experiment and further reduces the amount of starting material for a given sample [25]. Peptides generated by trypsin digestion of proteins from HCT116 cells under proteasome inhibition were enriched using the diglycine remnant antibody, followed by the labeling of these Gly-Gly peptides with ten-plex tandem mass tags (TMTs), where they were then combined and separated into fractions before being analyzed on a mass spectrometer. TMT labeling differentially labels multiple samples with isobaric tags, so that a labeled peptide has an identical mass regardless of its sample origin before fractionation [26]. One challenge of this approach is that the labeling with TMTs succeeds the Gly-Gly immunoprecipitation and small changes in sample handling could impact the results. However, the authors of the study reported a variability of less than 10% between biological and technical replicates—the latter consisting of the same biological sample being processed in parallel after the trypsin digest for the immunoprecipitation and TMT labeling steps—while quantifying 8000 ubiquitination sites using only 1 mg of starting material [25]. In addition to using less starting material per assessed condition, the ten-plex TMT allows for the quantitative analysis of more complex experiments due a higher level of multiplexing, as compared to typical SILAC duplex or triplex methods.

2.2.2 Improvements of the Gly-Gly Enrichment Method

Several other groups have improved and modified the method to increase the yield of the Gly-Gly pulldown. One such development is the immobilization of the anti-K-ε-GG antibody to beads by chemical cross-linking, established by Udeshi et al. [21]. The authors of this study found that cross-linking considerably diminishes the number of non-Gly-Gly peptides and antibody contaminants that are present in the final sample, which have been shown to have deleterious effects. Chemical cross-linking in combination with further fractionation leads to the identification of approximately 10,000 ubiquitination sites [21].

In a study looking at the cross talk between protein phosphorylation and ubiquitination, Swaney et al. utilized strong cation exchange (SCX) coupled with diGly enrichment to identify peptides containing multiple posttranslational modifications [27]. At a low pH, diGly peptides typically have four positive charges. The addition of negatively charged phosphates on these peptides reduces the overall charge. Using fractionation by SCX, ubiquitinated phosphopeptides (and unmodified peptides) are enriched in fractions eluted with a low salt concentration, whereas ubiquitinated peptides are eluted upon a higher salt concentration in later fractions [27]. In this study, Swaney et al. identified a large number of ubiquitinated proteins (1817), as well as 245 ubiquitinated phosphoproteins in *Saccharomyces cerevisiae*.

As previously discussed, the diGly motif mass of 114.04 kDa is used in the database searching algorithms in order to find the ubiquitination sites. However, these algorithms frequently generate false positives, and this leads to a significant amount of time spent doing manual data interpretation [28]. With the purpose of eliminating these false positives and enhancing the detection of the ubiquitin modification, Chicooree et al. developed an approach termed reductive methylation of ubiquitinated isopeptides (RUbI) [29]. This method uses formaldehyde and cyanoborohydride to demethylate both N-termini of diglycine branched isopeptides, adding a specific mass to the peptides. As such, upon collision-induced dissociation (CID) in the mass spectrometer, an abundant a_1 ion (m/z 62.09) and a highly specific b_2 ion (m/z 147.11) are generated, which are not present in the unmodified linear version of this peptide [29]. These two ions therefore provide an additional marker for diglycine branched peptides, which leads to an improvement in isopeptide assignment. Keeping the concept of generating diagnostic a_1 and b_2 ions for diGly peptides, the same research group improved upon their previous work by instead using mTRAQ labeling (nonisobaric stable isotope-labeled molecules that bind specifically to the N-terminus and side chain amines of proteins) to generate these ions followed by SWATH acquisition, which they termed MEDUSA (mass spectral enhanced detection of Ubls using SWATH acquisition) [30]. The ions generated by the mTRAQ labeling offer more diagnostic value than the RUbI and combined with SWATH, a recently developed data-independent acquisition method [31], allows for more accurate identification of modification-specific peptides than previous methods.

3 Identification and Quantification of Ubiquitin Chain Linkages

In order to determine the linkage specificity for any poly-ubiquitin chain conjugated to a protein, several proteomic approaches have been developed. Ubiquitin contains seven lysine residues (K6, K11, K27, K29, K33, K48, K63) as well as an N-terminal amino group (M1) from which additional ubiquitin molecules can be subsequently attached to form poly-ubiquitin chains. These chains can either be uniform and connected through the same residue at each linkage site (e.g., K48), have mixed linkages, or even be branched where a single ubiquitin within the chain is conjugated to two ubiquitin molecules [32]. Because the type of linkage(s) within a poly-ubiquitin chain dictates in large the biological fate of the conjugated proteins, the identification of the chain types can provide major insights into the roles of the modifications and the E2-E3 enzymes that generate them. For example, K48-linked chains typically signal proteins for degradation by the 26S proteasome, whereas K63 linkages have a role in vacuolar degradation or other non-proteolytic functions [33]. The large-scale methods using the anti-K-ε-GG antibody described earlier do not provide any information about the type of the ubiquitin chains attached to the identified proteins, as tryptic digestion only retains the diGly motif from the most proximal ubiquitin molecule attached to a protein. The section below will review some of the methods that have been developed to identify and quantify ubiquitin chain linkages.

3.1 Ubiquitin-AQUA Mass Spectrometry

Absolute quantification (AQUA) of ubiquitin is a method developed by Dr. Steven Gygi and colleagues to analyze which linkages are present within poly-ubiquitin chains and was first applied to characterize how cyclin B was ubiquitinated by the APC (anaphase-promoting complex) E3 ligase in vitro [34]. Upon the principle that the tryptic digestion of a ubiquitin-ubiquitin linkage also creates a unique diGly ubiquitin peptide, it is possible to determine which lysine residue(s) was/were conjugated in a given poly-ubiquitin chain. In order to perform AQUA, isotopically labeled ubiquitin standard peptides are generated (both with unmodified and modified lysine residues). Known amounts of these AQUA peptides are then added to the digested sample that is then analyzed by selected reaction monitoring (SRM) experiments. The spiked peptide standards co-elute alongside their experimental counterparts that can then be quantified in order to precisely measure the amount of each ubiquitin-ubiquitin linkage in the sample [34]. Since its introduction, this method has been improved such that a wider range of ubiquitin peptides can be monitored, including the N-terminus of ubiquitin that forms linear poly-ubiquitin chains [35]. AQUA provides a framework to characterize ubiquitin linkages from both reconstituted ubiquitination assays and complex samples derived from cells or tissues.

3.2 Modified Trypsinolysis and the Analysis of Branched Chains

Branched chains have been shown to promote the proteasomal degradation of substrates through the addition of K11-linked chains to shorter chains to enhance substrate recognition by the proteasome [36, 37]. Although AQUA has the ability to determine the composition of a poly-ubiquitin chain, it is unable to directly distinguish whether or not it is continuous or branched. A method was developed by Valkevich et al. that utilizes middle-down proteomics, which exploits minimal trypsinolysis under nondenaturing conditions to identify branched chains [38]. Instead of bottom-up proteomics where the extensive use of trypsin bestows only short peptides, the middle-down approach leaves the entire ubiquitin molecule intact apart from the last two glycine residues that remain associated to the conjugated lysine of the preceding polypeptide. In the case of a branched chain, the limited proteolysis generates a ubiquitin peptide that then contains two anchored diGly instead of one. By using this approach, Valkevich et al. characterized poly-ubiquitin chains produced in vitro by the bacterial effector E3 ligases NleL and IpaH9.8 to reveal that branch points occurred in 10% of the analyzed chain population [38]. More recently, ubiquitin enrichment combined with minimal trypsinolysis was employed to observe branched ubiquitin in cell culture, illustrating that this method can be applied to more complex samples [39].

In order to specifically detect and quantify K48-K63 branched chains, Dr. Teiji Tanaka and colleagues employed an elegant approach by combining AQUA with a ubiquitin replacement method (*see* also Subheading 3.5) [40]. The authors of the study expressed a ubiquitin variant in which they mutated the R54 to an alanine to prevent trypsin cleavage between K48 and K63. With this ubiquitin mutant, K48-K63 branched chains produce a longer peptide with diGly on both K48 and K63. Interestingly, it was estimated that about 20% of K63 linkages are K48-K63 branched chains in U2OS cells. Furthermore, the appearance of K48-K63 branched chains are markedly increased upon induction of the transcription factor NF-κB, in order to protect K63 chains from DUB activity [40].

3.3 Enrichment of Poly-ubiquitin Chains with Ubiquitin-Binding Domains

In combination with the development of tools for ubiquitin chain analysis, ubiquitin chain enrichment allowed for the study of specific lysine linkages. Enrichment of poly-ubiquitin chains with recombinant proteins that contain ubiquitin-associated domains (UBAs) and ubiquitin-interacting motifs (UIMs) was first used in combination with immobilized-metal affinity chromatography to further purify ubiquitinated proteins containing His-tagged ubiquitin for the identification of proteasome substrates [11, 12]. The approach was then further developed using tandem-repeated ubiquitin-binding entities (TUBEs) and tandem ubiquitin-interacting motifs (tUIMs). TUBEs consist of multiple UBAs and show superior binding to poly-ubiquitin in comparison with individual UBA domains [41]. Because of their strong affinities for poly-ubiquitinated proteins, TUBEs protect them from DUBs and the proteasome, acting as a "trap" to allow for biochemical analysis of these conjugates [41]. Depending on the type of UBA used, TUBEs have been designed to capture either all ubiquitin chains types [41] or to specifically target individual types including M1 [42], K29 and K33 [43], K48 [44], and K63 and K6 [45]. TUBEs are typically covalently bound to a purification tag that allows them and their bound counterparts to be pulled down with beads for further analysis by methods such as MS.

Similarly to TUBEs, tUIMs bind strongly to poly-ubiquitin, preventing their deubiquitination, and, when expressed in vivo, can act as inhibitors to the signaling pathways associated with that poly-ubiquitin chain linkage type [46]. Up to this point, only K48- and K63-tUIMs have been designed and successfully applied, for example, to elucidate protein components involved in K63-specific ubiquitination of targets at DNA double-stranded break sites

[47]. However, the design of tUIMs specific to other chains could be possible as their specificity is determined by the length of the linker between the UIMs [48]. More recently, a trypsin-resistant TUBE was also developed (TR-TUBE) and used in combination with the anti-K-ε-GG antibody for MS identification of ubiquitinated proteins [49]. The expression of a tagged TR-TUBE in cells protects the poly-ubiquitinated proteins from DUBs and proteasome degradation, as well as trypsin digestion in native conditions. The authors of the study found that by using TR-TUBE, they could more readily identify substrates of an overexpressed E3 ligase [49].

3.4 Pulldowns with Chain-Specific Antibodies and Affimers

There are several antibodies that are specific to different ubiquitin-ubiquitin linkages that can enrich for proteins conjugated to different poly-ubiquitin chains. As of the writing of this review, K48, K63, K27, and K11 antibodies are commercially available, and Chapter 24 of this book describes how several of these antibodies can be employed in conjunction with ubiquitin-AQUA. Recently, Dr. Michael Rape and colleagues used an engineered bispecific antibody to characterize K11-K48 branched chains. They found that in addition to cell cycle-regulated proteins, several misfolded nascent polypeptides were conjugated by the Ubr4 and Ubr5 E3 ligases for proteasome degradation [50].

The Komander lab has developed affimers, 12 kDa non-antibody scaffolds, that are highly specific for K6- and K33-/K11-linked ubiquitin chains [51]. While further validation of the K33 affimer is needed, the Komander lab was able to develop a MS-compatible protocol for the enrichment of K6-linked poly-ubiquitin in a cellular context to identify proteins modified with this chain type. Using the affimer, they achieved over a 100-fold enrichment of K6 linkages in their sample demonstrating the potential of this approach.

3.5 Ubiquitin Mutants

One can also explore the roles of linkage-specific poly-ubiquitin chains by inhibiting their formation and observing the effect on various cellular processes. A common way of doing this is known as the ubiquitin replacement strategy, designed by Xu et al., where all four mRNAs encoding ubiquitin in U2OS cells are depleted by RNAi and the simultaneous expression of a RNAi-resistant ubiquitin mutant is induced with tetracycline [52]. These mutants may contain an arginine residue in place of any specific lysine, preventing poly-ubiquitin chain formation. This strategy can be theoretically applied to all ubiquitin lysines. Several studies have used the ubiquitin replacement strategy followed by downstream MS analysis or pulldowns to investigate the roles of poly-ubiquitin topology in the cell [47, 52, 53].

Linear poly-ubiquitin chains, which are linked through M1, play important roles in inflammatory signaling and apoptotic cell death [54]. So far, the only known E3 ligase that can synthesize these chains is the linear ubiquitin assembly complex (LUBAC), and only a handful of substrates have been found [55]. The lack of substrates and the importance they play in disease highlights the need to identify new targets of linear poly-ubiquitin chains. As mentioned before, the diGly motif antibodies are unable to bind to peptides generated from linear ubiquitin after trypsin digestion. Kliza et al. created a lysine-less ubiquitin variant internally tagged with Strep-tag II (INT-Ub) that can only be incorporated into M1-linked poly-ubiquitin chains and can be easily enriched for [55]. They generated stable cell lines expressing INT-Ub variants and pulled down the lysates with Strep-Tactin for MS analysis. Several novel substrates were successfully identified and validated, illustrating that using an INT-Ub is an effective way of studying the role of linear poly-ubiquitin chains.

4 Identification of E3 Ligase Substrates

Up to this point, we have discussed various MS techniques employed to generally identify ubiquitination sites as well as to characterize poly-ubiquitin topology. These methods can be combined with pulldowns or inhibitory assays to study specific components of the UPS. In the next section, we will review several procedures developed for the identification of specific substrates of E3 ligases.

4.1 From the Ubiquitinome to E3 Substrates

The identification of E3 substrates within the ubiquitinome equates to searching for a needle in a haystack, as only a small number of ubiquitinated proteins are targeted by a given E3 ligase in the cell. Our lab performed some of the first studies in which the impact of the absence of an E3 ligase on the ubiquitinome was assessed. More specifically, we sought to determine which proteins are ubiquitinated after heat-shock by the yeast Hul5 and Rsp5 E3s [23, 56]. Heat-shock induces a marked increase of poly-ubiquitination in the cell, which facilitated the identification of many candidate substrates. Ubiquitinated proteins were either enriched by immobilized-metal affinity chromatography of His-tagged ubiquitin or with the anti-K-ε-GG antibody. Levels of ubiquitinated proteins or ubiquitination sites were compared between cells that expressed the wild-type E3 and mutant cells, in which the ligase was either deleted or inactive at a higher temperature. Using this approach, we identified over 100 candidate substrates of the Rsp5 E3 ligase that were ubiquitinated following heat-shock in wild-type but not mutant cells. Notably, several of these substrates contained short linear PY motifs located in structured domains that could only be recognized by the Rsp5 ligase upon misfolding [23]. We utilized a similar approach to determine which Rsp5-conjugated substrates are targeted by the Ubp2 and Ubp3 DUBs for the editing of their poly-ubiquitin chains, demonstrating that a similar methodology can be employed to identify DUB substrates [57].

The approach can also be used to identify smaller subsets of E3 substrates. Thomson JW et al. identified the DNA damage-inducible transcript 4 (DDIT4) as a HUWE1 substrate. They employed an inducible RNAi approach to downregulate the E3 ligase in differentially SILAC-labeled cells and then chemically inhibited the proteasome prior to the diGly enrichment. Among the approximately 2700 and 2800 ubiquitination sites quantified in two independent experiments, ubiquitination levels of only two peptides (corresponding to HUME1 and DDIT4) were found to be reduced in cells in which HUWE1 was not expressed. Turnover of DDIT4 was then further shown to be HUWE-dependent [58].

Major breakthroughs were achieved when similar approaches were employed to identify which proteins are targeted in the presence of thalidomide derivatives by the cereblon E3 ligase complex that is formed with the damaged DNA-binding protein 1 (DDB1), cullin 4A (CUL4A), and regulator of cullins 1 (ROC1). Thalidomide was first proscribed after it was found to cause birth defects in the late 1950s, but its derivatives pomalidomide and lenalidomide were recently introduced in the clinic by Celgene to treat multiple myeloma. These small molecules are thought to mediate the binding of key therapeutic targets to the cereblon E3 ligase leading to both their ubiquitination and degradation. Dr. Benjamin Ebert and colleagues used a combination of proteomics approaches to identify these targets in B cells [59]. Using SILAC and the anti-K-ε-GG antibody, they determined which proteins were more ubiquitinated in the presence of either lenalidomide or thalidomide. They also determined which proteins were (1) present at lower levels within the proteome after the addition of these small molecules (potentially due to proteasome degradation) and (2) bound to cereblon following affinity purification of the cereblon E3 ligase in the presence of the drugs. Using these complementary MS approaches, the authors found that both Ikaros (IKZF1) and Aiolos (IKZF3), which are B cell-specific transcription factors, were targeted for degradation by cereblon in the presence of the drugs [59]. In a follow-up study using similar approaches, the same research teams identified casein kinase 1A1 (CK1α) as another substrate targeted by cereblon in the presence of these small molecules [60]. More recently, An J. et al. used a pulse-SILAC approach to identify which proteins display, upon the addition lenalidomide, an increase in turnover [61]. Pulse-SILAC consists of switching the labeling (i.e., light to heavy) for several hours and then comparing the SILAC ratio for each protein. In this particular case, ubiquitination is not directly assessed. Instead, the increase in turnover for a given protein is revealed when there is relatively less pre-labeled light SILAC species upon the addition of the drug (RNAseq is also performed independently to exclude changes in transcription). Using the pulse-SILAC approach, the authors found that CK1α and the E3 ubiquitin-protein ligase ZFP91, a novel target of cereblon, displayed an increased turnover upon the addition of the drug in multiple experiments using HCT116 and Hek293 tissue culture cells [61]. These studies highlight how crucial MS approaches are to deciphering the inner workings of the UPS, as well as exploiting this system for human therapies.

4.2 Affinity Purification and Proximity Tagging

For many years, affinity purification of a ubiquitin ligase following by MS analysis remained the only proteomic approach available to identify E3 substrates. Dr. Michele Pagano and colleagues were, for instance, highly successful in using this approach to identify novel substrates of SKP1-CUL1-F-box (SCF) E3 ligases after performing co-immunoprecipitation experiments with the substrate-adaptor F-box proteins ßTrCP2, Fbxl3 and Fbxw7α [62–65]. However, due to the transient nature of E3 ligases with their substrates and the often short-lived nature of ubiquitinated proteins, it is typically difficult to identify interacting partners [66]. To address this problem, Dr. Wade Harper and colleagues developed parallel adaptor capture (PAC) proteomics where cells are subjected to proteasome inhibition with bortezomib and the cullin ring ligase (CRL) inhibitor MLN4924, which in turn increases the abundance of CRL and proteasomal substrates [67]. This buildup allows for a higher recovery of these substrates when subjected to subsequent affinity purification, MS, and analysis using *Comp*Pass (comparative proteomics analysis software suite). In this particular study, Tan et al. identified candidate targets for the leucine-rich repeat family of F-box proteins (FBXLs) that function with SCF, as well as validated one of the newly identified targets. By inhibiting both the proteasome and different E3 ligases, this procedure can be applied to a diverse subset of E3s to improve substrate identification if they are degraded by the proteasome [67]. Using a somewhat similar concept, pulldown of a ligase substrate receptor has also been done with a mutated version of an F-box protein that could not bind to the rest of the SCF complex (note that F-box proteins are the substrate receptor subunits of the SCF E3s). Expression of the mutant FBOX21 protein led to the identification of the EID1 (EP300-interacting inhibitor of differentiation 1) substrate following pulldown and label-free MS analysis [68].

An alternative to co-immunoprecipitation experiments is to fuse the bait protein to a biotin ligase causing all proteins in proximity to become biotinylated. This method is termed biotin identification (BioID) where these biotinylated proteins (potential interactors) can be extracted from a lysate using streptavidin beads and then analyzed by MS [69]. The advantage of BioID over traditional immunoprecipitation-based screens is that it can be applied to both insoluble proteins and weak interactors. BioID has been successfully applied within the UPS, for example, to identify SCF$^{\beta\text{-TrCP1/2}}$ E3 ligase substrates [70]. This concept can be applied across a large number of candidate proteins and is potentially a favorable method to screen for interacting partners within the UPS in vivo.

4.3 Engineered E3 Ligases

Several elegant proteomic approaches involve reengineering E3 ligases. In Zhuang M et al., a fusion protein was produced between an E3 substrate binding domain and Ubc12, the Nedd8 E2 conjugating enzyme [71]. This engineered E3 directs the conjugation of the rare ubiquitin-like protein Nedd8 to the ligase substrates, thereby facilitating their identification using MS. Using this approach, the authors identified numerous candidate substrates for the inhibitors of apoptosis proteins 1 (IAP1) E3 ligase [71].

A different approach was used by Mark KG et al. to trap E3 ligase substrates [72]. They fused a UBA domain to F-box adaptor proteins to identify substrates of several SCF E3 ligase complexes (recall that UBA domains have affinity for poly-ubiquitin chains). Following the formation of a poly-ubiquitin chain by SCF onto its substrates, the poly-ubiquitinated substrates are trapped to the F-box by having a higher affinity for the reengineered SCF ligase. These substrates can then be identified following co-immunoprecipitation and MS analysis. Using this approach, the authors found that the Saf1 F-box protein mediates the degradation of unprocessed lysosomal proteins in yeast [72].

The concept of trapping E3 substrates was also developed by O'Connor et al., using ubiquitin-activated interaction traps (UBAITs) to covalently link substrates to their cognate E3 ligases [73], and is also presented in Chapter 7 of this book. UBAITs consist of fusion proteins between the carboxyl end of an E3 and the amino terminus of ubiquitin that, through the ubiquitin cascade, trap target substrates to the E3. The fused ubiquitin is first activated similar to free ubiquitin by forming a thioester bond to either an E2, or in the case of HECT E3s, to the E3 itself. In principle, the fused ubiquitin is then conjugated to a substrate that becomes covalently linked to the engineered E3 ubiquitin. Following a pulldown, these E3-Ub substrates can be subjected to MS analysis and subsequent substrate identification. UBAITs have been validated with both HECT and RING E3 ligases from both humans and yeast, and several novel substrates were discovered in this study. Similar to biotinylated proximity labeling, the authors suggest that this method has the potential to be applied to other proteins besides E3 ligases [73].

Another tool that recently has been developed is E3 enzyme reversal, where an E3 ligase is fused to the catalytic domain of a DUB. This DUB fused to the E3 antagonizes the activity of the endogenous ubiquitin ligase counterpart in a dominant manner. It can stabilize the E3 substrates by preventing them from lysosomal or proteasomal degradation [74]. MacDonald et al. created a library of E3-DUBs composed of over 70 yeast ubiquitin ligases that can be inducibly expressed with the addition of copper to the media. Coupling these "anti-ligases" with other ubiquitin enrichment methods and MS analysis, one can observe the effect that a loss of an E3 has on the ubiquitinome in yeast. As such, the authors used their library as a discovery tool and identified several candidate substrates of the Pib1 E3 ligase involved in the ubiquitin-dependent vacuolar sorting [74].

5 Conclusion

In summary, there has been exciting and significant progress made to reveal the ubiquitinome using proteomic approaches over the past decade. Our understanding of the UPS through the analysis of ubiquitination sites, chain linkages, and system components will continue to expand through the development of new techniques and improvement upon existing ones. Continued proteomic-based research of the UPS remains crucial to the understanding of associated diseases such as various cancers and neurodegenerative disorders and will likely play a major role in identifying novel targets for therapeutic drugs.

Acknowledgments

The authors would like to thank all lab members for the discussions and Cristen Molzahn for the comments on the manuscript. They also acknowledge support from the Canadian Institutes of Health Research (CIHR) and British Columbia Proteomics Network (BCPN); TM is a MSFHR new investigator.

References

1. Cohen P, Tcherpakov M (2010) Will the ubiquitin system furnish as many drug targets as protein kinases? Cell 143(5):686–693. https://doi.org/10.1016/j.cell.2010.11.016
2. Huang X, Dixit VM (2016) Drugging the undruggables: exploring the ubiquitin system for drug development. Cell Res 26 (4):484–498. https://doi.org/10.1038/cr.2016.31
3. Weathington NM, Mallampalli RK (2014) Emerging therapies targeting the ubiquitin proteasome system in cancer. J Clin Invest 124(1):6–12. https://doi.org/10.1172/JCI71602
4. Komander D, Rape M (2012) The ubiquitin code. Annu Rev Biochem 81:203–229. https://doi.org/10.1146/annurev-biochem-060310-170328
5. Hershko A, Ciechanover A (1998) The ubiquitin system. Annu Rev Biochem 67:425–479. https://doi.org/10.1146/annurev.biochem.67.1.425
6. Wilkinson KD (2000) Ubiquitination and deubiquitination: targeting of proteins for degradation by the proteasome. Cell Dev Biol 11 (3):141–148. https://doi.org/10.1006/scdb.2000.0164
7. Rajalingam K, Dikic I (2016) SnapShot: expanding the ubiquitin code. Cell 164

(5):1074–1074.e1. https://doi.org/10.1016/j.cell.2016.02.019
8. Zhang Y, Fonslow BR, Shan B et al (2013) Protein analysis by shotgun/bottom-up proteomics. Chem Rev 113(4):2343–2394. https://doi.org/10.1021/cr3003533
9. Peng J, Schwartz D, Elias JE et al (2003) A proteomics approach to understanding protein ubiquitination. Nat Biotechnol 21 (8):921–926. https://doi.org/10.1038/nbt849
10. Lichti CF, Wildburger NC, Emmett MR et al (2014) Post-translational modifications in the human proteome. In: Genomics and proteomics for clinical discovery and development. Springer, Dordrecht, pp 101–136
11. Mayor T, Lipford JR, Graumann J et al (2005) Analysis of poly-ubiquitin conjugates reveals that the Rpn10 substrate receptor contributes to the turnover of multiple proteasome targets. Mol Cell Proteomics 4(6):741–751. https://doi.org/10.1074/mcp.M400220-MCP200
12. Mayor T, Graumann J, Bryan J et al (2007) Quantitative profiling of ubiquitylated proteins reveals proteasome substrates and the substrate repertoire influenced by the Rpn10 receptor pathway. Mol Cell Proteomics 6 (11):1885–1895. https://doi.org/10.1074/mcp.M700264-MCP200
13. Meierhofer D, Wang X, Huang L, Kaiser P (2008) Quantitative analysis of global ubiquitination in HeLa cells by mass spectrometry. J Proteome Res 7(10):4566–4576. https://doi.org/10.1021/pr800468j
14. Tagwerker C, Flick K, Cui M et al (2006) A tandem affinity tag for two-step purification under fully denaturing conditions: application in ubiquitin profiling and protein complex identification combined with in vivo cross-linking. Mol Cell Proteomics 5(4):737–748. https://doi.org/10.1074/mcp.M500368-MCP200
15. Xu G, Paige JS, Jaffrey SR (2011) Global analysis of lysine ubiquitination by ubiquitin remnant immunoaffinity profiling. Nat Biotechnol 28(8):868–873. https://doi.org/10.1038/nbt.1654
16. Kim W, Bennett EJ, Huttlin EL et al (2011) Systematic and quantitative assessment of the ubiquitin-modified proteome. Mol Cell 44 (2):325–340. https://doi.org/10.1016/j.molcel.2011.08.025
17. Skaug B, Chen ZJ (2010) Emerging role of ISG15 in antiviral immunity. Cell 143 (2):187–190. https://doi.org/10.1016/j.cell.2010.09.033
18. Wagner SA, Beli P, Weinert BT et al (2012) Proteomic analyses reveal divergent ubiquitylation site patterns in murine tissues. Mol Cell Proteomics 11(12):1578–1585. https://doi.org/10.1074/mcp.M112.017905
19. Wagner SA, Beli P, Weinert BT et al (2011) A proteome-wide, quantitative survey of in vivo ubiquitylation sites reveals widespread regulatory roles. Mol Cell Proteomics 10(10):M111.013284. https://doi.org/10.1074/mcp.M111.013284
20. Hornbeck PV, Zhang B, Murray B et al (2015) PhosphoSitePlus, 2014: Mutations, PTMs and recalibrations. Nucleic Acids Res 43(DI):D512–D520. https://doi.org/10.1093/nar/gku1267
21. Udeshi ND, Mertins P, Svinkina T, Carr SA (2013) Large-scale identification of ubiquitination sites by mass spectrometry. Nat Methods 8 (10):1950–1960. https://doi.org/10.1038/nprot.2013.120
22. Mann M (2006) Functional and quantitative proteomics using SILAC. Nat Rev Mol Cell Biol 7(12):952–959. https://doi.org/10.1038/nrm2067
23. Fang NN, Chan GT, Zhu M et al (2014) Rsp5/Nedd4 is the main ubiquitin ligase that targets cytosolic misfolded proteins following heat stress. Nat Cell Biol 16(12):1227–1237. https://doi.org/10.1038/ncb3054
24. Silva G, Finley D, Vogel C (2015) K63 polyubiquitination is a new modulator of the oxidative stress response. Nat Struct Mol Biol 22 (2):116–123. https://doi.org/10.1038/nsmb.2955
25. Rose CM, Isasa M, Ordureau A et al (2016) Highly multiplexed quantitative mass spectrometry analysis of ubiquitylomes. Cell Syst 3 (4):395–403. https://doi.org/10.1016/j.cels.2016.08.009
26. Rauniyar N, Yates JR (2014) Isobaric labeling-based relative quantification in shotgun proteomics. J Proteome Res 13(12):5293–5309. https://doi.org/10.1021/pr500880b
27. Swaney DL, Beltrao P, Starita L et al (2014) Global analysis of phosphorylation and ubiquitylation crosstalk in protein degradation. Nat Methods 10(7):676–682. https://doi.org/10.1038/nmeth.2519
28. Shi Y, Xu P, Qin J (2011) Ubiquitinated proteome: ready for global? Mol Cell Proteomics 10 (5):R110.006882. https://doi.org/10.1074/mcp.R110.006882
29. Chicooree N, Connolly Y, Tan C-T et al (2013) Enhanced detection of ubiquitin isopeptides. J Am Soc Mass Spectrom 24

(3):421–430. https://doi.org/10.1007/s13361-012-0538-0

30. Griffiths JR, Connolly Y, Griffiths JR et al (2014) Mass spectral enhanced detection of Ubls using SWATH of SUMO and ubiquitin-derived isopeptides. J Am Soc Mass Spectrom 25(5):767–777. https://doi.org/10.1007/s13361-014-0835-x
31. Gillet LC, Navarro P, Tate S et al (2012) Targeted data extraction of the MS/MS spectra generated by data-independent acquisition: a new concept for consistent and accurate proteome analysis. Mol Cell Proteomics 11(6): O111.016717. https://doi.org/10.1074/mcp.O111.016717
32. Yau R, Rape M (2016) The increasing complexity of the ubiquitin code. Nat Cell Biol 18 (6):579–586. https://doi.org/10.1038/ncb3358
33. Kaliszewski P, Zoładek T (2008) The role of Rsp5 ubiquitin ligase in regulation of diverse processes in yeast cells. Acta Biochim Pol 55 (4):649–662
34. Kirkpatrick DS, Hathaway NA, Hanna J et al (2006) Quantitative analysis of in vitro ubiquitinated cyclin B1 reveals complex chain topology. Nat Cell Biol 8(7):700–710. https://doi.org/10.1038/ncb1436
35. Phu L, Izrael-tomasevic A, Matsumoto ML et al (2011) Improved quantitative mass spectrometry methods for characterizing complex ubiquitin signals. Mol Cell Proteomics 10(5): M110.003756. https://doi.org/10.1074/mcp.M110.003756
36. Liu C, Liu W, Ye Y, Li W (2017) Ufd2p synthesizes branched ubiquitin chains to promote the degradation of substrates modified with atypical chains. Nat Commun 8:14274. https://doi.org/10.1038/ncomms14274
37. Meyer H, Rape M (2014) Enhanced protein degradation by branched ubiquitin chains. Cell 157(4):910–921. https://doi.org/10.1016/j.cell.2014.03.037
38. Valkevich EM, Sanchez NA, Ge Y, Strieter ER (2014) Middle-down mass spectrometry enables characterization of branched ubiquitin chains. Biochemistry 53(30):4979–4989. https://doi.org/10.1021/bi5006305
39. Crowe SO, Rana ASJB, Deol KK et al (2017) Ubiquitin chain enrichment middle-down mass spectrometry enables characterization of branched ubiquitin chains in cellulo. J Anal Chem 89(8):4428–4434. https://doi.org/10.1021/acs.analchem.6b03675
40. Ohtake F, Saeki Y, Ishido S et al (2016) The K48-K63 branched ubiquitin chain regulates NF-κB signaling. Mol Cell 64(2):251–266. https://doi.org/10.1016/j.molcel.2016.09.014
41. Hjerpe R, Aillet F, Lopitz-Otsoa F et al (2009) Efficient protection and isolation of ubiquitylated proteins using tandem ubiquitin-binding entities. EMBO Rep 10(11):1250–1258. https://doi.org/10.1038/embor.2009.192
42. Rahighi S, Ikeda F, Kawasaki M et al (2009) Specific recognition of linear ubiquitin chains by NEMO is important for NF-kB activation. Cell 136(6):1098–1109. https://doi.org/10.1016/j.cell.2009.03.007
43. Kristariyanto YA, Rehman SAA, Campbell DG et al (2015) K29-selective ubiquitin binding domain reveals structural basis of specificity and heterotypic nature of K29 polyubiquitin. Mol Cell 58(1):83–94. https://doi.org/10.1016/j.molcel.2015.01.041
44. Kristariyanto YA, Rehman SAA, Weidlich S et al (2017) A single MIU motif of MINDY-1 recognizes K48-linked polyubiquitin chains. EMBO Rep 18(3):392–402. https://doi.org/10.15252/embr.201643205
45. Kulathu Y, Akutsu M, Bremm A et al (2009) Two-sided ubiquitin binding explains specificity of the TAB2 NZF domain. Nat Struct Mol Biol 16(12):1328–1330. https://doi.org/10.1038/nsmb.1731
46. Sims JJ, Scavone F, Cooper EM et al (2012) Polyubiquitin-sensor proteins reveal localization and linkage- type dependence of cellular ubiquitin signaling. Nat Methods 9 (3):303–309. https://doi.org/10.1038/nmeth.1888
47. Thorslund T, Ripplinger A, Hoffmann S et al (2015) Histone H1 couples initiation and amplification of ubiquitin signalling after DNA damage. Nature 527(7578):389–393. https://doi.org/10.1038/nature15401
48. Sims JJ, Cohen RE (2009) Linkage-specific avidity defines the lysine 63-linked polyubiquitin-binding preference of Rap80. Mol Cell 33 (6):775–783. https://doi.org/10.1016/j.molcel.2009.02.011
49. Yoshida Y, Saeki Y, Murakami A et al (2015) A comprehensive method for detecting ubiquitinated substrates using TR-TUBE. Proc Natl Acad Sci 112(15):4630–4635. https://doi.org/10.1073/pnas.1422313112
50. Yau RG, Doerner K, Castellanos ER et al (2017) Assembly and function of heterotypic ubiquitin chains in cell-cycle and protein quality control. Cell 171(4):918–933.e20. https://doi.org/10.1016/j.cell.2017.09.040
51. Michel MA, Swatek KN, Hospenthal MK, Komander D (2017) Ubiquitin linkage-specific affimers reveal insights into K6-linked

ubiquitin signaling. Mol Cell 68(1):233–246.e5. https://doi.org/10.1016/j.molcel.2017.08.020

52. Xu M, Skaug B, Zeng W, Chen ZJ (2009) A ubiquitin replacement strategy in human cells reveals distinct mechanisms of IKK activation by TNFα and IL-1β. Mol Cell 36(2):302–314. https://doi.org/10.1016/j.molcel.2009.10.002
53. Ordureau A, Heo J-M, Duda DM et al (2015) Defining roles of PARKIN and ubiquitin phosphorylation by PINK1 in mitochondrial quality control using a ubiquitin replacement strategy. Proc Natl Acad Sci 112(21):6637–6642. https://doi.org/10.1073/pnas.1506593112
54. Rittinger K, Ikeda F (2017) Linear ubiquitin chains: enzymes, mechanisms and biology. Open Biol 7(4):170026. https://doi.org/10.1098/rsob.170026
55. Kliza K, Taumer C, Pinzuti I et al (2017) Internally tagged ubiquitin: a tool to identify linear polyubiquitin-modified proteins by mass spectrometry. Nat Methods 14(5):504–512. https://doi.org/10.1038/nmeth.4228
56. Fang N, Ng A, Measday V, Mayor T (2011) Hul5 HECT ubiquitin ligase plays a major role in the ubiquitylation and turn over of cytosolic misfolded proteins. Nat Cell Biol 13 (11):1344–1352. https://doi.org/10.1038/ncb2343
57. Fang NN, Zhu M, Rose A et al (2016) Deubiquitinase activity is required for the proteasomal degradation of misfolded cytosolic proteins upon heat-stress. Nat Commun 7:12907. https://doi.org/10.1038/ncomms12907
58. Thompson JW, Nagel J, Hoving S et al (2014) Quantitative Lys-∈–Gly-Gly (diGly) proteomics coupled with inducible RNAi reveals ubiquitin-mediated proteolysis of DNA damage-inducible transcript 4 (DDIT4) by the E3 Ligase HUWE1. J Biol Chem 289 (42):28942–28955. https://doi.org/10.1074/jbc.M114.573352
59. Krönke J, Udeshi ND, Narla A et al (2014) Lenalidomide causes selective degradation of IKZF1 and IKZF3 in multiple myeloma cells. Science 343(6168):301–305. https://doi.org/10.1126/science.1244851
60. Krönke J, Fink EC, Hollenbach PW et al (2015) Lenalidomide induces ubiquitination and degradation of CK1α in del(5q) MDS. Nature 523(7559):183–188. https://doi.org/10.1038/nature14610
61. An J, Ponthier CM, Sack R et al (2017) PSI-LAC mass spectrometry reveals ZFP91 as IMiD-dependent substrate of the CRL4 CRBN ubiquitin ligase. Nat Commun 8:15398. https://doi.org/10.1038/ncomms15398
62. Peschiaroli A, Dorrello NV, Guardavaccaro D et al (2006) SCFβTrCP-mediated degradation of claspin regulates recovery from the DNA replication checkpoint response. Mol Cell 23 (3):319–329. https://doi.org/10.1016/j.molcel.2006.06.013
63. Dorrello NV, Peschiaroli A, Guardavaccaro D et al (2006) S6K1- and bTRCP-mediated degradation of PDCD4 promotes protein translation and cell growth. Science 314 (5798):467–472. https://doi.org/10.1126/science.1130276
64. Ko HW, Jiang J, Edery I et al (2007) SCFFbxl3 controls the oscillation of the circadian clock by directing the degradation of cryptochrome proteins. Science 316(5826):900–904. https://doi.org/10.1126/science.1141194
65. Busino L, Millman SE, Scotto L et al (2012) Fbxw7α- and GSK3-mediated degradation of p100 is a pro-survival mechanism in multiple myeloma. Nat Cell Biol 14(4):375–385. https://doi.org/10.1038/ncb2463
66. Harper JW, Tan M-KM (2012) Understanding cullin-RING E3 biology through proteomics-based substrate identification. Mol Cell Proteomics 11(12):1541–1550. https://doi.org/10.1074/mcp.R112.021154
67. Tan MKM, Lim HJ, Bennett EJ et al (2013) Parallel SCF adaptor capture proteomics reveals a role for SCFFBXL17 in NRF2 activation via BACH1 repressor turnover. Mol Cell 52(1):9–24. https://doi.org/10.1016/j.molcel.2013.08.018
68. Watanabe K, Yumimoto K, Nakayama KI (2015) FBXO21 mediates the ubiquitylation and proteasomal degradation of EID1. Genes Cells 20(8):667–674. https://doi.org/10.1111/gtc.12260
69. Roux KJ, Kim DI, Burke B (2013) BioID: a screen for protein-protein interactions. Curr Protoc Protein Sci 74:Unit 19.23. https://doi.org/10.1002/0471140864.ps1923s74
70. Coyaud E, Mis M, Laurent EMN et al (2015) BioID-based identification of Skp cullin F-box (SCF) β-TrCP1/2 E3 ligase substrates. Mol Cell Proteomics 14(7):1781–1795. https://doi.org/10.1074/mcp.M114.045658
71. Zhuang M, Guan S, Wang H et al (2013) Substrates of IAP ubiquitin ligases identified with a designed orthogonal E3 ligase, the NEDDylator. Mol Cell 49(2):273–282. https://doi.org/10.1016/j.molcel.2012.10.022
72. Mark KG, Simonetta M, Maiolica A et al (2014) Ubiquitin ligase trapping identifies an

SCFSaf1 pathway targeting unprocessed vacuolar/lysosomal proteins. Mol Cell 53 (1):148–161. https://doi.org/10.1016/j.molcel.2013.12.003

73. O'Connor HF, Lyon N, Leung JW et al (2015) Ubiquitin-activated interaction traps (UBAITs) identify E3 ligase binding partners. EMBO Rep 16(12):1699–1712. https://doi.org/10.15252/embr.201540620

74. MacDonald C, Winistorfer S, Pope RM et al (2017) Enzyme reversal to explore the function of yeast E3 ubiquitin-ligases. Traffic 18 (7):465–484. https://doi.org/10.1111/tra.12485

Chapter 23

Ubiquitin diGLY Proteomics as an Approach to Identify and Quantify the Ubiquitin-Modified Proteome

Amit Fulzele and Eric J. Bennett

Abstract

Protein ubiquitylation is one of the most prevalent posttranslational modifications (PTM) within cells. Ubiquitin modification of target lysine residues typically marks substrates for proteasome-dependent degradation. However, ubiquitylation can also alter protein function through modulation of protein complexes, localization, or activity, without impacting protein turnover. Taken together, ubiquitylation imparts critical regulatory control over nearly every cellular, physiological, and pathophysiological process. Affinity purification techniques coupled with quantitative mass spectrometry have been robust tools to identify PTMs on endogenous proteins. A peptide antibody-based affinity approach has been successfully utilized to enrich for and identify endogenously ubiquitylated proteins. These antibodies recognize the Lys-ϵ-Gly-Gly (diGLY) remnant that is generated following trypsin digestion of ubiquitylated proteins, and these peptides can then be identified by standard mass spectrometry approaches. This technique has led to the identification of >50,000 ubiquitylation sites in human cells and quantitative information about how many of these sites are altered upon exposure to diverse proteotoxic stressors. In addition, the diGLY proteomics approach has led to the identification of specific ubiquitin ligase targets. Here we provide a detailed method to interrogate the ubiquitin-modified proteome from any eukaryotic organism or tissue.

Key words Ubiquitin, Proteomics, diGLY, Affinity purification, Mass spectrometry, SILAC

1 Introduction

Protein posttranslational modifications (PTMs) impart critical regulatory control on nearly every cellular process. PTMs diversify the proteome to such a degree that we will likely never realize the full extent of proteome complexity. Common posttranslational modifications include phosphorylation [1–6], glycosylation [7–11], ubiquitylation [12–20], nitrosylation [21], methylation [22], acetylation [23], and lipidation [24], all of which impact normal cell biology as well as pathogenesis. Therefore, identifying and understanding the function of individual PTMs are critical to understand cellular homeostasis.

PTMs modify a subpopulation of proteins and only a small portion of the total proteome; therefore, identification relies on

Thibault Mayor and Gary Kleiger (eds.), *The Ubiquitin Proteasome System: Methods and Protocols*, Methods in Molecular Biology, vol. 1844, https://doi.org/10.1007/978-1-4939-8706-1_23,

the development of efficient and robust enrichment approaches. A number of different strategies to enrich for ubiquitylated proteins have been developed, including utilizing overexpression of epitope-tagged versions of ubiquitin followed by affinity-based enrichment of ubiquitylated proteins [12, 25]. Another strategy uses purified ubiquitin-binding domains from select proteins to enrich for endogenously ubiquitylated proteins [26]. Each of these techniques has several advantages and drawbacks which have been reviewed previously [27–29]. While each approach can identify putative ubiquitylated proteins, they are less effective in identifying the precise sites of ubiquitylation on those proteins.

Since the initial peptide sequencing of ubiquitin and the demonstration that ubiquitin modifies lysine residues on substrates, it was appreciated that trypsin digestion of a ubiquitylated protein would generate peptides containing a characteristic diglycine (diGLY)-modified lysine residue that could be used to identify ubiquitylation sites. In fact, the existence of the remnant ubiquitin diGLY residue was first reported on histone H2A by A24 (later renamed ubiquitin) by Goldknopf and Busch in 1977 [30]. These diGLY-modified peptides were then easily distinguished by "bottom-up" mass spectrometry approaches, identifying ubiquitin-modified proteins along with the exact site of modification. However, the low abundance of ubiquitylated peptides compared to linear non-modified peptides in a whole-proteome digest makes the identification of diGLY-modified peptides difficult without enrichment steps. In 2003, Peng J et al. first attempted to study the ubiquitin-modified proteome based on the existence of the known ubiquitin remnant diGLY motif on ubiquitylated substrate peptides upon trypsinolysis [12]. This report pressed the need for development of new tools to capture these modifications.

Global analysis of the ubiquitin-modified proteome using mass spectrometry gained momentum when initial studies reported the development of antibody-based enrichment strategies that utilized antibodies capable of binding diGLY-modified peptides [18]. Thereafter, the generation of more robust ubiquitin remnant diGLY motif-specific antibodies allowed for the identification of more than 10,000 ubiquitylated (diGLY) peptides [19, 20]. It is critical to note that the C-terminal protein sequences of the ubiquitin-like proteins, Nedd8 and ISG15, are like ubiquitin and will leave identical diGLY-modified peptides upon trypsinolysis of Neddylated or ISGylated proteins. As such, identification of a diGLY-modified peptide does not, on its own, unequivocally identify a protein as being ubiquitylated. However, studies have shown that ~95% of all diGLY peptides identified using the diGLY-antibody enrichment approach arise from ubiquitylation versus Neddylation or ISGylation [19].

The diGLY proteomics approach now has become an indispensable tool to systematically interrogate protein ubiquitylation with site-level resolution [31–33]. The use of quantitative mass spectrometry approaches coupled with diGLY-antibody affinity enrichment steps has led to a more complete understanding of both the breadth of ubiquitylation and global alterations in protein ubiquitylation in response to an increasing variety of cell stimuli and stressors [34–43]. This approach has also proven useful to identify substrates for specific ubiquitin ligases [19, 38, 41, 43–49]. Another advantage of the diGLY-antibody affinity approach is that it can be applied toward the identification of ubiquitylated proteins from any human or murine primary tissue or with any eukaryote [37, 40, 50–54].

Here, we describe a SILAC-based quantitative proteomic method to identify differential ubiquitylation between two samples (Fig. 1). SILAC-based approaches are not required, and label-free as well as chemical isobaric-labeling approaches can also be used (*see* **Note 1**). Additional enrichment and/or fractionation steps can be used to improve the depth of the enrichment of diGLY-modified peptides depending on the exact experimental question [18, 20, 35, 39, 41, 42, 48, 51, 53–58]. Here, we limit our method description to a simple one-pot affinity strategy without extensive post-lysis or post-digestion fractionation steps.

2 Materials

2.1 Cell Culture Media and Lysis Buffer

The media may vary according to the cells used for the experiment. Here, we use standard DMEM which can be used with a variety of human cell types commonly grown in culture.

1. Light-SILAC media: DMEM lacking lysine and arginine, 10% dialyzed fetal bovine serum, 150 μg/mL L-lysine-2HCl (light), 85 μg/mL L-arginine-HCl (light), and 1% penicillin-streptomycin (Pen/Strep) (*see* **Note 2**). Remove 50 mL of DMEM from the 500 mL bottle, and add 50 mL of thawed FBS. Dissolve 150 mg of light L-lysine-2HCl and 85 mg of light L-arginine-HCl using 1 mL of DMEM media. Mix thoroughly and add it to the light media. Add Pen/Strep to avoid contamination during the cell culture.
2. Heavy-SILAC media: DMEM lacking lysine and arginine, 10% dialyzed FBS, 150 μg/mL $^{13}C_6$-$^{15}N_2$ L-lysine-2HCl (heavy; $^{13}C_6$, 99%; $^{15}N_2$, 99%; Cambridge Isotope Laboratories), 85 μg/mL $^{13}C_6$-$^{15}N_4$ L-arginine-HCl (heavy; $^{13}C_6$, 99%; $^{15}N_4$, 99%; Cambridge Isotope Laboratories), and 1% Pen/Strep. Remove 50 mL of DMEM media lacking lysine and arginine from the 500 mL bottle, and add 50 mL of thawed dialyzed FBS. Dissolve 150 mg of heavy L-lysine-2HCl and 85 mg of heavy L-arginine-HCl using 1 mL of DMEM media. Mix thoroughly, and add it to the heavy media and add Pen/Strep.

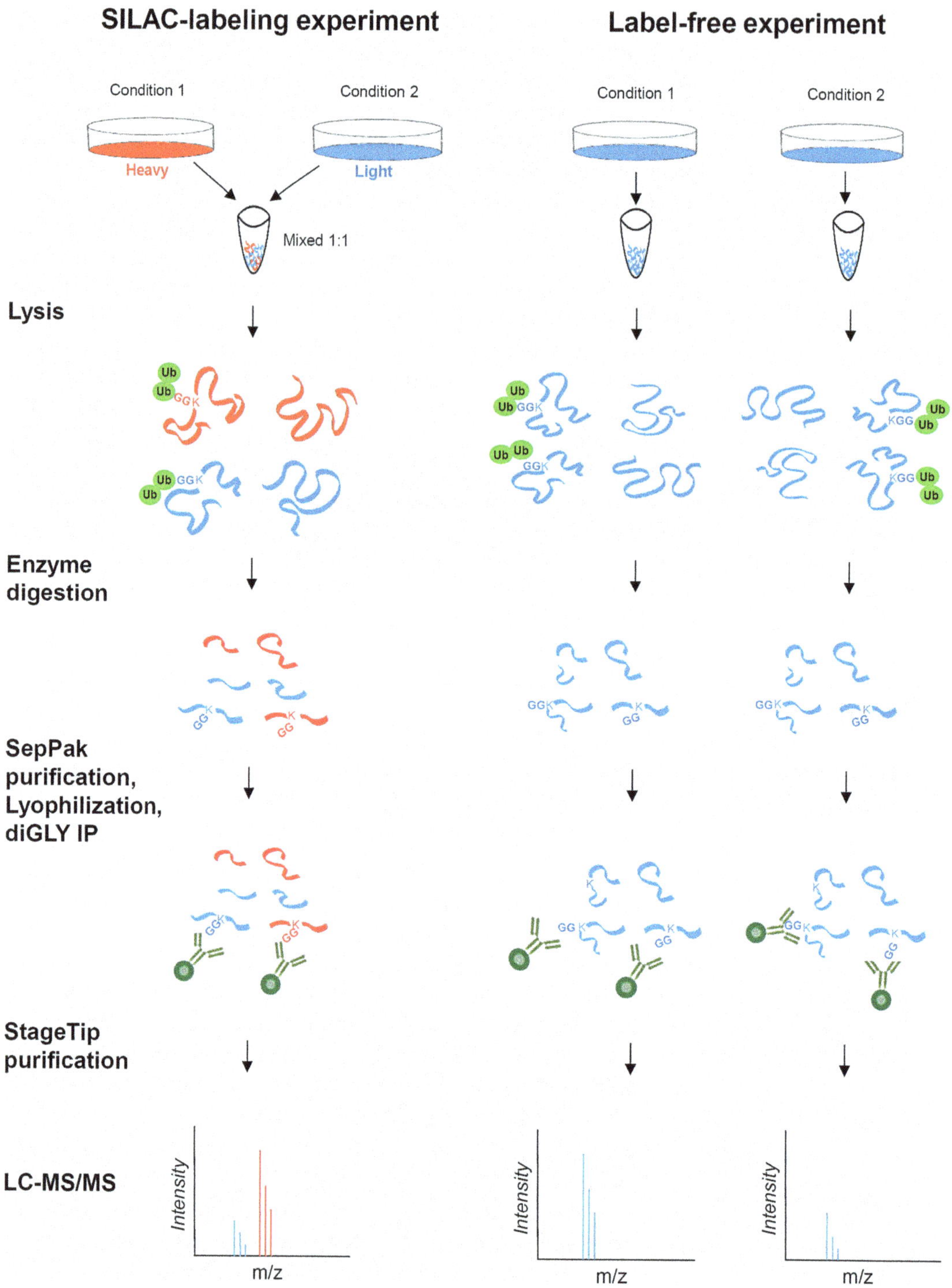

Fig. 1 Workflow schematic for diGLY proteomics

3. Sterile phosphate-buffered saline (PBS).
4. 0.25% trypsin EDTA (cell culture grade).
5. Lysis buffer: 8 M urea, 150 mM NaCl, 50 mM Tris–HCl (pH 8), complete protease inhibitor, 1 mM sodium fluoride (NaF), 1 mM β-glycerophosphate (β-Gly), 1 mM sodium orthovanadate (NaV), and 5 mM N-ethylmaleimide (NEM). Dissolve NEM in ethanol and prepare fresh before addition to the lysis buffer (*see* **Notes 3–5**).

2.2 Protein Digestion

1. LysC protease enzyme (2 AU; Wako): For 2 AU vial of LysC, resuspend in 400 μL of HPLC grade water for a final concentration of 0.005 AU/μL, approximately 2 mg/mL. Make 20 μL aliquots in 0.5 mL tubes and store at −80 °C.
2. Trypsin protease enzyme (TPCK treated; Sigma): Prepare the stock concentration of 0.1 mg/mL in a 50 mM ammonium bicarbonate buffer, and store at −80 °C (*see* **Note 6**).
3. 1 M $CaCl_2$ stock solution.

2.3 Peptide Desalting

1. SepPak™ tC18 reverse-phase column (waters): For 30 mg of a protein digest, a 500 mg SepPak™ tC18 column is recommended [59]. This protocol uses 3 cc volume capacity cartridges (*see* **Note** 7). If using the other cartridge sizes, the volume of solutions should be adapted accordingly (i.e., for 6 cc, tC18 cartridge, use 6 mL of the volume of the solution during each wash step).
2. 100% acetonitrile (ACN): Use HPLC-grade acetonitrile for preparing the solutions.
3. 50% acetonitrile (ACN) and 0.5% acetic acid (HAcO).
4. 0.1% trifluoroacetic acid (TFA).
5. 0.4% trifluoroacetic acid (TFA).
6. 0.5% acetic acid (HAcO).

2.4 diGLY IP

1. Ubiquitin remnant motif (K-Ɛ-GG) antibody or PTMScan® ubiquitin remnant motif (K-Ɛ-GG) kit (Cell Signaling Technology) (*see* **Note 8**).
2. Protein-A Plus Ultralink™ Resin (Thermo Fisher, 53,142).
3. Phosphate-buffered saline (PBS).
4. HPLC-grade water.
5. Immunoaffinity purification (IAP) buffer: The composition of 10× IAP buffer is 500 mM MOPS-NaOH (pH 7.5), 100 mM Na_2HPO_4, and 500 mM NaCl. IAP buffer can be stored at 4 °C for up to 1 month. We typically make the 10× IAP fresh each time.

2.5 Antibody Cross-Linking (Optional)

1. Cross-linking buffer: 0.2 M Triethanolamine, pH 8.2 (F.W. 149.19, *d* = 1.124 g/mL).
2. Cross-linking agent: Dimethyl pimelimidate dihydrochloride (DMP). Prepare fresh.
3. Blocking buffer: 0.1 M Ethanolamine, pH8.2 (M.W. 61.08 g/mol).
4. Antibody cross-linked bead storage buffer: 0.1% Tween-20, 0.02% sodium azide in PBS.

2.6 Stage-Tip Peptide Purification

1. Stage tips, Empore™ C18-embedded membrane (Sigma-Aldrich): Punch a small hole through the four layers of Empore C18 material discs, and mount it at the narrow end of an unautoclaved 200 μL pipette tip.
2. Methanol, HPLC grade.
3. 80% acetonitrile (ACN), 5% formic acid (FA).
4. 5% formic acid.
5. 5% acetonitrile (ACN), 5% formic acid (FA).

2.7 Equipment

1. Sonicator.
2. 37 °C incubator.
3. Lyophilizer.
4. Speed vacuum centrifuge.
5. Nano-HPLC.
6. Fused silica column (15 cm, 100 μm ID) packed with C18 material (porous spherical silica, 1.9 μM; pore diameter to surface area, 120 $A^0/300$ m^2/g; %carbon, 15%; Dr. Maisch GmbH).
7. Mass spectrometer.

2.8 Cells

1. We typically use eight 15 cm dishes of HCT116 cells per experiment (four plates for heavy-labeled cells, four plates for light-labeled cells). This corresponds to ~2 × 10^8 cells and ~30 mg of total protein upon lysis (*see* **Note 9**). The heavy cell population is typically treated with a cell stressor or is altered in other ways (e.g., knockout of an E3 ligase gene).

3 Methods

3.1 Cell/Tissue Harvest and Lysis

1. Once the cells are ready to harvest, remove the cells from 15 cm dishes with 4 mL of 0.25% trypsin EDTA (cell culture grade) per dish.

2. Add 30 mL of cold SILAC media (add appropriate heavy or light media depending upon the treatment condition) to a single dish.
3. Resuspend the cells, and move to the next dish to collect all the heavy or light cell populations. Alternatively, the cells can be scraped and collected in 5 mL cold PBS for each dish.
4. Collect the separated heavy and light cells by centrifugation (300 RCF, 4 °C, 5 min). Decant and discard the supernatant.
5. Resuspend the cell pellets in 20 mL of cold PBS. Make sure the cell pellet is well resuspended to get the accurate cell count.
6. Count the cells using hemocytometer or by using an automated cell counter. Mix equal amounts of heavy and light cells (by cell number) into a single 50 mL tube. Save approximately 100 μL or more of the unmixed heavy and light cell populations to perform Western blot analysis if necessary (*see* **Note 10**).
7. Collect the combined heavy and light cells by centrifugation (300 RCF, 4 °C, 5 min). Decant and discard the supernatant.
8. Cell pellets can be stored at −80 °C at this stage.
9. Add 1.5 to 2mL of freshly prepared urea lysis buffer (depending on the pellet size to be mixed) to the frozen cell pellets. As the cell pellet thaws, resuspend the pellet completely (*see* **Notes 4** and **11**).
10. Sonicate with a 30% intensity (8 mV) setting using a microtip-fitted sonicator. Sonicate three times for 10 s each with 30 s rest on ice between the cycles.
11. Centrifuge the cell lysate at 20,000 RCF for 15 min at 4 °C to pellet the insoluble material.
12. Collect and transfer the supernatant to a new 15 mL conical tube.

3.2 Protein Digestion

1. Determine the protein concentration of the cell lysates using an appropriate total protein detection assay method (i.e., BCA or Bradford assay).
2. Dilute the lysate with equal volume of lysis buffer without urea or protease inhibitors to bring the final urea concentration to 4 M.
3. Add LysC to the lysates at the final concentration of 10 ng/μL.
4. Incubate at 37 °C for 1–2 h with end-over-end rotation.
5. Dilute the LysC-digested samples to 1 M urea final concentration using 50 mM Tris–HCl, pH 8.
6. Add $CaCl_2$ to final concentration of 1 mM.
7. Add Sigma trypsin (T1426 TPCK treated) at the ratio of 1:100; enzyme: substrate.

8. Incubate overnight at 37 °C.
9. Stop the digestion by adding TFA to a final concentration of 0.4%. Verify the pH is less than 2.0 using pH determination strips. Add enough TFA to bring down the pH if in case it is above 2.
10. Centrifuge at 300 RCF for 15 min at room temperature. Collect the supernatant.

3.3 Peptide Desalting

1. The peptide desalting steps are done essentially as described previously [59]. Wash and condition the C18 SepPak™ cartridge three times using 3 mL of ACN for 3 cc capacity cartridges (use vacuum) (*see* **Note 7**).
2. Wash with 3 mL of 50% ACN and 0.5% HAcO to clear any unwanted material bound to the cartridges (use vacuum).
3. Equilibrate three times with 3 mL of 0.1% TFA (use vacuum).
4. Load the digested cell lysates in 0.4% TFA (gravity flow).
5. Wash and desalt four times with 3 mL of 0.1% TFA (use vacuum).
6. Wash (to remove TFA) with 1 mL of 0.5% HAcO (gravity flow).
7. Elute the peptides bound to the C18 cartridges two times with 3 mL of 50% ACN and 0.5% HAcO (gravity flow). Use a new 15 cm conical tube to collect the peptide eluate. To increase the throughput, the C18 cartridge can be placed in a 15 mL conical tube and spun at 200 RCF to increase the flow rate (*see* **Note 7**).
8. At this point, collect ~250 μg of the digested and desalted peptides in a separate tube to be used for analysis of the total proteome by mass spectrometry.
9. Freeze the eluate with liquid N_2 or store at −80 °C for 2–4 h, and lyophilize for a minimum of 2 days. The lyophilized peptides appear as white (sometimes yellowish) fluffy powder. The lyophilization step should remove all the residual acid from the peptide sample, and the peptides should be completely dried prior to diGLY IP.

3.4 Affinity Purification with diGLY Antibody

The subsequent steps of the diGLY IP should be done on ice or in the cold room.

1. To prepare the antibody/protein-A resin, pre-equilibrate the protein-A beads by washing five times in 1 mL of cold PBS followed by two washes in 2× IAP buffer. One IP reaction uses 20 μL of dry resin (40 μL of the 1:1 slurry is needed) (*see* **Note 8**).
2. After the final wash, resuspend the resin as a 1:1 slurry in 2× IAP buffer.

3. Add 60 μg of the diGLY antibody per 20 μL of prepared protein-A beads and couple overnight (end-over-end rotation at 4 °C). Alternatively, the antibody can be cross-linked to the resin (*see* **Note 12**).
4. Resuspend the lyophilized peptides in 1.3 mL of 2× IAP buffer. Shaking and sonication can help dissolve peptides. Check pH of solution (*see* **Note 13**). The pH should be around 7–7.5 (use the pH paper and compare peptide solution to 2× IAP buffer alone).
5. Clear solution by centrifugation, 30 min at 20,000 RCF at 4 °C (a sizable pellet may be present, but most of the peptides will be in solution).
6. Transfer the peptide solution to the tube with the antibody-coupled beads, and incubate for 1–2 h rotating at 4 °C.
7. Centrifuge at 300 RCF, 4 °C, and collect the supernatant. Save the supernatant for additional IP reactions or as a backup if IP fails (*see* **Note 14**).
8. Wash the beads four times with 1 mL of 2× IAP buffer. Rotate at 4 °C for 10 min between each wash.
9. Wash with 1 mL cold HPLC-grade water. Rotate at 4 °C for 10 min.
10. Elution 1: Add 55 μL of 5% formic acid, mix (tap bottom of tube lightly), and let it stand at room temperature for 10 min. Collect eluate in a new 1.5 mL Eppendorf tube.
11. Elution 2: Add 45 μL of 5% formic acid, mix, incubate 10 min at room temperature, and combine it with the first elution.

3.5 Antibody Cross-Linking (Optional)

The cross-linking of the ubiquitin remnant Lys-ϵ-Gly-Gly motif-specific antibody has been reported to increase the number of detectable diGLY sites due to the reduced deleterious effects that antibody-derived contaminants have on the enrichment [55, 60]. The cross-linking of antibody is an optional step to save the antibody reagents for subsequent immune enrichments. The shelf life of the cross-linked antibody is not well-studied (*see* **Note 12**). The cross-linking procedure is adapted and modified from a protocol available from New England BioLabs.

1. After diGLY-antibody coupling to the protein-A beads (as per diGLY IP protocol procedure), wash two times with PBS. Resuspend resin as a 1:1 slurry with PBS.
2. Add 1 mL of cross-linking buffer to the protein-A immobilized antibody, and resuspend thoroughly. Mix with end-over-end rotation at room temperature for 10 min.
3. Centrifuge at 300 RCF and remove the supernatant.
4. Repeat **steps 2–3** one more time.

5. Resuspend in 1 mL cross-linking buffer containing 25 mM DMP. Mix thoroughly, and incubate at room temperature for 45 min with agitation.
6. Centrifuge at 300 RCF and remove the supernatant.
7. Add 1 mL blocking buffer and resuspend completely.
8. Centrifuge at 300 RCF and remove the supernatant.
9. Add 1 mL blocking buffer, resuspend, and incubate for 1 h at room temperature with agitation.
10. Centrifuge at 300 RCF and remove the supernatant.
11. Wash the beads with 1 mL PBS.
12. Repeat PBS wash two more times.
13. After the final PBS wash, proceed with the diGLY protocol, add the reconstituted peptides in 1.3 mL 2× IAP buffer, and check the pH (it should be around 7).
14. After the final elution (as per diGLY protocol) of the peptides in 5% formic acid, wash the beads with 1 mL, 5% formic acid 3 times, and then with 1 mL PBS 5 times.
15. Resuspend the beads in 100 μL PBS, 0.1% Tween-20, and 0.02% sodium azide for long-term storage at 4 °C.

3.6 Stage-Tip Purification of Peptides

Samples are cleaned up for mass spectrometry analysis using in-house prepared C18 stage tips, as described previously [61].

1. Puncture a hole in the cap of a 1.5 mL microcentrifuge tube (unautoclaved) to allow for placement of a 200 μl micropipette. Condition the stage tip containing four layers of Empore C18-embedded membrane with 40 μL of methanol. Centrifuge at 550 RCF for 1 min (*see* **Note 15**).
2. Wash stage tip with 40 μL of 80% acetonitrile and 5% formic acid. Centrifuge at 550 RCF for 1 min.
3. Wash three times with 40 μL of 5% formic acid. Centrifuge at 550 RCF for 1 min.
4. Load the diGLY IP eluate on the stage tip. Centrifuge at 550 RCF for 1 min. Pass the IP eluate through the tip twice.
5. Wash three times with 40 μL of 5% formic acid. Centrifuge at 550 RCF for 1 min.
6. Transfer the stage tip to the new punctured 1.5 mL microcentrifuge tube.
7. Elute the peptides with 40 μL of 80% acetonitrile and 5% formic acid. Repeat two times.
8. Evaporate the solvent using a speed vacuum centrifuge.
9. Resuspend the dried peptides in 14 μL of 5% formic acid and 5% acetonitrile. Centrifuge the tubes. Transfer the peptides to an autosampler vial or other mass spec compatible vial.

Centrifuge the sample again (sample can be stored in 4 °C for short-term or at −20 °C for long-term storage before using it for MS analysis).

3.7 Mass Spectrometry Analysis

The choice of the nano-HPLC and the mass spectrometer depends on the availability of the instrument for the user. High-resolution, high-speed instruments are ideal for acquiring data of high quality and quantity. Here, the information is provided for the EASY-nLC1000 and the Q-Exactive mass spectrometer instrument (Thermo Fisher) (*see* **Note 16**). Solvent A contains 0.1% formic acid in water, and solvent B contains 0.1% formic acid in acetonitrile. We typically inject 3 μL of the sample per run, and run each sample in triplicate (Fig. 2).

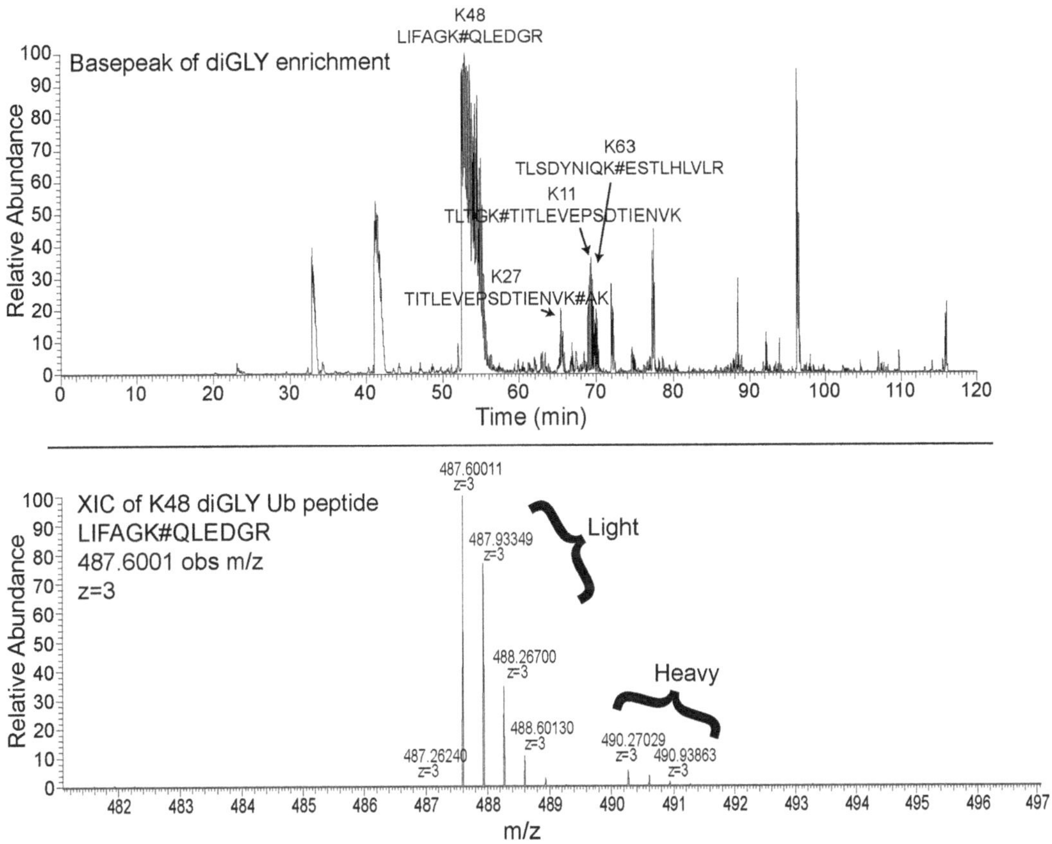

Fig. 2 Example of typical base peak ion chromatogram for a diGLY-antibody enriched sample. Top: Heavy (K8 only)-labeled cells were treated with a ubiquitin E1 enzyme inhibitor and the proteasome inhibitor MG132. Heavy cells were mixed with light cells treated with MG132 only prior to cell lysis and subsequent sample processing for diGLY-based proteomic analysis. Shown is the base peak ion chromatogram for the diGLY-antibody enriched sample. Arrows signify prominent ion peaks that correspond to a subset of peptides from abundant ubiquitin-ubiquitin linkages. Bottom: Extracted ion chromatogram for the K48-linked Ub peptide ($z + 3$). The sequence of the peptide is shown with # representing a diGLY-modified lysine residue. Note that the ion intensity of the heavy ion pair is significantly lower than the corresponding light pair due to E1 enzyme inhibition

3.7.1 LC-MS-MS Parameters

1. We run a 2 h gradient and instrument method for diGLY-enriched samples.
2. Peptides are eluted from the C18 column with the following gradient: 100 min, 2–30% ACN gradient; 5 min, 30–60% ACN gradient; 5 min, 60–95% ACN gradient; and 5 min, 95–0% gradient; with a final 5 min, isocratic step in solvent A. Total run time of 120 min at a flow rate of 250 nL/min.
3. Collect the MS/MS data in a data-dependent fashion using a top 10 method with a full MS mass range from 300 to 1750 m/z, 70,000 resolution, and an AGC target of 3e6.
4. Set the MS2 scans to trigger on when an ion intensity threshold of 1e5 reaches with a maximum injection time of 250 ms.
5. Set the normalized collision energy setting to 22 for peptides fragmentation.
6. A dynamic exclusion time of 40 s is used, and the peptide match setting is disabled.
7. Singly charged ions, charge states above 8, and unassigned charge states are excluded.

3.7.2 Peptide and Protein Identification and Quantification

1. Subsequent instrument files containing the raw MS/MS data can be processed with a variety of MS data processing pipelines including the Trans-Proteomic Pipeline, MaxQuant, or vendor-specific software packages.
2. Use the suitable search algorithm such as SEQUEST, OMSSA, or MASCOT to search MS/MS spectra against a concatenated target-decoy database comprised of forward and reversed sequences from the appropriate sequence database.
3. Typical search parameters used are as follows: 10 ppm precursor ion tolerance and 0.01 Da fragment ion tolerance; trypsin (1 1 KR P) is set as the enzyme; allow up to three missed cleavages; dynamic modifications of 15.99491 Da on methionine (oxidation), 114.04293 Da on lysine (diGLY), 42.010564 Da on peptide N-term (acetylation), and static modification of 125.047679 Da on cysteines (NEM alkylation).
4. Filter the peptide matches to a peptide false discovery rate of 1% (*see* **Notes 17** and **18**).
5. Perform appropriate quantitative measurement (MS1 precursor area or max signal comparison, peptide counting, or MS2-based quantification). Exclude all peptides with a C-terminal diGLY-modified lysine residue.

4 Notes

1. A SILAC-based approach can be used for experiments in Saccharomyces cerevisiae as well [36, 37, 48, 56]. Label-free quantitative approaches have also been used to examine the ubiquitin-modified proteome with good success. Further, tandem-mass-tagged approaches that label peptides after diGLY-antibody-based enrichment has been used to quantify changes to the ubiquitin-modified proteome [53]. However, special care must be taken with both label-free and post-IP labeling approaches to ensure consistent lysis, sample digestion, and IP conditions across both control and experimental samples. Small deviations in sample handling between experiments can result in quantitative differences between peptide amounts resulting in sample handling rather than for biological reasons.
2. SILAC media using only heavy lysine can be used as well. All peptides of interest (i.e., ubiquitylated peptides) will have a lysine residue which will carry the heavy label. Thus, usage of heavy arginine is not required. However, we routinely use input peptides to the diGLY-IP to measure changes in total protein abundance. For this, labeling with both lysine and arginine can be helpful although not essential. If lysine-only labeling is used, we discard all non-lysine containing peptides during quantitative analysis of both the ubiquitin-modified and total proteome.
3. All reagents used should be of highest mass spectrometry grade quality.
4. Urea lysis buffer should be prepared fresh before every experiment.
5. N-Ethylmaleimide (NEM) is a cysteine-alkylating agent that will inhibit cellular thiol-dependent deubiquitylating enzymes upon lysis. Addition of fresh NEM (or other cysteine-alkylating agents) is critical as omission can lead to loss of ubiquitylated peptides upon lysis (even in 8 M urea). NEM is readily soluble in ethanol, and it should be prepared fresh before addition to the lysis buffer.
6. We typically do not use the highest-quality mass spectrometry grade trypsin during sample preparation for subsequent diGLY enrichment. The recommended starting protein amount is more than 20 mg, and the use of mass spectrometry grade trypsin to facilitate proper digestion of this amount of protein would add additional reagent costs to the experiment.
7. The SepPak™ cartridges are available with varying volume holding capacities. The 3 cc tC18 cartridge allows for convenient nesting on 15 mL conical tubes for centrifugation in the

last step of the peptide desalting procedure to extract maximum volume from the cartridge.

8. Lucerna also sells a diGLY antibody that has been successfully used in many studies [20, 34, 51, 62]. However, the clear majority of published studies have used the cell signaling antibody, and we have used it exclusively [19, 35–44, 46–58, 60, 63–65]. As such, the subsequent protocol is based on our experience using the diGLY antibody available from Cell Signaling Technology. The ubiquitin remnant motif (K-ε-GG) antibody, when purchased as part of the PTMScan kit, typically arrives already conjugated to the resin. We prefer to use the unconjugated antibody without the kit. As such, this protocol includes the procedure for antibody coupling to the beads (both with and without cross-linking). If you are using the antibody pre-coupled to the resin as part of the PTMScan kit, then skip to **step 4** in the affinity purification section.
9. We recommend using ~30 mg of starting material (tissue lysate, worm extract, etc.). Other studies report using less starting material with good outcomes; we have obtained good data from as little as 10 mg. However, 30 mg of starting material is recommended to obtain the depth of coverage of ubiquitin-modified peptides usually desired.
10. Heavy- and light-labeled samples can also be mixed 1:1 according to total protein after lysis.
11. Try to use the minimum possible volume of lysis buffer for cell lysis, so when the samples are diluted to decrease the urea concentration, prior to lysC and trypsin digestion, the protein concentration remains high enough for efficient digestion.
12. Previous studies have reported greater numbers of ubiquitin-modified peptides identified after cross-linking the antibody to the resin [55] (Table 1). We have not observed this result and have obtained similar results from un-cross-linked and cross-linked resins. We have successfully reused cross-linked resin after acid elution as long as the resin is reused within 2–3 weeks.
13. After the lyophilized peptides are resuspended in the 2× IAP buffer, it is very important to check the pH of the dissolved peptides. If the pH is below 7, it indicates that trace amounts of TFA and acetic acid used in the desalting steps are still present in the sample, and the lyophilization was not efficient. To avoid such a case, it is recommended to lyophilize the sample for more than 2 days.
14. To identify and quantify larger numbers of diGLY-modified peptides, we have previously done sequential IPs from a single sample (i.e., the flow through of the first diGLY IP is applied to a second round of IP using a new aliquot of antibody-coupled

Table 1
Table describing results of key previously published diGLY proteomics studies

Published study PMID (ref #)	12872131 [12]	20639865 [18]	21139048	21987572 [45]	21906983 [19]	21890473 [20]	21963094 [44]
Authors	Peng J et al.	Xu G et al.	Danielsen JM et al.	Lee KA et al.	Kim W et al.	Wagner SA et al.	Emanuele MJ et al.
Year	2003	2010	2011	2011	2011	2011	2011
Tissues/cells	*S. cerevisiae*	HEK293T cells	U2OS, HEK293T cells	HeLa TREx cells	HCT116, HEK293T	HEK293T	HeLa cells
Labeling		SILAC	SILAC	SILAC	SILAC	SILAC	SILAC
Ub sites identified	110	374	753	1800	19,000	11,054	9957
Ub proteins identified	72	236	471	900	5000	4273	2814
IP/pre-IP enrichment	Ni-NTA affinity chromatography of His-tagged Ub	(1) Ni-NTA affinity chromatography of His-tagged Ub and (2) diGLY IP	Streptavidin affinity chromatography of Strep-HA-tagged Ub	diGLY-antibody IP (antibody 3925), (Cell Signaling Technology)	diGLY-antibody IP (Cell Signaling Technology)	diGLY-antibody IP (Lucerna)	PTMScan ubiquitin remnant antibody beads (Cell Signaling Technology) diGLY-antibody IP
Post-enrichment fractionation	SCX					IEF	
Remarks	First report of Ub-modified peptide identification using the signature diGLY remnant for identification by MS	Reported the use of diGLY affinity purification strategy for immuno-enrichment of the diGLY-modified peptides	Study reports the promiscuity of lysine ubiquitylation at the site level	Study used peptide- and protein-level approaches to enrich for ubiquitinated proteins in the presence and absence of the HRD1 ubiquitin ligase	Reported the identification and quantification of a large number of diGLY-modified peptides and their response to proteasome inhibition and translation inhibition. Also identified substrates for CRL ligases	Reported the identification and quantification of a large number of diGLY-modified peptides and their response to proteasome inhibition. Characterized possible cross talk with acetylation	Identified proteins regulated by CRL ligases

(continued)

Table 1 (continued)

Published study PMID (ref #)	22871113 [50]	23000965 [34]	22505724 [35]	22790023 [51]	22817900 [63]	23266961 [55]	23749301 [56]	23716602 [36]
Authors	Na CH et al.	Povlsen LK et al.	Udeshi ND et al.	Wagner SA et al.	Beltrao P et al.	Udeshi N D et al.	Swaney DL et al.	Ng AH et al.
Year	2012	2012	2012	2012	2012	2013	2013	2013
Tissues/cells	Rat brain tissue	Human U2OS cells	Jurkat E6–1 cells	Tissues from C57BL/6 mice	*S. cerevisiae*	Jurkat E6–1 cells	*S. cerevisiae*	*S. cerevisiae*
Labeling		SILAC	SILAC			SILAC	SILAC	SILAC
Ub sites identified	1786	6700	3300 peptides	20,085	2500	20,004 sites, 3143 peptides	5465	1321
Ub proteins identified	921						1307	240
IP/pre-IP enrichment	(1) SCX, (2) PTMScan ubiquitin remnant antibody beads (Cell Signaling Technology)	diGLY IP (Lucerna, clone GX41)	(1) SCX, (2) PTMScan ubiquitin remnant antibody beads (Cell Signaling Technology)	diGLY-antibody IP (clone GX41) and PTMScan ubiquitin remnant antibody beads (Cell Signaling Technology), diGLY-antibody IP	(1) His-tag purification of histidine tagged Ub and (2) PTMScan ubiquitin remnant antibody beads (Cell Signaling Technology)	(1) Basic pH reversed-phase separation pre-IP, (2) PTMScan ubiquitin remnant antibody beads cross-linked with DMP (Cell Signaling Technology) diGLY-antibody IP	(A) (i) cobalt-NTA affinity chromatography for His-tagged purification, (ii) PTMScan ubiquitin remnant antibody beads (Cell Signaling Technology); (B) (i) SCX, (ii) PTMScan ubiquitin remnant antibody beads (Cell Signaling Technology)	PTMScan ubiquitin remnant antibody beads (Cell Signaling Technology)
Post-enrichment fractionation				SCX				
Remarks	Study utilized the antibody enrichment strategy for analyzing the ubiquitome from rat brain	Study demonstrated the widespread ultraviolet-regulated ubiquitylation of known DDR components and many proteins not previously implicated in this response	Examined the effects of proteasome inhibition and deubiquitinase inhibition by PR-619 on ubiquitination sites	Study demonstrated the enrichment of endogenous ubiquitylated peptides from murine tissue lysates and their subsequent identification by mass spectrometry	This study identified 2500 ubiquitylation sites for *S. cerevisiae*	Study described a number of enhancements to the diGLY-based enrichment method	Study developed two methods to identify protein isoforms that are both phosphorylated and ubiquitylated in the yeast *Saccharomyces cerevisiae*, identifying 466 proteins with 2100 phosphorylation sites co-occurring with 2189 ubiquitylation sites	Study shows that upon heat shock, the distinct populations of structured and intrinsically disordered proteins are prone to ubiquitylation

Published study PMID (ref #)	23503661 [46]	23749302 [57]	25093938	25078903 [48]	24961812 [48]	25147182 [64]	25278611 [58]
Authors	Sarraf SA et al.	Mertins P et al.	Akimov V et.al.	Tong Z et al.	Iesmantavicius V et al.	Thompson JW et al.	Theurillat JP et al.
Year	2013	2013	2014	2014	2014	2014	2014
Tissues/cells	HCT116^{PARKIN}, HeLa^{PARKIN}, HCT116, SH-SY5Y	HeLa S3 cells	StUbEx-stable expressing HeLa cells	*S. cerevisiae*	*S. cerevisiae*	BT-549 cells stably transduced with an inducible HUWE1 shRNA	LHMAR cells
Labeling	SILAC	SILAC	SILAC	SILAC	SILAC	SILAC	SILAC
Ub sites identified	6934 sites in HCT116^{PARKIN}, 3286 sites in HeLa^{PARKIN}, 6149 sites in HCT116, 3450 sites in SH-SY5Y cells	15,408		3116	2299	(A) 2735 peptides in condition 1 and 2806 peptides in condition 2; (B) 6614 peptides in MG132-treated cells and 4157 peptides in DMSO-treated cells	7181 Ub peptides
Ub proteins identified	1993 proteins in HCT116^{PARKIN}, 1220 proteins in HeLa^{PARKIN}, 1927 proteins in HCT116, 1329 proteins in SH-SY5Y cells		3400	1111			
IP/pre-IP enrichment	diGLY-antibody IP (Cell Signaling Technology)	(1) Basic pH reversed-phase separation pre-IP, (2) PTMScan ubiquitin remnant antibody beads (Cell Signaling Technology)	Ni-Sepharose beads	(1) Ni-NTA affinity chromatography of His-tagged Ub, (2) PTMScan ubiquitin remnant antibody beads (Cell Signaling Technology)	PTMScan ubiquitin remnant antibody beads (Cell Signaling Technology)	(A) PTMScan ubiquitin remnant antibody beads (Cell Signaling Technology), (B) (i) bRPLC, (ii) PTMScan ubiquitin remnant antibody beads	(1) Basic pH reversed-phase (bRP) chromatography, (2) diGLY-antibody IP (Cell Signaling Technology) cross-linked using DMP
Post-enrichment fractionation		pH-gradient fractionation		SCX			SCX
Remarks	Study elucidated the ubiquitylation site specificity and topology of PARKIN-dependent target modification in response to mitochondrial depolarization	Reported the integrated analysis of protein expression, phosphorylation, ubiquitylation, and acetylation by serial enrichment	Used the StUbEx strategy to identify Ub-modified peptides	diGLY proteomics approach identified 3116 ubiquitylation sites, including 10 sites in Tul1 candidate substrates	Study reported the parallel quantification of ubiquitylation, phosphorylation, and proteome changes in rapamycin-treated yeast cells	Study implemented an inducible loss-of-function approach in combination with quantitative diGly proteomics to find novel Huwe1 substrates	Study analyzed changes in the ubiquitin landscape induced by prostate cancer-associated mutations of SPOP, an E3 ubiquitin ligase substrate-binding protein

(continued)

Table 1 (continued)

Published study PMID (ref #)	26038114 [62]	26051181 [39]	26051182 [40]	26131937 [41]	27211868 [65]	27185884 [42]	27667366 [53]	28665616 [54]
Authors	Satpathy S et al.	Elia A E et al.	Higgins R et al.	Kronke J et al.	Fiskin E et al.	Gendron JM et al.	Rose C M et al.	Sap KA et al.
Year	2015	2015	2015	2015	2016	2016	2016	2017
Tissues/cells	A20 cells	HeLa cells	HCT116 cells *Drosophila* S2 cells *S. cerevisiae*	KG-1 cells	Human HCT116 and HeLa cells infected with salmonella or left untreated	HCT116 cells	HTC116 cells, mice tissues	*Drosophila melanogaster* Schneider's line 2 cells
Labeling	SILAC	SILAC	SILAC	SILAC	SILAC	SILAC	Ten-plex tandem mass tags (TMTs)	SILAC
Ub sites identified	Quantified 6059	33,500	5893	13,061	11,606	9000	16,842 in HCT116 cells, 16,925 in mice tissues	14,018
Ub proteins identified					3824	3000		
IP/pre-IP enrichment	diGLY-antibody IP (Lucerna, clone GX41)	(1) SCX (for nuclear extracts), (2)	diGLY-antibody IP (Cell Signaling Technology)	(1) Basic pH reversed-phase separation pre-IP, (2) PTMScan ubiquitin remnant antibody beads (Cell Signaling Technology)	diGLY-antibody IP (Cell Signaling Technology)	(1) Cell fractionation using differential centrifugation, (2) diGLY-antibody IP (Cell Signaling Technology)	(1) diGLY-antibody IP (Cell Signaling Technology), (2) basic pH reversed-phase separation	(1) Fractionation by HILIC, (2) PTMScan ubiquitin remnant antibody beads (Cell Signaling Technology)
Post-enrichment fractionation					SCX			
Remarks	Study illustrated the power of multilayered proteomic analyses for discovering novel BCR signaling	Systematically examined alterations to the Ub-modified proteome upon induction of DNA damage responses	Study identified the evolutionarily conserved, site-specific regulatory ubiquitylation of 40S ribosomal proteins upon UPR activation and translation inhibition	Study demonstrated that lenalidomide induces the ubiquitination of casein kinase 1A1 (CK1a) by the E3 ubiquitin ligase CUL4-RBX1-DDB1-CRBN	Study observed ~5–10% of all quantified diGLY sites being over twofold regulated, corresponding to Salmonella invasion-induced changes in multiple cellular processes	Study demonstrated that combining Ub proteomics with subcellular fractionation can effectively separate degradative and regulatory ubiquitylation events on distinct protein populations	Study reports TMT-multiplexing strategy to quantify ubiquitin-modified peptides and reveals PINK1- and PARKIN-dependent ubiquitylation events during early and late mitophagy	Study reports Ub-modified proteome changes upon proteasome inactivation both by chemical inhibitors and by dsRNA-mediated knockdown of specific subunits in *Drosophila* S2 cells

resin). While this can achieve a greater number of identified peptides, it comes at the cost of using more diGLY antibody per experiment.

15. During the stage-tip purification of the peptides, make sure that the C18 material does not dry down during all the centrifugation steps. Adjust the centrifugation speed if required.
16. Individual LC and mass spectrometer settings will need to be adjusted based on the instrumentation available. The settings listed here are meant to serve as an example and are not to be considered the only suitable instrument method for data collection (Fig. 2).
17. We typically observe that between 30 and 80% of all identified peptides contain a diGLY-modified lysine residue. Enrichment is typically higher when using samples treated with a proteasome inhibitor.
18. We typically identify ~10,000 total diGLY peptides (4000 unique peptides) in a single sample shot in triplicate. Treatment of cells with proteasome inhibitor will result in a greater number of identified peptides (25,000 total diGLY-modified peptides, 10,000 unique peptides) (Fig. 2, Table 1).

Acknowledgments

We thank Marilyn Leonard and Danielle Garshott for providing a critical reading of this manuscript and Ruoyu (Lulu) Li for assistance in making Fig. 1. This work was supported by the NIH (DP2-GM119132, PGM085764) (E.J.B).

References

1. Mann M, Ong S-E, Grønborg M, Steen H, Jensen ON, Pandey A (2002) Analysis of protein phosphorylation using mass spectrometry: deciphering the phosphoproteome. Trends Biotechnol 20(6):261–268
2. Steen H, Küster B, Fernandez M, Pandey A, Mann M (2001) Detection of tyrosine phosphorylated peptides by precursor ion scanning quadrupole TOF mass spectrometry in positive ion mode. Anal Chem 73(7):1440–1448
3. Thalassinos K, Grabenauer M, Slade SE, Hilton GR, Bowers MT, Scrivens JH (2008) Characterization of phosphorylated peptides using traveling wave-based and drift cell ion mobility mass spectrometry. Anal Chem 81(1):248–254
4. Carr SA, Huddleston MJ, Annan RS (1996) Selective detection and sequencing of phosphopeptides at the femtomole level by mass spectrometry. Anal Biochem 239(2):180–192
5. Ficarro SB, McCleland ML, Stukenberg PT, Burke DJ, Ross MM, Shabanowitz J, Hunt DF, White FM (2002) Phosphoproteome analysis by mass spectrometry and its application to Saccharomyces cerevisiae. Nat Biotechnol 20 (3):301–305
6. Ptacek J, Devgan G, Michaud G, Zhu H (2005) Global analysis of protein phosphorylation in yeast. Nature 438(7068):679
7. Zhang H, Xiao-jun L, Martin DB, Aebersold R (2003) Identification and quantification of

N-linked glycoproteins using hydrazide chemistry, stable isotope labeling and mass spectrometry. Nat Biotechnol 21(6):660

8. Dell A, Morris HR (2001) Glycoprotein structure determination by mass spectrometry. Science 291(5512):2351–2356
9. Apweiler R, Hermjakob H, Sharon N (1999) On the frequency of protein glycosylation, as deduced from analysis of the SWISS-PROT database. Biochim Biophys Acta 1473(1):4–8
10. Brownlee M (1995) Advanced protein glycosylation in diabetes and aging. Annu Rev Med 46(1):223–234
11. Spiro RG (2002) Protein glycosylation: nature, distribution, enzymatic formation, and disease implications of glycopeptide bonds. Glycobiology 12(4):43R–56R
12. Peng J, Schwartz D, Elias JE, Thoreen CC, Cheng D, Marsischky G, Roelofs J, Finley D, Gygi SP (2003) A proteomics approach to understanding protein ubiquitination. Nat Biotechnol 21(8):921–926. https://doi.org/10.1038/nbt849
13. Scheffner M, Nuber U, Huibregtse JM (1995) Protein ubiquitination involving an E1-E2-E3 enzyme ubiquitin thioester cascade. Nature 373(6509):81
14. Kirkpatrick DS, Denison C, Gygi SP (2005) Weighing in on ubiquitin: the expanding role of mass-spectrometry-based proteomics. Nat Cell Biol 7(8):750–757. https://doi.org/10.1038/ncb0805-750
15. Xu P, Peng J (2006) Dissecting the ubiquitin pathway by mass spectrometry. Biochim Biophys Acta 1764(12):1940–1947. https://doi.org/10.1016/j.bbapap.2006.09.004
16. Xu P, Duong DM, Seyfried NT, Cheng D, Xie Y, Robert J, Rush J, Hochstrasser M, Finley D, Peng J (2009) Quantitative proteomics reveals the function of unconventional ubiquitin chains in proteasomal degradation. Cell 137(1):133–145. https://doi.org/10.1016/j.cell.2009.01.041
17. Bennett EJ, Rush J, Gygi SP, Harper JW (2010) Dynamics of cullin-RING ubiquitin ligase network revealed by systematic quantitative proteomics. Cell 143(6):951–965. https://doi.org/10.1016/j.cell.2010.11.017
18. Xu G, Paige JS, Jaffrey SR (2010) Global analysis of lysine ubiquitination by ubiquitin remnant immunoaffinity profiling. Nat Biotechnol 28(8):868–873. https://doi.org/10.1038/nbt.1654
19. Kim W, Bennett EJ, Huttlin EL, Guo A, Li J, Possemato A, Sowa ME, Rad R, Rush J, Comb MJ, Harper JW, Gygi SP (2011) Systematic and quantitative assessment of the ubiquitin-modified proteome. Mol Cell 44(2):325–340. https://doi.org/10.1016/j.molcel.2011.08.025
20. Wagner SA, Beli P, Weinert BT, Nielsen ML, Cox J, Mann M, Choudhary C (2011) A proteome-wide, quantitative survey of in vivo ubiquitylation sites reveals widespread regulatory roles. Mol Cell Proteomics 10(10): M111 013284. https://doi.org/10.1074/mcp.M111.013284
21. Jaffrey SR, Erdjument-Bromage H, Ferris CD, Tempst P, Snyder SH (2001) Protein S-nitrosylation: a physiological signal for neuronal nitric oxide. Nat Cell Biol 3(2):193
22. Clarke S (1993) Protein methylation. Curr Opin Cell Biol 5(6):977–983
23. Choudhary C, Kumar C, Gnad F, Nielsen ML, Rehman M, Walther TC, Olsen JV, Mann M (2009) Lysine acetylation targets protein complexes and co-regulates major cellular functions. Science 325(5942):834–840
24. Ichimura Y, Kirisako T, Takao T, Satomi Y (2000) A ubiquitin-like system mediates protein lipidation. Nature 408(6811):488
25. Mayor T, Deshaies RJ (2005) Two-step affinity purification of multiubiquitylated proteins from Saccharomyces cerevisiae. Methods Enzymol 399:385–392. https://doi.org/10.1016/S0076-6879(05)99026-5
26. Aillet F, Lopitz-Otsoa F, Hjerpe R, Torres-Ramos M, Lang V, Rodríguez MS (2012) Isolation of ubiquitylated proteins using tandem ubiquitin-binding entities. In: Ubiquitin family modifiers and the proteasome: reviews and protocols, pp 173–183
27. Iconomou M, Saunders DN (2016) Systematic approaches to identify E3 ligase substrates. Biochem J 473(22):4083–4101. https://doi.org/10.1042/BCJ20160719
28. Beaudette P, Popp O, Dittmar G (2016) Proteomic techniques to probe the ubiquitin landscape. Proteomics 16(2):273–287. https://doi.org/10.1002/pmic.201500290
29. Ordureau A, Munch C, Harper JW (2015) Quantifying ubiquitin signaling. Mol Cell 58(4):660–676. https://doi.org/10.1016/j.molcel.2015.02.020
30. Goldknopf IL, Busch H (1977) Isopeptide linkage between nonhistone and histone 2A polypeptides of chromosomal conjugate-protein A24. Proc Natl Acad Sci 74(3):864–868
31. Bustos D, Bakalarski CE, Yang Y, Peng J, Kirkpatrick DS (2012) Characterizing ubiquitination sites by peptide-based immunoaffinity enrichment. Mol Cell Proteomics 11

(12):1529–1540. https://doi.org/10.1074/mcp.R112.019117

32. Sylvestersen KB, Young C, Nielsen ML (2013) Advances in characterizing ubiquitylation sites by mass spectrometry. Curr Opin Chem Biol 17(1):49–58. https://doi.org/10.1016/j.cbpa.2012.12.009
33. Carrano AC, Bennett EJ (2013) Using the ubiquitin-modified proteome to monitor protein homeostasis function. Mol Cell Proteomics 12(12):3521–3531
34. Povlsen LK, Beli P, Wagner SA, Poulsen SL, Sylvestersen KB, Poulsen JW, Nielsen ML, Bekker-Jensen S, Mailand N, Choudhary C (2012) Systems-wide analysis of ubiquitylation dynamics reveals a key role for PAF15 ubiquitylation in DNA-damage bypass. Nat Cell Biol 14(10):1089–1098. https://doi.org/10.1038/ncb2579
35. Udeshi ND, Mani D, Eisenhaure T, Mertins P, Jaffe JD, Clauser KR, Hacohen N, Carr SA (2012) Methods for quantification of in vivo changes in protein ubiquitination following proteasome and deubiquitinase inhibition. Mol Cell Proteomics 11(5):148–159
36. Ng AH, Fang NN, Comyn SA, Gsponer J, Mayor T (2013) System-wide analysis reveals intrinsically disordered proteins are prone to ubiquitylation after misfolding stress. Mol Cell Proteomics 12(9):2456–2467. https://doi.org/10.1074/mcp.M112.023416
37. Iesmantavicius V, Weinert BT, Choudhary C (2014) Convergence of ubiquitylation and phosphorylation signaling in rapamycin-treated yeast cells. Mol Cell Proteomics 13(8):1979–1992
38. Kronke J, Udeshi ND, Narla A, Grauman P, Hurst SN, McConkey M, Svinkina T, Heckl D, Comer E, Li X, Ciarlo C, Hartman E, Munshi N, Schenone M, Schreiber SL, Carr SA, Ebert BL (2014) Lenalidomide causes selective degradation of IKZF1 and IKZF3 in multiple myeloma cells. Science 343(6168):301–305. https://doi.org/10.1126/science.1244851
39. Elia AE, Boardman AP, Wang DC, Huttlin EL, Everley RA, Dephoure N, Zhou C, Koren I, Gygi SP, Elledge SJ (2015) Quantitative proteomic atlas of ubiquitination and acetylation in the DNA damage response. Mol Cell 59(5):867–881. https://doi.org/10.1016/j.molcel.2015.05.006
40. Higgins R, Gendron JM, Rising L, Mak R, Webb K, Kaiser SE, Zuzow N, Riviere P, Yang B, Fenech E, Tang X, Lindsay SA, Christianson JC, Hampton RY, Wasserman SA, Bennett EJ (2015) The unfolded protein response triggers site-specific regulatory ubiquitylation of 40S ribosomal proteins. Mol Cell 59(1):35–49. https://doi.org/10.1016/j.molcel.2015.04.026
41. Kronke J, Fink EC, Hollenbach PW, MacBeth KJ, Hurst SN, Udeshi ND, Chamberlain PP, Mani DR, Man HW, Gandhi AK, Svinkina T, Schneider RK, McConkey M, Jaras M, Griffiths E, Wetzler M, Bullinger L, Cathers BE, Carr SA, Chopra R, Ebert BL (2015) Lenalidomide induces ubiquitination and degradation of CK1alpha in del(5q) MDS. Nature 523(7559):183–188. https://doi.org/10.1038/nature14610
42. Gendron JM, Webb K, Yang B, Rising L, Zuzow N, Bennett EJ (2016) Using the ubiquitin-modified proteome to monitor distinct and spatially restricted protein homeostasis dysfunction. Mol Cell Proteomics 15(8):2576–2593. https://doi.org/10.1074/mcp.M116.058420
43. Sundaramoorthy E, Leonard M, Mak R, Liao J, Fulzele A, Bennett EJ (2017) ZNF598 and RACK1 regulate mammalian ribosome-associated quality control function by mediating regulatory 40S ribosomal ubiquitylation. Mol Cell 65(4):751–760 e754. https://doi.org/10.1016/j.molcel.2016.12.026
44. Emanuele MJ, Elia AE, Xu Q, Thoma CR, Izhar L, Leng Y, Guo A, Chen YN, Rush J, Hsu PW, Yen HC, Elledge SJ (2011) Global identification of modular cullin-RING ligase substrates. Cell 147(2):459–474. https://doi.org/10.1016/j.cell.2011.09.019
45. Lee KA, Hammerle LP, Andrews PS, Stokes MP, Mustelin T, Silva JC, Black RA, Doedens JR (2011) Ubiquitin ligase substrate identification through quantitative proteomics at both the protein and peptide levels. J Biol Chem 286(48):41530–41538. https://doi.org/10.1074/jbc.M111.248856
46. Sarraf SA, Raman M, Guarani-Pereira V, Sowa ME, Huttlin EL, Gygi SP, Harper JW (2013) Landscape of the PARKIN-dependent ubiquitylome in response to mitochondrial depolarization. Nature 496(7445):372–376. https://doi.org/10.1038/nature12043
47. Fang NN, Chan GT, Zhu M, Comyn SA, Persaud A, Deshaies RJ, Rotin D, Gsponer J, Mayor T (2014) Rsp5/Nedd4 is the major ubiquitin ligase that targets cytosolic misfolded proteins upon heat-stress. Nat Cell Biol 16(12):1227
48. Tong Z, Kim M-S, Pandey A, Espenshade PJ (2014) Identification of candidate substrates for the Golgi Tul1 E3 ligase using quantitative diGly proteomics in yeast. Mol Cell Proteomics 13(11):2871–2882

49. Garzia A, Jafarnejad SM, Meyer C, Chapat C, Gogakos T, Morozov P, Amiri M, Shapiro M, Molina H, Tuschl T (2017) The E3 ubiquitin ligase and RNA-binding protein ZNF598 orchestrates ribosome quality control of premature polyadenylated mRNAs. Nat Commun 8:16056
50. Na CH, Jones DR, Yang Y, Wang X, Xu Y, Peng J (2012) Synaptic protein ubiquitination in rat brain revealed by antibody-based ubiquitome analysis. J Proteome Res 11 (9):4722–4732. https://doi.org/10.1021/pr300536k
51. Wagner SA, Beli P, Weinert BT, Scholz C, Kelstrup CD, Young C, Nielsen ML, Olsen JV, Brakebusch C, Choudhary C (2012) Proteomic analyses reveal divergent ubiquitylation site patterns in murine tissues. Mol Cell Proteomics 11(12):1578–1585. https://doi.org/10.1074/mcp.M112.017905
52. Buckley SM, Aranda-Orgilles B, Strikoudis A, Apostolou E, Loizou E, Moran-Crusio K, Farnsworth CL, Koller AA, Dasgupta R, Silva JC, Stadtfeld M, Hochedlinger K, Chen EI, Aifantis I (2012) Regulation of pluripotency and cellular reprogramming by the ubiquitin-proteasome system. Cell Stem Cell 11 (6):783–798. https://doi.org/10.1016/j.stem.2012.09.011
53. Rose CM, Isasa M, Ordureau A, Prado MA, Beausoleil SA, Jedrychowski MP, Finley DJ, Harper JW, Gygi SP (2016) Highly multiplexed quantitative mass spectrometry analysis of ubiquitylomes. Cell Syst 3(4):395–403 e394. https://doi.org/10.1016/j.cels.2016.08.009
54. Sap KA, Bezstarosti K, Dekkers DHW, Voets O, Demmers JAA (2017) Quantitative proteomics reveals extensive changes in the ubiquitinome after perturbation of the proteasome by targeted dsRNA-mediated subunit knockdown in drosophila. J Proteome Res 16 (8):2848–2862. https://doi.org/10.1021/acs.jproteome.7b00156
55. Udeshi ND, Svinkina T, Mertins P, Kuhn E, Mani D, Qiao JW, Carr SA (2013) Refined preparation and use of anti-diglycine remnant (K-ε-GG) antibody enables routine quantification of 10,000 s of ubiquitination sites in single proteomics experiments. Mol Cell Proteomics 12(3):825–831
56. Swaney DL, Beltrao P, Starita L, Guo A, Rush J, Fields S, Krogan NJ, Villén J (2013) Global analysis of phosphorylation and ubiquitylation cross-talk in protein degradation. Nat Methods 10(7):676–682
57. Mertins P, Qiao JW, Patel J, Udeshi ND, Clauser KR, Mani DR, Burgess MW, Gillette MA, Jaffe JD, Carr SA (2013) Integrated proteomic analysis of post-translational modifications by serial enrichment. Nat Methods 10 (7):634–637. https://doi.org/10.1038/nmeth.2518
58. Theurillat JP, Udeshi ND, Errington WJ, Svinkina T, Baca SC, Pop M, Wild PJ, Blattner M, Groner AC, Rubin MA, Moch H, Prive GG, Carr SA, Garraway LA (2014) Prostate cancer. Ubiquitylome analysis identifies dysregulation of effector substrates in SPOP-mutant prostate cancer. Science 346 (6205):85–89. https://doi.org/10.1126/science.1250255
59. Villen J, Gygi SP (2008) The SCX/IMAC enrichment approach for global phosphorylation analysis by mass spectrometry. Nat Protoc 3(10):1630–1638. https://doi.org/10.1038/nprot.2008.150
60. Udeshi ND, Mertins P, Svinkina T, Carr SA (2013) Large-scale identification of ubiquitination sites by mass spectrometry. Nat Protoc 8 (10):1950–1960. https://doi.org/10.1038/nprot.2013.120
61. Rappsilber J, Mann M, Ishihama Y (2007) Protocol for micro-purification, enrichment, pre-fractionation and storage of peptides for proteomics using StageTips. Nat Protoc 2 (8):1896
62. Satpathy S, Wagner SA, Beli P, Gupta R, Kristiansen TA, Malinova D, Francavilla C, Tolar P, Bishop GA, Hostager BS, Choudhary C (2015) Systems-wide analysis of BCR signalosomes and downstream phosphorylation and ubiquitylation. Mol Syst Biol 11(6):810. https://doi.org/10.15252/msb.20145880
63. Beltrao P, Albanese V, Kenner LR, Swaney DL, Burlingame A, Villen J, Lim WA, Fraser JS, Frydman J, Krogan NJ (2012) Systematic functional prioritization of protein posttranslational modifications. Cell 150(2):413–425. https://doi.org/10.1016/j.cell.2012.05.036
64. Thompson JW, Nagel J, Hoving S, Gerrits B, Bauer A, Thomas JR, Kirschner MW, Schirle M, Luchansky SJ (2014) Quantitative Lys--Gly-Gly (diGly) proteomics coupled with inducible RNAi reveals ubiquitin-mediated proteolysis of DNA damage-inducible transcript 4 (DDIT4) by the E3 ligase HUWE1. J Biol Chem 289(42):28942–28955. https://doi.org/10.1074/jbc.M114.573352
65. Fiskin E, Bionda T, Dikic I, Behrends C (2016) Global analysis of host and bacterial ubiquitinome in response to salmonella typhimurium infection. Mol Cell 62(6):967–981. https://doi.org/10.1016/j.molcel.2016.04.015

Chapter 24

Interpreting the Language of Polyubiquitin with Linkage-Specific Antibodies and Mass Spectrometry

Marissa L. Matsumoto, Erick R. Castellanos, Yi Jimmy Zeng, and Donald S. Kirkpatrick

Abstract

Posttranslational modification of cellular proteins by ubiquitin serves a variety of functions. Among the multitude of ubiquitin substrates, ubiquitin itself is the most prevalent. For many years, the direct detection of polyubiquitin chains attached to cellular substrates was not practical, with cell biologists relegated to indirect approaches involving site-directed mutagenesis or in vitro biochemistry. Recent advances in two technologies—polyubiquitin linkage-specific antibodies and mass spectrometry proteomics, have overcome that limitation. Using one or both of these, the direct analysis of polyubiquitin chain linkages on cellular substrate proteins may be performed. This paper describes the complimentary nature of linkage-specific antibodies and mass spectrometry proteomics for the characterization of complex ubiquitin signals using lessons learned in early development of both technologies.

Key words Ubiquitin, Polyubiquitin chains, Linkage-specific antibodies, Mass spectrometry, Ub-AQUA

1 Introduction

The covalent attachment of ubiquitin to proteins within eukaryotic cells has proven to be as pervasive as its name suggests [1]. From controlling protein degradation and organelle turnover to regulating protein localization and enzyme activities, the importance of ubiquitin in directing dynamic cellular processes continues to be elucidated. At the heart of these controls is a complex cellular language, encoded by spatially organized clusters of ubiquitin [2] and read out by receptors bearing a diverse array of ubiquitin-binding domains (UBD) [3].

Ubiquitin is classically conjugated via its C-terminus to one or more lysine residues on a substrate protein through the actions of a multistep cascade involving E1-activating, E2-conjugating, and E3 ligase enzymes. Ubiquitin-like (UBL) proteins including Nedd8, Sumo, and ISG15 function in a similar manner as protein

Thibault Mayor and Gary Kleiger (eds.), *The Ubiquitin Proteasome System: Methods and Protocols*, Methods in Molecular Biology, vol. 1844, https://doi.org/10.1007/978-1-4939-8706-1_24,

posttranslational modifications. In addition to targeting lysines, evidence exists for the attachment of ubiquitin to substrate N-termini and residues including Cys, Ser, and Thr. A critical element of the ubiquitin language is the formation of polyubiquitin chains where ubiquitin molecules are conjugated to one or more lysines (or the N-terminus) of another ubiquitin molecule. Ubiquitin contains seven lysine residues, and depending on the nature of this ubiquitin-ubiquitin attachment, chain structures can be formed that bear mixtures of linkages and/or branching. Additional complexity emerges from the formation of hybrid signals where ubiquitin is attached to lysines of another UBL protein. While early investigations into ubiquitin biochemistry uncovered the importance of homogenous polyubiquitin chains, a growing body of evidence highlights the structural and biochemical diversity of ubiquitin signals that can be assembled, as well as circumstances when heterogeneous chains regulate cellular events [4–6]. Answering questions about the functional relevance of complex ubiquitin signals requires analytical methods to detect and quantify structural isomers of polyubiquitin in a substrate-specific manner.

The advancement of tools for characterizing mono-, multi-, and polyubiquitination of protein substrates has helped to elucidate the various functions of protein ubiquitination over the past two decades. Prominent among these have been polyubiquitin linkage-specific antibodies [7–12] and mass spectrometry methods [13–16], which together facilitate the direct analysis of the ubiquitin forms within a sample. This technical paper seeks to articulate how these two analytical technologies complement one another and can be used to interrogate questions about the forms of ubiquitin bound to individual protein substrates.

2 Materials

2.1 Ubiquitin Linkage-Specific Antibodies

The following methods section focuses on lessons learned in the development and utilization of linkage-specific antibodies generated at Genentech [7, 10–12]. In addition to these four reagent antibodies (Table 1), two additional publications from Tokunaga, F. et al. [9] and Wang, H. et al. [8] describe a linear/M1 polyubiquitin-specific antibody and a K63 linkage-specific antibody (clone HWA4C4), respectively. Besides the published reagents, linkage-specific antibodies are also commercially available from multiple sources, although the origin and performance of these reagents are less apparent. Given our familiarity with the Genentech developed reagents, we have limited our discussion to those in Table 1. These antibodies can be requested through Genentech's material transfer agreement (MTA) program (https://www.gene.com/scientists/mta).

Table 1
Linkage-specific antibodies generated at Genentech

Linkage specificity	Clone name	Source and references
Linear/M1-linked polyubiquitin	1F11/3F5/Y102L	Genentech, Inc. [11]
K11-linked polyubiquitin	2A3/2E6	Genentech, Inc. [10]
K48-linked polyubiquitin	Apu2.07	Genentech, Inc. [7]
K63-linked polyubiquitin	Apu3.A8	Genentech, Inc. [7]

An important consideration is that polyubiquitin linkage-specific antibodies have several design features that are not typical among reagent antibodies. Genentech polyubiquitin linkage-specific antibodies were generated by phage display selection of antibody Fab fragments against full-length diubiquitins of defined linkage. Crystal structure determination of the anti-linear/M1 and anti-K63 specific Fabs in complex with their respective antigens revealed that, rather than binding only to the linkage peptide, these antibodies recognize conformational epitopes made up of both ubiquitin subunits presented in a linkage-specific orientation [7, 11]. Although crystal structures of anti-K11 and anti-K48 Fabs in complex with their respective diubiquitins have not been reported, similar phage display approaches were used to select these Fabs. This coupled with their analytical performance provides evidence that they likewise interact with conformation-specific epitopes on folded ubiquitin protein. This concept stands in contrast with most reagent antibodies that are generated by immunization of animals, often with a short peptide from the protein of interest. Due of a lack of secondary and tertiary structure in most short peptide antigens, the resulting antibodies frequently recognize linear epitopes within their target proteins that are naturally suited for detection of denatured proteins by Western blot.

Despite recognizing conformational epitopes, linkage-specific antibodies surprisingly bind to polyubiquitin chains by Western blot, under conditions where proteins are typically denatured. Because ubiquitin is known to be a well-folded protein and because a conformational epitope is required for binding, we speculate that polyubiquitin chains either resist denaturation by SDS-PAGE or undergo refolding upon transfer to a membrane. Consistent with this notion, we find that adherence to specific Western blotting conditions is required to maintain antibody specificity. In immunoprecipitation experiments, conformational flexibility of polyubiquitin chains in solution [17–20] creates an opportunity for cross-reactivity with chains of other linkages that needs to be carefully controlled. Strict adherence to immunoprecipitation protocols is essential to maintaining the desired specificity. For step-by-step protocol instructions, please refer to a previous Methods in Molecular Biology chapter [12] with additional details and considerations described below. A new addition since that time is the anti-linear/M1 polyubiquitin antibody. This antibody requires more stringent conditions (i.e., higher urea concentration) during immunoprecipitation, as noted.

2.2 Western Blot with Linkage-Specific Antibodies

1. $PBST_{0.05}$: PBS with 0.05% Tween20 (*see* **Note 1**).
2. Western blot blocking buffer: 5% (w/v) skim milk powder in $PBST_{0.05}$ (*see* **Note 1**).
3. Wet transfer apparatus.
4. Linkage-specific and anti-human secondary antibodies (*see* Table 1 and **Notes 2** and **3**).

2.3 Immunoprecipitation with Linkage-Specific Antibodies

1. Lysis buffer: 8 M urea, 20 mM Tris (pH 7.5), 135 mM NaCl, 1% Triton X-100, 10% glycerol, 1.5 mM $MgCl_2$, 10 μL/mL of 100× Halt™ protease and phosphatase inhibitor cocktail (Thermo Fisher Scientific), 5 mM EDTA, and 2 mM *N*-ethylmaleimide (NEM) (*see* **Note 4**).
2. Immunoprecipitation buffer: 20 mM Tris (pH 7.5), 135 mM NaCl, 1% Triton X-100, 10% glycerol, 1.5 mM $MgCl_2$, 10 μL/mL of 100× Halt™ protease and phosphatase inhibitor cocktail, 5 mM EDTA, and 2 mM NEM (*see* **Note 4**).
3. Wash buffer for anti-K11, anti-K48, anti-K63 antibodies: 4 M urea, 20 mM Tris (pH 7.5), 135 mM NaCl, 1% Triton X-100, 10% glycerol, and 1.5 mM $MgCl_2$.
4. Wash buffer for anti-linear/M1 antibody: 7 M urea, 20 mM Tris (pH 7.5), 135 mM NaCl, 1% Triton X-100, 10% glycerol, and 1.5 mM $MgCl_2$.
5. Linkage-specific antibody (*see* Table 1).
6. Protein A-coated magnetic beads or protein A-conjugated resin.

2.4 Isotopically Labeled Peptides for Mass Spectrometry

Synthetic, isotopically labeled internal standard peptides have been the primary reagents employed in mass spectrometry experiments to characterize total ubiquitin levels, polyubiquitin chain linkage abundance, and the presence of emerging ubiquitin signals such as phosphorylation of ubiquitin on S65 [15, 21]. The battery of tryptic peptides in Table 2 below spans the entirety of the ubiquitin protein sequence with the exception of the final four amino acids. Redundancy is built in to cover loci susceptible to incomplete trypsin proteolysis (especially K33, but also K48 and K63), involved in polyubiquitin chains, or affected by posttranslational modification. Central to this repertoire of peptides are the K-GG peptides representing the seven isopeptide-linked ubiquitin chains (K6, K11, K27, K29, K33, K48, K63). In addition, the peptide

Table 2
List of peptides for mass spectrometry

Peptide		Isotope label			Heavy				Light			
Name	Sequence	+13C	+15N	Δmass	MH+	M2H+	M3H+	M4H+	MH+	M2H+	M3H+	M4H+
MQIF	MQIFVK	5	1	6.0138	771.44658	386.22693			765.43280	383.22004		
MQIFox	M@QIFVK	5	1	6.0138	787.44148	394.22438			781.42770	391.21749		
M1	GGMQIFVK	5	1	6.0138	885.48948	443.24838			879.47570	440.24149		
M1ox	GGM@QIFVK	5	1	6.0138	901.48438	451.24583			895.47060	448.23894		
K6	MQIFVK*TLTGK	6	1	7.0171	1386.78863	693.89796	462.93440		1379.77150	690.38939	460.59535	
K6ox	M@QIFVK*TLTGK	6	1	7.0171	1402.78353	701.89541	468.26603		1395.76640	698.38684	465.92699	
K11	TLTGK*TITLEVEPSDTIENVK	5	1	6.0138	2408.27988	1204.64358	803.43148	602.81997	2402.26610	1201.63669	801.42689	601.3165
TITLE	TITLEVEPSDTIENVK	5	1	6.0138	1793.94108	897.47418	598.65188		1787.92730	894.46729	596.64729	
K27	TITLEVEPSDTIENVK*AK	5	1	6.0138	2107.11608	1054.06168	703.04355		2101.10230	1051.05479	701.03895	
K29	AK*IQDK	6	1	7.0171	823.47453	412.24091			816.45740	408.73234		
K29-EG	AK*IQDKEGIPPDQQR	5	1	6.0138	1842.96998	921.98863	614.99485		1836.95620	918.98174	612.99025	
IQ-EG	IQDKEGIPPDQQR	5	1	6.0138	1529.79498	765.40113	510.60318		1523.78120	762.39424	508.59859	
EGI	EGIPPDQQR	5	1	6.0138	1045.53048	523.26888			1039.51670	520.26199		
K33	IQDK*EGIPPDQQR	5	1	6.0138	1643.83798	822.42263	548.61751		1637.82420	819.41574	546.61292	
LIF	LIFAGK	9	1	10.0272	658.43508	329.72118			648.40790	324.70759		
QLE	QLEDGR	6	1	7.0171	724.36973	362.68851			717.35260	359.17994		
LI-QL	LIFAGKQLEDGR	6	1	7.0171	1353.75983	677.38356	451.92480		1346.74270	673.87499	449.58575	
K48	LIFAGK*QLEDGR	6	1	7.0171	1467.80273	734.40501	489.93910		1460.78560	730.89644	487.60005	
TLS	TLSDYNIQK	6	1	7.0171	1088.56953	544.78841			1081.55240	541.27984		
EST	ESTLHLVLR	6	1	7.0171	1074.63793	537.82261			1067.62080	534.31404		
TL-ES	TLSDYNIQKESTLHLVLR	6	1	7.0171	2137.17243	1069.08986	713.06233	535.04857	2130.15530	1065.58129	710.72329	533.2943
K63	TLSDYNIQK*ESTLHLVLR	6	1	7.0171	2251.21543	1126.11136	751.07666	563.55932	2244.19830	1122.60279	748.73762	561.8050
pS65	ES#TLHLVLR	6	1	7.0171	1154.60426	577.80577	385.53961		1147.58713	574.29721	383.20056	
pS65-TL-ES	TLSDYNIQKES#TLHLVLR	6	1	7.0171	2217.13876	1109.07302	739.71777	555.04015	2210.12163	1105.56446	737.37873	553.28587
K63-pS65	TLSDYNIQK*EsTLHLVLR	6	1	7.0171	2331.18173	1166.09451	777.73210	583.55089	2324.16460	1162.58594	775.39305	581.7966

The table shows the predicted precursor ion masses for the isotopically labeled peptides (Heavy) as well as the corresponding unlabeled peptides (Light) measured in a trypsin-digested sample. The charge states most often used for quantification in standard LC-MS applications are highlighted. The isotopically labeled residues are indicated by red underline. For peptides containing methionine residues, masses are shown for the oxidized (M@) and unoxidized sequences. The isopeptide-linked diglycine remnant, representing the position of covalent ubiquitin attachment, is denoted by K* and named according to the lysine number. Similarly, serine phosphorylation is indicated by S#

representing linear polyubiquitin bears diglycine attached to the N-terminus of M1 [15]. The diglycine remnant of each of these mimics the covalently attached C-terminus of ubiquitin that remains linked upon trypsin digestion [13, 22]. Additional reagents could be prepared for other less well-characterized events such as phosphorylation of S57 [13], acetylation of K6/K48 [23], deamidation of Q40 [24], or ADP/phospho-ribosylation of R42 [25, 26]. For each internal standard peptide, a single amino acid is replaced with an isotopically labeled counterpart. While many amino acids are amenable to isotopic labeling, the preference has been to introduce labels through leucine, valine, and proline due to their frequency within sequences and the fact that they impart >6 Da additional mass.

2.5 Mass Spectrometry Reagents and Buffers

1. NuPAGE 4–12% Bis-Tris protein gels, 1.5 mm, 10-well.
2. 20× NuPAGE MOPS SDS running buffer.
3. 4× NuPAGE LDS sample buffer.
4. Gel destain solution: 50 mM ammonium bicarbonate (AMBIC), 50% acetonitrile (ACN).
5. SimplyBlue SafeStain.

6. 100% acetonitrile (ACN).
7. Vacuum concentrator (e.g., SpeedVac).
8. In-gel digestion buffer: 50 mM AMBIC, 5% ACN, pH 8.0, chilled to 4 °C (*see* **Note 5**).
9. Sequencing Grade Modified Trypsin (20 μg aliquot in 50 μL; Promega).
10. Extraction Buffer: 50% ACN, 5% formic acid (FA).
11. Peptide resuspension buffer: 10% ACN, 5% FA, 0.01% hydrogen peroxide (*see* **Note 6**).
12. Peptide stock solutions: 25 pmol/μL for a single peptide diluted in 25% ACN/0.1% FA.
13. Ub-AQUA Peptide Mix: 1 pmol/μL for each peptide diluted in 25% ACN/0.1% FA (*see* **Note 7**).

3 Methods

3.1 Western Blot with Linkage-Specific Antibodies

1. Run a biochemical ubiquitination reaction, cell lysate, or immunoprecipitation reaction on an SDS-PAGE gel, and transfer to nitrocellulose at 30 V for 1–2 h using a traditional wet transfer apparatus (*see* **Note 8**). As a control for specificity we highly recommend running pure polyubiquitin chains of a known linkage on the gel. High-quality polyubiquitin chains of all linkages are commercially available.
2. Incubate the blot in the Western blot blocking buffer for 1 h at room temperature with gentle shaking.
3. Incubate the blot with 1 μg/mL linkage-specific antibody diluted in the Western blot blocking buffer for 1 h at room temperature with gentle shaking (*see* **Note 9**).
4. Wash the blot three times in $PBST_{0.05}$.
5. Incubate in a 1:10,000 dilution of anti-human secondary antibody for 1 h at room temperature with gentle shaking (*see* **Note 2**).
6. Wash the blot three times in $PBST_{0.05}$, once in PBS and develop (*see* **Note 3**).

3.2 Immunoprecipitation with Linkage-Specific Antibodies

1. Wash cells in PBS and lyse cell pellet in roughly two volumes of lysis buffer containing 8 M urea. If lysate is viscous, sonication can be performed to shear DNA.
2. Dilute lysates with immunoprecipitation buffer. For the anti-K11, anti-K48, and anti-K63 linkage-specific antibodies, a final concentration of 4 M urea is required to maintain antibody specificity. For the anti-linear/M1 linkage-specific antibody, a final concentration of 7 M urea is required for specificity (*see* **Note 10**).

3. Preclear lysates with protein A-coated magnetic beads or protein A-conjugated resin.
4. Remove beads and immunoprecipitate with the linkage-specific antibody of interest.
5. Capture antibody-protein complexes with protein A beads/resin.
6. Wash beads/resin five times with the appropriate wash buffer (anti-K11, anti-K48, and anti-K63 wash buffer containing 4 M urea or the anti-linear/M1 wash buffer containing 7 M urea) followed by PBS.
7. Transfer beads/resin to a new tube during the last wash, capture/pellet, and remove PBS.
8. Elute by resuspending the beads in SDS-PAGE sample buffer, heat, and run on an SDS-PAGE gel.
9. Carry out a Western blot to detect your protein of interest.

3.3 Mass Spectrometry: Sample Preparation and Analysis

A range of mass spectrometry methods have proven useful for quantifying polyubiquitin chain linkages. Included among these are selected/multiple reaction monitoring (SRM/MRM) [14], high-resolution MS1 precursor ion quantification [15], and parallel reaction monitoring (PRM) [27]. Recent work suggests the possibility that targeted studies involving isobaric tags and targeted multiplexing are also options for these studies [28].

As the attachment of ubiquitin imparts approximately 8 kDa of additional mass to the substrate, SDS-PAGE is particularly valuable in separating unmodified substrates from mono- and multi-ubiquitinated species, as well as substrates bearing polyubiquitin chains [29]. Separation permits the assessment of polyubiquitin linkage type as a function of molecular weight, implicitly providing information about the nature of ubiquitin attachment both proximal and distal to the substrate protein.

A critical part of any such study is the complete digestion of ubiquitin molecules using trypsin into peptides that can be detected by the mass spectrometer. Achieving complete and consistent in-gel trypsin digestion across a batch of samples requires that the protocol be experimentally optimized. The protocol provided here has been optimized for analysis of ubiquitinated proteins separated by 1.5-mm-thick NuPAGE 4–12% Bis-Tris protein gels.

In-gel trypsin digestion for quantitative analysis of ubiquitin linkages:

1. Dilute protein samples containing ubiquitinated protein in 4× LDS sample buffer. Load samples into lanes of a 10-well, 1.5 mm thick 4–12% Bis-Tris gel (*see* **Note 11**).
2. Separate proteins by electrophoresis in 1× MOPS running buffer, diluted from 20× with ddH_2O.

3. Remove gel from cassette and rinse with ddH_2O. Coomassie stain gel with SimplyBlue SafeStain following manufacturer's protocol. Perform an initial destaining with ddH_2O to remove excess Coomassie reagent and reveal stained protein bands.
4. Excise gel regions containing proteins of interest, and transfer into 1.5 mL Eppendorf tubes (Fig. 1).
5. To further destain the gel pieces containing Coomassie stained bands, add sufficient gel destain solution to fully cover gel pieces (10× gel volume). Perform gentle agitation for 30 min.
6. Remove destain solution and replace with 5× gel volume 100% ACN to dehydrate gel pieces. Perform gentle agitation for 30 min. Dehydrated gel pieces turn opaque white in color.
7. After dehydrating, remove ACN and place Eppendorf tube with gel pieces into Labconco SpeedVac for 30 min to dry (*see* **Note 12**).
8. While gel pieces are destaining/drying, prepare trypsin digestion solution on ice by diluting Sequencing Grade Modified

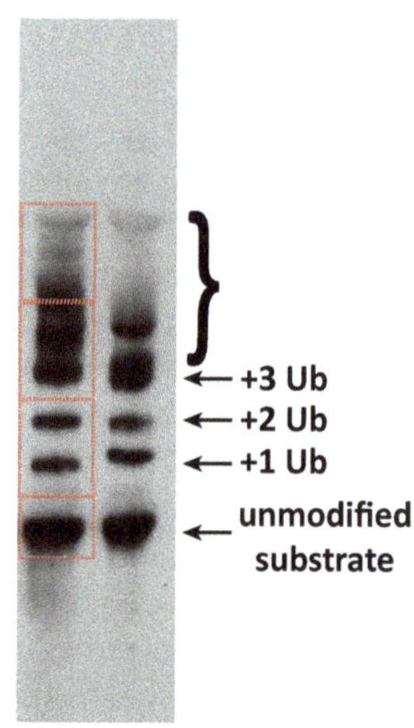

Fig. 1 Analysis of ubiquitin linkage profile from SDS-PAGE gels. The goal of an analysis is to dissect the region containing unmodified substrate and the region containing the first two ubiquitin molecules from the region where smearing begins. Polyubiquitin chains are typically abundant in the region above discrete Ub-substrate bands. The discrete bands are comprised primarily of multi-mono ubiquitin modifiers, making the stoichiometric relationship of polyubiquitin to substrate low. These bands are ideally suited for identifying sites of ubiquitination on the substrate. Keeping excised gel regions near equally sized and performing an analysis on a region where little/no ubiquitin signal is expected (i.e., unmodified substrate) allow for assessment of background signal in the analysis. Of note, Western blots such as this are run in parallel to the Coomassie stained gel used for in-gel trypsin digestion

Trypsin to a concentration of 20 ng/μL by adding 950 μL in-gel digestion buffer to 20 μg aliquot of trypsin (50 μL).

9. Add ice-cold trypsin digestion solution to dry gel pieces, and allow them to reswell on ice for 2 h. Sufficient volume should be added to fully cover gel pieces and ensure they remain covered throughout the duration of the reswelling period (*see* **Note 13**).
10. Incubate in-gel trypsin digests overnight at 37 °C (12–16 h).
11. Quench in gel digests with 0.5× gel volume of Extraction Buffer.
12. Add Ub-AQUA peptide mixture to samples (*see* **Notes 14** and **15**) and briefly vortex sample to mix. Centrifuge sample for 1 min at 13,000 × *g*.
13. Immediately remove sample (with Ub-AQUA peptides and extraction buffer) from gel pieces using a gel loading tip, and transfer to a clean tube (*see* **Note 16**).
14. Add 1× gel volume of extraction buffer to the gel pieces, perform quick centrifuge, and add extracted sample to tube containing original extraction. Repeat this step one additional time for a total of three extractions.
15. Add 1× gel volume of 100% ACN to the gel pieces, perform a quick centrifuge, and add peptides to tube containing original extraction.
16. Tube (or autosampler vial; *see* **Note 16**) containing digested peptides, Ub-AQUA mix, and extraction buffer should be dried to completion in SpeedVac. The length of this step can vary between a few minutes and multiple hours (*see* **Note 17**).
17. Resuspend dry peptide sample in 5–10 μL peptide resuspension buffer. Incubate for 30 min at room temperature prior to injection on the mass spectrometer to allow time for oxidation of methionine-containing peptides.

3.4 Examples of Applications

The methods required for effective characterization of a complex ubiquitin signal on a protein substrate depend on the origin and purity of the ubiquitinated substrate, as well as the nature of the ubiquitin signal. This section seeks to guide an experimentalist new to the ubiquitin field through these decision-making steps to decide what approach[es] to utilize. Subsequently, methods are provided for characterization of ubiquitin signals on a range of ubiquitin-conjugated proteins from the simple reaction products with near homogeneous linkages to more complex substrates derived from cells.

3.4.1 Validating Biochemically Synthesized Polyubiquitin Chains

Purified polyubiquitin chains are widely commercially available and can be readily synthesized in the lab using purified enzymes and in vitro biochemical reactions [30]. Free polyubiquitin chains are commonly used in deubiquitination assays to assess linkage specificity of deubiquitinase enzymes and in binding assays to determine specificity of proteins containing ubiquitin-binding domains (UBDs) such as ubiquitin-associated domains (UBA), ubiquitin-interacting motifs (UIM), and ubiquitin binding in ABIN and NEMO (UBAN) domains. Due to the promiscuous nature of some E2 and E3 enzymes used in generating these free polyubiquitin chains, in vitro ubiquitination reactions frequently contain small amounts of unintended linkages as contaminants. Biochemically synthesized polyubiquitin chains should be validated to ensure they contain the specific linkage of interest, as mixed linkages can alter the outcome of certain in vitro assays. Furthermore, when studying the specificity of an E2-E3 enzyme combination, it can be valuable to decipher the nature of polyubiquitin linkage (s) synthesized by different enzyme variants [31]. To determine the identity of the linkages present in a reaction and measure their abundance, two complementary methods can be employed:

1. Use linkage-specific antibodies to probe the reaction by Western blot. This is a qualitative technique that will allow for detection of linear/M1, K11, K48, and K63 linkages, even when present at sub-stoichiometric levels. For detection of the rare linkages and for quantitative determination of the amount of each linkage present, mass spectrometry is a more appropriate method to employ.
2. Use mass spectrometry analysis with Ub-AQUA peptide mixture to probe reaction products when the goal is to detect and quantify all linkages present in the biochemical reaction, including minor amounts of contaminant linkages.

As an example, we reference the characterization of the diubiquitin species generated by an in vitro reaction with Ube1, Ube2S, and ubiquitin [10]. Ube2S is an E2 ubiquitin-conjugating enzyme that is reported to synthesize exclusively K11-linked polyubiquitin chains [32]. The diubiquitin species can be detected by Western blot using the K11 polyubiquitin linkage-specific antibody, and the use of ubiquitin-AQUA peptide mass spectrometry confirms that the reaction yields 99.1% K11-linked ubiquitin, with minor amounts of K48 and K63 linkages and trace amounts of K6, K27, K29, and K33. Similar characterization of M1-, K48-, and K63-linked dimers has been carried out both by immunoblotting with linkage-specific antibodies and ubiquitin-AQUA analysis—the latter demonstrating greater than 99% purity of the linkage of interest [30].

3.4.2 Deciphering the Ubiquitin Linkages Attached to a Single Substrate

Whether elucidating the function of a protein of interest or deciphering the mechanism by which it is regulated, determining the polyubiquitin chain linkages attached to a biologically meaningful substrate can provide valuable information about the consequences of ubiquitination. To characterize the forms of ubiquitin linked to an individual substrate, linkage-specific antibody detection and mass spectrometry serve as complementary techniques and in combination provide a balance of efficiency and thoroughness in interrogation of posttranslational modifications of substrates by polyubiquitin. Immunoprecipitation with a substrate-specific antibody has proven a reliable method when coupled with linkage-specific immunoblotting [5, 10] or mass spectrometry [33] to identify and quantify the types of polyubiquitin linkages attached to a substrate of interest. Conversely, immunoprecipitation with a linkage-specific antibody under denaturing conditions followed by a Western blot for the substrate protein has been used to demonstrate direct modification of substrates with specific linkages [7, 33–35]. Furthermore, immunoprecipitation with a linkage-specific antibody coupled with Ub-AQUA mass spectrometry has been used to demonstrate mixed, branched, or multiple types of homogenous chains attached to substrate proteins [7, 10, 11, 15]. Such approaches can be used on more complex lysates and organellar preparations as well [36, 37]. From either an in vitro biochemical reaction or a whole cell lysate, linkage-specific antibodies paired with Ub-AQUA mass spectrometry can be used to characterize both the substrate of interest and the polyubiquitin chains attached to the substrate:

1. Use the linkage-specific antibody to immunoprecipitate polyubiquitin chains of a given linkage under denaturing conditions, and then perform a Western blot on the reaction with an antibody against the substrate protein. The use of denaturing conditions in the immunoprecipitations should disrupt any protein complexes, allowing detection of substrates directly modified with polyubiquitin chains of a specific linkage. Conversely, immunoprecipitate under denaturing conditions with an antibody against the known substrate, followed by Western blot with a linkage-specific antibody, to confirm substrate modification with a given polyubiquitin chain linkage.
2. After performing an immunoprecipitation with an antibody against the substrate, Ub-AQUA mass spectrometry can be used to identify and quantify the linkages present. Similarly after immunoprecipitation with a linkage-specific antibody, mass spectrometry can be used to quantify all linkages in the immunoprecipitation to determine the presence of mixed, branched, or multiple homogeneous chains attached to the substrate of interest.

As an example, we reference the ubiquitination reaction of Ube1, UbcH5c, and MuRF1, where the E3 ligase MuRF1 auto-ubiquitinates itself [10]. The polyubiquitin linkages synthesized by MuRF1 can be immunoprecipitated with linkage-specific antibodies and quantified by Ub-AQUA mass spectrometry methods. Immunoprecipitation with the K11 linkage-specific antibody under denaturing conditions resulted in enrichment of not only K11 linkages, but also significant amounts of K48 and K63 linkages. The use of the K11R mutant ubiquitin eliminates the linkage-specific antigen from being generated and, as predicted, abrogates the enrichment of all linkages upon immunoprecipitation with the anti-K11 antibody. This result provides direct demonstration that the antibody is uniquely specific for K11-linked polyubiquitin chains. Further, this provided direct evidence that MuRF1 generates a mixture of linkages on an individual substrate, either as branched/mixed polyubiquitin chains or parallel homogenous polyubiquitin chains of different linkages.

4 Notes

1. TBS buffer can be used in place of PBS without affecting specificity of the antibodies.
2. The Genentech linkage-specific antibodies are formatted as human IgG1 isotypes; therefore, an anti-human secondary antibody should be used.
3. We generally use HRP-conjugated secondary antibodies and develop with chemiluminescence by either film or a digital blot scanner such as the LI-COR C-DiGit. Although not systematically tested, the authors see no reason why near-infrared fluorescence detection of labeled antibodies would affect specificity.
4. Both cell lysis and immunoprecipitation buffers can be made ahead of time; however, the Halt protease and phosphatase inhibitors and NEM should be added fresh immediately before use.
5. Trypsin digestion buffer must be chilled to 4 °C prior to diluting the enzyme to avoid the protease digesting itself prematurely. For similar reasons, gel pieces must be kept on ice during reswelling so that trypsin enzyme can penetrate the gel matrix and retain its activity.
6. Oxidation is performed to convert the inherently mixed population of methionine and methionine sulfoxide (oxMet; +16 Da) into a unified population in the form of oxMet. Oxidation occurs when performic acid, formed when hydrogen peroxide and formic acid are combined, is incubated with Met-containing peptides. Note that commercially available

hydrogen peroxide is most often prepared by vendors at 30% w/w solution in water.

7. Peptide stock solutions and Ub-AQUA peptide mixes can be stored at −80 °C. Although individual peptides in mixture are present at 1 pmol/μL, the aggregate concentration when all peptides are added to the mix will be >20 pmol/μL.
8. The use of a traditional wet transfer apparatus at 30 V is critical to allow the polyubiquitin chains to either maintain a folded state or to properly refold upon transfer. Transfer for 1 h is recommended for short polyubiquitin chains, whereas transfer for 2 h is recommended for longer chains of high molecular weight. The use of a semidry transfer or rapid dry transfer system like the Thermo Fisher Scientific iBlot can result in nonspecific binding with our antibodies. We find that nitrocellulose works best to maintain antibody specificity. Transfer to PVDF can result in nonspecific binding to other linkages.
9. Incubation at room temperature for 1 h with the linkage-specific antibody is critical to maintain specificity. We find that incubation overnight at 4 °C can lead to nonspecific binding to other linkages.
10. The anti-linear/M1 polyubiquitin antibody displays some cross-reactivity with K63-linked chains in solution. The close spatial proximity of Lys63 and Met1 in the tertiary structure of ubiquitin results in flexible chains that can potentially adopt similar conformations. Extensive optimization of the immunoprecipitation reactions with the anti-linear antibody demonstrated that the use of 7 M urea prevents this cross-reactivity with K63-linked chains [11].
11. For precisely run gels, load no more than 40 μL (sample + sample buffer) into each lane of a 10-well, 1.5 mm thick gel. Maximum loading capacity is ~50 μL.
12. Effectively drying gel pieces with ACN desiccation and SpeedVac drying steps is critical to maximizing the efficiency of in-gel trypsin digestion. Incompletely dry gel pieces are less efficient in allowing trypsin enzyme to permeate, negatively impacting digestion efficiency.
13. The length of gel reswelling in this protocol is substantially longer than typical in-gel digestion protocols. While successful digestion is possible with shorter times, empirical testing showed that consistent, maximal digestion efficiency was reached under these conditions.
14. The amount of Ub-AQUA peptides to be added varies depending on the nature of the sample. A typical range is from 50 fmol to 5 pmol per peptide per sample. If the Ub-AQUA peptide mixture must be diluted down from 1 pmol/μL, this should be

done in 25% ACN/0.1% FA immediately before addition to digested sample. At peptide concentrations below 1 pmol/μL, peptide losses can occur. Experience suggests that such losses occur differentially for each peptide in the mixture, adding variability to quantitative data.

15. For peptide mixes including incompletely digested Ub-AQUA peptides (i.e., LI-QL, TL-ES), the order of addition is critical. If Ub-AQUA peptide mix is added to an active digest (prior to quenching with extraction buffer), these peptides will be digested. Digestion of labeled internal standard will generate two peptides—one from the labeled end and one from the unlabeled side. Signals from these two unintended digestion products will interfere with expected signals for the corresponding heavy or light peptides.
16. If total volume of the extracted sample is <100 μL, extracted peptides can be added directly to an autosampler injection vial rather than an Eppendorf tube.
17. Excessive drying times for extracted peptide samples can arise when the liquid-air interface is disrupted by contaminants introduced during sample handling. Experience suggests that freezing extracted peptide sample at −80 °C prior to placing in SpeedVac can improve performance of this drying step.

Acknowledgments

All authors are employees of Genentech Inc. and shareholders of the Roche group.

References

1. Varshavsky A (2012) The ubiquitin system, an immense realm. Annu Rev Biochem 81:167–176. https://doi.org/10.1146/annurev-biochem-051910-094049
2. Komander D, Rape M (2012) The ubiquitin code. Annu Rev Biochem 81:203–229. https://doi.org/10.1146/annurev-biochem-060310-170328
3. Husnjak K, Dikic I (2012) Ubiquitin-binding proteins: decoders of ubiquitin-mediated cellular functions. Annu Rev Biochem 81:291–322. https://doi.org/10.1146/annurev-biochem-051810-094654
4. Emmerich CH, Ordureau A, Strickson S et al (2013) Activation of the canonical ikk complex by k63/m1-linked hybrid ubiquitin chains. Proc Natl Acad Sci U S A 110:15247–15252. https://doi.org/10.1073/pnas.1314715110
5. Meyer H-J, Rape M (2014) Enhanced protein degradation by branched ubiquitin chains. Cell 157:910–921. https://doi.org/10.1016/j.cell.2014.03.037
6. Ohtake F, Saeki Y, Ishido S, Kanno J, Tanaka K (2016) The K48-K63 branched ubiquitin chain regulates NF-κB signaling. Mol Cell 64:251–266. https://doi.org/10.1016/j.molcel.2016.09.014
7. Newton K, Matsumoto ML, Wertz IE et al (2008) Ubiquitin chain editing revealed by polyubiquitin linkage-specific antibodies. Cell 134:668–678. https://doi.org/10.1016/j.cell.2008.07.039
8. Wang H, Matsuzawa A, Brown SA et al (2008) Analysis of nondegradative protein ubiquitylation with a monoclonal antibody specific for lysine-63-linked polyubiquitin. Proc Natl

Acad Sci U S A 105:20197–20202. https://doi.org/10.1073/pnas.0810461105

9. Tokunaga F, Sakata S, Saeki Y et al (2009) Involvement of linear polyubiquitylation of NEMO in NF-kappaB activation. Nat Cell Biol 11:123–132. https://doi.org/10.1038/ncb1821
10. Matsumoto ML, Wickliffe KE, Dong KC et al (2010) K11-linked polyubiquitination in cell cycle control revealed by a K11 linkage-specific antibody. Mol Cell 39:477–484. https://doi.org/10.1016/j.molcel.2010.07.001
11. Matsumoto ML, Dong KC, Yu C et al (2012) Engineering and structural characterization of a linear polyubiquitin-specific antibody. J Mol Biol 418:134–144. https://doi.org/10.1016/j.jmb.2011.12.053
12. Newton K, Matsumoto ML, Ferrando RE et al (2012) Using linkage-specific monoclonal antibodies to analyze cellular ubiquitylation. Methods Mol Biol 832:185–196. https://doi.org/10.1007/978-1-61779-474-2_13
13. Peng J, Schwartz D, Elias JE et al (2003) A proteomics approach to understanding protein ubiquitination. Nat Biotechnol 21:921–926. https://doi.org/10.1038/nbt849
14. Kirkpatrick DS, Hathaway NA, Hanna J et al (2006) Quantitative analysis of in vitro ubiquitinated cyclin b1 reveals complex chain topology. Nat Cell Biol 8:700–710. https://doi.org/10.1038/ncb1436
15. Phu L, Izrael-Tomasevic A, Matsumoto ML et al (2011) Improved quantitative mass spectrometry methods for characterizing complex ubiquitin signals. Mol Cell Proteomics 10: M110.003756. https://doi.org/10.1074/mcp.M110.003756
16. Kaiser SE, Riley BE, Shaler TA et al (2011) Protein standard absolute quantification (PSAQ) method for the measurement of cellular ubiquitin pools. Nat Methods 8:691–696. https://doi.org/10.1038/nmeth.1649
17. Castañeda CA, Kashyap TR, Nakasone MA, Krueger S, Fushman D (2013) Unique structural, dynamical, and functional properties of K11-linked polyubiquitin chains. Struct Lond Engl 1993(21):1168–1181. https://doi.org/10.1016/j.str.2013.04.029
18. Castañeda CA, Chaturvedi A, Camara CM, Curtis JE, Krueger S, Fushman D (2016) Linkage-specific conformational ensembles of non-canonical polyubiquitin chains. Phys Chem Chem Phys 18:5771–5788. https://doi.org/10.1039/c5cp04601g
19. Liu Z, Gong Z, Jiang W-X et al (2015) Lys63-linked ubiquitin chain adopts multiple conformational states for specific target recognition. eLife 4. https://doi.org/10.7554/eLife.05767
20. Varadan R, Assfalg M, Haririnia A, Raasi S, Pickart C, Fushman D (2004) Solution conformation of lys63-linked di-ubiquitin chain provides clues to functional diversity of polyubiquitin signaling. J Biol Chem 279:7055–7063. https://doi.org/10.1074/jbc.M309184200
21. Ordureau A, Münch C, Harper JW (2015) Quantifying ubiquitin signaling. Mol Cell 58:660–676. https://doi.org/10.1016/j.molcel.2015.02.020
22. Kirkpatrick DS, Weldon SF, Tsaprailis G, Liebler DC, Gandolfi AJ (2005) Proteomic identification of ubiquitinated proteins from human cells expressing his-tagged ubiquitin. Proteomics 5:2104–2111. https://doi.org/10.1002/pmic.200401089
23. Ohtake F, Saeki Y, Sakamoto K et al (2015) Ubiquitin acetylation inhibits polyubiquitin chain elongation. EMBO Rep 16:192–201. https://doi.org/10.15252/embr.201439152
24. Cui J, Yao Q, Li S et al (2010) Glutamine deamidation and dysfunction of ubiquitin/nedd8 induced by a bacterial effector family. Science 329:1215–1218. https://doi.org/10.1126/science.1193844
25. Qiu J, Sheedlo MJ, Yu K et al (2016) Ubiquitination independent of E1 and E2 enzymes by bacterial effectors. Nature 533:120–124. https://doi.org/10.1038/nature17657
26. Bhogaraju S, Kalayil S, Liu Y et al (2016) Phosphoribosylation of ubiquitin promotes serine ubiquitination and impairs conventional ubiquitination. Cell 167:1636–1649.e13. https://doi.org/10.1016/j.cell.2016.11.019
27. Ordureau A, Heo J-M, Duda DM et al (2015) Defining roles of parkin and ubiquitin phosphorylation by Pink1 in mitochondrial quality control using a ubiquitin replacement strategy. Proc Natl Acad Sci U S A 112:6637–6642. https://doi.org/10.1073/pnas.1506593112
28. Erickson BK, Rose CM, Braun CR et al (2017) A strategy to combine sample multiplexing with targeted proteomics assays for high-throughput protein signature characterization. Mol Cell 65:361–370. https://doi.org/10.1016/j.molcel.2016.12.005
29. Seyfried NT, Xu P, Duong DM, Cheng D, Hanfelt J, Peng J (2008) Systematic approach for validating the ubiquitinated proteome. Anal Chem 80:4161–4169. https://doi.org/10.1021/ac702516a
30. Dong KC, Helgason E, Yu C et al (2011) Preparation of distinct ubiquitin chain reagents of high purity and yield. Struct Lond Engl

1993(19):1053–1063. https://doi.org/10.1016/j.str.2011.06.010

31. Bosanac I, Phu L, Pan B et al (2011) Modulation of K11-linkage formation by variable loop residues within UbcH5a. J Mol Biol 408:420–431. https://doi.org/10.1016/j.jmb.2011.03.011
32. Wickliffe KE, Lorenz S, Wemmer DE, Kuriyan J, Rape M (2011) The mechanism of linkage-specific ubiquitin chain elongation by a single-subunit E2. Cell 144:769–781. https://doi.org/10.1016/j.cell.2011.01.035
33. Dynek JN, Goncharov T, Dueber EC et al (2010) C-IAP1 and UbcH5 promote K11-linked polyubiquitination of RIP1 in TNF signalling. EMBO J 29:4198–4209. https://doi.org/10.1038/emboj.2010.300
34. Goncharov T, Niessen K, de Almagro MC et al (2013) OTUB1 modulates c-IAP1 stability to regulate signalling pathways. EMBO J 32:1103–1114
35. de Almagro MC, Goncharov T, Newton K, Vucic D (2015) Cellular IAP proteins and Lubac differentially regulate necrosome-associated RIP1 ubiquitination. Cell Death Dis 6:e1800. https://doi.org/10.1038/cddis.2015.158
36. Cunningham CN, Baughman JM, Phu L et al (2015) USP30 and parkin homeostatically regulate atypical ubiquitin chains on mitochondria. Nat Cell Biol 17:160–169. https://doi.org/10.1038/ncb3097
37. Ordureau A, Sarraf SA, Duda DM et al (2014) Quantitative proteomics reveal a feedforward mechanism for mitochondrial parkin translocation and ubiquitin chain synthesis. Mol Cell 56:360–375. https://doi.org/10.1016/j.molcel.2014.09.007

Chapter 25

Dissecting Dynamic and Heterogeneous Proteasome Complexes Using In Vivo Cross-Linking-Assisted Affinity Purification and Mass Spectrometry

Xiaorong Wang and Lan Huang

Abstract

Protein-protein interactions are essential for protein complex formation and function. Affinity purification coupled with mass spectrometry (AP-MS) is the method of choice for studying protein-protein interactions at the systems level under different physiological conditions. Although effective in capturing stable protein interactions, transient, weak, and/or dynamic interactors are often lost due to extended procedures during conventional AP-MS experiments. To circumvent this problem, we have recently developed XAP (in vivo cross-linking (*X*)-assisted *a*ffinity *p*urification)-MS strategy to better preserve dynamic protein complexes under native lysis conditions. In addition, we have developed XBAP (in vivo cross-linking (*X*)-assisted *b*imolecular tandem *a*ffinity *p*urification)-MS method by incorporating XAP with bimolecular affinity purification to define dynamic and heterogeneous protein subcomplexes. Here we describe general experimental protocols of XAP- and XBAP-MS to study dynamic protein complexes and their subcomplexes, respectively. Specifically, we present their applications in capturing and identifying proteasome dynamic interactors and ubiquitin receptor (UbR)-proteasome subcomplexes.

Key words Protein-protein interaction, AP-MS, XAP, XBAP, Dynamic interactions, Proteasome complexes, Ubiquitin receptor-proteasome subcomplexes

1 Introduction

Protein-protein interactions (PPIs) that establish the "protein sociology" in different cellular contexts play key regulatory roles in nearly all biological processes including development, cell homeostasis, growth, and proliferation. Aberrant protein-protein interactions lead to various human diseases including cancer [1, 2]. Modulation of protein-protein interaction represents an emerging therapeutic paradigm, and protein interaction interfaces describe a new class of attractive targets for drug development. Therefore, understanding protein interacting network is critical not only for a comprehensive understanding of protein function and regulation but also provides potential targets for better

Thibault Mayor and Gary Kleiger (eds.), *The Ubiquitin Proteasome System: Methods and Protocols*, Methods in Molecular Biology, vol. 1844, https://doi.org/10.1007/978-1-4939-8706-1_25,

therapeutics. Several methodologies including yeast two-hybrid system, protein microarray, fluorescence imaging, and affinity purification-mass spectrometry have been developed for studying protein-protein interactions. Owing to new developments in sample preparation strategies, mass spectrometry technologies, and bioinformatics tools, affinity purification coupled with mass spectrometry (AP-MS) has become the method of choice for studying protein-protein interactions at the systems level with great sensitivity, accuracy, versatility, and speed [3–8]. While AP-MS has proven effective and powerful in capturing stable interactions, weakly bound interactors are often lost due to extended biochemical manipulation under native purification conditions. Given the complex nature of PPIs, alternative strategies would be needed to enable the capture of all types of interactions. To this end, we have previously developed the QTAX (*q*uantitative analysis of *t*andem *a*ffinity purified in vivo cross-linked (*X*) protein complexes) method to capture a wide range of interactions in a single experiment [9]. One of the key components in the QTAX strategy is in vivo formaldehyde (FA) cross-linking of living cells to freeze protein-protein interactions prior to cell lysis, thus allowing better preservation of the dynamic proteome of protein complexes during subsequent affinity purification [9]. The second key component in the QTAX method is the integration of the HB (His-Bio) tag for tandem affinity purification of protein complexes, which is versatile and can be used for protein purification under both native and fully denatured conditions [9–13]. In combination with quantitative mass spectrometry, the QTAX strategy has been successfully employed to capture and identify stable and transient/weak interactors of protein complexes including proteasomes [9, 14]. While the composition of protein complexes resulting from denaturing purification can be readily determined, subunit stoichiometry can be misrepresented due to variance in protein cross-linking efficiency, leading to varying protein absolute abundances. Therefore, to isolate functional protein assemblies for proteomic analysis, native purification would be preferred to maintain the integrity of protein complexes. Previously, it has been shown that mild cross-linking can be coupled with native purification to preserve protein interactions [15–17]. In order to preserve the dynamic proteome of protein complexes and maintain complex intactness during affinity purification under native conditions, we have developed a new AP-MS strategy, namely, XAP (in vivo cross-linking (*X*)-assisted *a*ffinity *p*urification)-MS by incorporating mild in vivo formaldehyde (<0.1%) cross-linking of intact cells with HB-tag-based single-step affinity purification under native conditions [18, 19] (Fig. 1a). The XAP method has been successfully applied to reliably capture dynamic/weak interactors of human proteasome complexes including a known proteasome interactor Ecm29, which enabled us to investigate and determine the evolutionarily

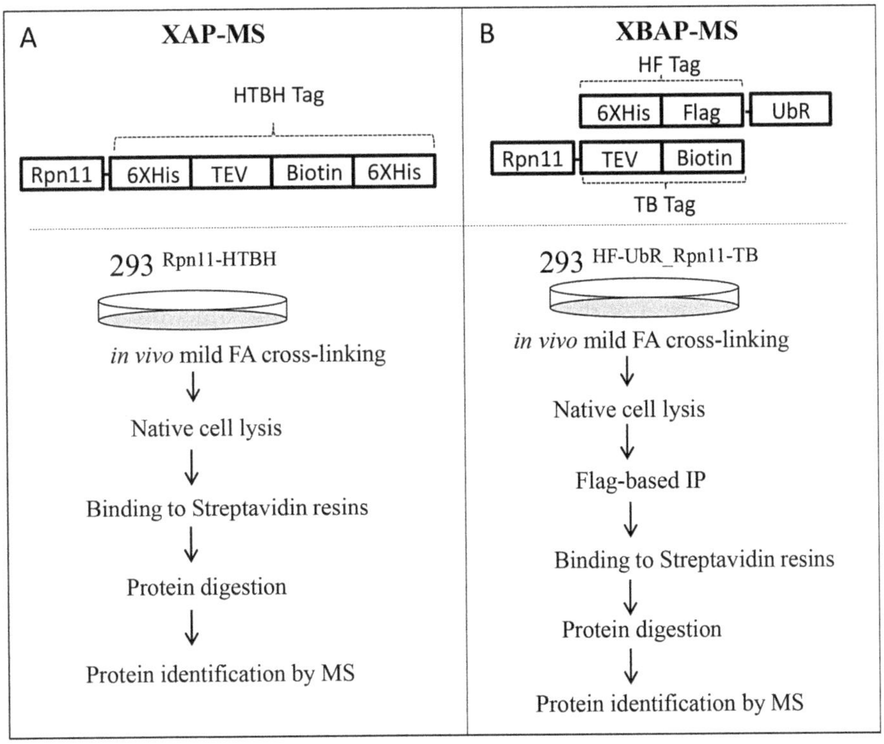

Fig. 1 The general schemes for (**a**) XAP- and (**b**) XBAP-MS strategies. While one single bait is tagged for affinity purification in the XAP experiment, two baits are selected for tagging in the XBAP experiment

conserved functional role of Ecm29 in regulating 26S proteasome disassembly mediated by oxidative stress [19]. In comparison to the QTAX strategy [9], XAP utilizes significantly less amount of formaldehyde (0.05 vs. 1%) for in vivo cross-linking [18, 19]. Our results have further demonstrated that such mild in vivo cross-linking allows effective stabilization of protein structures and interacting networks without altering the integrity and functionality of protein complexes [18]. In addition, limited formaldehyde cross-linking does not interfere with native cell lysis, subsequent purification, and mass spectrometric analysis.

It is well known that proteasomes exists in cells as dynamic and heterogeneous populations. To dissect dynamic ubiquitin receptor-proteasome subcomplexes, we have developed the XBAP (in vivo cross-linking (*X*)-assisted *b*imolecular tandem *a*ffinity *p*urification) method to enable the effective purification of these subcomplexes and to define their subunit compositions [18] (Fig. 1b). In comparison to XAP, XBAP requires tagging of two constituents of a given complex, thus allowing the isolation of protein subcomplexes containing the two selected baits. To this end, the XBAP strategy employs two unique tags, i.e., HF (His-FLAG) and TB (Tev-biotin) tags for two distinct baits, permitting bimolecular purification not only under native conditions but also under fully denaturing conditions [18]. Importantly, the employment of two tagged baits enables the dissection of heterogeneous protein complexes into subpopulations. With the XBAP strategy, we have effectively identified seven UbR-proteasome subcomplexes and quantitatively compared their subunit compositions and relative abundance based on label-free quantitative mass spectrometry [18]. Collectively, the development of the XAP- and XBAP-MS strategies enhance our ability to define dynamic and heterogeneous protein complexes in more detail, thus facilitating the determination of their structure, regulation, and function. Here we describe how the two new AP-MS strategies are applied to study proteasome complexes.

2 Materials

2.1 Cell Culture

1. A HEK293 cell line stably expressing C-terminal HTBH-tagged Rpn11: 293 $^{\text{Rpn11-HTBH}}$ (*see* **Note 1**).
2. A HEK293 cell lines stably expressing the His-Flag-UbR and Rpn11-TB: 293 $^{\text{HF-UbR\_Rpn11-TB}}$ (*see* **Note 2**).
3. Culture medium: DMEM medium supplemented with 10% fetal bovine serum and 1% penicillin/streptomycin.

2.2 HB-Tag-Based Affinity Purification for XAP and XBAP Experiments

1. 1× protease inhibitor cocktail: 1 μg/mL phenylmethylsulfonyl fluoride, leupeptin, aprotinin, and pepstatin. Make a 100× stock solution for phenylmethylsulfonyl fluoride in 100% isopropanol, store at 4 °C, and 1000× stock solutions for leupeptin, aprotinin, and pepstatin, store at −20 °C.
2. 1× phosphatase inhibitor cocktail: 5 mM NaF, 0.1 mM Na_3VO_4, 2.5 mM $Na_4P_2O_7$, and 1 mM EDTA. Make a 10× stock solution; store at −20 °C.
3. 1× trypsin–EDTA solution for cell culture.
4. 1× PBS.
5. 2.5 M glycine (pH 7).
6. Lysis buffer A: 100 mM NaCl, 50 mM sodium phosphate, 10% glycerol, 5 mM ATP, 1 mM DTT, 5 mM $MgCl_2$, 1× protease inhibiter cocktail, 1× phosphatase inhibitor cocktail, 0.5% IGEPAL CA-630, and pH 7.5. Make the buffer right before the experiment.
7. 20 gauge needles.
8. Streptavidin agarose resin (Thermo Scientific).
9. TBS buffer: 50 mM Tris–HCl (pH 7.4), 100 mM NaCl, 5 mM ATP, and 10% glycerol.
10. Anti-FLAG M2 affinity gel (Sigma).

11. 3× FLAG peptides (Sigma).
12. Microspin column (Bio-Rad).
13. Siliconized tubes (Axygen).

2.3 Protein Digestion and LC-MS/MS Analysis

1. 50 mM NH_4HCO_3.
2. 50 mM NH_4HCO_3 in 8 M urea.
3. Trifluoroacetic acid (TFA).
4. 1 μg/μL sequencing-grade endopeptidase LysC stock solution.
5. 0.4 μg/μL sequencing-grade trypsin stock solution: dissolve 20 μg of trypsin in 50 μL of 1 mM TFA (*see* **Note 3**).
6. 10% formic acid.
7. OMIX C18 tip (Agilent Technologies) (*see* **Note 4**).
8. SpeedVac.
9. EASY NanoLC 1000 system (ThermoFisher Scientific).
10. Mass spectrometer: LTQ-Orbitrap XL MS (ThermoFisher Scientific).
11. Nano LC solvent A: 2% acetonitrile, 0.1% formic acid in H_2O.
12. Nano LC solvent B: 98% acetonitrile, 0.1% formic acid in H_2O.
13. Beads extraction buffer: 25% acetonitrile, 0.1% formic acid.
14. Gel loading tip.

2.4 Database Searching for Protein Identification and Quantification

1. LC-MS/MS data extraction: instrument specific scripts from the manufacturer.
2. Protein identification and quantitation software: Protein Prospector (University of California, San Francisco).

3 Methods

3.1 Cell Culture and In Vivo Formaldehyde Treatment

1. Culture HEK93$^{Rpn11\text{-}HTBH}$ or HEK293 $^{HF\text{-}UbR\_Rpn11\text{-}TB}$ cells to (90%) confluence in complete DMEM medium.
2. Wash the cells with 1× PBS once.
3. Treat the cells 0.05% FA for 10 min in PBS at 37 °C in the incubator (*see* **Note 5**).
4. Quench with a final concentration of 0.125 M glycine for 5 min.
5. Wash the cells once in 1× PBS to remove the formaldehyde and glycine.

3.2 HB-Tag-Based Affinity Purification for XAP

1. Detach cells with trypsin-EDTA solution, and wash them once with 1× PBS buffer.
2. Collect cell pellets, and lyse cells using lysis buffer A by pushing the lysate through a 20 gauge needle 15 times. Centrifuge the

lysates at maximum speed of a microcentrifuge for 15 min to remove cell debris.

3. Incubate the supernatant with 25 μL bead volume of streptavidin resin per plate for the desired amount of time at 4 °C (*see* **Note 6**).
4. Wash the streptavidin beads with 50 bed volumes of lysis buffer A without protease and phosphatase inhibitors (*see* **Note 7**).
5. Wash the beads with ten bead volumes of TBS buffer.

3.3 Bimolecular Tandem Affinity Purification Strategies for XBAP

1. Follow the same procedure for cell lysis as **steps 1** and **2** in Subheading 3.2.
2. Incubate the supernatant with anti-FLAG M2 affinity gel at 4 °C for 3 h; add 20 μL bead volume of FLAG affinity gel per 150 mm plate cells (*see* **Note 8**).
3. Wash the FLAG affinity gel with 30 bead volumes of lysis buffer A without protease and phosphatase inhibitors.
4. Wash the FLAG affinity gel with 15 bed volumes of TBS buffer.
5. Elute the bound proteins with 150 μg/mL of 3× FLAG peptide in TBS buffer by incubating at 4 °C for 30 min. Discard the beads.
6. Incubate the resulting eluent with streptavidin resin at 4 °C for 2 h. Use 10 μL bead volume of streptavidin resin per 150 mm plate cells (*see* **Note 9**).
7. Wash the streptavidin resin with 50 bed volumes of TBS buffer.

3.4 Protein Digestion and LC-MS/MS Analysis

To avoid keratin contamination in your samples, you need to wear a hairnet, sleeves, and clean gloves for the following procedure. To minimize sample loss, bound protein samples are directly subjected to subsequent LysC/trypsin digestion and mass spectrometric analysis. Each purification experiment needs to be repeated once to obtain two biological replicates to assess reproducibility.

1. Add one bead volume of 8 M urea with 50 mM ammonium bicarbonate buffer to streptavidin resin, then add 1 μL of endopeptidase LysC stock solution to the protein complex, and incubate at 37 °C for 4 h (*see* **Note 10**).
2. Decrease urea concentration to <1.5 M by adding an adequate volume of 50 mM NH_4HCO_3. Add trypsin to a final concentration of 5–10 ng/μL, and incubate overnight at 37 °C (*see* **Note 11**).
3. To recover digested peptides, add 10% formic acid to a final concentration of 1% to the sample to stop the digestion. Mix and spin at 100 × *g* for 1 min.
4. Take off the supernatant and transfer to a clean siliconized tube with a gel loading tip.

5. Add half bead volume of bead extraction buffer to the beads, shake gently for 5 min, spin the sample, and collect the supernatant.
6. Repeat **step 7** two more times.
7. Pool all the supernatant from **steps 6–8** together. Dry it in a SpeedVac.
8. Add 0.1% formic acid to dissolve the pellet. Desalt the pellet with Agilent OMIX C18 tips.
9. Analyze peptide mixtures by LC-MS/MS using nanoflow reverse phase liquid chromatography (EASY NanoLC 1000 system) coupled online to a LTQ-Orbitrap XL MS. Each cycle of an MS/MS experiment includes one MS scan in FT mode (350–1400 m/z, resolution of 60,000 at m/z 400) followed by data-dependent MS2 scans in the LTQ with normalized collision energy at 35% for the top ten peaks. Elute peptides starting with nanoLC solvent A with a linear gradient 0–35% nanoLC solvent B for 80 min at a flow of 250 nl/min. LC-MS/MS is operated in an information-dependent mode in which each full MS analysis is followed by ten MS/MS acquisitions where the three most abundant peptide molecular ions are dynamically selected for collision induced dissociation (CID) to generate tandem mass spectra (*see* **Note 12**).

3.5 Protein Identification Using Protein Prospector (See Note 13)

1. Obtain monoisotopic masses of both parent ions and corresponding fragment ions, parent ion charge states, and ion intensities from the MS/MS by using Raw_Extract script from Xcalibur v2.4.
2. Use the Batch-tag program within Protein Prospector for database searching. Select trypsin as the enzyme, and set the maximum number of missed tryptic cleavage sites as 2. Chemical modifications such as protein amino-terminal acetylation, methionine oxidation, amino-terminal pyroglutamine, and deamidation of asparagine residues are selected as variable modifications. These modifications, except for protein amino-terminal acetylation, need to be chosen because of their frequent occurrence during sample preparation. Set the mass accuracy for parent ions and fragment ions as ±20 ppm and 0.6 Da, respectively. Any annotated protein databases such as SwissProt and UniProt can be used for database searching. A concatenated database composed of a normal and its random database can be generated in Protein Prospector for database searching. Because we purify the samples from human cell lines, *Homo sapiens* is selected as the restricted species.
3. General protein identification is based on at least two peptides with an expectation value cutoff of 0.01.

4 Notes

1. The HTBH-tag consists of two hexahistidine tags, a TEV cleavage site, and a signal sequence for in vivo biotinylation, which allows efficient purification of proteasome complexes in a single step by binding to streptavidin resins and specific elution by cleavage with TEV protease [11].
2. The His-FLAG tag consists of one hexahistidine tag and three FLAG tags. TB tag consists of a TEV cleavage site and a signal sequence for in vivo biotinylation, which allows efficient purification of proteasome complexes in a single step by binding to streptavidin resins and specific elution by cleavage with TEV protease.
3. Make fresh 1 mM TFA each time from a 100 mM TFA stock solution.
4. Any C18-based desalting tips and cartridges should work.
5. If the cells are not adherent, you can do the treatment in a tube with gentle agitation.
6. For the HTBH-tag, use 25 μl of streptavidin beads per 150 mm plate of HEK293 cells for maximum specific binding efficiency with minimal background binding. Purification efficiency should be followed by western blot analysis. The optimal binding for proteasome complexes to streptavidin beads is 2 h. The binding efficiency decreases when the incubation time decreases.
7. Effective washing steps can be achieved in micro-columns from Bio-Rad, for example, to minimize the bead loss.
8. For the FLAG-tag, use 20 μl bead volume of FLAG affinity gel per 150 mm plate of HEK293 cells for maximum specific binding efficiency with minimal background binding. Purification efficiency should be evaluated by western blot analysis. The optimal binding for UbR complex to FLAG beads is 3 h. The binding efficiency decreases when the incubation time decreases.
9. Since streptavidin resin are used for the second step binding in the XBAP experiment, the amount of beads required is less. For our bait, 10 μl bead volume per 150 mm plate cells is optimal.
10. LysC digestion can proceed from 4 h to overnight.
11. Trypsin digestion can proceed from 8 h to overnight.
12. For MS instruments with fast scanning rates such as the LTQ-Orbitrap, top ten peaks can be sequenced in each LC-MS/MS acquisition cycle.
13. You can access protein prospector from the following website: http://prospector.ucsf.edu/prospector/mshome.htm. Other database searching tools such as MaxQuant, Mascot, proteome discoverer, etc. can also be used.

Acknowledgments

This work was supported by National Institutes of Health grants RO1GM074830 to L.H, R01GM106003 to L.H. and S. R.

References

1. Ryan DP, Matthews JM (2005) Protein-protein interactions in human disease. Curr Opin Struct Biol 15(4):441–446. https://doi.org/10.1016/j.sbi.2005.06.001
2. Kar G, Gursoy A, Keskin O (2009) Human cancer protein-protein interaction network: a structural perspective. PLoS Comput Biol 5 (12):e1000601
3. Gingras AC, Gstaiger M, Raught B, Aebersold R (2007) Analysis of protein complexes using mass spectrometry. Nat Rev Mol Cell Biol 8 (8):645–654
4. Ryan CJ, Cimermancic P, Szpiech ZA, Sali A, Hernandez RD, Krogan NJ (2013) High-resolution network biology: connecting sequence with function. Nat Rev Genet 14 (12):865–879. https://doi.org/10.1038/nrg3574
5. Lambert JP, Ivosev G, Couzens AL, Larsen B, Taipale M, Lin ZY, Zhong Q, Lindquist S, Vidal M, Aebersold R, Pawson T, Bonner R, Tate S, Gingras AC (2013) Mapping differential interactomes by affinity purification coupled with data-independent mass spectrometry acquisition. Nat Methods 10 (12):1239–1245. https://doi.org/10.1038/nmeth.2702
6. Collins BC, Gillet LC, Rosenberger G, Rost HL, Vichalkovski A, Gstaiger M, Aebersold R (2013) Quantifying protein interaction dynamics by SWATH mass spectrometry: application to the 14-3-3 system. Nat Methods 10 (12):1246–1253. https://doi.org/10.1038/nmeth.2703
7. Sowa ME, Bennett EJ, Gygi SP, Harper JW (2009) Defining the human deubiquitinating enzyme interaction landscape. Cell 138 (2):389–403
8. Kaake RM, Wang X, Huang L (2010) Profiling of protein interaction networks of protein complexes using affinity purification and quantitative mass spectrometry. Mol Cell Proteomics 9 (8):1650–1665
9. Guerrero C, Tagwerker C, Kaiser P, Huang L (2006) An integrated mass spectrometry-based proteomic approach: quantitative analysis of tandem affinity-purified in vivo cross-linked protein complexes (QTAX) to decipher the 26 S proteasome-interacting network. Mol Cell Proteomics 5(2):366–378. https://doi.org/10.1074/mcp.M500303-MCP200
10. Tagwerker C, Flick K, Cui M, Guerrero C, Dou Y, Auer B, Baldi P, Huang L, Kaiser P (2006) A tandem affinity tag for two-step purification under fully denaturing conditions: application in ubiquitin profiling and protein complex identification combined with in vivocross-linking. Mol Cell Proteomics 5 (4):737–748
11. Wang X, Chen CF, Baker PR, Chen PL, Kaiser P, Huang L (2007) Mass spectrometric characterization of the affinity-purified human 26S proteasome complex. Biochemistry 46 (11):3553–3565
12. Fang L, Wang X, Yamoah K, Chen PL, Pan ZQ, Huang L (2008) Characterization of the human COP9 signalosome complex using affinity purification and mass spectrometry. J Proteome Res 7(11):4914–4925
13. Fang L, Kaake RM, Patel VR, Yang Y, Baldi P, Huang L (2012) Mapping the protein interaction network of the human COP9 signalosome complex using a label-free QTAX strategy. Mol Cell Proteomics 11(5):138–147. https://doi.org/10.1074/mcp.M111.016352
14. Guerrero C, Milenkovic T, Przulj N, Kaiser P, Huang L (2008) Characterization of the proteasome interaction network using a QTAX-based tag-team strategy and protein interaction network analysis. Proc Natl Acad Sci U S A 105 (36):13333–13338
15. Subbotin RI, Chait BT (2014) A pipeline for determining protein-protein interactions and proximities in the cellular milieu. Mol Cell Proteomics 13(11):2824–2835. https://doi.org/10.1074/mcp.M114.041095
16. Smart SK, Mackintosh SG, Edmondson RD, Taverna SD, Tackett AJ (2009) Mapping the local protein interactome of the NuA3 histone acetyltransferase. Protein Sci 18 (9):1987–1997. https://doi.org/10.1002/pro.212
17. Bousquet-Dubouch MP, Baudelet E, Guerin F, Matondo M, Uttenweiler-Joseph S, Burlet-Schiltz O, Monsarrat B (2009) Affinity purification strategy to capture human endogenous

proteasome complexes diversity and to identify proteasome-interacting proteins. Mol Cell Proteomics 8(5):1150–1164

18. Yu C, Yang Y, Wang X, Guan S, Fang L, Liu F, Walters KJ, Kaiser P, Huang L (2016) Characterization of dynamic UbR-proteasome subcomplexes by in vivo cross-linking (X) assisted bimolecular tandem affinity purification (XBAP) and label-free quantitation. Mol Cell Proteomics 15(7):2279–2292. https://doi.org/10.1074/mcp.M116.058271

19. Wang X, Chemmama IE, Yu C, Huszagh A, Xu Y, Viner R, Block SA, Cimermancic P, Rychnovsky SD, Ye Y, Sali A, Huang L (2017) The proteasome-interacting Ecm29 protein disassembles the 26S proteasome in response to oxidative stress. J Biol Chem 292(39):16310–16320. https://doi.org/10.1074/jbc.M117.803619

Index

Thibault Mayor and Gary Kleiger (eds.), *The Ubiquitin Proteasome System: Methods and Protocols* , Methods in Molecular Biology, vol. 1844, https://doi.org/10.1007/978-1-4939-8706-1, © Springer Science+Business Media, LLC, part of Springer Nature 2019

Lightning Source UK Ltd.
Milton Keynes UK
UKHW05n1303270918
329493UK00005B/102/P

9 781493 987054